Proceedings
of the
Steklov Institute of Mathematics

edited by
I. G. Petrovskiĭ

Number 107 (1971)

Equations in free semigroups

by

Ju. I. Hmelevskiĭ

Providence, Rhode Island
AMERICAN MATHEMATICAL SOCIETY
1976

Академия Наук
Союза Советских Социалистических Республик

ТРУДЫ

ордена Ленина

МАТЕМАТИЧЕСКОГО ИНСТИТУТА имени В. А. СТЕКЛОВА
CVII

Ответственный редактор
академик И. Г. ПЕТРОВСКИЙ

УРАВНЕНИЯ В СВОБОДНОЙ ПОЛУГРУППЕ

Ю. И. ХМЕЛЕВСКИЙ

Издательство „Наука"

Translated by G. A. Kandall

AMS (MOS) subject classifications (1970). Primary 20M05; Secondary 10B35.

ABSTRACT. This monograph is devoted to an investigation of equations and systems of equations in a free semigroup with a finite number of generators. Such equations are also called word equations. Algorithms are constructed for solving equations in three unknowns and systems of equations in two unknowns. The structure of the solutions of equations is investigated.

Word equations are connected with Diophantine equations. Namely, to each word equation it is possible to assign a Diophantine equation in such a way that there exists a one-to-one correspondence between the solutions of both equations.

The investigation of word equations may be useful in the study of equations in free groups, a problem having great significance for the elementary theory of free groups.

This monograph is intended for advanced undergraduates, graduate students and scientists interested in algorithmic problems.

Library of Congress Cataloging in Publication Data CIP

Khmelevskiĭ, IUriĭ Il'ich.
Equations in free semigroups.

(Proceedings of the Steklov Institute of Mathematics ; no. 107 (1971)
Translation of Urvneniia v svobodnoĭ polugruppe.
Bibliography: p. 270.
1. Semigroups. 2. Diophantine analysis. I. Title.
II. Series: Akademiia nauk SSSR. Mathematicheskiĭ institut. Proceedings ; no. 107 (1971)
QA1.A413 no. 107 (1971) [QA171] 510'.8s [512'.2]
ISBN 0-8218-3007-4 76-4883

PROCEEDINGS OF THE STEKLOV INSTITUTE OF MATHEMATICS IN THE ACADEMY OF SCIENCES OF THE USSR

(Труды математического института им. В. А. Стеклова, т. CVII, 1971)

TABLE OF CONTENTS

INTRODUCTION

1. This monograph deals with equations in a free semigroup with a finite number of generators or, more briefly, with word equations, a term suggested by A. A. Markov. A system of word equations with n unknowns $x_1, \cdots, x_n$ and coefficient alphabet $\{a_1, \cdots, a_k\}$ is a system of equalities

$$\Phi_1 = \Psi_1,\quad \cdots\cdots \quad \Phi_N = \Psi_N, \tag{1}$$

where $N \geqq 1$ and $\Phi_1, \cdots, \Phi_N, \Psi_1, \cdots, \Psi_N$ are words in the alphabet $\{a_1, \cdots, a_k, x_1, \cdots, x_n\}$. A solution of the system of equations (1) is any list $(X_1, \cdots, X_n)$ of words in the alphabet $\{a_1, \cdots, a_k\}$ such that as a result of substituting the words $X_1, \cdots, X_n$ for the unknowns $x_1, \cdots, x_n$ in the left- and right-hand sides of each equality in (1) one obtains words which are graphically equal.

The study of word equations was begun by A. A. Markov. He considered systems of equations in two unknowns and constructed an algorithm recognizing the existence of solutions of such systems. In the first part of this present monograph (Chapter II) we construct an algorithm recognizing the existence of solutions of systems of word equations in three unknowns.

In considering arbitrary systems of word equations various reductions are possible. For example (see Proposition 1.28), for any system of equations in n unknowns it is possible to construct a single equation in n unknowns which is equivalent to the system. On the other hand, any system of word equations reduces to a system in which each equation has the form $\alpha = \beta\gamma$ or $\alpha = \beta$, where α, β and γ are letters of the alphabet $\{a_1, \cdots, a_k, x_1, \cdots, x_n\}$. This is well illustrated in the following example. The equation $x_1a_1x_2x_3 = x_4x_5$ has a solution if and only if there is a solution of the system

$$\begin{aligned} x_1a_1 &= y_1, \\ y_1x_2 &= y_3, \\ y_3x_3 &= y_4, \\ x_4x_5 &= y_5, \\ y_4 &= y_5. \end{aligned}$$

In the second part of this monograph (Chapter III) we construct an algorithm recognizing the existence of solutions of systems consisting of word equations in two unknowns (the whole system may contain any number of unknowns).

To denote a product of (semigroup) factors depending on the index i we use the symbol $[\ \ ]_{i=m}^{n}$. More precisely, by induction, if there is given a sequence of words $A_m, \cdots, A_n$ (when $m > n$ the sequence is empty), then $[A_i]_{i=m}^{n}$ denotes the empty word if $m > n$ and the word $A_m[A_i]_{i=m+1}^{n}$ if $m \leqq n$. By means of this symbol any word equation can be written in the form

$$A_0[\xi_i A_i]_{i=1}^{n} = B_0[\eta_i B_i]_{i=1}^{m},$$

where $n, m \geqq 0$, $\xi_1, \cdots, \xi_n$, $\eta_1, \cdots, \eta_m$ are unknowns, and $A_0, \cdots, A_n$, $B_0, \cdots, B_m$ are words in the alphabet $\mathit{И}$.[1] The words $A_0, \cdots, A_n$, $B_0, \cdots, B_m$ are called the coefficients of the equation. A word equation is called coefficient-free if all its coefficients are empty words.

In the third and final part of this monograph (Chapter IV) we consider coefficient-free word equations.

2. The problem of solving word equations is related to the well-known problem of solving Diophantine equations. This relation is based on a result of Nielsen (see [1] or [13]) which states that the second-order matrices with natural elements and determinant equal to 1 form a free semigroup Π with two generators. Such generators are the matrices

$$A = \begin{bmatrix} 1 & 1 \\ 0 & 1 \end{bmatrix}, \quad B = \begin{bmatrix} 1 & 0 \\ 1 & 1 \end{bmatrix}.$$

Using this result, for any system $\mathfrak{A}$ of word equations with a two-letter coefficient alphabet we can construct a system of Diophantine equations $\mathfrak{A}^*$ such that there is a one-to-one correspondence between the solutions of $\mathfrak{A}$ and the natural solutions of $\mathfrak{A}^*$.

We will show how to construct the system $\mathfrak{A}^*$.

To each word Φ in the alphabet $\{a, b, x_1, \cdots, x_n\}$ we assign a second-order matrix $\gamma(\Phi)$ by the following inductive rule:

1) $\gamma(a) = A$, $\gamma(b) = B$, $\gamma(x_i) = \begin{bmatrix} x_{i1} & x_{i2} \\ x_{i3} & x_{i4} \end{bmatrix}$,

where the x_{ij} are unknowns which assume natural values.

2) If $\Phi = \xi_1 \cdots \xi_k$, where $k \geqq 1$ and $\xi_1, \cdots, \xi_k$ are letters of the alphabet $\{a, b, x_1, \cdots, x_n\}$, then $\gamma(\Phi) = \gamma(\xi_1) \times \cdots \times \gamma(\xi_k)$, where $\times$ is the symbol for matrix multiplication. To the empty word is assigned the identity matrix. For any word Φ in the alphabet $\{a, b, x_1, \cdots, x_n\}$ the matrix $\gamma(\Phi)$ has the form $\begin{bmatrix} P_1 & P_2 \\ P_3 & P_4 \end{bmatrix}$, where the P_i are polynomials in the x_{ij} with natural coefficients. If now there is given a system of word equations (1), the matrices $\gamma(\Phi_i)$ have the form $\begin{bmatrix} P_{i1} & P_{i2} \\ P_{i3} & P_{i4} \end{bmatrix}$ and the matrices $\gamma(\Psi_i)$ the form $\begin{bmatrix} Q_{i1} & Q_{i2} \\ Q_{i3} & Q_{i4} \end{bmatrix}$, where the P_{ij} and Q_{ij} are polynomials with natural coefficients in the x_{ij}. To the system (1) there corresponds the system of Diophantine equations $P_{ij} = Q_{ij}$ $(i = 1, \cdots, N;\ j = 1, \cdots, 4)$, $x_{i1}x_{i4} - x_{i2}x_{i3} = 1$ $(i = 1, \cdots n)$. This is the system $\mathfrak{A}^*$.

If P is a word in the alphabet $\{a, b\}$ and $\gamma(P) = \begin{bmatrix} a_1 & a_2 \\ a_3 & a_4 \end{bmatrix}$, we put $\gamma_i(P) = a_i$ $(i = 1, \cdots, 4)$. If M is a matrix in Π and has the representation $M = A_1 \times \cdots \times A_k$ in terms of the generating matrices, put $\bar{\gamma}(M) = a_1 \cdots a_k$, where $a_i = a$ if $A_i = A$ and $a_i = b$ if $A_i = B$.

Nielsen's result implies the following. If $(X_1, \cdots, X_n)$ is a solution of the system $\mathfrak{A}$, the set $x_{ij} = \gamma_j(X_i)$ satisfies the system $\mathfrak{A}^*$. If the set $\{x_{ij}\}$, $1 \leqq i \leqq n$, $1 \leqq j \leqq 4$, satisfies $\mathfrak{A}^*$, then the set $(X_1, \cdots, X_n)$, where X_i is defined by $X_i = \bar{\gamma}\left(\begin{bmatrix} x_{i1} & x_{i2} \\ x_{i3} & x_{i4} \end{bmatrix}\right)$, satisfies $\mathfrak{A}$.

Thus the solution of word equations may be of arithmetic interest. This

[1] Here and henceforth the letter $\mathit{И}$ denotes the coefficient alphabet.

is especially true since, as was proved by Ju. V. Matijasevič, a general algorithm for solving Diophantine equations does not exist.

3. This monograph consists of four chapters.

Chapter I is of an auxiliary nature and contains information needed for the succeeding chapters. It consists of three sections. In §1 there are collected a number of technical facts which will be used throughout the entire paper. In §2 we consider so-called exponential equations, which play an important role in the study of word equations. In §3 the method used in this monograph is described in general terms and the fundamental concepts and definitions are introduced. The reader may begin with §3, referring to the auxiliary material of §§1 and 2 when it becomes necessary.

Chapter II deals with word equations in three unknowns. We construct an algorithm recognizing whether or not an arbitrary given equation in three unknowns has a solution.

In Chapter III we consider systems consisting of word equations in two unknowns (W_2-systems). We construct an algorithm recognizing whether or not an arbitrary given W_2-system has a solution. In §§11 and 12 of this chapter we obtain a description of the general solution of word equations in two unknowns, which is of interest in its own right.

Chapter IV is devoted to coefficient-free equations. The main result concerns equations in three unknowns. We prove that the general solution of such equations can be expressed by means of a finite number of formulas containing only the (semigroup) operations of multiplication and raising to a power. We also prove that for equations in four unknowns such a representation can no longer be obtained.

The results of Chapter II were first stated in [**2**], those of Chapter III in [**3**], and those of Chapter IV in [**4**].

Throughout this entire paper we use the symbolism and propositions of mathematical logic (see, for example, [**5**]). In particular, we use the symbols & (conjunction), $\vee$ (disjunction), $\urcorner$ (negation), $\supset$ (implication), $\Leftrightarrow$ (equivalence), $\exists$ (existential quantifier) and $\forall$ (universal quantifier).

Propositions are given two numbers. The first indicates the number of the section, the second the number of the proposition within that section.

I would like to take this opportunity to express my deep gratitude to my teacher, A. A. Markov, under whose guidance the main results were obtained, and also to S. I. Adjan for his constant interest and encouragement.

CHAPTER I

PRELIMINARIES

§1. Auxiliary propositions

In this section we prove a number of auxiliary propositions which will be used throughout the entire work. The starting point for the proof of these propositions is the theorem on solutions of coefficient-free word equations in two unknowns (1.1).

By analogy with numbers, the symbol A^m denotes the word $A \cdots A$ (m times) if $m > 0$ and the empty word if $m = 0$. We say that a word P is a power of a word A if $P = A^m$ for some $m \geqq 0$. A word P is called simple if P is not representable in the form $P = S^a$, where $a > 1$. Note that the empty word Λ is not simple, since $\Lambda = \Lambda^a$ for any $a \geqq 0$. Hence any simple word is nonempty.

It is obvious that any word P is representable in the form $P = S^a$, where $a \geqq 0$ and S is a simple word. Moreover, if $P = \Lambda$, then $a = 0$; if P is simple, then $a = 1$; and if P is nonempty and is not simple, then $a > 1$. Thus any word is a power of a simple word.

The number of letters in a word is called the length of the word. The length of a word P is denoted by the symbol $[P^\partial$. The length of the empty word is taken to be zero.

A word equation is called trivial (nontrivial) if its left-hand side is graphically equal (unequal) to its right-hand side. We may assume that the left- and right-hand sides of any nontrivial equation begin with different letters. It is always possible to arrange this by cancelling the equation on the left and, if one of the sides is empty, by multiplying on the right by a suitable letter.

Let us consider a nontrivial coefficient-free word equation in two unknowns, say x and y. It has the form

$$x\Phi(x,y) = y\Psi(x,y), \tag{*}$$

where Φ and Ψ are words in the alphabet $\{x,y\}$. We prove the following assertion.

1.1. *If a pair of words (X, Y) is a solution of a nontrivial coefficient-free equation in two unknowns, there exist integers $a, b \geqq 0$ and a simple word S such that $X = S^a$ and $Y = S^b$.*

PROOF (BY INDUCTION ON $[XY^\partial$). If $[XY^\partial = 0$, then $X = Y = \Lambda = S^0$, where S is simple, and the assertion is true. Suppose $n > 0$ and assume that the assertion is true whenever $[XY^\partial < n$. Now suppose that $[XY^\partial = n$ and the pair (X, Y) is a solution of (*). Then we have the equality in the alphabet $\mathcal{U}$

$$X\Phi(X,Y) = Y\Psi(X,Y). \tag{1}$$

Without loss of generality, we consider the case $[X^\partial \geqq [Y^\partial$. Then (1) implies that there exists a word Z such that $X = YZ$. But then from (1) we have

$YZ\Phi(YZ, Y) = Y\Psi(YZ, Y)$. Cancelling this equality on the left by Y, we have that $Z\Phi(YZ, Y) = \Psi(YZ, Y)$, so that the pair (Y, Z) satisfies the word equation

$$z\Phi(yz, y) = \Psi(yz, y). \tag{2}$$

If $\Psi = \Lambda$, then $Z = \Lambda$; hence $X = Y$ and the assertion is obvious. If $\Psi \neq \Lambda$, the right-hand side of (2) begins with y, so that (2) is a nontrivial coefficient-free equation in two unknowns. If $Y = \Lambda$, then $Y = X^0$ and the assertion is obvious. And if $Y \neq \Lambda$, then $[X^\partial > [Z^\partial$, $[YZ^\partial < [YX^\partial$ and the induction assumption is applicable to the pair (Y, Z), whence we obtain that $Z = S^c$ and $Y = S^b$ for some $b, c \geqq 0$ and simple word S. Then $X = YZ = S^{b+c}$. The proposition is proved.

We note some special cases of Proposition 1.1.

1.2. *If $XY = YX$, then $X = S^a$ and $Y = S^b$, where $a, b \geqq 0$ and S is a simple word.*

1.2′. *If AB is a simple word and $AB = BA$, then either $A = \Lambda$ or $B = \Lambda$.*

This follows from 1.2 and the definition of a simple word.

1.3. *Suppose S and T are simple words and $S^a = T^b$, $a \geqq 1$, $b \geqq 1$. Then $S = T$.*

This follows from 1.1 and the definition of a simple word.

A word P is called a beginning (end) of a word Q if there exists a word R such that $Q = PR$ $(Q = RP)$.

We write $P \angle Q$ if P is a beginning of Q.

1.4. *If S is a simple word and $PS \angle S^a$, then P is a power of S.*

Proof. Since $PS \angle S^a$, we have $PSV = S^a$ for some V. This equality implies that there exist $k \geqq 0$ and words A and B such that $P = S^k A$ and $S = AB$. Then $(AB)^k AABV = (AB)^a$. Since AB is a simple word, we have $AB \neq \Lambda$, $k < a$. Cancelling the last equality on the left by $(AB)^k A$, we obtain $ABV = (BA)^{a-k-1}B$. If $a - k - 1 = 0$, then $A = \Lambda$, and hence $P = S^k$. If $a - k - 1 > 0$, then, clearly, $AB = BA$ and by 1.2′ either $A = \Lambda$ or $B = \Lambda$. If $A = \Lambda$, then $P = S^k$. If $B = \Lambda$, then $A = S$ and $P = S^{k+1}$. The proposition is proved.

1.4′. *Suppose $PA \angle A^n$. Then there exist $a, b \geqq 0$ and a simple word S such that $A = S^a$ and $P = S^b$.*

Proof. If $A = \Lambda$, the assertion is obvious. If $A \neq \Lambda$, write A in the form $A = S^c$, where $c \geqq 1$ and S is a simple word, and apply 1.4.

1.4″. *Suppose $PAB \angle B(AB)^n$, $P \neq \Lambda$ and AB is a simple word. Then there exists $c \geqq 0$ such that $P = B(AB)^c$.*

Proof. Let us first prove that $[P^\partial \geqq [B^\partial$. We argue by contradiction. If $[P^\partial < [B^\partial$, then $B = PC$ and $APC \angle C(APC)^n$ for some $C \neq \Lambda$. Since $P \neq \Lambda$, then $n > 0$ and $APC = CAP$. This is impossible by 1.2′, since $C \neq \Lambda$, $P \neq \Lambda$ and APC is a simple word. We have proved that $[P^\partial \geqq [B^\partial$. Then $P = BC$, $CAB \angle (AB)^n$ and, by 1.4, C is a power of AB: $C = (AB)^c$; hence $P = B(AB)^c$, as required.

The following group of propositions is related to the equality $A^n V_1 = B^m V_2$. We prove that under certain specific conditions on n, m, $[A^\partial$ and $[B^\partial$ the words A and B are powers of the same word.

1.5. *Suppose $A^n V_1 = B^2 V_2$, where $[A^\partial \leqq [B^\partial \leqq (n-1)[A^\partial$. Then there exist $a, b \geqq 0$ and a simple word S such that $A = S^a$ and $B = S^b$. In particular, if A and B are simple words, then $A = B$.*

PROOF. The assertion is trivial if $[A^\partial = [B^\partial = (n-1)[A^\partial$. There remains the case $[A^\partial < [B^\partial < (n-1)[A^\partial$. In this case there exist an integer k, $0 < k < n-1$, and words P and Q such that

$$B = A^k P, \qquad A = PQ. \tag{1}$$

Substituting for A and B the expressions (1) in the equality $A^n V_1 = B^2 V_2$ and cancelling the resulting equality on the left by $(PQ)^k P$, we have $(QP)^{n-k-1} Q V_1 = (PQ)^k P$. Since $0 < k < n-1$, the last equality implies that $QP = PQ$. In view of 1.2, the words Q and P are powers of a simple word S, and the assertion being proved follows from (1).

1.6. *Suppose $A^n V_1 = B^m V_2$ and $n[A^\partial, m[B^\partial \geqq [AB^\partial$. Then A and B are powers of the same simple word. In particular, if A and B are simple words, then $A = B$.*

PROOF. We restrict ourselves to the case $[B^\partial \geqq [A^\partial > 0$. From the estimates for n and m it follows that $[B^\partial \leqq (n-1)[A^\partial$, $m \geqq 2$. Now apply 1.5. The proposition is proved.

1.7. *Suppose $A^n V_1 = B^m V_2$, where $n, m > 1$, and $V_1 \angle A$, $V_2 \angle B$. Then A and B are powers of the same word. If A and B are simple words, then under the additional assumption $V_1, V_2 \neq \Lambda$ we have $A = B$, $n = m$ and $V_1 = V_2$.*

PROOF. Without loss of generality, we consider the case $[B^\partial \geqq [A^\partial$. Clearly, $[B^\partial \leqq n[A^\partial$. Since $V_1 \angle A$, on multiplying $A^n V_1 = B^m V_2$ by a suitable end of the word A we obtain $A^{n+1} = B^m V$. We have $[A^\partial \leqq [B^\partial \leqq n[A^\partial$, $m \geqq 2$. In view of 1.5 we obtain $A = S^a$ and $B = S^b$, where $a, b \geqq 0$ and S is a simple word. If A and B are simple words, then $A = B$. Using the assumptions $V_1 \angle A$, $V_2 \angle B$ and $V_1, V_2 \neq \Lambda$, we obtain $n = m$ and $V_1 = V_2$. The proposition is proved.

Any beginning of a word P other than P itself is called a proper beginning of P. We write $Q \mathrel{\underset{\cdot}{\angle}} P$ if Q is a proper beginning of P.

1.7′. *Suppose $A^n V_1 = B^m V_2$, where A and B are simple words, $n, m > 1$, $V_1 \mathrel{\underset{\cdot}{\angle}} A$ and $V_2 \mathrel{\underset{\cdot}{\angle}} B$. Then $A = B$, $V_1 = V_2$ and $n = m$.*

The proof is analogous to that of 1.7.

We say that a word S is a cyclic permutation of a word T if there exist words A and B such that $S = AB$ and $T = BA$.

1.8. *A cyclic permutation of a simple word is a simple word.*

PROOF (BY CONTRADICTION). Assume that AB is a simple word but BA is not simple. Then $BA = T^a$, where $a > 1$ and T is a simple word ($BA \neq \Lambda$ in view of the simplicity of AB). The equality $BA = T^a$ implies that there exist k, $0 \leqq k < a$, and P and Q such that $B = T^k P$, $T = PQ$ and $A = QT^{a-k-1}$.

Then $AB=(QP)^a$, which, since $a>1$, contradicts the simplicity of AB. The proposition is proved.

We now consider the equality $PS^nV_1=QT^mV_2$, where S and T are simple words. We will prove that if $n[S^\partial$ and $m[T^\partial$ are sufficiently large compared to $[S^\partial$, $[T^\partial$ and $|[P^\partial-[Q^\partial|$, then the words S and T are obtained from one another by a cyclic permutation.

1.9. *Suppose S and T are simple words, $[P^\partial \geqq [Q^\partial$, and*

(1) $$PS^nV_1=QT^mV_2,$$

(2) $$n[S^\partial \geqq [ST^\partial, \qquad m[T^\partial \geqq [T^2S^\partial+[P^\partial-[Q^\partial.$$

Then there exist $k\geqq 0$, A and B such that $S=AB$, $T=BA$ and $P=QT^kB$.

PROOF. Since $[P^\partial \geqq [Q^\partial$, it follows from (1) and (2) that there exist an integer k, $0\leqq k<m$, and words A and B such that

(3) $$P=QT^kB, \quad T=BA.$$

Cancelling (1) on the left by P, we obtain

(4) $$S^nV_1=(AB)^{m-k-1}AV_2.$$

From (3) we have the estimate $k[AB^\partial \leqq [P^\partial-[Q^\partial$. From (2) we have $(m-1)[T^\partial \geqq [ST^\partial+[P^\partial-[Q^\partial \geqq [ST^\partial+k[AB^\partial$. Hence $(m-k-1)[AB^\partial \geqq [SAB^\partial$. We have $n[S^\partial \geqq [SAB^\partial$. Furthermore, since T is a simple word and $T=BA$, by 1.8 the word AB is simple. From (4), in view of 1.6, it follows that $S=AB$. Our assertion now follows from (3).

1.9′. *Suppose S and T are simple words and $V_1S^nP=V_2T^m$, where $n[S^\partial \geqq [ST^\partial$ and $m[T^\partial \geqq [T^2SP^\partial$. Then there exist $k\geqq 0$, A and B such that $S=AB$, $T=BA$ and $P=A(BA)^k$.*

The proof is analogous to that of 1.9 (with $Q=\Lambda$), only instead of beginnings one considers ends.

An important role in the sequel is played by the question of when word functions (i.e. functions whose values are words in some alphabet) assume simple values. For example, this situation arises when applying 1.5–1.7 to the case where the bases A and B depend on parameters.

We say that the words X and Y commute if we have a graphical equality $XY=YX$. It follows from 1.2 that X and Y commute if and only if they are powers of the same word. If X and Y do not commute, i.e. $XY\neq YX$, then we write $\mathscr{E}(X,Y)$ (read "X and Y do not commute"). Obviously, if $\mathscr{E}(X,Y)$, then $X\neq\Lambda$ and $Y\neq\Lambda$.

In Propositions 1.10–1.13 we prove the simplicity of certain words depending on parameters and made up of noncommuting words.

1.10. *Suppose $\mathscr{E}(X,Y)$ and $[X^\lambda Y^\partial \geqq 2[XY^\partial$. Then $X^\lambda Y$ is a simple word.*

PROOF (BY CONTRADICTION). Assume that the hypothesis of the theorem is satisfied but that the word $X^\lambda Y$ is not simple. Then there exist an integer $a>1$ and a simple word S such that

(1) $$X^\lambda Y=S^a.$$

Taking into account that $a>1$ and $[X^\lambda Y^\partial \geqq 2[XY^\partial$, we have the chain of inequalities

$$(a-1)\,[S^\partial \geqq (a/2)\,[S^\partial = [X^\lambda Y^\partial/2 \geqq [XY^\partial \geqq [X^\partial,$$

$$(\lambda-1)\,[X^\lambda = \tfrac{1}{2}\cdot 2(\lambda-1)\,[X^\partial \geqq \tfrac{1}{2}((\lambda-1)\,[X^\partial + [XY^\partial)$$
$$= \tfrac{1}{2}\,[X^\lambda Y^\partial = (a/2)\,[S^\partial \geqq [S^\partial;$$

hence $a[S^\partial$, $\lambda[X^\partial \geqq [XS^\partial$. It follows from (1), in view of 1.6 and the simplicity of S, that X is a power of S. But then Y is also a power of S, which contradicts the condition $\mathscr{E}(X, Y)$. We have obtained the required contradiction. The proposition is proved.

1.11. *Suppose* $\mathscr{E}(X, Y)$ *and* $[(X^\lambda Y)^\mu X^\partial \geqq 2[XYX^\partial$, $\lambda > 0$, $\mu > 0$. *Then the word* $(X^\lambda Y)^\mu X$ *is simple.*

Proof. The condition $\mathscr{E}(X, Y)$ implies $\mathscr{E}(X, YX)$. This is easy to see by contradiction. Hence if $\mu = 1$, then the validity of our assertion follows from 1.10 (XY now plays the role of Y in 1.10).

Consider the case $\mu > 1$. Here the proof will be carried out by contradiction. Assume that the word $(X^\lambda Y)^\mu X$ is not simple. Then there exist an integer $a > 1$ and a simple word S such that

$$(1) \qquad (X^\lambda Y)^\mu X = S^a \quad (a, \mu > 1).$$

1) Suppose that $[X^\lambda Y^\partial \geqq [S^\partial$. It follows from (1) that $[X^\lambda Y^\partial \leqq (a-1)\,[S^\partial$. Hence 1.5 is applicable to (1) (S and $X^\lambda Y$ now play the roles of A and B in 1.5); hence $X^\lambda Y$ is a power of S. Then, as is evident from (1), X and Y are powers of S, which contradicts the condition $\mathscr{E}(X, Y)$. Thus the case $[X^\lambda Y^\partial \geqq [S^\partial$ is impossible.

2) Suppose that

$$(2) \qquad [X^\lambda Y^\partial < [S^\partial.$$

If $\mu > 2$ and $[S^\partial > (\mu - 1)\,[X^\lambda Y^\partial$, then we have the chain of inequalities $a[S^\partial \geqq 2[S^\partial > 2(\mu-1)\,[X^\lambda Y^\partial \geqq (\mu+1)\,[X^\lambda Y^\partial \geqq [(X^\lambda Y)^\mu X^\partial$, which contradicts (1). Consequently, if $\mu > 2$, then $[S^\partial \leqq (\mu-1)\,[X^\lambda Y^\partial$, so that $[X^\lambda Y^\partial < [S^\partial \leqq (\mu-1)\,[X^\lambda Y^\partial$. Then it follows from (1) by 1.5 ($X^\lambda Y$ and S now play the roles of A and B in 1.5) that $X^\lambda Y$ is a power of S, which, together with (1), contradicts the condition $\mathscr{E}(X, Y)$. It remains to consider the case $\mu = 2$. In this case it follows from (1), (2) that $a = 2$. But (1) implies that there exist A and B such that $X = AB$, $[A^\partial = [B^\partial$ and $S = X^\lambda YA = BX^{\lambda-1}YX$. From these equalities we obtain in succession $A = B$, $X = A^2$, $AY = YA$ and $XY = A^2Y = YA^2 = YX$, which contradicts the condition $\mathscr{E}(X, Y)$. From the contradictions obtained we conclude that case 2) is also impossible. The proposition is proved.

1.12. *Suppose* $\mathscr{E}(X, Y)$ *and* λ, μ, $\nu \geqq 1$. *Then the word* $P = ((X^\lambda Y)^\mu X)^\nu X^\lambda Y$ *is simple.*

Proof. The reasoning employed in 1.11 shows that if $\mathscr{E}(X, Y)$, $\mu > 0$ and $\nu > 1$, then the word $(X^\mu Y)^\nu X$ is simple (when considering the case $\mu > 1$ in 1.11 we did not use the fact that $[(X^\lambda Y)^\mu X^\partial \geqq 2[XYX^\partial$). On the other hand, $\mathscr{E}(X, Y)$ implies $\mathscr{E}(X^\lambda Y, X)$ (this is easy to prove by contradiction). This implies the simplicity of P when $\nu > 1$. And if $\nu = 1$, P has the form $P = (X^\lambda Y)^\mu X X^\lambda Y$. By what has been proved, its cyclic permutation $(X^\lambda Y)^{\mu+1}X$ is a simple word. In view of 1.8, P is simple. The proposition is proved.

1.13. *Suppose S is a simple word, $\mathscr{E}(S, C)$, $Q \neq C$, $\nu > 4$ and $\sigma[S^\partial \geqq \max\{2[SC^\partial, [Q^\partial\}$. Then the word $(S^\sigma C)^\nu S^\sigma Q$ is simple.*

PROOF (BY CONTRADICTION). Assume that the word $(S^\sigma C)^\nu S^\sigma Q$ is not simple. Then there exist $a > 1$ and a simple word T such that

$$(S^\sigma C)^\nu S^\sigma Q = T^a. \tag{1}$$

In view of 1.10, our assumptions imply that the word $S^\sigma C$ is simple. Comparing the lengths of both sides of (1), we obtain the estimates $[S^\sigma C^\partial \leqq (a-1)[T^\partial$ and $[T^\partial \leqq (\nu - 1)[S^\sigma C^\partial$.

In view of 1.6, these estimates and (1) imply that $S^\sigma C = T$. Then it is evident from (1) that $S^\sigma Q$ is a power of $S^\sigma C$. By our assumptions, $0 < [S^\sigma Q^\partial < 2[S^\sigma C^\partial$. Hence $S^\sigma Q = S^\sigma C$, which contradicts the condition $Q \neq C$. We have obtained the required contradiction. The proposition is proved.

1.14. *Suppose S and T are simple words, $\mathscr{E}(S, C)$ and $\mathscr{E}(T, D)$. Suppose also that $\sigma[S^\partial, t[T^\partial \geqq 2[SCTD^\partial$ and*

$$(S^\sigma C)^3 V_1 = (T^t D)^3 V_2. \tag{1}$$

Then $S = C$ and $S^\sigma C = T^t D$.

PROOF. Applying Proposition 1.6 to (1), we obtain $S = T$. The equality (1) takes the form

$$(S^\sigma C)^3 V_1 = (S^t D)^3 V_2. \tag{2}$$

Let us estimate the lengths of the words $S^\sigma C$ and $S^t D$. If $t[S^\partial \geqq [S^\sigma CS^\partial$, then (2) implies that $CS \angle S^{t-\sigma}$ and, by 1.4, C is a power of S, which contradicts the condition $\mathscr{E}(S, C)$. From this contradiction we conclude that $t[S^\partial < [S^\sigma CS^\partial$. Similarly, $\sigma[S^\partial < [S^t DS^\partial$. Taking into account the lower estimates for $\sigma[S^\partial$ and $t[T^\partial$, we now have

$$[S^\sigma C^\partial < [S^t DSC^\partial \leqslant 2\,[S^{t\partial} \leqslant 2\,[S^t D^\partial,$$

$$[S^t D^\partial < [S^\sigma CSD^\partial \leqslant 2\,[S^{\sigma\partial} \leqslant 2[S^\sigma C^\partial.$$

Hence $3[S^\sigma C^\partial$, $3[S^t D^\partial \geqq [S^\sigma CS^t D^\partial$. By 1.10, the words $S^\sigma C$ and $S^t D$ are simple. Now from (2), in view of 1.6, we conclude that $S^\sigma C = S^t D$, so that $S^\sigma C = T^t D$. The proposition is proved.

1.15. *Suppose S is a simple word, $\mathscr{E}(S,C)$ and*

$$AS^\sigma CS^\sigma V_1 = BS^\sigma CS^\sigma V_2, \tag{1}$$

$$(\sigma - 1)[S^\partial \geqq |\,[A^\partial - [B^\partial|. \tag{2}$$

Then $A = B$, $V_1 = V_2$.

PROOF. It suffices to prove that $[A^\partial = [B^\partial$. We will prove this by contradiction. Assume that $[A^\partial \neq [B^\partial$. Without loss of generality, we suppose that $[A^\partial > [B^\partial$. Then (1) implies that there exists $D \neq \Lambda$ such that $A = BD$. Then from (1) we have

$$DS^\sigma CS^\sigma V_1 = S^\sigma CS^\sigma V_2. \tag{3}$$

Let us estimate $[DS^\partial$ from above. We have $[D^\partial = [A^\partial - [B^\partial \leqq (\sigma - 1)[S^\partial$, so that $[DS^\partial \leqq \sigma[S^\partial$. Then (3) implies that $DS \angle S^\sigma$ and, by 1.4, D is a power of S. Since $D \neq \Lambda$, we have $D = S^a$ for some $a > 0$. Now cancelling (3) on the left by S^σ, we obtain

$$S^a CS^\sigma V_1 = CS^\sigma V_2. \tag{4}$$

Since $D = S^a$ and $[D^\partial \leqq (\sigma - 1)[S^\partial$, we conclude that $a < \sigma$. Then (4) implies that $S^a C = CS^a$. We have $a > 0$, so that the pair (S, C) satisfies a coefficient-free equation in two unknowns. In view of 1.1, this contradicts the condition $\mathscr{E}(S, C)$. The equality $[A^\partial = [B^\partial$, and hence also Proposition 1.15, is proved.

The following assertion is the starting point for the method of studying word equations used in this monograph. Its formulation brings to mind division of natural numbers (X being the dividend, Y the divisor, t the quotient and A the remainder).

1.16. *Suppose $a > 0$, $Y \neq \Lambda$ and $XV_1 = Y^a X V_2$. Then there exist uniquely an integer $t \geqq 0$ and a word A such that $X = Y^t A$ and $A \underset{\cdot}{\angle} Y$.*

PROOF. We proceed by induction on the length of X.

If $[X^\partial < [Y^\partial$, then, obviously, $X \underset{\cdot}{\angle} Y$ and the assertion is true with $t = 0$ and $A = X$.

Suppose $n \geqq [Y^\partial$, and assume that the assertion being proved is true whenever $[X^\partial < n$. Now suppose that $[X^\partial = n$. Since $n \geqq [Y^\partial$, we have $[X^\partial \geqq [Y^\partial$. Hence $Y \angle X$, so that there exists (uniquely) an X_1 such that $X = YX_1$. Cancelling the equality $XV_1 = Y^a X V_2$ on the left by Y, we obtain $X_1 V_1 = Y^a X_1 V_2$. Since $Y \neq \Lambda$ and $X = YX_1$, it follows that $[X_1^\partial < [X^\partial$, i.e. $[X_1^\partial < n$. Hence the induction assumption is applicable to the equality $X_1 V_1 = Y^a X_1 V_2$, so there exist uniquely an integer $t \geqq 0$ and a word A such that $X_1 = Y^t A$, $A \underset{\cdot}{\angle} Y$. Then $X = Y^{t+1} A$. The proposition is proved.

1.16′. *Suppose $a > 0$, $X \neq \Lambda$, $Y \neq \Lambda$ and $XV_1 = Y^a X V_2$. Then there exist uniquely an integer $t \geqq 0$ and a word A such that $X = Y^t A$ and $\Lambda \neq A \angle Y$.*

The proof is the same as that of 1.16, except that as the basis of the induction we use $0 < [X^\partial \leqq [Y^\partial$ and in the induction step $n > [Y^\partial$.

1.16″. *Suppose $a > 0$, $X \neq \Lambda$, $Y \neq \Lambda$ and $XV_1 = Y^a X V_2$. Then there exist $t \geqq 0$ and words A and B such that $Y = AB$, $X = (AB)^t A$ and $A \neq \Lambda$.*

This is a rephrasing of 1.16′.

1.17. *Suppose $X^a Y V_1 = Y^b X V_2$, $a > 0$ and $b > 0$. Then $XY = YX$.*

PROOF. The assertion is trivial if $[X^\partial = [Y^\partial$ or if one of the words X or Y is empty. Suppose $X, Y \neq \Lambda$ and $[X^\partial \neq [Y^\partial$. Without loss of generality, we may assume that $[X^\partial > [Y^\partial$. By 1.16″ we have $Y = AB$ and $X = (AB)^t A$ for $t \geqq 0$ and some A and B. Since $[X^\partial > [Y^\partial$, we have $t > 0$. Substituting the expressions obtained for X and Y in place of X and Y in the equality $X^a Y V_1 = Y^b X V_2$ and cancelling the resulting equality on the left by $(AB)^t A$, we obtain

(1) $$((AB)^t A)^{a-1} ABV_1 = (BA)^b V_2.$$

Since $b > 0$ and $t > 0$, we have from (1) that $AB = BA$; hence it follows from 1.2 that A and B are powers of a simple word. Our assertion now follows from the equalities $Y = AB$ and $X = (AB)^t A$. The proposition is proved.

The following group of propositions deals with the solution of word equations in one unknown.

When considering word equations we may assume that the left- and right-hand sides of an equation contain occurrences of the unknowns (it suffices to multiply an equation on the right by an unknown). Furthermore, we may restrict ourselves to equations which are not cancellable on the left. Under these assumptions any word equation in one unknown has the form[2)]

(*) $$A\,[xA_i]_{i=1}^n = [xB_i]_{i=1}^m,$$

where $n, m \geqq 1$; $A, A_1, \cdots, A_n$, $B_1, \cdots, B_m$ are the coefficients and $A \neq \Lambda$. It follows from 1.16 that if X is a solution of (*), then $X = A^\lambda B$, where $\lambda \geqq 0$ and $B \angle A$. We obtain an equality

(**) $$P_0[A^\lambda P_i]_{i=1}^n = Q_0[A^\lambda Q_i]_{i=1}^m,$$

where $n, m \geqq 1$; A, P_i and Q_i are words in the alphabet И; and $A \neq \Lambda$. We will prove that if (**) holds for a sufficiently large λ, then it holds for any $\lambda \geqq 0$.

1.18. *Suppose $A \neq \Lambda$. Let*

$$k = \min\{n, m\}, \qquad d = \max_{0 \leq i \leq k}\{\,|\,[P_0 \cdots P_i^\partial - [Q_0 \cdots Q_i^\partial\,|\,\}, \qquad G = d/[A^\partial.$$

If (**) holds for some $\lambda \geqq G + 1$, then it holds for any $\lambda \geqq 0$.

PROOF. Suppose (**) holds and $\lambda \geqq G+1$. Comparing the lengths of the left- and right-hand sides of (**), we obtain $n = m = k$. We will now prove 1.18 under the assumption that the word A is simple. We proceed by induction on k.

Suppose $k = 1$. Then (1) has the form $P_0 A^\lambda P_1 = Q_0 A^\lambda Q_1$. Without loss of generality, we may assume that $[P_0^\partial \geqq [Q_0^\partial$. Then $P_0 = Q_0 R$ and $RA^\lambda P_1 = A^\lambda Q_1$, for some R. From the estimate for λ it follows that $[RA^\partial \leqq \lambda[A^\partial$, and then from the latter equality we have $RA \angle A^\lambda$; hence, by 1.4, R is a power of A. We have $R = A^c$ and $Q_1 = A^c P_1$, where $c \geqq 0$. Hence $P_0 = Q_0 A^c$, $Q_1 = A^c P_1$ and the equality $P_0 A^\lambda P_1 = Q_0 A^\lambda Q_1$ holds for any $\lambda \geqq 0$. The assertion is proved for $k = 1$.

Suppose $l > 1$ and the assertion is valid whenever $k < l$. We will prove it for $k = l$. We restrict ourselves to the case $[P_0^\partial \geqq [Q_0^\partial$. Arguing as in the case $k = 1$, we have for some $c \geqq 0$ that $P_0 = Q_0 A^c$ and $A^c P_1[A^\lambda P_i]_{i=2}^n = Q_1[A^\lambda Q_i]_{i=2}^n$. Obviously, $[A^c P_1^\partial - [Q_1^\partial = [P_0 P_1^\partial - [Q_0 Q_1^\partial$ and, as is easily seen, the induction assumption is applicable to the next to last equality. Applying it and taking into account that $P_0 = Q_0 A^c$, we see that the assertion being proved is true.

Thus, the assertion is true if A is a simple word. In the general case, since $A \neq \Lambda$, we put $A = S^a$, where $a \geqq 1$ and S is a simple word, and consider the equality

---

[2)] The symbol $[\ \ ]_{i=m}^n$ was defined in §1 of the Introduction.

(1) $$P_0[S^\mu P_i]_{i=1}^n = Q_0[S^\mu Q_i]_{i=1}^m.$$

Now if (**) holds, then (1) holds with $\mu = a\lambda$, and $\mu = a\lambda \geqq a(G+1) = ad/[A^\partial + a = d/[S^\partial + a \geqq d/[S^\partial + 1$. By what has been proved, (1) holds for any $\mu \geqq 0$. Then (**) holds for any $\lambda \geqq 0$. The proposition is proved.

Now consider the equation (*). The number $M = \max_{i,j}\{n, m[A_i^\partial, [B_j^\partial, [A^\partial\}$ is called the characteristic of (*).

1.19. *Suppose a word equation in one unknown of characteristic M has a solution of length $\geqq M^2 + 3M$. Then the equation has a solution of length $\leqq M$.*

PROOF. Let X be a solution of (*). Then for some V_1 and V_2 we have $AXV_1 = XV_2$; hence, in view of 1.16, there exist an integer $\lambda \geqq 0$ and a word B such that $X = A^\lambda B$, $B \angle A$. We have in the alphabet **И** the equality

(1) $$A[A^\lambda B A_i]_{i=1}^n = [A^\lambda B B_i]_{i=1}^m.$$

For the quantities d, G of 1.18 we have the estimates

$$d = \max_{0 \leqq i \leqq k}\{|[AA_1 \cdots A_i^\partial - [Q_1 \cdots Q_i^\partial|\} \leqq [A^\partial + M^2, \quad G = d/[A^\partial \leqq 1 + M^2/[A^\partial.$$

Since by hypothesis $[X^\partial \geqq 3M + M^2$ and $X = A^\lambda B$, $B \angle A$, it follows that $\lambda \geqq (2M + M^2)/[A^\partial \geqq 1 + G$. By 1.18, equality (1) is satisfied for any $\lambda \geqq 0$. But, obviously, if $\lambda \geqq 0$ satisfies (1), then $X = A^\lambda B$ is a solution of (*). Hence $X = B$ is the desired solution of (*). The proposition is proved.

1.20. *There exists an algorithm recognizing the solvability of word equations in one unknown.*

PROOF. The algorithm consists of going through the words in the alphabet **И** of length $\leqq M^2 + 3M$, where M is the characteristic of the equation. If among these words there is a solution of the equation, the equation has a solution, but if not, then by 1.19 the equation has no solution.

It is sometimes convenient to consider equalities not in a free semigroup, but in the free group whose positive alphabet is the alphabet **И**. In 1.21 and 1.22 the symbol $\amalg$ denotes equality in the free group whose positive alphabet is **И**.

1.21. *Suppose $AS^\lambda B = CS^\lambda D$, $(\lambda - 1)[S^\partial \geqq |[A^\partial - [B^\partial|$ and S is a simple word. Then there exists an integer d such that $A \amalg CS^d$ and $S^d B \amalg D$.*

This assertion was proved, in essence, in the analysis of the case $k = 1$ in 1.18.

1.22. *Suppose $S^a X S^\lambda V_1 \amalg Y S^\mu V_2$, where X, Y, S, V_1, V_2 are words in the alphabet **И**, S is a simple word, $\mathscr{L}(X,S)$, $\mathscr{L}(Y,S)$, and $(\lambda - 2)[S^\partial \geqq [Y^\partial$ and $(\mu - 2)[S^\partial \geqq [X^\partial$. Then there exists an integer c such that $S^a X \amalg Y S^c$.*

PROOF. We carry out the argument not in the group, but, as usual, in the semigroup. Without loss of generality, we consider the case $a \geqq 0$. We have in the alphabet **И** the equality

(1) $$S^a X S^\lambda V_1 = Y S^\mu V_2.$$

We will first prove that $[Y^\partial \geqq (a-1)[S^\partial$. We argue by contradiction. If

$[Y^\partial < (a-1)[S^\partial$, then $Y = S^kA$ and $S = AB$ for certain A, B and k, $0 \leqq k < a-1$. Cancelling (1) on the left by Y, we obtain $B(AB)^{a-1-k}\cdots = (AB)^\mu V_2$. Since $a-1-k$, $\mu > 0$, we have $BA = AB$ and, by 1.2, the words A and B are powers of some T. Then S and Y are powers of T, which contradicts the condition $\mathscr{E}(Y,S)$. We have proved that $[Y^\partial \geqq (a-1)[S^\partial$.

Now if $[S^aX^\partial \geqq [Y^\partial$, then $S^aX = YR$ and $RS^\lambda V_1 = S^\mu V_2$ for some R. We have $[R^\partial = a[S^\partial + [X^\partial - [Y^\partial \leqq [SX^\partial \leqq (\mu - 1)[S^\partial$. Then $RS \angle S^\mu$ and, by 1.4, $R = S^c$ and $S^aX = YS^c$ for some $c \geqq 0$.

If $[S^aX^\partial \leqq [Y^\partial$, we have $Y = S^aXR$ and $S^\lambda V_1 = RS^\mu V_2$ for some R. Since $[R^\partial \leqq [Y^\partial \leqq (\lambda - 1)[S^\partial$, it follows that $RS \angle S^\lambda$, $R = S^c$ and $Y = S^aXS^c$, i.e. $YS^{-c} \amalg S^aX$, where $c \geqq 0$. The proposition is proved.

1.23. *Suppose $XYZ = ZYX$ and $X, Z \neq \Lambda$. Then there exist integers $a, b, c \geqq 0$ and words U, V such that $X = (UV)^aU$, $Y = V(UV)^b$, $Z = (UV)^cU$ and the word UV is simple.*

PROOF. Suppose $XYZ = ZYX$. Then $XYZY = ZYXY$, and XY and ZY commute. By 1.2 we have $XY = S^n$ and $ZY = S^m$, where $n, m \geqq 0$ and S is a simple word. Since $X, Z \neq \Lambda$, we see that $n, m > 0$, and hence, obviously, $X = (UV)^aU$, $S = UV$, $Y = V(UV)^b$ and $Z = (UV)^cU$ for some $a, b, c \geqq 0$ and U, V. The proposition is proved.

1.24. *Suppose $XYZ = ZXY$ and $XY \neq \Lambda$. Then there exist $a, b, c \geqq 0$ and words U, V such that $X = (UV)^aU$, $Y = V(UV)^b$ and $Z = (UV)^c$.*

PROOF. In view of 1.2, $XYZ = ZXY$ implies that $XY = S^d$ and $Z = S^c$ for some $c, d \geqq 0$ and simple word S. Since $XY \neq \Lambda$, we have $d > 0$; hence $XY = S^d$ implies that there exist $a, b \geqq 0$ and words U, V such that $S = UV$, $X = (UV)^aU$ and $Y = V(UV)^b$. The proposition is proved.

1.25. *Suppose $XYZ = ZXY = YZX$. Then X, Y and Z are powers of the same word.*

PROOF. If $XY = \Lambda$, the assertion is trivial. If $XY \neq \Lambda$, then by 1.24 we have $X = (UV)^aU$, $Y = V(UV)^b$ and $Z = (UV)^c$ for some U, V and a, b, $c \geqq 0$. Hence $XYZ = (UV)^{a+b+c+1}$, $YZX = (VU)^{a+b+c+1}$. Since $XYZ = YZX$, we have $(UV)^{a+b+c+1} = (VU)^{a+b+c+1}$, and so $UV = VU$. We now apply 1.2. The proposition is proved.

1.26. *Suppose XA and T are simple words, $[X^\partial > [A^\partial$ and*

$$(XA)^\mu X = T^\lambda V, \qquad \mu > 0,\ \lambda > 1,$$

where $\Lambda \neq V \angle T$ and either $[V^\partial > [A^\partial$ or $\lambda > [A^\partial + 1$. Then $XA = T$, $X = V$ and $\lambda = \mu$.

PROOF. If $\mu > 1$, the assertion follows from 1.7. The case $\mu = 1$, as we will now prove, is impossible.

Suppose $\mu = 1$. In this case we have

$$XAX = T^\lambda V, \qquad \lambda > 1,\ \Lambda \neq V \angle T. \tag{1}$$

We will prove that $[T^\partial < [XA^\partial < \lambda[T^\partial$. The inequality $[T^\partial < [XA^\partial$ obviously follows from (1). We will prove by contradiction that $[XA^\partial < \lambda[T^\partial$. Assume that $[XA^\partial \geqq \lambda[T^\partial$. Then (1) implies that $[X^\partial \leqq [V^\partial$. Since $V \angle T$, we see that

$[V^\partial \leqq [T^\partial$. Hence $[X^\partial \leqq [V^\partial \leqq [T^\partial$. But since by hypothesis $[A^\partial < [X^\partial$, we get $[XA^\partial < 2[T^\partial$, which contradicts the assumption $[XA^\partial \geqq \lambda[T^\partial$, since $\lambda > 1$. Hence we have proved that $[T^\partial < [XA^\partial < \lambda[T^\partial$. Then (1) implies that there exist words P, Q and an integer a, $0 < a \leqq \lambda - 1$, such that $XA = T^aP$, $T = PQ$ and $P \neq \Lambda$. Now cancelling (1) on the left by XA, we obtain $X = Q(PQ)^{\lambda-a-1}V$; hence $XA = Q(PQ)^{\lambda-a-1}VA$. But $XA = T^aP$. Hence

$$(PQ)^aP = Q(PQ)^{\lambda-a-1}VA. \tag{2}$$

We will prove that the right-hand side of (2) begins with the word QP. This is obvious if $a < \lambda - 1$. Suppose $a = \lambda - 1$. Then

$$(PQ)^{\lambda-1}P = QVA, \qquad P \neq \Lambda, \quad \lambda > 1. \tag{3}$$

We will prove that $[V^\partial \geqq [P^\partial$. Indeed, by hypothesis either $[V^\partial > [A^\partial$ or $\lambda > [A^\partial + 1$. If $\lambda > [A^\partial + 1$, then the inequality $[V^\partial \geqq [P^\partial$ obviously follows from (3). Consider the case $[V^\partial > [A^\partial$. It follows from (3) that $[VA^\partial \geqq 2[P^\partial$, and since $[V^\partial > [A^\partial$, we have $2[V^\partial > 2[P^\partial$ and $[V^\partial > [P^\partial$. We have proved that $[V^\partial \geqq [P^\partial$. But $P \angle T$ and by hypothesis $V \angle T$. Hence $P \angle V$, so the right-hand side of (3) begins with the word QP. From (3) we obtain $PQ = QP = T$. Since $P \neq \Lambda$, by 1.2′ we must have $Q = \Lambda$. But then $T = P$ and $XA = T^{a+1}$, which contradicts the simplicity of the word XA, since $a > 0$. Thus the case $\mu = 1$ is impossible. The proposition is proved.

1.27. *Suppose $m, n > 1$, $[XP_i]_{i=1}^n = S^m$, $[X^\partial \geqq [S^\partial > \max_{1 \leqq i \leqq n}\{[P_i^\partial\}$ and S is a simple word. Then $P_1 = \cdots = P_n$.*

Proof. Our assumptions imply that there exist $k > 0$, A and B such that $X = S^kA$ and $S = AB$. The given equality takes the form $[S^kAP_i]_{i=1}^n = S^m$ or, after cancelling on the left and multiplying on the right by S^k, the form $[AP_iS^k]_{i=1}^n = S^m$. From this last equality we obtain by an obvious induction using 1.4 that the words $AP_1, \cdots, AP_n$ are powers of S. The hypothesis $[S^\partial > \max\{[P_1^\partial, \cdots, [P_n^\partial\}$ implies that the exponents on all these powers are the same; hence $P_1 = \cdots = P_n$. The proposition is proved.

In conclusion, we consider the reduction of a system of equations to a single equation.

Suppose $\mathfrak{A}$ and $\mathfrak{B}$ are systems of word equations in n unknowns. The systems $\mathfrak{A}$ and $\mathfrak{B}$ are called equivalent if any solution of $\mathfrak{A}$ is a solution of $\mathfrak{B}$, and conversely.

1.28. *For any system of word equations in n unknowns it is possible to construct a word equation in n unknowns which is equivalent to the system.*

Proof.[3] It suffices to prove the assertion for a system consisting of two equations. Suppose a and b are letters of the alphabet **И**.[4] We will prove that the system $\Phi_1 = \Phi_2$, $\Psi_1 = \Psi_2$ is equivalent to the equation $\Phi_1 a \Phi_2 \Phi_1 b \Phi_2 = \Psi_1 a \Psi_2 \Psi_1 b \Psi_2$. This is a consequence of the following lemma.

Lemma. *Suppose that X_1, X_2, Y_1 and Y_2 are words in the alphabet* **И** *and $X_1 a X_2 X_1 b X_2 = Y_1 a Y_2 Y_1 b Y_2$. Then $X_1 = Y_1$ and $X_2 = Y_2$.*

---

[3] This proof is due to A. B. Livčak. Another construction is given in [3].

[4] Here and henceforth we assume that the coefficient alphabet contains at least two letters. If it consists of one letter, solutions will be powers of this letter and word equations obviously reduce to linear Diophantine equations.

Indeed, if words with an even number of letters are equal, then "halves" of these words are equal. Hence $X_1aX_2 = Y_1AY_2$ and $X_1bX_2 = Y_1bY_2$. If X_1 were longer than Y_1, then for some U and V we would have $X_1 = Y_1aU$, $X_1 = Y_1bV$, $aU = bV$ and $a = b$, which is impossible. Similarly, Y_1 is not longer than X_1. Hence the lengths of X_1 and Y_1 are equal. But then $X_1 = Y_1$ and $X_2 = Y_2$. The lemma is proved.

Proposition 1.28 is also proved.

§2. Exponential equations

In solving word equations it is necessary to consider equations whose left- and right-hand sides are words in the alphabet И which depend on natural exponents. We call such equations exponential word equations or, more briefly, exponential equations. More precisely, exponential equations are defined in the following manner. Variables whose values are natural numbers are called natural variables. Let us now define parametric words. The inductive definition of parametric words has the following form: 1) if φ is a word in the alphabet И, then φ is a parametric word; 2) if φ is a parametric word and λ a natural variable, then $(\varphi)^\lambda$ is a parametric word; 3) if φ_1 and φ_2 are parametric words, then $\varphi_1\varphi_2$ is a parametric word. In 2) instead of a natural variable λ there may appear a function λ of natural variables taking integral values.

We write $\varphi(t_1, \cdots, t_k)$ if the list $t_1, \cdots, t_k$ of variables contains all of the variables occurring in the parametric word φ (φ may depend fictitiously on certain of the t_i). Suppose there is given a parametric word $\varphi(t_1, \cdots, t_k)$ and let $\bar{t}_1, \cdots, \bar{t}_k$ be values of the variables $t_1, \cdots, t_k$. Then the "exponents" λ occurring in the parametric word take integral values. If all these values are nonnegative, the result of substituting into φ the numbers $\bar{t}_1, \cdots, \bar{t}_k$ for the unknowns $t_1, \cdots, t_k$ is a word in the alphabet И. We denote this word by $\varphi(\bar{t}_1, \cdots, \bar{t}_k)$ and say that $\varphi(\bar{t}_1, \cdots, \bar{t}_k)$ is defined. If even one exponent is negative, then $\varphi(\bar{t}_1, \cdots, \bar{t}_k)$ is not defined.

By exponential equations we mean equations whose left- and right-hand sides are parametric words. A solution of an exponential equation $\varphi(t_1, \cdots, t_k) = \psi(t_1, \cdots, t_k)$ is any set $\bar{t}_1, \cdots, \bar{t}_k$ of values of the unknowns $t_1, \cdots, t_k$ such that $\varphi(\bar{t}_1, \cdots, \bar{t}_k)$ and $\psi(\bar{t}_1, \cdots, \bar{t}_k)$ are defined and we have in the alphabet И the equality $\varphi(\bar{t}_1, \cdots, \bar{t}_k) = \psi(\bar{t}_1, \cdots, \bar{t}_k)$.

Exponential equations were first considered by Lyndon [6]. Lyndon studied exponential equations in free groups, but his results are also valid for exponential word equations. He proved that any parametric word in a free group is equivalent to the disjunction of a finite number of so-called normal parametric words and thereby deduced that any exponential equation in a free group is equivalent to some Diophantine equation. Similarly, for any exponential word equation $\mathfrak{A}(t_1, \cdots, t_k)$ it is possible to construct a Diophantine equation $P(t_1, \cdots, t_k) = 0$ which is equivalent to $\mathfrak{A}$. Conversely, as is easily seen, any Diophantine equation is equivalent to some exponential word equation. For example, the Diophantine equation $x^2 - y^3 = 1$ is equivalent to the exponential equation $(a^x)^x = ((a^y)^y)^y a$. Hence the general problem of solving exponential word equations is equivalent to the problem of solving Diophantine equations and, consequently, is algorithmically unsolvable. Lyndon, however, showed that if the left- and right-hand sides of an exponential equation are normal parametric words with linear exponents, such an equation is equivalent to a system composed of linear equations and inequalities. The nonlinearity in the general case is related to the fact that in reducing to normal form the unknowns are multiplied. For example, the word $((a^\lambda b)^\mu a^\sigma b)^\nu$ is normal in the region $\lambda \neq \sigma$, but in the region $\lambda = \sigma$ it is equivalent to the normal word $(a^\lambda b)^{\nu(\mu+1)}$ with nonlinear exponent.

The peculiarity of exponential equations encountered in this monograph is that in reducing them to normal form the unknowns are not multiplied, even though we consider both two- and three-stage equations. This is achieved by imposing conditions on the bases of the powers and on the exponents. It is possible that this is always the situation in the case of exponential equations arising in the solution of word equations.

Our investigation of exponential equations differs from that of Lyndon. Since we are constructing an algorithm, only linearizable exponential equations are suitable and our main concern is the problem of linearization. In this regard, the main question is how an exponential equation and its equivalent system of linear Diophantine equations are related. In applications the bases of powers are not constants, but depend on various parameters, in particular, word parameters. Obtaining uniform estimates with respect to such bases for the coefficients of the Diophantine equations is necessary for the solution of various problems (see §§4, 8 and 22 below).

In this section we consider a special case of exponential equations, the so-called one-stage exponential equations. These are equations whose left- and right-hand sides are one-stage parametric words. These words, by definition, have the form

$$P_0[S_i^{\lambda_i} P_i]_{i=1}^n, \tag{*}$$

where $n \geqq 0$; $P_0, \cdots, P_n$, $S_1, \cdots, S_n$ are words in the alphabet И; $S_1, \cdots, S_n$ are nonempty words; and $\lambda_1, \cdots, \lambda_n$ are linear functions of natural variables with positive integral coefficients and integral free term. In the sequel these functions will be called α-polynomials.

Any one-stage exponential equation has the form

$$P_0[A_i^{\lambda_i} P_i]_{i=1}^n = Q_0[B_i^{\mu_i} Q_i]_{i=1}^m, \tag{2.1}$$

where $n, m \geqq 0$; P_i, Q_j are words in the alphabet И; A_i, B_j are nonempty words in the alphabet И; λ_i, μ_j are α-polynomials. The words $P_0, \cdots, P_n$, $Q_0, \cdots, Q_m$ are called the coefficients and the words $A_1, \cdots, A_n$, $B_1, \cdots, B_m$ the bases of equation (2.1).

We are interested in describing the solution set of a one-stage exponential equation. Our results yield an algorithm recognizing whether or not a given one-stage exponential equation has a solution.

It is convenient to divide the study of one-stage exponential equations into two parts. We first consider the so-called reduced exponential equations, and then show that the general case depends on the reduced case.

A parametric word (*) is called reduced if the following conditions are satisfied:

1) $S_1, \cdots, S_n$ are simple words.

2) For any i, $1 \leqq i \leqq n$, if $P_{i-1} \neq \Lambda$, then the last letters of the words P_{i-1} and S_i are distinct.

3) For any i, $1 \leqq i < n$, the word S_i is not a beginning of the word $P_i S_{i+1}^c$, where c is the integral part of $[S_i^\partial / [S_{i+1}^\partial$.

An exponential equation is called reduced if its left- and right-hand sides are reduced parametric words. We will show how one-stage exponential equations depend on reduced equations.

Suppose $\mathscr{P}(t_1, \cdots, t_k)$ is some predicate. Parametric words $\varphi(t_1, \cdots, t_k)$ and $\psi(t_1, \cdots, t_k)$ are called equivalent in the region $\mathscr{P}(t_1, \cdots, t_k)$ if for any natural numbers $\bar{t}_1, \cdots, \bar{t}_k$ such that $\mathscr{P}(\bar{t}_1, \cdots, \bar{t}_k)$ we have $\varphi(\bar{t}_1, \cdots, \bar{t}_k) = \psi(\bar{t}_1, \cdots, \bar{t}_k)$. Now suppose that $\mathfrak{A}(t_1, \cdots, t_k)$ and $\mathfrak{B}(t_1, \cdots, t_k)$ are exponential equations. We call them equivalent in the region $\mathscr{P}(t_1, \cdots, t_k)$ if any set $(\bar{t}_1, \cdots, \bar{t}_k)$ of natural numbers such that $\mathscr{P}(\bar{t}_1, \cdots, \bar{t}_k)$ which satisfies one of the equations $\mathfrak{A}$ or $\mathfrak{B}$ also satisfies the other. It is clear that if the words φ and ψ are equivalent to the words φ_1 and ψ_1 in some region, then the equation $\varphi = \psi$ is equivalent in this region to the equation $\varphi_1 = \psi_1$.

Throughout this section the letters $A, B, C, D, P, Q, R, S, T$ with or without indices denote words in the alphabet **И**. If $A = S^a$, where S is a simple word, then S is called the period of A.

2.1. *For any parametric word $\varphi = P_0[A_i^{\lambda_i} P_i]_{i=1}^n$ it is possible to construct a reduced parametric word which is equivalent to φ in the region $\lambda_1, \cdots, \lambda_n \geqq N$, where $N = 1 + 3\max_{1 \leqq i \leqq n-1}\{a_i/a_{i+1}\}$, a_i being the length of the period of A_i.*

Proof. The number n is called the height of the parametric word φ. We will prove the assertion by induction on the height. Since $A_1, \cdots, A_n \neq \Lambda$, we have $A_i = S_i^{b_i}$, where $b_i \geqq 1$ and S_i is a simple word. The word φ is equivalent to the word $\varphi_1 = P_0[S_i^{b_i\lambda_i} P_i]_{i=1}^n$, which satisfies condition 1). To guarantee condition 2), it suffices in φ_1 to "shift" each S_i as far as possible to the left. More explicitly, if the word $AS^\lambda B$, $A \neq \Lambda$, does not satisfy 2), then it is equivalent to a word $A'S'^\lambda B'$, where A' is obtained from A by deleting the last letter, B' from B by multiplying on the left by the last letter of S, and S' from S by a cyclic permutation of the first letter. Continuing the shift to the left as far as possible, we obtain a word $PT^\lambda Q$ which is equivalent to $AS^\lambda B$ and satisfies 2). Here T is a cyclic permutation of S, since by 1.8 condition 1) is also satisfied. Carrying out these transformations in φ_1 for $i = 1, \cdots, n$ and taking into account that in transforming the segment $P_t' S_t^{b_t\lambda_t} P_{t+1}$ the previously transformed part $[P_{i-1} S_i^{b_i\lambda_i}]_{i=1}^{t-1}$ is not affected, we obtain a word φ_2 which is equivalent to φ and satisfies 1) and 2). It remains to satisfy 3). For $n = 1$ condition 3) is satisfied trivially. Assume that $n > 1$ and φ satisfies 1) and 2), but not 3). Suppose 3) is first violated, advancing from left to right, when $i = t$, i.e. $A_t \angle P_t A_{t+1}^c$, where c is the integral part of $[A_t^\partial / [A_{t+1}^\partial$. Split off from the left of P_t as many powers as possible of A_t: $P_t = A_t^a P$, where $a \geqq 0$ and P does not begin with A_t. The word $A_t^{\lambda_t} P_t$ is equivalent to the word $A_t^{\lambda_t + a} P$. If A_t is not a beginning of PA_{t+1}^c, then the segment $A_t^{\lambda_t + a} P A_{t+1}^{\lambda_{t+1}}$ satisfies conditions 1)–3). Replacing in φ the segment $A_t^{\lambda_t} P_t$ by the segment $A_t^{\lambda_t + a} P$, we obtain a word φ' which is equivalent to φ and satisfies conditions 1), 2), and in which 3) can be violated only when $i > t$. We now apply induction on t.

Consider the case $A_t \angle PA_{t+1}^c$. Let b be the largest integer such that $A_t^b \angle PA_{t+1}^{3c+1}$. We have $A_t = PQ$ and $A_{t+1}^{3c+1} = (QP)^{b-1}QV$. Since φ satisfies 1) and 2), the words A_{t+1}, A_t and QP are simple. Hence if $(b-1)[QP^\partial \geqq [A_{t+1}QP^\partial$, then by 1.5 we have $A_{t+1} = QP$. Then the word $\psi = A_t^{\lambda_t} P_t A_{t+1}^{\lambda_{t+1}}$ is equivalent to the word $\psi_1 = A_t^{\lambda_t + \lambda_{t+1} + a} P$. Replacing in φ the segment ψ by the segment ψ_1, we obtain a word equivalent to φ of height $< n$. Now use the induction assumption. Consider the case $(b-1)[QP^\partial < [A_{t+1}QP^\partial$. Since $A_t = PQ$,

$$[(QP)^{b-1}Q^\partial < [A_{t+1}A_tA_t^\partial \leqq (2c+1)[A_{t+1}^\partial.$$

Hence $A_{t+1}^{3c+1} = (QP)^{b-1}QRA_{t+1}^d$, where $d \geqq c$, and R is an end of A_{t+1}, i.e. $A_{t+1} = TR$. By definition of the number b the word RA_{t+1}^d does not begin with A_t. But then, obviously, $(RT)^c$ does not begin with A_t. The word $\psi = A_t^{\lambda_t}P_tA^{\lambda_{t+1}}$ is equivalent in the region $\lambda_{t+1} \geqq N$ to the word

$$\psi_1 = A_t^{\lambda_t+a+b}(RT)^{\lambda_{t+1}-3c-1+d}R,$$

which satisfies conditions 1) $-$3). Replacing in φ the word ψ by ψ_1, we obtain a word φ_1' which is equivalent to φ in the region $\lambda_1, \cdots, \lambda_n \geqq N$ and satisfies 1) and 2), and in which 3) can be violated only when $i > t$. We now apply induction on t. The proposition is proved.

REMARK. From the above procedure for reducing the word $P_0[A_i^{\lambda_i}P_i]_{i=1}^n$ it follows that if the periods of the words $A_1, \cdots, A_n$ have the same length, then equivalence to the reduced word is obtained in the region $\lambda_1, \cdots, \lambda_n > 0$. It suffices to note that in the second paragraph of the proof we have $[A_t^\partial = [A_{t+1}^\partial$. Hence the reference to 1.5 is not needed and the two subcases are $b > 1$, $b = 1$. Furthermore, $c = 1$, and the exponents $3c+1$ and d are replaced in the text by 2 and 1.

Let us define the characteristic and height of a one-stage exponential equation. Suppose there is given an equation (2.1). By the height of the equation we mean the number $h = \max\{n, m, 8\}$. By the characteristic we mean the number

$$H = \frac{\max\{[A_i^\partial, [B_j^\partial, [Q_i^\partial, [P_j^\partial\}}{\min\{[A_i^\partial, [B_j^\partial\}},$$

where max and min are taken over all possible values of the indices i and j. We have $h \geqq 8$ and $H \geqq 1$. In view of 2.1, any one-stage exponential equation $\mathfrak{A}(\lambda_1, \cdots, \lambda_n)$ of characteristic H is equivalent in the region $\lambda_1, \cdots, \lambda_n \geqq 3H+1$ to a reduced equation. Note that in reducing an equation the characteristic and height are, in general, altered. For reduced equations the following assertion is true.

2.2. *Suppose there is given a reduced equation*

$$(*) \qquad P_0[S_i^{\lambda_i}P_i]_{i=1}^n = Q_0[T_i^{\mu_i}Q_i]_{i=1}^m$$

of characteristic H. If (*) *has a solution $\lambda_1, \cdots, \lambda_n, \mu_1, \cdots, \mu_m \geqq 4H$, then $n = m$, $P_0 = Q_0$, $P_i = Q_i$, $S_i = T_i$ and $\lambda_i = \mu_i$, $i = 1, \cdots, n$.*

PROOF. We proceed by induction on $n+m$. If $n+m = 0$, the assertion is trivial. If one of the numbers n or m is greater than zero, so is the other, because of the estimates for λ_i and μ_j. Hence suppose $n, m > 0$. Without loss of generality, consider the case $[P_0^\partial \geqq [Q_0^\partial$. By 1.9 (whose applicability is guaranteed by the estimates for λ_1 and μ_1) for some t, A and B we have $S_1 = AB$, $T_1 = BA$ and $P_0 = Q_0(BA)^tB$. Since the equation is reduced, we must have $B = \Lambda$ and $t = 0$; hence $P_0 = Q_0$ and $S_1 = T_1$. We will prove that $\lambda_1 = \mu_1$. Suppose, for example, that $\lambda_1 > \mu_1$. After cancellation we have $S_1^{\lambda_1-\mu_1}P_1[S_i^{\lambda_i}P_i]_{i=2}^n = Q_1[T_i^{\mu_i}Q_i]_{i=2}^m$. Since the equation is reduced, the word Q_1 cannot begin with S_1. Consequently, $S_1 = Q_1R$, where $R \neq \Lambda$. Then $m > 1$

and we have $R \cdots = T_2^{\mu_2} \cdots$. Since $\mu_2 \geqq 4H$, we see that $R \angle T_2^{\mu_2}$. Hence $T_1 = S_1 \angle Q_1 T_2^{\mu_2}$, which contradicts condition 3) in the definition of a reduced word. This contradiction shows that the case $\lambda_1 > \mu_1$ is impossible. Similarly, the case $\lambda_1 < \mu_1$ is impossible. Hence $\lambda_1 = \mu_1$. Then $P_1[S_i^{\lambda_i} P_i]_{i=2}^n = Q_1[T_i^{\mu_i} Q_i]_{i=2}^m$ and the assertion follows from the induction assumption. The proposition is proved.

Remark. It is easy to see that 2.2 remains true if the condition $\lambda_1, \cdots, \lambda_n, \mu_1, \cdots, \mu_m \geqq 4H$ is replaced by the condition

$$\min\{\lambda_i[S_i^\partial, \mu_j[T_j^\partial\} \geqq 4 \max\{[P_i^\partial, [Q_j^\partial, [S_i^\partial, [T_j^\partial\},$$

where min and max are taken over all possible values of the indices i, j.

We will need the concept of a lineo-Diophantine relation (LDR). By definition, any LDR has the form

$$\bigvee_{i=1}^{N} [\omega_i(x_1, \ldots, x_n) \,\&\, \mathop{\&}_{j=1}^{n} (x_i > N_{ij})],$$

where $N \geqq 0$, $\omega_1, \cdots, \omega_n$ are systems of linear Diophantine equations, N_{ij} is a rational number and $x_1, \cdots, x_n$ are natural unknowns. When $N = 0$ the disjunction is assumed to be false. Instead of the symbols $>$ may appear the symbols $\geqq$.

2.3. *Any one-stage exponential equation is equivalent to some LDR.*

Proof. We proceed by induction on the number k of unknowns occurring in the equation. When $k = 0$ the assertion is trivial. Suppose $k > 0$ and the assertion is true for equations in less than k unknowns, and suppose that $\mathfrak{A}$ is an exponential equation in the k unknowns $t_1, \cdots, t_k$. Let H denote the characteristic of $\mathfrak{A}$. By 2.1, $\mathfrak{A}$ is equivalent in the region $t_1, \cdots, t_k \geqq 3H + 1$ to a reduced equation $\mathfrak{A}^*$. Let H^* be the characteristic of $\mathfrak{A}^*$ and let $H' = \max\{H, H^*\}$. Let Ω denote the region $t_1, \cdots, t_k \geqq 8H'$. In the region $\urcorner\Omega$ we have the equivalence

$$\mathfrak{A}(t_1, \ldots, t_k) \Leftrightarrow \bigvee_{i=1}^{k} \bigvee_{a=0}^{8H'} [\mathfrak{A}(t_1, \ldots, t_{i-1}, a, t_{i+1}, \ldots, t_k) \,\&\, (t_i = a)].$$

By means of the induction assumption we conclude that in the region $\urcorner\Omega$ the equation $\mathfrak{A}$ is equivalent to some LDR $\mathscr{D}_1$. Consider the region Ω. The equation $\mathfrak{A}^*$ has the form 2.2(*), where the λ_i and μ_j are α-polynomials in the unknowns $t_1, \cdots, t_k$. It follows from the reduction procedure that in Ω the inequalities λ_i, $\mu_j > 8H' - 3H - 1 \geqq 4H' \geqq 4H^*$ are satisfied. By 2.2, the equation $\mathfrak{A}^*$ in Ω either has no solution or is equivalent to the system $\lambda_i = \mu_i$, $i = 1, \cdots, n$. In other words, $\mathfrak{A}^*$ in the region Ω is equivalent to some LDR $\mathscr{D}_2$. The equation $\mathfrak{A}$ is equivalent to the LDR $\mathscr{D}_1 \vee \mathscr{D}_2$. The proposition is proved.

2.3′. *There exists an algorithm recognizing whether or not a given one-stage exponential equation has a solution.*

Proof. There exists an algorithm recognizing whether or not a given LDR has a solution. This follows from well-known results on the decidability of elementary arithmetic without the multiplication rule (see [7]). Our assertion now follows from 2.3.

Propositions 2.3 and 2.3′ answer the questions posed at the beginning of the section concerning the structure of the solutions and the decision algorithm for one-stage exponential equations. A more subtle question is that of estimating a "critical" region for an equation, i.e. a region in which the equation must have a solution, if it has solutions. This estimate depends on the height and characteristic of the equation, but, in general, not on these alone. It is of great importance to know for which equations there exists an estimate depending only on the characteristic and height. In Propositions 2.4–2.7 we consider certain one-stage exponential equations of a special form (in general, nonreduced), for which there exists an estimate of a critical region depending only on the height and characteristic of the equation.

2.4. *Suppose $\varphi(x_1, \cdots, x_n)$ and $\psi(x_1, \cdots, x_n)$ are words in the alphabet $И \cup \{x_1, \cdots, x_n\}$.*[5] *Consider the exponential equation*

$$\varphi(A_1^{\lambda_1}, \cdots, A_n^{\lambda_n}) = \psi(A_1^{\lambda_1}, \cdots, A_n^{\lambda_n}), \tag{*}$$

where $\lambda_1, \cdots, \lambda_n$ are natural unknowns and $A_1, \cdots, A_n$ are simple words of the same length. Let h and H be the height and characteristic of the equation. Then in the region $\lambda_1, \cdots, \lambda_n \geqq 5H$ the equation (*) is equivalent to a system $\mathscr{D}$ of linear Diophantine equations whose coefficients are in absolute value $\leqq h$ and whose free terms are of absolute value $\leqq 2hH$. Each positive solution of the system $\mathscr{D}$ will be a solution of (*).

PROOF. In accordance with 2.1 we construct a reduced exponential equation which, in view of the remark following the proof of 2.1, is equivalent to (*) in the region $\lambda_1, \cdots, \lambda_n > 0$. It will have the form

$$P_0[S_i^{\alpha_i} P_i]_{i=1}^k = Q_0[T_i^{\beta_i} Q_i]_{i=1}^m, \tag{1}$$

where $k, m \geqq 0$; P_i and Q_j are coefficients, S_i and T_j are cyclic shifts of the words $A_1, \cdots, A_n$, and the exponents α_i and β_j are α-polynomials in the unknowns $\lambda_1, \cdots, \lambda_n$. Denote by M and M_1 the maximum coefficient length in equations (*) and (1), and by H_1 the characteristic of (1). Denote the length of the words $A_1, \cdots, A_n$ by d. From the reduction procedure (see the proof of 2.1) it follows that $M_1 \leqq M + d$. Since $[A_1^\partial = \cdots = [A_n^\partial = d$, for the characteristics H and H_1 we have

$$H = \frac{\max\{M, d\}}{d}, \qquad H_1 = \frac{\max\{M_1, d\}}{d}.$$

Since $M_1 \leqq M + d$, we have $H_1 \leqq H + 1$. Also, it follows from the reduction procedure that each of the exponents α_i and β_j has the form $c_1\lambda_1 + \cdots + c_n\lambda_n + b$, where $c_1, \cdots, c_n$ and b are integers, $0 \leqq c_i \leqq h$, $i = 1, \cdots, n$, $b \leqq (h+1)M/d$, and, moreover, $c_1 + \cdots + c_n \geqq 1$ and, in view of the above-mentioned remark, $b \geqq -1$. Since $b \geqq -1$ and $c_1 + \cdots + c_n \geqq 1$, in the region $\lambda_1, \cdots, \lambda_n \geqq 5H$ we have the inequalities $\alpha_i, \beta_j \geqq 5H - 1$ and, consequently, $\alpha_i, \beta_j \geqq 4H_1$, since $H_1 \leqq H + 1$ and $H \geqq 1$. By 2.2, either equation (1) has no solution in the region $\lambda_1, \cdots, \lambda_n \geqq 5H$ or we have the relations $k = l$, $P_0 = Q_0$, $P_i = Q_i$ and $S_i = T_i$, $i = 1, \cdots, k$. In the first case we may take as the desired system $\mathscr{D}$ the equation $\lambda_1 = -1$, which has no natural solutions.

---

[5] А ∪ Б denotes the union of the alphabets А and Б.

In the second case the system $\alpha_i - \beta_i = 0$, $i = 1, \cdots, k$, will serve. The necessary estimates for the coefficients and free terms follow from the above estimates for c_i and b if we take into account that $(h+1)M/d \leqq (h+1)H \leqq 2hH$, since $h \geqq 2$. The proposition is proved.

2.4′. *Suppose $\mathfrak{A}$ is the exponential equation in two unknowns $\varphi(S^\lambda, T^\mu) = \psi(S^\lambda, T^\mu)$, where $\varphi(x,y)$ and $\psi(x,y)$ are words in the alphabet $\mathbf{H} \cup \{x,y\}$; λ and μ are natural unknowns, and S and T are simple words, $[S^\partial = [T^\partial$. Suppose the height and characteristic of $\mathfrak{A}$ are h and H. Then the following assertions are true.*

1) *$\mathfrak{A}$ is equivalent in the region $\lambda, \mu > 4h^2H$ to an equation $a\lambda + b\mu = c$, where a, b, c are integers, $|a|, |b| \leqq h$ and $|c| \leqq 2hH$. Any positive solution of the equation $a\lambda + b\mu = c$ satisfies $\mathfrak{A}$.*

2) *if $\mathfrak{A}$ has a solution in the region $\lambda, \mu > 4h^2H$, then it has a positive solution (λ, μ) such that $\lambda, \mu \leqq 4hH$.*

PROOF OF 1). By 2.4, when $n = 2$ the equation $\mathfrak{A}$ is equivalent in the region $\lambda, \mu \geqq 5H$ to a system of linear Diophantine equations

$$a_i\lambda + b_i\mu = c_i, \qquad i = 1, \cdots, N;\ N \geqq 1, \tag{1}$$

where $|a_i|, |b_i| \leqq h$ and $|c_i| \leqq 2hH$. Here any positive solution of (1) satisfies $\mathfrak{A}$. But if the system (1) has large solutions, it is equivalent to one of its own equations. Indeed, for the case $N = 2$, if the determinant Δ of the system is different from zero, then by Cramer's rule we obtain

$$|\lambda| = \frac{|c_1b_2 - b_1c_2|}{\Delta} \leqslant 4h^2H,$$

since Δ is an integer and $|\Delta| \geqq 1$. Consequently, if the system (1) with $N = 2$ has a solution in which $\lambda > 4h^2H$, then $\Delta = 0$ and one equation can be obtained from the other by multiplying it by some number. Hence we may exclude from the system (1) all equations except one, which is the desired equation.

PROOF OF 2). In view of 1) it suffices to prove that if $|a|, |b|, |c| \leqq M$ and the Diophantine equation

$$a\lambda + b\mu = c \tag{2}$$

has a solution, then it has a positive solution such that $\lambda, \mu \leqq 2M$. Let us prove this assertion. It is obvious if one of the numbers a or b (or both) is equal to zero, and also if a and b have the same signs. Consider the case $ab < 0$. The criterion for the solvability of (2) is (see [8], p. 113) that c should be divisible by the greatest common divisor d of a and b. If this is the case, then dividing equation (2) by d we obtain $a_1\lambda + b_1\mu = c_1$, where a_1 and b_1 are relatively prime. We have $a_1b_1 < 0$. Hence either $a_1c_1 \leqq 0$ or $b_1c_1 \leqq 0$. Without loss of generality, we consider the case $a_1c_1 \leqq 0$. Then $b_1c_1 \geqq 0$. Changing, if necessary, the signs in (2), we may assume that $a_1 > 0$. Then $b_1 < 0$. It is known (see [8], p. 109) that the equation $a_1\lambda + b_1\mu = c_1$ has a solution (λ_0, μ_0) such that $0 < \lambda_0 \leqq |b_1|$ (λ_0 is the smallest positive solution of the congruence $a_1\lambda \equiv c_1 \pmod{b_1}$). We have for μ_0 the expression $\mu_0 = (a_1\lambda_0 - c_1)/-b_1$. Since $b_1 < 0$, $c_1 \leqq 0$, $a_1 > 0$ and $0 < \lambda_0 < |b_1|, |a_1|, |c_1| \leqq M$, it follows that $0 < \mu_0 \leqq 2M$. The solution (λ_0, μ_0) is the desired solution of (2). The assertion concerning equation (2), hence also statement 2), is proved. Thus the proposition is proved.

In Propositions 2.4 and 2.4′ the bases of the equations were assumed to be simple words. In the general case, where the bases are not assumed to be simple, the analogues of these assertions do not hold. More precisely, in the analogues of 2.4 and 2.4′ the constants bounding the coefficients of the Diophantine equations and the critical region depend not only on the height and characteristic, but also on the periodicity exponents of the bases (if $A = S^a$, where S is a simple word, then the number a is called the periodicity exponent of the word A). The elimination of this dependence is of essential importance for the construction of our algorithms, since in applications the bases, and hence also their periodicity exponents, depend on various parameters.

In the following propositions (2.5 – 2.7) the bases are not assumed to be simple.

2.5. *Suppose $\varphi(x_1, \cdots, x_n)$ and $\psi(x_1, \cdots, x_n)$ are words in the alphabet $\mathbf{H} \cup \{x_1, \cdots, x_n\}$. Consider the exponential equation*

$$(*) \qquad \varphi(A^{\lambda_1}, \cdots, A^{\lambda_n}) = \psi(A^{\lambda_1}, \cdots, A^{\lambda_n}).$$

where $\lambda_1, \cdots, \lambda_n$ are natural unknowns and A is a word in the alphabet $\mathbf{H}$. Let h and H be the height and characteristic of (*). Then in the region $\lambda_1, \cdots, \lambda_n \geqq 5H$ the equation (*) is equivalent to a system of linear Diophantine equations whose coefficients are in absolute value $\leqq h$ and whose free terms are in absolute value $\leqq 2hH$.

Proof. Represent A as a power of a simple word: $A = S^a$, where $a \geqq 1$ and S is simple. Let $\mathfrak{A}$ denote the equation obtained from (*) by substituting $S^{a\lambda_i}$ for A^{λ_i}. The equations $\mathfrak{A}$ and (*) are equivalent. Denote by h_1 and H_1 the height and characteristic of $\mathfrak{A}$, and by M and M_1 the maximum coefficient length of (*) and $\mathfrak{A}$. We have $h = h_1$, $M = M_1$, $H = \max\{M, [A^\partial\}/[A^\partial$, $H_1 = \max\{M_1, [S^\partial\}/[S^\partial$, and $H_1 \leqq aH$. Let Ω and Ω_1 be the regions defined by the inequalities $\lambda_1, \cdots, \lambda_n \geqq 5H$ and $a\lambda_1, \cdots, a\lambda_n \geqq 5H_1$. Since $H_1 \leqq aH$, we have $\Omega \subset \Omega_1$. By 2.4, the equation $\mathfrak{A}$ in the region Ω_1 is equivalent to a system $\mathscr{D}$ of equations $ac_{i1}\lambda_1 + \cdots + ac_{in}\lambda_n = d_i$, $i = 1, \cdots, k$, where the c_{ij} and d_i are integers, $|c_{ij}| \leqq h_1$ and $|d_i| \leqq 2h_1H_1$. Since $\Omega \subset \Omega_1$, the equivalence holds in Ω. If even one of the d_i is not divisible by a, then the system $\mathscr{D}$, hence also the equation (*), has no solution in Ω and as the desired system we may take $\lambda_1 = \cdots = \lambda_n = 1$, which also has no solution in Ω. If all of the d_i are divisible by a, then the desired system will be $c_{i1}\lambda_1 + \cdots + c_{in}\lambda_n = d_i/a$, $i = 1, \cdots, k$, since $h_1 = h$ and $|d_i/a| \leqq 2h_1H_1/a \leqq 2hH$. The proposition is proved.

2.5′. *Consider the exponential equation $\varphi(A^\lambda, A^\mu) = \psi(A^\lambda, A^\mu)$, where $\varphi(x,y)$ and $\psi(x,y)$ are words in the alphabet $\mathbf{H} \cup \{x,y\}$, λ and μ are natural unknowns, and A is a word in the alphabet $\mathbf{H}$. Let h and H be the height and characteristic of this equation. If the equation has a solution $\lambda, \mu > 4h^2H$, then it has a positive solution $\lambda, \mu \leqq 4hH$.*

Proof. This follows from 2.4′ in the same way that 2.5 follows from 2.4.

2.6. *Consider the exponential equation $\varphi(A^\lambda, B^\mu) = \psi(A^\lambda, B^\mu)$, where $\varphi(x,y)$ and $\psi(x,y)$ are words in the alphabet $\mathbf{H} \cup \{x,y\}$, and λ and μ are natural unknowns. Let h and H be the height and characteristic of this equation, and suppose*

$A = S^a$ and $B = T^b$, where S and T are simple words and $[S^\partial \neq [T^\partial$. If the equation has a solution $\lambda, \mu \geqq 2hH$, then it is satisfied for any $\lambda, \mu \geqq 2hH$.

This assertion is a consequence of the following more general one when $n = 1$.

2.6′. *Suppose $\varphi(x, x_1, \cdots, x_n)$ and $\psi(x, x_1, \cdots, x_n)$ are words in the alphabet $\text{И} \cup \{x, x_1, \cdots, x_n\}$. Consider the exponential equation*

$$(*) \qquad \varphi(A^\lambda, B^{\lambda_1}, \cdots, B^{\lambda_n}) = \psi(A^\lambda, B^{\lambda_1}, \cdots, B^{\lambda_n}),$$

where $\lambda, \lambda_1, \cdots, \lambda_n$ are natural unknowns and A and B are words in the alphabet И. Let h and H be the height and characteristic of the equation, and suppose $A = S^a$ and $B = T^b$, where S and T are simple words and $[S^\partial \neq [T^\partial$. If the equation has a solution $\overline{\lambda}, \overline{\lambda}_1, \cdots, \overline{\lambda}_n \geqq 2hH$, then the set $(\lambda, \overline{\lambda}_1, \cdots, \overline{\lambda}_n)$ satisfies $(*)$ for any $\lambda \geqq 2hH$.

PROOF. Construct in accordance with 2.1 a reduced equation equivalent to $(*)$ in the region $\lambda, \lambda_1, \cdots, \lambda_n \geqq 3H + 1$. It will have the form

$$(1) \qquad P_0[S_i^{\alpha_i} P_i]_{i=1}^k = Q_0[T_i^{\beta_i} Q_i]_{i=1}^m,$$

where $k, m \geqq 0$; P_i and Q_j are the coefficients; S_i and T_j are cyclic shifts of the words S and T; and the exponents α_i and β_j are α-polynomials in the unknowns $\lambda, \lambda_1, \cdots, \lambda_n$. The reduction of $(*)$ involves the following transformations: 1) replacing A^λ and B^{λ_i} by $S^{a\lambda}$ and $T^{b\lambda_i}$; 2) cyclic shifts to the left of the words S and T; 3) splitting off powers of these shifts from coefficients and powers standing to the right; 4) coalescing powers with the same bases. Since $[S^\partial \neq [T^\partial$, S and T do not have the same cyclic shifts. Hence shifts of S can be coalesced only with shifts of S, and shifts of T with shifts of T. It follows that exponents where the bases are shifts of S have the form $pa\lambda + q$, and where shifts of T, the form $a_1 b\lambda_1 + \cdots + a_n b\lambda_n + r$, where p, q, r and a_i are integers and $p \geqq 1$, $a_i \geqq 0$, $a_1 + \cdots + a_n \geqq 1$, $q \geqq -3e - 1$ and $r \geqq -3/e - 1$, where $e = [T^\partial / [S^\partial$. If L denotes the maximum coefficient length of the equation $\varphi = \psi$, then $q \leqq (h+1)L/[S^\partial$, since q occurs only when splitting off powers of shifts of S from the coefficients. Put $M = \max\{[A^\partial, [B^\partial, L\}$, $N = \max\{[S^\partial, [T^\partial\}$, $M_1 = \max\{[P_i^\partial, [Q_j^\partial, N\}$. It follows from the reduction procedure that $\max\{[P_i^\partial, [Q_j^\partial\} \leqq L + N$; hence $M_1 \leqq L + N$.

We now assume that $(*)$ has a solution $\overline{\lambda}, \overline{\lambda}_1, \cdots, \overline{\lambda}_n \geqq 2hH$. The exponents α_i and β_j take values $\overline{\alpha_i}$ and $\overline{\beta}_j$. Let us estimate the products $\overline{\alpha_i}[S_i^\partial$ and $\overline{\beta}_j[T_j^\partial$. By definition of the characteristic, $H = M/\min\{[A^\partial, [B^\partial\}$. Hence $H[A^\partial$, $H[B^\partial \geqq M$. Also, $h \geqq 8$, by definition of the height. For exponents on shifts of S we have

$$\begin{aligned}(pa\overline{\lambda} + q)\,[S^\partial &= p\overline{\lambda}[A^\partial + q[S^\partial \geqq \overline{\lambda}[A^\partial - 3[T^\partial - [S^\partial \\ &\geqq 2hH[A^\partial - 4N \geqq 2hM - 4N \geqq h\max\{[A^\partial, [B^\partial\} + hL - 4N \\ &\geqq hN + hL - 4N \geqq 4(N + L) \geqq 4M_1.\end{aligned}$$

Similarly, if γ is an exponent on a shift of T, we have $\overline{\gamma}[T^\partial \geqq 2hH[B^\partial - 4N \geqq [M_1$. We obtain that equation (1) has a solution $(\overline{\alpha_1}, \cdots, \overline{\alpha_k}, \overline{\beta}_1, \cdots, \overline{\beta}_m)$ such that $\overline{\alpha_i}[S_i^\partial, \overline{\beta}_j[T_j^\partial \geqq 4M_1$. In view of the remark following the proof of 2.2, we have $k = m$, $P_0 = Q_0$, $P_i = Q_i$, $S_i = T_i$ and $\overline{\alpha_i} = \overline{\beta}_i$, $i = 1, \cdots, k$. It follows that any solution $(\lambda, \lambda_1, \cdots, \lambda_n)$ of the system $\alpha_i = \beta_i$, $i = 1, \cdots, k$, will be a solution of (1). We will show that if S_i is a shift of S, then the equation

$\alpha_i - \beta_i = 0$ has the form $0 = 0$. Indeed, in view of what has been said above about exponents, the equation $\alpha_i - \beta_i = 0$ in this case has the form $ca\lambda + d = 0$, where $|d|\,[S^\partial \leqq 4N + (h+1)L \leqq (h+1+4)M < 2hM$. On the other hand, if $c \neq 0$, then $|c|a\overline{\lambda}[S^\partial \geqq \overline{\lambda}[A^\partial \geqq 2hH[A^\partial \geqq 2hM > |\,d|\,[S^\partial$. Since $ca\overline{\lambda} + d = 0$, we have $c = d = 0$, as required. Hence the system $\alpha_i - \beta_i = 0$, $i = 1, \cdots, k$, does not contain λ, and therefore the set $(\lambda, \overline{\lambda}_1, \cdots, \overline{\lambda}_n)$ satisfies it for any $\lambda \geqq 0$. The assertion being proved now follows from the fact that equations (*) and (1) are equivalent in the region $\lambda, \lambda_1, \cdots, \lambda_n \geqq 2hH$. The proposition is proved.

2.7. *Consider the exponential equation in one unknown*

$$(*) \qquad P_0[A^\lambda P_i]_{i=1}^n = Q_0[A^\lambda Q_i]_{i=1}^m,$$

where $n, m \geqq 0$; A, P_i and Q_j are words in the alphabet $\mathrm{И}$, $A \neq \Lambda$. *Put* $k = \min\{n, m\}$, $M = \max_{1\leq i\leq k}\{\,|\,[P_0 \cdots P_i^\partial - [Q_0 \cdots Q_i^\partial|\}$, $G = M/[A^\partial$. *If the equation* (*) *has a solution* $\lambda \geqq G+1$, *then it is satisfied for any* $\lambda \geqq 0$. *Put* $N = 2hH$, *where h and H are the height and characteristic of* (*). *If* (*) *has a solution* $\lambda \geqq N$, *then it is satisfied for any* $\lambda \geqq 0$.

Proof. The first part is a rephrasing of 1.18. The second follows from the first, since $2hH \geqq G+1$.

We use both the estimate $\lambda \geqq 2hH$ and the more precise estimate $\lambda \geqq G+1$. Where the estimate $\lambda \geqq 2hH$ results in large numbers we prefer to use $\lambda \geqq G+1$. However, in nontrivial cases G requires more laborious calculations. We also call G a characteristic. Thus for equations in one unknown we have two characteristics and we distinguish them by saying the characteristic G or the characteristic H.

One-stage exponential equations have various applications to the solution of word equations. Solving one-stage exponential equations is equivalent to the problem of solving the following system of word equations:

$$\varphi(x_1, \cdots, x_n) = \psi(x_1, \cdots, x_n), \qquad S_i x_i = x_i S_i, \quad i = 1, \cdots, n,$$

where φ and ψ are words in the alphabet $\mathrm{И} \cup \{x_1, \cdots, x_n\}$, while $S_1, \cdots, S_n$ are simple words in the alphabet И. By 1.2, all solutions of the equation $S_i x_i = x_i S_i$ are given by $x_i = S_i^{\lambda_i}$, where λ_i is a natural unknown. We arrive at the one-stage exponential equation

$$\varphi(S_1^{\lambda_1}, \cdots, S_n^{\lambda_n}) = \psi(S_1^{\lambda_1}, \cdots, S_n^{\lambda_n}).$$

An important application of one-stage equations involves the notion of a periodic word. A word $A^t A_1$, where $t \geqq 0$ and $A_1 \angle A$, is called a periodic word with period A. The determination of solutions of an equation $\mathfrak{A}(x_1, \cdots, x_n)$ in a region, where $x_1, \cdots, x_n$ are periodic words with period of bounded length (each x_i has its own period), obviously reduces to solving a finite set of one-stage exponential equations.

The appearance of natural exponents in the solution of word equations goes back to 1.16. This proposition is fundamental to the method of investigating word equations presented in this monograph. It leads to the consideration of exponential equations whose bases depend on various parameters (including word parameters).

We now consider certain two-stage exponential equations.

2.8. *Consider the exponential equation*

$$(*)\qquad [(S^\sigma C)^{\lambda_i} S^\sigma A_i]_{i=1}^n = [(S^\sigma C)^{\mu_i} S^\sigma B_i]_{i=1}^m,$$

where $n, m \geqq 1$; σ, λ_i *and* μ_j *are natural unknowns*; S, C, A_i *and* B_j *are words in the alphabet* $\mathbf{И}$; S *is simple, and* $\mathscr{L}(S, C)$. *Either* (*) *has no solution in the region*

$$(1)\qquad \sigma [S^\partial + [C^\partial \geqslant \max_{i,j} \{[A_i^\partial, [B_j^\partial\} + 2[SC^\partial;\ \lambda_1, \ldots, \lambda_n, \mu_1, \ldots, \mu_m \geqslant 3,$$

or there exist sequences $(n_1, \cdots, n_k)$, $(m_1, \cdots, m_k)$ *and* $(\epsilon_1, \cdots, \epsilon_k)$ *of integers satisfying the following conditions*:

1) $k > 0$, $1 \leqq n_1 < \cdots < n_k = n$, $1 \leqq m_1 < \cdots < m_k = m$; $|\epsilon_1|, \cdots, |\epsilon_k| \leqq 1$, $\epsilon_k = 0$.

2) $B_i = C$ *for any* i *different from* $m_1, \cdots, m_k$.

3) (*) *is equivalent in the region* (1) *to the system*

$$(2)\qquad \lambda_1 + \cdots + \lambda_{n_i} - (\mu_1 + \cdots + \mu_{m_i}) = m_i - n_i + \epsilon_i \qquad (i = 1, \cdots, k).$$

PROOF. We will first prove an auxiliary assertion. Assume that all of the words A_i and B_j are different from C. We will prove that if (*) has a solution $(\sigma, \lambda_1, \cdots, \lambda_n, \mu_1, \cdots, \mu_m)$ in the region (1), then $n = m$ and there exists a sequence $\epsilon_1, \cdots, \epsilon_n$ such that $|\epsilon_1|, \cdots, |\epsilon_n| \leqq 1$, $|\epsilon_n| = 0$ and (*) is equivalent to the system

$$(3)\qquad \lambda_1 + \cdots + \lambda_i - (\mu_1 + \cdots + \mu_i) = \epsilon_i \qquad (i = 1, \cdots, n).$$

We proceed by induction on $k = \min\{n, m\}$. By hypothesis, $k \geqq 1$.

Suppose that (*) has in the region (1) a solution $(\sigma, \lambda_1, \cdots, \lambda_n, \mu_1, \cdots, \mu_m)$. In the alphabet $\mathbf{И}$ we have

$$(4)\qquad (S^\sigma C)^{\lambda_1} S^\sigma A_1 [(S^\sigma C)^{\lambda_i} S^\sigma A_i]_{i=2}^n = (S^\sigma C)^{\mu_1} S^\sigma B_1 [(S^\sigma C)^{\mu_i} S^\sigma B_i]_{i=2}^m.$$

Put $\epsilon_1 = \lambda_1 - \mu_1$. We will prove that $|\epsilon_1| \leqq 1$. Without loss of generality, we consider the case $\epsilon_1 \geqq 0$. If $\epsilon_1 > 1$, then from (4) we have $(S^\sigma C)^{\epsilon_1} S^\sigma A_1 \cdots = S^\sigma B_1 [\cdots]_{i=2}^m$. Comparing the lengths of the left- and right-hand sides of this last equality, we conclude that $m > 1$, whence for some V_1 and V_2 we have $S^\sigma CS^\sigma CS^\sigma V_1 = S^\sigma B_1 S^\sigma CS^\sigma V_2$, and therefore $CS^\sigma CS^\sigma V_1 = B_1 S^\sigma CS^\sigma V_2$. It now follows by 1.15 (which is obviously applicable) that $B_1 = C$, which contradicts the assumption $B_1 \neq C$. This contradiction implies that $\epsilon_1 \leqq 1$. The inequality $|\epsilon_1| \leqq 1$ is proved.

Suppose $k = 1$. Comparing the lengths of the left- and right-hand sides of (4) and taking into account that $|\epsilon_1| \leqq 1$, we obtain $n = m = 1$, $\lambda_1 = \mu_1$, $A_1 = B_1$ and $\epsilon_1 = 0$. The sequence $\epsilon_1 = 0$ is the desired one.

Now suppose $k > 1$. Without loss of generality, we consider the case $\epsilon_1 \geqq 0$. Cancelling (4) on the left by $(S^\sigma C)^{\mu_1} S^\sigma$, we obtain

$$(5)\qquad (CS^\sigma)^{\varepsilon_1} A_1 (S^\sigma C)^2 V_1 = B_1 (S^\sigma C)^{\varepsilon_1} (S^\sigma C)^2 V_2$$

for certain V_1 and V_2. We will prove that $(CS^a)^{\epsilon_1} A_1 = B_1 (S^a C)^{\epsilon_1}$ for any $a \geqq 0$. We have

$$|\,[(CS^\sigma)^{\varepsilon_1} A_1^\partial - [B_1 (S^\sigma C)^{\varepsilon_1\,\partial}| = |\,[A_1^\partial - [B_1^\partial| < (\sigma - 1)[S^\partial.$$

Hence from (5), in view of 1.15, we obtain $(CS^\sigma)^{\epsilon_1} A_1 = B_1 (S^\sigma C)^{\epsilon_1}$. If $\epsilon_1 = 0$, then $A_1 = B_1$. If $\epsilon_1 = 1$, then $CS^\sigma A_1 = B_1 S^\sigma C$. If we had $\epsilon_1 = -1$, then we would have had $A_1S^\sigma C = CS^\sigma B_1$. By 1.21, ϵ_1 is uniquely determined by the words S, C, A_1 and B_1, and the relation $(CS^\sigma)^{\epsilon_1}A_1 = B_1(S^\sigma C)^{\epsilon_1}$ holds for any $\sigma \geqq 0$. Then (*) is equivalent to the system

$$\text{(6)}\qquad \begin{gathered} \lambda_1 - \mu_1 = \epsilon_1, \\ (S^\sigma C)^{\lambda_2+\epsilon_1}S^\sigma A_2\,[(S^\sigma C)^{\lambda_i}S^\sigma A_i]_{i=3}^{n} = [(S^\sigma C)^{\mu_i}S^\sigma B_i]_{i=2}^{m}. \end{gathered}$$

We have $\epsilon_1 \geqq 0$. Applying the induction assumption to (6.1), we obtain that $n = m$ and there exists a sequence $\epsilon_2, \cdots, \epsilon_n$ such that $|\epsilon_i| \leqq 1$, $\epsilon_n = 0$ and (6.1) is equivalent to the system

$$\lambda_2 + \epsilon_1 + \lambda_3 + \cdots + \lambda_k - (\mu_2 + \cdots + \mu_k) = \epsilon_k \qquad (k = 2, \cdots, n).$$

Hence the system (6), so also the equation (*), is equivalent to the system (3). The auxiliary assertion is proved.

Suppose the equation (*) is given. We define the sequences $(n_1, \cdots, n_k)$ and $(m_1, \cdots, m_q)$ as the increasing sequences of indices for which the words A and B with these indices differ from C, and we put $n_0 = m_0 = 0$. Then (*) can be written in the form

$$\text{(7)}\qquad [(S^\sigma C)^{\alpha_i}S^\sigma A_{n_i}]_{i=1}^{k} = [(S^\sigma C)^{\beta_i}S^\sigma B_{m_i}]_{i=1}^{q},$$

where $\alpha_i = \lambda_{n_{i+1}+1} + \cdots + \lambda_{n_i} + n_i - n_{i-1} - 1$ and $\beta_i = \mu_{m_{i+1}+1} + \mu_{m_i} + m_i - m_{i-1} - 1$. It is obvious that in the region (1) we have $\alpha_i, \beta_j \geqq 3$. Now applying the auxiliary assertion proved above we obtain that if (*) has a solution in the region (1), then $k = q$ and (7) is equivalent to the system

$$\text{(8)}\qquad \alpha_1 + \cdots + \alpha_i - (\beta_1 + \cdots + \beta_i) = \epsilon_i \qquad (i = 1, \cdots, k),$$

where $|\epsilon_1|, \cdots, |\epsilon_k| \leqq 1$ and $\epsilon_k = 0$. Substituting in (8) for α_i and β_j their expressions in terms of λ and μ, we obtain the system (2). The sequences $(\epsilon_1, \cdots, \epsilon_k)$, $(n_1, \cdots, n_k)$ and $(m_1, \cdots, m_k)$ are the desired ones. The proposition is proved.

2.9. *The equation* 2.8 (*) *is equivalent to some LDR in the region* $\lambda_1, \cdots, \lambda_n, \mu_1, \cdots, \mu_m \geqq 3$.

PROOF. For large σ (we have in mind the estimate 2.8 (1)), the assertion follows from 2.8. For small σ, 2.8 (*) is a one-stage exponential equation with unknowns $\lambda_1, \cdots, \lambda_n$, $\mu_1, \cdots, \mu_m$, and the assertion follows from 2.3. The proposition is proved.

2.10. *Consider the equation* $\Phi(T_1, T_2) = \Psi(T_1, T_2)$, *where* $\Phi(x, y)$ *and* $\Psi(x, y)$ *are words in the alphabet* $\mathfrak{A} \cup \{x, y\}$, $T_1 = (S^\sigma C)^\lambda S^\sigma R$ *and* $T_2 = (S^\sigma C)^\mu S^\sigma R$, *while* σ, λ *and* μ *are positive unknowns*, S, C *and* R *are words in the alphabet* $\mathfrak{A}$, S *is simple, and* $\mathscr{L}(S, C)$. *The equation* $\Phi(T_1, T_2) = \Psi(T_1, T_2)$ *is equivalent to some LDR.*

PROOF. The equation $\Phi(T_1, T_2) = \Psi(T_1, T_2)$ is an exponential equation of the form 2.8 (*), where λ_i and μ_j are equal to λ or μ. In the region $\lambda, \mu \geqq 3$ the assertion follows from 2.8. It remains to prove it in the region $(\lambda < 3) \vee (\mu < 3)$. It suffices, obviously, to prove it when one of the unknowns λ or μ is fixed. Let us fix λ. Denote equation 2.8 (*) with λ fixed by $\mathfrak{A}(\sigma, \mu)$. Now consider this equation $\mathfrak{A}(\sigma, \mu)$. For fixed σ it is a one-stage exponential equation with one unknown μ. The characteristic G of this equation is bounded by a constant which does not depend on σ. This follows from the fact that σ occurs in the base of a power with exponent μ. In view of 2.7, this implies that there exists a constant N which does not depend on σ such that for any σ we have the implication $(\exists \mu \geqq N)\mathfrak{A}(\sigma,\mu) \supset (\forall \mu \geqq N)\mathfrak{A}(\sigma,\mu)$. We obtain the equivalence $\forall \sigma\mu[\mathfrak{A}(\sigma,N), \mu \geqq N \Leftrightarrow \mathfrak{A}(\sigma,\mu), \mu \geqq N]$.[6] Our assertion follows from the equivalences[7]

$$\mathfrak{A}(\sigma,\mu) \Leftrightarrow [\mathfrak{A}(\sigma,\mu), \mu < N] \vee [\mathfrak{A}(\sigma,\mu), \mu \geqq N]$$
$$\Leftrightarrow \bigvee_{k=0}^{N-1} [\mathfrak{A}(\sigma,k), \mu = k] \vee [\mathfrak{A}(\sigma,N), \mu \geqslant N],$$

since each of the equations $\mathfrak{A}(\sigma,k)$, $k = 0,1,\cdots,N$, is a one-stage exponential equation with the one unknown σ. The proposition is proved.

§3. Transformations and images

In this section we describe in general terms the method used in this monograph for studying word equations. The starting point of this method is Proposition 1.16.

We will consider word equations whose left- and right-hand sides begin with different unknowns. The general case reduces easily to this one (see Proposition 4.10). Suppose there is given a word equation in n unknowns $x_1,\cdots,x_n$ of the form $x\cdots = y\cdots$, where x and y are distinct unknowns and the dots denote the unwritten ends of the left- and right-hand sides of the equation. Multiplying, if necessary, this equation on the right by x, we can write it in the form $x\cdots = Dx\cdots$, where D is nonempty and does not contain x. If $(X_1,\cdots,X_n)$ is a solution of this equation, then in the alphabet $\mathbf{И}$ we have an equality $XV_1 = AXV_2$. It is natural to apply Proposition 1.16 to this equality. In this connection, it is convenient to pass from the usual word equations of the form $\Phi = \Psi$ to "one-sided" word equations.

A one-sided word equation is an expression of the form $x\Phi \rightarrow y\Psi$ or $x\Phi \leftarrow y\Psi$, where x and y are distinct unknowns, and Φ and Ψ are words in the alphabet $\mathbf{И} \cup \mathbf{X}$.[8] Suppose $x\Phi \rightarrow y\Psi$ is a one-sided equation in the n unknowns $x_1,\cdots,x_n$. By a solution of the equation $x\Phi \rightarrow y\Psi$ we mean any solution $(X_1,\cdots,X_n)$ of the equation $x\Phi = y\Psi$ such that $X_1,\cdots,X_n$ are nonempty words and $[X^\partial > [Y^\partial$. A solution of the equation $x\Phi \leftarrow y\Psi$ is defined analogously, with $[X^\partial > [Y^\partial$ replaced by $[Y^\partial > [X^\partial$. We have the obvious equivalence

$$(x\Phi = y\Psi)$$
$$\Leftrightarrow (x\Phi \rightarrow y\Psi) \vee (x\Phi \leftarrow y\Psi) \vee \bigvee_{i=1}^{n} (x_i = \Lambda, x\Phi = y\Psi) \vee (x = y, \Phi = \Psi).$$

In view of this equivalence the decision problem for word equations in n unknowns of the form $x\cdots = y\cdots$ reduces to the decision problems for one-sided equations in n unknowns and word equations in $n-1$ unknowns.

Let us define some concepts which arise when applying 1.16 to one-sided word equations.

---

6) In addition to the symbol & we also use the symbol , (comma) to denote a conjunction.

7) Here and henceforth, free variables appearing in the left- and right-hand sides of equivalences and implications have a universal interpretation.

8) Here and below, $\mathbf{X}$ denotes the alphabet of unknowns.

Consider a one-sided equation in the n unknowns $x_1, \cdots, x_n$. We assume that the equation has the form $x_1\Phi \to Dx_1\Psi$, where D is nonempty and does not contain x_1. Expressing D more precisely, we can write the equation in the form

$$x_1\Phi(x_1, \cdots, x_n) \to [\xi_i A_i]_{i=1}^k x_1 \Psi(x_1, \cdots, x_n), \tag{3.1}$$

where $k \geqq 1$; $\xi_1, \cdots, \xi_k$ are unknowns different from x_1; $A_1, \cdots, A_k$ are coefficients; and Φ and Ψ are words in the alphabet $\mathbf{Н} \cup \mathbf{X}$.

Let us define transformations and images of equation (3.1).

We will distinguish between finite and infinite transformations and images. Any equality of the form

$$x_1 = ([\xi_i A_i]_{i=1}^k)^t [\xi_i A_i]_{i=1}^r \xi_{r+1} A, \tag{3.2}$$

where $t \geqq 0$, $0 \leqq r < k$, $A \angle A_{r+1}$ and also $[\xi_i A_i]_{i=1}^r \xi_{r+1} A \not\angle [\xi_i A_i]_{i=1}^k$, is called a finite transformation of equation (3.1). The equation obtained as a result of substituting the right-hand side of (3.2) for x_1 in (3.1) is called the finite image of (3.1) under the transformation (3.2).

Any equality of the form

$$x_1 = ([\xi_i A_i]_{i=1}^k)^t [\xi_i A_i]_{i=1}^r y, \tag{3.3}$$

where $t \geqq 0, 0 \leqq r < k, kt + r > 0$ and y is an unknown different from $x_1, \cdots, x_n$, is called an infinite transformation of (3.1). Let $\tilde{x}_1$ denote the right-hand side of (3.3). The equation

$$y\Phi(\tilde{x}_1, x_2, \ldots, x_n) \leftarrow [\xi_i A_i]_{i=r+1}^k [\xi_i A_i]_{i=1}^r y \Psi(\tilde{x}_1, x_2, \ldots, x_n) \tag{3.4}$$

is called the infinite image of (3.1) under the transformation (3.3). The unknown y will be called a "new" unknown, and the number t appearing in (3.2), (3.3), the parameter of the transformations (3.2), (3.3).

Finite images of one-sided word equations in n unknowns are word equations in $n-1$ unknowns. Infinite images of one-sided equations in n unknowns are one-sided word equations in n unknowns. Any one-sided word equation has infinitely many finite and infinite images.

The following lemma is obtained by applying 1.16 to equation (3.1).

3.1. *The words $X_1, \cdots, X_n$ will also be denoted by $\bar{x}_1, \cdots, \bar{x}_n$. The following assertions are true.*

a) *Suppose $(X_1, \cdots, X_n)$ is a solution of equation* (3.1). *Then one and only one of the following two statements holds:*

1) *There exist uniquely integers $t \geqq 0$, r, $0 \leqq r < k$, and a beginning A of the coefficient A_{r+1} such that*

$$X_1 = ([\bar{\xi}_i A_i]_{i=1}^k)^t [\bar{\xi}_i A_i]_{i=1}^r \bar{\xi}_{r+1} A. \tag{1}$$

The set $(X_2, \cdots, X_n)$ is a solution of the image of equation (3.1) *under the transformation* (3.2).

2) *There exist uniquely integers $t \geqq 0$, r, $0 \leqq r < k$, and a word Y such that*

$$X_1 = ([\bar{\xi}_i A_i]_{i=1}^k)^t [\bar{\xi}_i A_i]_{i=1}^r Y. \tag{2}$$

The set $(Y, X_2, \cdots, X_n)$ is a solution of the image of equation (3.1) *under the transformation* (3.3).

b) *For any solution* $(Y, X_2, \cdots, X_n)$ *whatever of the image of equation* (3.1)) *under the transformation* (3.3), *the set* $(X_1, \cdots, X_n)$, *where* X_1 *is defined by* (2), *is a solution of equation* (3.1). *For any solution* $(X_2, \cdots, X_n)$ *whatever of the image of equation* (3.1) *under the transformation* (3.2), *the set* $(X_1, \cdots, X_n)$, *where* X_1 *is defined by* (1), *is a solution of equation* (3.1).

PROOF. a) Suppose $(X_1, \cdots, X_n)$ is a solution of (3.1). We have in the alphabet **И** the equality

$$(3) \qquad X_1\Phi(X_1, \ldots, X_n) = [\overline{\xi}_i A_i]_{i=1}^{k} X_1 \Psi(X_1, \ldots, X_n).$$

Because of the arrow in (3.1) we have $[X_1^{\partial} > [\overline{\xi}_1^{\partial} > 0$. In particular, $[\overline{\xi}_i A_i]_{i=1}^{k} \neq \Lambda$. By 1.16, there exist uniquely an integer $t \geqq 0$ and words U and V such that

$$(4) \qquad X_1 = ([\overline{\xi}_i A_i]_{i=1}^{k})^t U,$$

$$(5) \qquad [\overline{\xi}_i A_i]_{i=1}^{k} = UV; \quad V \neq \Lambda.$$

Let us consider (5). There are two possible cases.

1) A coefficient is "cut". There exist r, $0 \leqq r < k$, and a beginning A of the word A_{r+1} such that $U = [\overline{\xi}_i A_i]_{i=1}^{r} \overline{\xi}_{r+1} A$. The condition $V \neq \Lambda$ implies that $U \mathrel{\underset{\cdot}{\angle}} [\overline{\xi}_i A_i]_{i=1}^{k}$. Since $\overline{\xi}_1, \cdots, \overline{\xi}_k \neq \Lambda$, the number r and the word A are uniquely determined.

2) An unknown is "cut". There exist r, $0 \leqq r < k$, and a word Y such that $U = [\overline{\xi}_i A_i]_{i=1}^{r} Y$, where $Y \neq \Lambda$ and $Y \mathrel{\underset{\cdot}{\angle}} \overline{\xi}_{r+1}$. It is obvious that the number r and the word Y are uniquely determined.

In the first case we obtain the expression (1) for X_1 and in the second X_1 has the form (2). Since $\Lambda \neq Y \mathrel{\underset{\cdot}{\angle}} \overline{\xi}_{r+1}$, cases 1) and 2) are mutually exclusive. It is obvious that in the first case we have 1). We will prove that in the second we have 2).

First note that $kt + r > 0$, as otherwise we would have $t = r = 0$, $X_1 = Y$ and $Y \mathrel{\underset{\cdot}{\angle}} \overline{\xi}_1$, which is impossible since $[X_1^{\partial} > [\overline{\xi}_1^{\partial}$. Substituting the right-hand side of (2) for X_1 in (3) and cancelling the resulting equation on the left by $([\overline{\xi}_i A_i]_{i=1}^{k})^t [\overline{\xi}_i A_i]_{i=1}^{r}$, we obtain

$$(6) \qquad Y\Phi(\widetilde{X}_1, X_2, \ldots, X_n) = [\overline{\xi}_i A_i]_{i=r+1}^{k} [\overline{\xi}_i A_i]_{i=1}^{r} Y \Psi(\widetilde{X}_1, X_2, \ldots, X_n),$$

where $\widetilde{X}_1$ denotes the right-hand side of (2). Since $0 < [Y^{\partial} < [\overline{\xi}_{r+1}^{\partial}$, it is clear from (2) and (6) that the set $(Y, X_2, \cdots, X_n)$ satisfies the image of equation (3.1) under the transformation (3.3), so that 2) holds. Thus the assertion a) is proved.

Reversing the argument, we verify b). The lemma is proved.

Lemma 3.1 is involved in all of the subsequent machinery used in solving equations in three unknowns.

It follows from the lemma that any solution $\Sigma = (X_1, \cdots, X_n)$ of (3.1) uniquely determines a transformation π of this equation and a solution Σ' of the image of (3.1) under π. Conversely, if $\mathfrak{A}'$ is an image of an equation $\mathfrak{A}$, then to any solution Σ' of $\mathfrak{A}'$ there corresponds a solution Σ of $\mathfrak{A}$. In this case we say that the solution Σ generates the solution Σ', and the solution Σ' induces the solution Σ (or the solution Σ is induced by the solution Σ'). The solution Σ is obtained from Σ' by the formulas which carry $\mathfrak{A}$ into $\mathfrak{A}'$.

A solution of an equation $\mathfrak{A}$ is called finite (infinite) if the transformation of $\mathfrak{A}$ generated by this solution is a finite (infinite) transformation.

Now consider a pair $(\mathfrak{A},\Sigma)$, where $\mathfrak{A}$ is a one-sided word equation and Σ is a solution of this equation. As we have already stated, any such pair uniquely determines an image $\mathfrak{A}'$ of $\mathfrak{A}$ and a solution Σ' of this image. In this case we say that the pair $(\mathfrak{A}',\Sigma')$ is the image of the pair $(\mathfrak{A},\Sigma)$. A sequence of pairs $(\mathfrak{A}_0,\Sigma_0),\cdots,(\mathfrak{A}_k,\Sigma_k)$ is called a chain of pairs if for any i, $1 \leqq i \leqq k$, the pair $(\mathfrak{A}_i,\Sigma_i)$ is the image of the pair $(\mathfrak{A}_{i-1},\Sigma_{i-1})$. We say that a chain of pairs $(\mathfrak{A}_0,\Sigma_0),\cdots,(\mathfrak{A}_k,\Sigma_k)$ is generated by the pair $(\mathfrak{A}_0,\Sigma_0)$ or, more briefly, is a chain of the pair $(\mathfrak{A}_0,\Sigma_0)$.

A sequence $\mathfrak{A}_0,\cdots,\mathfrak{A}_k$ of word equations is called a chain of images if for any i, $1 \leqq i \leqq k$, the equation $\mathfrak{A}_i$ is an image of the equation $\mathfrak{A}_{i-1}$. A chain of images $\mathfrak{A}_0,\cdots,\mathfrak{A}_k$ is called an infinite chain if for any i, $1 \leqq i \leqq k$, the equation $\mathfrak{A}_i$ is an infinite image of the equation $\mathfrak{A}_{i-1}$. We say that a chain $\mathfrak{A}_0,\cdots,\mathfrak{A}_k$ is generated by the equation $\mathfrak{A}_0$ or, more briefly, is a chain of the equation $\mathfrak{A}_0$. If $\mathfrak{A}_0,\cdots,\mathfrak{A}_k$ is an infinite chain of $\mathfrak{A}_0$, then $\mathfrak{A}_k$ is called an infinite image of order k of $\mathfrak{A}_0$.

A chain of pairs $(\mathfrak{A}_0,\Sigma_0),\cdots,(\mathfrak{A}_k,\Sigma_k)$ is called complete if $\mathfrak{A}_0,\cdots,\mathfrak{A}_k$ is an infinite chain and Σ_k is a finite solution of $\mathfrak{A}_k$. Lemma 3.1 implies that any pair generates a unique complete chain of pairs.

We say that a solution Σ of an equation $\mathfrak{A}$ is a solution of rank k if the complete chain of the pair $(\mathfrak{A},\Sigma)$ consists of $k+1$ pairs. Solutions of rank zero are the finite solutions. Obviously, an equation $\mathfrak{A}$ has a solution of rank $\geqq k$ if and only if at least one infinite image of order k of $\mathfrak{A}$ has a solution. It is also obvious that solutions of different ranks are distinct.

Suppose there is given a transformation (3.3) of equation (3.1). We say that the unknown x_1 is affected and the unknowns $x_2,\cdots,x_n$ are not affected in (3.3). Suppose $\mathfrak{A}_0,\cdots,\mathfrak{A}_k$ is an infinite chain and $\pi_1,\cdots,\pi_k$ a sequence of transformations, where π_i carries $\mathfrak{A}_{i-1}$ into $\mathfrak{A}_i$. We say that an unknown ξ is affected (not affected) in the chain $\mathfrak{A}_0,\cdots,\mathfrak{A}_k$ if ξ is affected in at least one (not affected in any) of the transformations $\pi_1,\cdots,\pi_k$.

The exponent t on the right-hand side of (1), (2) in 3.1 is called the height of the solution $(X_1,\cdots,X_n)$. The height of a solution is numerically equal to the parameter of the transformation generated by this solution.

Lemma 3.1 implies that if a one-sided equation has a solution, then at least one image of this equation has a solution and, conversely, if some image of an equation has a solution, then the equation itself has a solution. Knowing the structure of the solutions of the images of an equation, we can also determine the structure of the solutions of the equation itself. Thus, the study of one-sided equations reduces to the study of the images of these equations. However, in order to use this reduction it is necessary to consider an infinite number of images. Is it possible to somehow limit the problem to a consideration of only a finite number of images? This is the so-called problem of eliminating the parameter of a transformation. We will state this problem more precisely.

We first generalize the notion of a parametric word. Namely, we add to the inductive definition of parametric word in §2 the following condition: 1′. If φ is a word unknown, then φ is a parametric word. By a word unknown we mean an unknown whose values are words in the alphabet $\mathfrak{N}$.

Consider the equation

$$(*)\qquad \Phi(x_1, \cdots, x_n, t) = \Psi(x_1, \cdots, x_n, t),$$

where $x_1, \cdots, x_n$ are word unknowns, t is a natural unknown, and Φ and Ψ are parametric words. If in the words Φ and Ψ we substitute for t a natural number, we obtain a word equation in the n unknowns $x_1, \cdots, x_n$. We say that the parameter t in equation (*) is eliminable if for some $N \geqq 0$ we have the equivalence

$$\exists x_1 \ldots x_n t\,(\Phi = \Psi) \Leftrightarrow \bigvee_{i=0}^{N} \exists x_1 \ldots x_n\,[\Phi(x_1, \ldots, x_n, i) = \Psi(x_1, \ldots, x_n, i)].$$

(1)

In other words, the parameter t in (*) is eliminable if the decidability of (*) (with respect to $x_1, \cdots, x_n, t$) implies the decidability of this equation for small t. It follows that if the parameter t in (*) is eliminable, the decidability of (*) with respect to $x_1, \cdots, x_n, t$ reduces to the decidability of a finite number of word equations. If the equivalence (1) holds, we say that N is an elimination bound for the parameter in (*).

Note that (1) is equivalent to the implication

$$\exists x_1 \ldots x_n t\,(\Phi = \Psi) \supset \bigvee_{i=0}^{N} \exists x_1 \ldots x_n\,[\Phi(x_1, \ldots, x_n, i) = \Psi(x_1, \ldots, x_n, i)],$$

since the reverse implication is trivial.

We will also use the notion of conditional elimination. Suppose $\mathscr{P}(x_1, \cdots, x_n, t)$ is a predicate. We say that the parameter in (*) is eliminable under the condition $\mathscr{P}$ (in the region $\mathscr{P}$) if for some $N \geqq 0$ we have the implication

$$\exists x_1 \ldots x_n t\,[(\Phi = \Psi) \,\&\, \mathscr{P}] \supset \bigvee_{i=0}^{N} \exists x_1 \ldots x_n\,[\Phi(x_1, \ldots, x_n, i) = \Psi(x_1, \ldots, x_n, i)].$$

Once again, N is called an elimination bound for the parameter in (*) in the region $\mathscr{P}$.

In investigating the structure of the solutions of word equations we use another form of parametric elimination. We say that the parameter in (*) is strongly eliminable if for some $N \geqq 0$ and arbitrary $x_1, \cdots, x_n$ we have the implication

$$(2)\qquad (\exists t \geqq N)\,(\Phi = \Psi) \supset (\forall t \geqq N)\,(\Phi = \Psi).$$

In other words, the parameter in (*) is strongly eliminable if any set $(X_1, \cdots, X_n)$ which is a solution of (*) for some $t \geqq N$ satisfies (*) for any $t \geqq N$. Obviously, strong eliminability of a parameter implies simple eliminability. The converse is false, as we shall see by an example in §21. The number N in (2) is called a strong elimination bound for the parameter in (*).

The definitions of elimination formulated above for "two-sided" equations of the form $\Phi = \Psi$ carry over verbatim to one-sided equations. It suffices to replace $\Phi = \Psi$ everywhere by $x\Phi \to y\Psi$, where x and y are distinct unknowns from the list $x_1, \cdots, x_n$.

We now turn to parametric elimination in transformations of word equations. Denote by $\mathfrak{A}_{t,r}$ the image of equation (3.1) under the transformation (3.3), and by $\mathfrak{A}_{t,r,A}$ the image of (3.1) under the transformation (3.2). We say that a number N is an elimination bound (strong elimination bound) for the parameter in the transformations of equation (3.1) if N is an elimination bound (strong elimination bound) for the parameter t in each of the equations $\mathfrak{A}_{t,r}$ and $\mathfrak{A}_{t,r,A}$ (for all r and A). In this regard we assume (a formal detail related to the one-sidedness of equation (3.1)) that in the equation $\mathfrak{A}_{t,r}$ with $r=0$ the parameter t takes on only positive values, i.e. the equivalence (1) has the form

$$\exists x_2 \ldots x_n y t\, (\mathfrak{A}_{t,0} \,\&\, t>0) \Leftrightarrow \bigvee_{i=1}^{N} \exists x_2 \ldots x_n y \mathfrak{A}_{i,0}.$$

Similarly, if $r=0$ and $A=\Lambda$, then in the equation $\mathfrak{A}_{t,r,A}$ the parameter t is taken to be positive. We say that the parameter is eliminable (strongly eliminable) in the transformations of $\mathfrak{A}$ if there exists an elimination bound (strong elimination bound) for the parameter in the transformations of $\mathfrak{A}$. It follows from the definitions of eliminability and from 3.1 that if the parameter is eliminable in the transformations of a one-sided equation, the decidability of this equation reduces to the decidability of a finite number of images of this equation. More precisely, we have in mind the equivalence

$$\exists x_1 \ldots x_n \mathfrak{A} \Leftrightarrow \bigvee_{t=0}^{N} \bigvee_{r=0}^{k-1} [\exists x_2 \ldots x_n y \mathfrak{A}_{t,r} \vee \bigvee_{A \angle A_{r+1}} \exists x_2 \ldots x_n \mathfrak{A}_{t,r,A}],$$

where $\mathfrak{A}$ is the equation (3.1) and N an elimination bound for the parameter in the transformations of $\mathfrak{A}$.

We also define eliminability of a parameter in transformations of an equation in some region. Suppose $\mathscr{P}(x_1, \cdots, x_n)$ is an n-place predicate. We say that a number N is an elimination bound for the parameter in transformations of $\mathfrak{A}$ in the region $\mathscr{P}(x_1, \cdots, x_n)$ if N is an elimination bound for the parameter t in the region $\mathscr{P}(\overline{x}_1, x_2, \cdots, x_n)$ in the equation $\mathfrak{A}_{t,r,A}$ and in the region $\mathscr{P}(\tilde{x}_1, x_2, \cdots, x_n)$ in the equation $\mathfrak{A}_{t,r}$ for all r, A. We say that the parameter is eliminable in transformations of $\mathfrak{A}$ in the region $\mathscr{P}(x_1, \cdots, x_n)$ if there exists an elimination bound for the parameter in transformations of $\mathfrak{A}$ in the region $\mathscr{P}(x_1, \cdots, x_n)$. It follows from these definitions and from 3.1 that if the parameter is eliminable in transformations of $\mathfrak{A}$ in the region $\mathscr{P}$, the decidability of $\mathfrak{A}$ in $\mathscr{P}$ reduces to the decidability of a finite number of images of $\mathfrak{A}$ in the corresponding regions.

We have already observed that if the parameter is eliminable in transformations of an equation, then the decidability of this equation reduces to the decidability of a finite number of its images. In §4 we prove the eliminability of the parameter in transformations of equations in three unknowns. Hence for any one-sided equation in three unknowns the decidability of this equation reduces to the decidability of a finite number of its images. Here the interest lies in the infinite images, since the finite images are equations in two unknowns. The infinite images are one-sided equations in three unknowns, and the discussion is applicable to them. We come to study infinite chains of equations in three unknowns. The method used in this monograph for investigating word equations consists of studying properties of an equation $\mathfrak{A}_k$ in an infinite chain $\mathfrak{A}_0, \cdots, \mathfrak{A}_k$ as $k \rightarrow \infty$.

In the same way we can attempt to study equations in more than three unknowns. But for equations in four unknowns the eliminability of the parameter of transformations remains an open question.

CHAPTER II

EQUATIONS IN THREE UNKNOWNS

In this chapter we construct an algorithm recognizing whether or not a given word equation in three unknowns has a solution.

The investigation of equations in three unknowns is carried out in accordance with the scheme set forth in §3. We first prove (under certain easily removed restrictions) the eliminability of the parameter in transformations of equations in three unknowns (§4). Here the exponential equations of §2 play a major role. Now, however, the bases of powers and the coefficients depend on x, y and z. The problem is to obtain "uniform" (independent of x, y and z) upper estimates for the exponents. Having proved the eliminability of the parameter, it is possible to use the device of transformations. Instead of equations we consider their images of sufficiently large order. In this way the solution of arbitrary equations in three unknowns is successfully reduced to the solution of special so-called basic equations (§§5 and 6). Then we must widen the circle of operations which we apply to equations: to the operation of taking an image we add certain others. This leads to the important concept of a tree of an equation.

Basic equations are considered in §§7–10. The reason for distinguishing basic equations and their "basic" role is explained by the following. A basic equation is an equation of the form $xA(y)z\Phi(x,y,z)\rightarrow zB(y)x\Psi(x,y,z)$, where A and B are words in the alphabet $\mathbf{H}\cup\{y\}$ containing the same number ($\geqq 0$) of occurrences of the letter y and where $[A^\partial=[B^\partial$. In view of the last condition, this basic equation is equivalent to the system consisting of the equations $xAz\rightarrow zBx$ and $\Phi=\Psi$. The equation $xAz\rightarrow zBx$ can be studied in great detail. This allows us in solving the basic equation $xAz\Phi\rightarrow zBx\Psi$ to obtain information about the solutions from the "prefix" $xAz\rightarrow zBx$ and then to take into account the influence of the "tail" $\Phi=\Psi$.

§4. Eliminability of the parameter of transformations

We first prove a general assertion on elimination.

4.1. *Consider an equation in n unknowns*

$$x_1\Phi(x_1,\cdots,x_n)\rightarrow D(x_2,\cdots,x_n)x_1\Psi(x_1,\cdots,x_n). \tag{1}$$

If each of the unknowns $x_2,\cdots,x_n$ occurs in D, then the parameter is strongly eliminable in transformations of (1).

PROOF. We restrict ourselves to infinite transformations of (1) (the proof is analogous for finite transformations). Any infinite transformation of (1) has the form $x_1=D^tAy$, where $t\geqq 0$, A is a proper beginning of the word D, and y is a new unknown. The infinite images of (1) have the form

$$y\Phi(D^tAy,x_2,\cdots,x_n)\leftarrow BAy\Psi(D^tAy,x_2,\cdots,x_n), \tag{2}$$

where B is defined by the graphical equality $AB=D$. We must prove that the parameter t in (2) is strongly eliminable. Let us do this. Suppose $(Y, X_2, \cdots, X_n)$ is a solution of (2) for a fixed t. Let A_0, B_0 and D_0 denote the results of substituting the words $Y, X_2, \cdots, X_n$ for the unknowns $y, x_2, \cdots, x_n$ in the words A, B and D. We have that $AB = D$, BA begins with one of the letters $x_2, \cdots, x_n$ and each of these letters occurs in D. Therefore,

$$(3) \qquad [D_0^\partial \geqq \max\{[A_0^\partial, [B_0^\partial, [X_2^\partial, \cdots, [X_n^\partial, [Y^\partial\} > 0.$$

Consider the exponential equation in one unknown

$$(4) \qquad Y\Phi(D_0^t A_0 Y, X_2, \cdots, X_n) = B_0 A_0 Y \Psi(D_0^t A_0 Y, X_2, \cdots, X_n).$$

The eliminability of t in (2) becomes evident if we apply Proposition 2.7 to (4) and use the fact that in view of the estimates (3) the characteristic G of 2.7 is bounded by a constant which does not depend on the solution $(Y, X_2, \cdots, X_n)$. The proposition is proved.

4.2. *In transformations of any word equation in two unknowns the parameter is strongly eliminable.*

This follows from 4.1, since the unknown x_2 in this case certainly occurs in D (D begins with it).

The main difficulty is the elimination of the parameter in the case where not all of the unknowns $x_2, \cdots, x_n$ occur in D. For equations in three unknowns this happens when one of the letters x_2 or x_3 occurs in D and the other does not. Then the equation has the form (with different notation)

$$(*) \qquad x\Phi(x, y, z) \to \mathscr{A}(y) x \Psi(x, y, z),$$

where $\mathscr{A}$ is a word in the alphabet $\mathbf{H} \cup \{y\}$.

Multiplying both sides of the equation on the right by a suitable word in the alphabet $\mathbf{H} \cup \mathbf{X}$, we may assume that each of the unknowns x, y, z occurs on both sides of the equation (and as many times as needed) and that the equation has at least one nonempty coefficient. In particular, the unknown z occurs on both sides of the equation. Consider the maximal beginnings of the left- and right-hand sides of the equation which do not contain z. Call them the prefixes. The equation (*) is called indecomposable if the number of occurrences of at least one of the unknowns x or y in the prefixes of the left- and right-hand sides is different.

We first consider indecomposable equations.

Let us define a relation τ between words and natural numbers. Let X be a word in the alphabet $\mathbf{H}$, and $M > 0$ an integer. By definition, we write $\tau(X, M)$ if there exist a number $t \geqq 0$ and words S and S_1 such that $X = S^t S_1$, $S_1 \angle S$ and $[S^\partial \leq M$. Obviously, if $M > 0$ and $[X^\partial \leqq M$, then $\tau(X, M)$, so that $\rceil \tau(X, M)$ implies $[X^\partial > M$. It is also obvious that if $\tau(X, M)$ and $Y \angle X$, then $\tau(Y, M)$. It is also obvious that in this definition we may assume that S is simple, since any word is a power of a simple word.

We define an n-place predicate τ by means of the equivalence

$$\tau(x_1, \ldots, x_n, M) \Leftrightarrow \mathop{\&}_{i=1}^{n} \tau(x_i, M).$$

The problem of solving equations satisfying the condition τ with a fixed M is the problem of solving one-stage exponential equations. As regards the solutions not satisfying the condition τ, in their determination the parameter of transformations can be assumed to be eliminable.

Before proving this assertion, we define projections of a word. If P is a word in the alphabet $\mathbf{A} \cup \mathbf{Б}$, then the projection of P on the alphabet $\mathbf{A}$ is the word obtained if we replace each letter of the alphabet $\mathbf{Б}$ in the word P by the empty word.

4.3. *Suppose* (*) *is an indecomposable equation. Denote by* l_1 *and* l_2 *the lengths of the projections of the left- and right-hand sides of* (*) *on the alphabet* $\{x, y, z\}$, *and by* k_1 *and* k_2 *the lengths of these projections on the alphabet* $\mathbf{И}$. *Put* $l = \max\{l_1, l_2, 8\}$ *and* $M = \max\{k_1, k_2, 1]$. *In transformations of* (*) *the parameter is eliminable under the condition* $\neg\tau(x, y, z; 2lM)$. *The number* $(2l)^{10}$ *is an elimination bound.*

PROOF. We will prove the eliminability of the parameter in infinite transformations. The proof for finite transformations is carried out in a similar way.

Any infinite transformation of (*) has the form $x = \mathscr{A}^{\lambda}\mathscr{A}_1 v$, where $\lambda \geqq 0$, $\mathscr{A}_1 \angle \mathscr{A}$, and v is a new unknown. Correspondingly, any infinite image of (*) has the form

$$v\Phi(\mathscr{A}^{\lambda}\mathscr{A}_1 v, y, z) \to y^{\mathscr{P}} v\Psi(\mathscr{A}^{\lambda}\mathscr{A}_1 v, y, z), \tag{1}$$

where $y^{\mathscr{P}}$ is a cyclic permutation of $\mathscr{A}$. If $\mathscr{A}_1 = \Lambda$, then $\lambda > 0$. We henceforth assume that if $\mathscr{A}_1 = \Lambda$, then the exponent λ takes on only positive values.

We have to prove that the parameter λ is eliminable in (1) under the condition $\neg\tau(\mathscr{A}^{\lambda}\mathscr{A}_1 v, y, z; 2lM)$. Let $\mathfrak{A}(v, y, z, \lambda)$ denote the predicate defined by equation (1). Then the assertion being proved can be written as the implication

$$\exists vyz\lambda\,[\mathfrak{A}(v, y, z, \lambda)\,\&\,\neg\tau(\mathcal{A}^{\lambda}\mathcal{A}_1 v, y, z; 2lM)] \supset \bigvee_{i=0}^{C} \exists vyz\mathfrak{A}(v, y, z, i), \tag{2}$$

where C is a natural number. We will prove that (2) holds with $C = (2l)^{10}$.

Assume that the left-hand side of (2) is satisfied. Then there exists a solution $(V, Y, Z, \bar{\lambda})$ of (1) such that $\neg\tau(\mathscr{A}(Y)^{\bar{\lambda}}\mathscr{A}_1(Y)\,V, Y, Z; 2lM)$. We must prove that at least one of the equations $\mathfrak{A}(v, y, z, i)$, $0 \leqq i \leqq C$, has a solution. This is true, obviously, if $\bar{\lambda} \leqq (2l)^{10}$. We may assume, then, that $\bar{\lambda} > (2l)^{10}$.

Our entire argument depends on the properties of solutions of one-stage exponential equations established in 2.4–2.7. The most important of these properties are the upper estimates, in terms of the height and characteristic of an equation, of a region in which the equation has a solution, if it has solutions.

We regard equation (1) with fixed v, y and z as an exponential equation in the one unknown λ. Equation (1) differs from equation (2) of 4.1 in that now the unknown z does not occur in the word $\mathscr{A}$. Hence the characteristic of (1), which depends on the ratio $\partial(z)/\partial(A)$,[9] is not bounded, as in 4.1,

[9] If $\varphi(\xi_1, \cdots, \xi_n)$ is a word function (the definition follows 1.9′), then $\partial(\varphi)$ denotes the length function: the value of $\partial(\varphi)$ on the set $(\bar{\xi}_1, \cdots, \bar{\xi}_n)$ is equal to the length of the word $\varphi(\bar{\xi}_1, \cdots, \bar{\xi}_n)$.

by an absolute constant (i.e. a constant which does not depend on the solution (v,y,z)). It is necessary to express z from (1).

Let us rewrite (1). Selecting in the left- and right-hand sides of (1) the left-most occurrences of z, and also the occurrences of $\mathscr{A}^\lambda$ in x in the words Φ and Ψ (the occurrences are in the form $*\mathscr{A}^\lambda*\mathscr{A}_1 v$ (see [9], Chapter I, §4.2)), we write this equation as

$$(3) \qquad \mathscr{P}_0\,[\mathscr{A}^\lambda\mathscr{P}_i]_{i=1}^n\, z\Phi_1\,(\mathscr{A}^\lambda\mathscr{A}_1 v, y, z) \leftarrow \mathscr{Q}_0\,[\mathscr{A}^\lambda\mathscr{Q}_i]_{i=1}^m\, z\Psi\,(\mathscr{A}^\lambda\mathscr{A}_1 v, y, z),$$

where $n,m \geqq 0$; $\mathscr{P}_i$ and $\mathscr{Q}_j$ are words in the alphabet $\boldsymbol{И}\cup\{v,y\}$, and Φ_1 and Ψ_1 are ends of the words Φ and Ψ. Since $(V,Y,Z,\bar\lambda)$ is a solution of (3), in the aphabet $\boldsymbol{И}$ we have the equality

$$(4) \qquad P_0\,[A^{\bar\lambda}P_i]_{i=1}^n\, Z\Phi_1\,(A^{\bar\lambda}A_1V, Y, Z) = Q_0\,[A^{\bar\lambda}Q_i]_{i=1}^m\, Z\Psi_1\,(A^{\bar\lambda}A_1V, Y, Z),$$

where P_i, Q_j, A and A_1 are the results of substituting the words Y and V for the unknowns y and v in the words $\mathscr{P}_i$, $\mathscr{Q}_j$, $\mathscr{A}$ and $\mathscr{A}_1$. By the indecomposability of (*) two cases are possible: 1) $n=m$, but the words $\mathscr{P}_0\cdots\mathscr{P}_n$ and $\mathscr{Q}_0\cdots\mathscr{Q}_m$ do not contain the letter y the same number of times; 2) $n\neq m$. Let us consider each of these cases.

Case 1. Suppose $n=m$. We will assume that there are k $(k\geqq 1)$ more occurrences of y in $\mathscr{P}_0\cdots\mathscr{P}_n$ than in $\mathscr{Q}_0\cdots\mathscr{Q}_m$. Then, taking into account the structure of the words $\mathscr{P}_i$ and $\mathscr{Q}_j$, we obtain $[P_0\cdots P_n^\partial - [Q_0\cdots Q_m^\partial = k[Y^\partial + d$, where d is the difference between the number of letters in the alphabet $\boldsymbol{И}$ in the prefixes of the left- and right-hand sides of (*). We have $|d|\leqq M$. From the conditions $\urcorner\tau(Y,2lM)$, $l\geqq 1$, $M\geqq 1$, we have $[Y^\partial > 2lM > M$. But then $k[Y^\partial + d > 0$. It now follows from (4) that there exists a word $B\neq\Lambda$ such that

$$[B^\partial = k[Y^\partial + d, \qquad P_0[A^{\bar\lambda}P_i]_{i=1}^n = Q_0[A^{\bar\lambda}Q_i]_{i=1}^m B,$$
$$BZ\Phi_1(A^{\bar\lambda}A_1V, Y, Z) = Z\Psi_1(A^{\bar\lambda}A_1V, Y, Z).$$

Since $B\neq\Lambda$, in view of 1.16 it follows from the last equality that there exist an integer $\bar\mu\geqq 0$ and a word B_1 such that $Z = B^{\bar\mu}B_1$ and $B_1\angle B$. We have obtained the desired expression for the component Z. We can now reduce equation (1) to an exponential equation in two unknowns.

Let us define words $f(x,y,z)$ and $g(x,y,z)$ in the alphabet $\boldsymbol{И}\cup\{x,y,z\}$. Put (see (1)) $f(x,y,z) = V\Phi(x,y,z)$ and $g(x,y,z) = y\mathscr{P}\,V\Psi(x,y,z)$. Consider the exponential equation in two unknowns

$$(5) \qquad f(A^\lambda A_1 V, Y, B^\mu B_1) = g(A^\lambda A_1 V, Y, B^\mu B_1).$$

The coefficients of (5) are determined by the triple (V,Y,Z). It follows from the definition of the words f and g that the pair $(\bar\lambda,\bar\mu)$ satisfies (5), and also that if (λ_0,μ_0) is a solution of (5), the quadruple $(V,Y,Z_0 = B^{\mu_0}B_1, \lambda=\lambda_0)$ satisfies the predicate $\mathfrak{A}(v,y,z,\lambda)$. To prove our assertion it suffices to show that (5) has a solution (λ_0,μ_0) such that $\lambda_0\leqq(2l)^{10}$. Let us demonstrate this.

We estimate $[A^\partial$ and $[B^\partial$. We have the lower estimates $[A^\partial\geqq[Y^\partial$ and $[B^\partial = k[Y^\partial + d \geqq [Y^\partial - M > 3M$, since $[Y^\partial > 4M$ in view of the conditions $\urcorner\tau(Y,2lM)$, $l>1$. We have the upper estimates $[A^\partial\leqq(l-1)\,[Y^\partial + M\leqq l[Y^\partial$ and $[B^\partial\leqq(l-1)\,[Y^\partial + M\leqq l[Y^\partial$. Clearly $[B_1^\partial\leqq[B^\partial$ and $[A^\partial\geqq[A_1V^\partial$. We consider separately the cases $\bar\mu\leqq 24l^3$ and $\bar\mu > 24l^3$.

If $\bar{\mu} \leqq 24l^3$, consider the equation in the one unknown λ

(6) $$f(A^\lambda A_1 V, Y, B^{\bar{\mu}} B_1) = g(A^\lambda A_1 V, Y, B^{\bar{\mu}} B_1).$$

Let h, H and N denote the height, characteristic and maximum coefficient length of the exponential equation (6). Using the inequalities $[A^\partial \geqq [Y^\partial$, $[B_1^\partial \leqq [B^\partial \leqq l[Y^\partial$ and $[A_1 V^\partial \leqq [A^\partial$, we obtain

$$\begin{aligned} N &\leqq M + [A_1 V^\partial + (l-1)([Y^\partial + [B^{\bar{\mu}} B_1^\partial) \\ &\leqq M + [A^\partial + (l-1)\,[A^\partial + (l-1)(\bar{\mu}+1)\,[B^\partial \\ &\leqq (l+1)\,[A^\partial + (l-1)25l^3[A^\partial \leqq 25l^4[A^\partial, \end{aligned}$$

and

$$H = \max\{N, [A^\partial\}/[A^\partial \leqq 25l^4.$$

It is also obvious that $h \leqq l$. Hence $2hH \leqq 50l^5$. We have that $\bar{\lambda} > (2l)^{10} > 50l^5$ is a solution of (6). Applying Proposition 2.7 to this equation, we see that (6) has a solution λ_0, $0 < \lambda_0 \leqq (2l)^{10}$. The pair $(\lambda_0, \bar{\mu})$ is the desired solution of (5).

There remains the case $\bar{\mu} > 24l^3$. Here we represent A and B as powers of simple words:

(7) $$A = S^\sigma, \qquad B = T^\tau,$$

where S and T are simple words, $\sigma > 0$ and $\tau > 0$. There are two possible subcases.

Subcase a). $[S^\partial \neq [T^\partial$. Let H, h and N denote the characteristic, height and maximum coefficient length of equation (5). Estimating N from above, we obtain $N \leqq M + (l-1)\,[Y^\partial + \max\{[A_1 V^\partial, [B_1^\partial\} \leqq l[Y^\partial + \max\{[A^\partial, [B^\partial\} \leqq 2l[Y^\partial$. Hence $N \leqq 2l[Y^\partial$. Using this inequality, the upper and lower estimates for $[A^\partial$ and $[B^\partial$, and the inequality $[Y^\partial > 3M$, we have for the characteristic H

$$H = \frac{\max\{N, [A^\partial, [B^\partial\}}{\min\{[A^\partial, [B^\partial\}} \leqslant \frac{2l\,[Y^\partial}{[Y^\partial - M} = \frac{2l}{1 - \dfrac{M}{[Y^\partial}} < \frac{2l}{\frac{2}{3}} = 3l.$$

Regarding the height h we clearly have $h \leqq l$. Hence $2hH \leqq 6l^2$. Equation (5) has a solution $(\bar{\lambda}, \bar{\mu})$, where $\bar{\lambda}, \bar{\mu} \geqq 6l^2$. By 2.6 equation (5) is satisfied for any $\lambda, \mu \geqq 0$. The pair $(\lambda_0 = 1, \bar{\mu})$ is the desired solution of (5).

Subcase b). $[S^\partial = [T^\partial$. Here we consider the exponential equation in the two unknowns α and β

(8) $$f(S^\alpha A_1 V, Y, T^\beta B_1) = g(S^\alpha A_1 V, Y, T^\beta B_1).$$

It has a solution $\bar{\alpha} = \sigma\bar{\lambda}$, $\bar{\beta} = \tau\bar{\mu}$. Let H, h and N be the characteristic, height and maximum coefficient length of (8). Equations (8) and (5) have the same coefficients. Hence $N \leqq 2l[Y^\partial$. Moreover, we have $[S^\partial = [T^\partial$. Hence for the characteristic H we have

$$H = \frac{\max\{N, [S^\partial, [T^\partial\}}{\min\{[S^\partial, [T^\partial\}} \leqslant \max\left\{\frac{2l\,[Y^\partial}{[S^\partial}, 1\right\} = \max\left\{\frac{2l\sigma\,[Y^\partial}{[A^\partial}, 1\right\} \leqslant 2l\sigma.$$

Similarly, $H \leqq \max\{2l[Y^\partial/[T^\partial, 1\} = \max\{2l\tau[Y^\partial/[B^\partial, 1\} \leqq 3l\tau$. Regarding the height h we have $h \leqq l$. We now estimate the solution $(\bar{\alpha}, \bar{\beta})$. We have $\bar{\alpha} = \sigma\bar{\lambda} > \sigma 24l^3 \geqq 8h^2H$, $\bar{\beta} = \tau\bar{\mu} > \tau 24l^3 \geqq 8h^2H$. Hence by 2.4′ there exists an equation $a\alpha+b\beta+c=0$, where a, b and c are integers, $|a|$, $|b| \leqq h$ and $|c| \leqq 4hH$, which is satisfied by the pair $(\bar{\alpha},\bar{\beta})$ and any positive solution (α,β) of which satisfies (8). But then in view of (7) any solution (λ,μ) of the equation

$$(9) \qquad a\sigma\lambda+b\tau\mu=c$$

satisfies (5).

It now remains for us to prove that (9) has a positive solution (λ_0,μ_0) in which $\lambda_0 \leqq (2l)^{10}$. The usual argument does not work, since the coefficients of (9) depend on σ and τ. However, as we will now show, the exponents σ and τ are connected by the relation $k\sigma=q\tau$, where q is the number of occurrences of the letter y in the word A. This allows us to estimate from above in the usual way the smallest solution of (9).

Let us prove that $k\sigma=q\tau$. Since $[S^\partial=[T^\partial$, we have from (7)

$$(10) \qquad [A^\partial=\sigma[S^\partial, \qquad [B^\partial=\tau[S^\partial.$$

On the other hand, $[A^\partial=q[Y^\partial+\delta$, where δ is the number of letters of the alphabet $\mathbf{H}$ in the word $\mathscr{A}$. We also have $[B^\partial=k[Y^\partial+d$. Therefore, comparing $[A^\partial$ and $[B^\partial$, we obtain $k[A^\partial-q[B^\partial=k\delta-qd$. From this and (10) we have $(k\sigma-q\tau)[S^\partial=k\delta-qd$. Dividing by $[S^\partial$, we have

$$(11) \qquad k\sigma-q\tau=(k\delta-qd)/[S^\partial.$$

Obviously, $|k\delta-qd| \leqq 2lM$. We will prove by contradiction that $[S^\partial > 2lM$. Indeed, suppose $[S^\partial \leqq 2lM$. Then from (7) and the relation $A_1V \angle A$ we obtain $A^{\bar{\lambda}}A_1V=S^tS_1$, where $t \geqq 0$ and $S_1 \angle S$, i.e. $\tau(A^{\bar{\lambda}}A_1V,2lM)$. Also, since $y \angle \mathscr{A}$ (see (*)), $Y \angle A$. Then (7) implies $\tau(Y,2lM)$. Finally, from (7) and the equality $Z=B^{\bar{\mu}}B_1$ we obtain $Z=T^\nu T_1$, where $\nu \geqq 0$ and $T_1 \angle T$, so that $\tau(Z,2lM)$. These relations contradict the condition

$$\urcorner\, \tau(A^{\bar{\lambda}}A_1V,Y,Z;2lM).$$

Thus we have proved that $[S^\partial > 2lM$. Then the right-hand side of (11) is a proper fraction and the left-hand side an integer. Hence the right-hand side is equal to zero and we have $k\sigma=q\tau$, as required.

Let us transform (9). Multiplying (9) by q, we obtain the equation $aq\sigma\lambda+bq\tau\mu=cq$. Since $q\tau=k\sigma$, the equation takes the form $\sigma(aq\lambda+bk\mu)=cq$. After dividing by σ we obtain

$$(12) \qquad aq\lambda+bk\mu=cq/\sigma.$$

But $|c| \leqq 4hH$. Therefore, $|c/\sigma| \leqq 4Hh/\sigma \leqq 8l\sigma l/\sigma \leqq 8l^2$, $|cq/\sigma| \leqq 8l^3$. It is also obvious that $|aq|,|bk| \leqq l^2$. Now recall that any solution of (12) (equation (9) modified) satisfies (5). Equation (12) has the solution $(\bar{\lambda},\bar{\mu})$. Hence in view of what was said in §2 on Diophantine equations, (12) has a solution (λ_0,μ_0) such that $\lambda_0 \leqq 16l^3 < (2l)^{10}$. This pair is the desired solution of (5). Our assertion is proved in Case 1.

Case 2. $n \neq m$. Without loss of generality, we will assume that $n > m$. Let d_1 and d_2 denote the number of letters of the alphabet $\mathbf{H}$ in the words $\mathscr{P}_0\cdots\mathscr{P}_m$ and $\mathscr{Q}_0\cdots\mathscr{Q}_m$. Put $d=d_1-d_2$. Then (4) implies that there exist words D_1 and D_2 such that

$$(13) \qquad P_0\,[A^{\bar{\lambda}}P_i]_{i=1}^m D_1 = Q_0\,[A^{\bar{\lambda}}Q_i]_{i=1}^m D_2,$$

$$(14)\qquad D_2\,[A^{\bar\lambda}P_i]_{i=m+1}^{n} Z\Phi_1\,(A^{\bar\lambda}A_1V, Y, Z) = D_1 Z\Psi_1\,(A^{\bar\lambda}A_1V, Y, Z),$$

where $[D_1^\partial = -d$, $D_2 = \Lambda$ if $d \leqq 0$, and $D_1 = \Lambda$, $[D_2^\partial = d$ if $d \geqq 0$.

Let us express Z from equation (14). If $D_1 = \Lambda$, then Z is expressed from (14) by means of 1.16. If $D_1 \neq \Lambda$, then first cancel (14) on the left by D_1. Note that $D_2 = \Lambda$ if $D_1 \neq \Lambda$. Furthermore, we have $[D_1^\partial = |d| \leqq M < [Y^\partial \leqq [A^\partial$ and $\bar\lambda > 0$. Hence (14) implies that $A = D_1 Q$ for some Q. Now cancelling (14) on the left by D_1 we have $QA^{\bar\lambda-1}P_{m+1}[A^{\bar\lambda-1}AP_i]_{i=m+2}^{n}Z\Phi_1(A^{\bar\lambda}A_1V, Y, Z) = Z\Psi_1(A^{\bar\lambda}A_1V, Y, Z)$. If $D_1 = \Lambda$, represent $A^{\bar\lambda}$ in the form $A^{\bar\lambda-1}A$. Then (14) can be represented in the form

$$(15)\qquad C_0\,[A^{\bar\lambda-1}C_i]_{i=1}^{n-m} Z\Phi_1\,(A^{\bar\lambda}A_1V, Y, Z) = Z\Psi_1\,(A^{\bar\lambda}A_1V, Y, Z),$$

where $C_0, \cdots, C_{n-m}$ are words in the alphabet $\mathbf{H}$ and where $\max\{[C_i^\partial\} \leqq [A^\partial + \max\{|d|, [P_{m+1}^\partial, \cdots, [P_n^\partial\}$. If we view (14) and (15) as exponential equations in the one unknown $\bar\lambda$, they will be equivalent in the region $\bar\lambda > 0$.

Put $D(\lambda) = C_0[A^{\lambda-1}C_i]_{i=1}^{n-m}$. It follows from (15) and 1.16 that there exist $\bar t \geqq 0$ and a word U such that $Z = (D(\bar\lambda))^{\bar t}U$ and $U \angle D(\bar\lambda)$. But then from (15) we have

$$(16)\qquad D\,(\bar\lambda)\,U\Phi_1\,(A^{\bar\lambda}A_1V, Y, D\,(\bar\lambda)^{\bar t}\,U) = U\Psi_1\,(A^{\bar\lambda}A_1V, Y, D\,(\bar\lambda)^{\bar t}\,U).$$

Consider the equation $P_0[A^\lambda P_i]_{i=1}^{m}D_1 = Q_0[A^\lambda Q_i]_{i=1}^{m}D_2$. For the height h and characteristic H of this equation we have the estimates $h \leqq l-1$ and

$$H \leqslant \frac{\max\{[P_i^\partial, [Q_j^\partial, [A^\partial\} + M}{[A^\partial} \leqslant \frac{[A_1V^\partial + 2M + (l-1)\,[y^\partial}{[A^\partial} \leqslant 1 + l,$$

whence we conclude that $2hH \leqq 2l^2$. The number $\bar\lambda > 2l^2$ satisfies this equation. Hence by 2.7 the equation is satisfied for any $\lambda \geqq 0$. We have proved that (13) holds for any $\bar\lambda \geqq 0$.

Consider the equation in the three unknowns u, λ and t

$$(17)\qquad D(\lambda)u\Phi_1(A^\lambda A_1V,\ Y,\ (D\,(\lambda))^t u) = u\Psi_1\,(A^\lambda A_1V,\ Y,\ (D\,(\lambda))^t u).$$

As is evident from (16), equation (17) has the solution $(\bar\lambda, \bar t, U)$. Since (13) is satisfied for any $\bar\lambda \geqq 0$, by comparing (1), (4), (13) and (17) we conclude that if a triple (λ_0, t_0, U_0) satisfies (17), then the quadruple $(V, Y, Z_0 = (D(\lambda_0))^{t_0}U_0, \lambda_0)$ satisfies (1), i.e. the predicate $\mathfrak{A}(v, y, z, \lambda)$. Hence it suffices to prove that (17) has a solution (λ_0, t_0, U_0) in which $\lambda_0 \leqq (2l)^{10}$. Let us prove this.

Denote the left- and right-hand sides of (17) by $f(\lambda, t, u)$ and $g(\lambda, t, u)$. Consider the exponential equation

$$(18)\qquad f(\bar\lambda, t, U) = g(\bar\lambda, t, U).$$

This is an equation in the one unknown t. Let us estimate the characteristic G (see the remark following the proof of 2.7) of this equation. We have $\bar\lambda - 1 > 3l$, $[A_1V^\partial \leqq [A^\partial$, $[D(\bar\lambda)^\partial \geqq (\bar\lambda - 1)\,[A^\partial$, $[U^\partial \leqq [D(\bar\lambda)^\partial$ and $[A^\partial \geqq [Y^\partial \geqq M$. Hence

$$G \leqslant \frac{M + (l-1)([A^{\bar\lambda}A_1V^\partial + [Y^\partial) + (l-1)\,[U^\partial + [D(\bar\lambda)^\partial}{[D\,(\bar\lambda)^\partial} \leqslant \frac{[A^\partial + (l-1)(\bar\lambda+2)\,[A^\partial}{[D\,(\bar\lambda)^\partial} + l$$
$$= l + \frac{(l-1)(\bar\lambda-1)\,[A^\partial}{[D\,(\bar\lambda)^\partial} + \frac{(3l-2)\,[A^\partial}{[D\,(\bar\lambda)^\partial} \leqslant 2l - 1 + \frac{(\bar\lambda-1)\,[A^\partial}{[D\,(\bar\lambda)^\partial} \leqslant 2l.$$

In view of 2.7, if equation (18) has a solution $t > 2l$, then it is satisfied for any $t \geqq 0$. Now (18) has a solution $t = \bar{t}$. Hence it has a solution $t = t_0 \leqq 2l$. Consider the equation in the two unknowns λ and u

$$(19) \qquad f(\lambda, t_0, u) = g(\lambda, t_0, u).$$

It is sufficient to prove that (19) has a solution (λ_0, U_0) in which $\lambda_0 \leqq (2l)^{10}$. Note that the pair $(\bar{\lambda}, U)$ satisfies (19). We have $U \angle D(\bar{\lambda})$. Consider two cases, depending on how the word U cuts the occurrences of $A^{\bar{\lambda}-1}$ in the word $D(\bar{\lambda})$.

Case a). $U = C_0[A^{\bar{\lambda}-1}C_i]_{i=1}^{s} A^{\bar{\lambda}-1}P$, where $0 \leqq s < n - m$ and $P \angle C_{s+1}$. Put $U(\lambda) = C_0[A^{\lambda-1}C_i]_{i=1}^{s} A^{\lambda-1}P$. Consider the equation

$$(20) \qquad f(\lambda, t_0, U(\lambda)) = g(\lambda, t_0, U(\dot{\lambda})).$$

This is an exponential equation in the one unknown λ. For the height h and characteristic H of this equation we have $h \leqq l - 1 + [t_0(n - m) + s + 1](l - 1) \leqq (l - 1)(1 + (n - m)(t_0 + 1)) \leqq (l - 1)(1 + (l - 1)(2l + 1)) \leqq 2l^3$, $\max\{[P_i^{\partial}\} \leqq (l - 1)[Y^{\partial} + M \leqq l[Y^{\partial} \leqq l[A^{\partial}$,

$$H \leqslant \frac{\max\{[C_i^{\partial}\} + [AA_1V^{\partial} + (l-1)[Y^{\partial} + [P^{\partial} + M}{[A^{\partial}} \leqslant 3l + 4 \leqslant 4l.$$

Thus, $2Hh \leqq 16l^4$. Since equation (20) is satisfied for $\lambda = \bar{\lambda} \geqq 16l^4$, by 2.7 it is satisfied for any $\lambda \geqq 0$, in particular for $\lambda_0 = 0$. But then the pair ($\lambda_0 = 0$, $U_0 = U(\lambda_0)$) is the desired solution of (19).

Case b). $U = C_0[A^{\bar{\lambda}-1}C_i]_{i=1}^{s} A^{\bar{\mu}}P$, where $0 \leqq s < n - m$, $0 \leqq \bar{\mu} < \bar{\lambda} - 1$ and $P \angle A$. Put $U(\lambda, \mu) = C_0[A^{\lambda-1}C_i]_{i=1}^{s} A^{\mu}P$ and consider the equation

$$(21) \qquad f(\lambda, t_0, U(\lambda, \mu)) = g(\lambda, t_0, U(\lambda, \mu)).$$

This is an exponential equation in the two unknowns λ and μ. For the characteristic H and the height h of this equation we have the same estimates as in Case a); that is, $H \leqq 4l$ and $h \leqq 2l^3$, so that $4h^2H \leqq 64l^7$. The pair $(\bar{\lambda}, \bar{\mu})$ satisfies this equation, and $\bar{\lambda} > 64l^7$. If $\bar{\mu} > 64l^7$, then by 2.5′ the equation has a solution (λ_0, μ_0) in which $\lambda_0 \leqq (2l)^{10}$. But then the pair $(\lambda_0, U_0 = U(\lambda_0, \mu_0))$ is the desired solution of (19). There remains the case where $\bar{\mu} \leqq 64l^7$. Consider the equation in one unknown

$$(22) \qquad f(\lambda, t_0, U(\lambda, \bar{\mu})) = g(\lambda, t_0, U(\lambda, \bar{\mu})).$$

It is satisfied when $\lambda = \bar{\lambda}$. Since $\bar{\mu} \leqq 64l^7$, we obtain for the height h and characteristic H of this equation that $h \leqq 2l^3$ and $H \leqq 65l^7$. Since (22) has a solution $\bar{\lambda} > (2l)^{10} \geqq 2hH$, by 2.7 it has a solution λ_0, $0 < \lambda_0 \leqq (2l)^{10}$. Then the pair $(\lambda_0, U_0 = U(\lambda_0, \bar{\mu}))$ is the desired solution of (19). The proposition is proved.

We now consider decomposable equations.

Let Φ be a word in the alphabet $\mathbf{И} \cup \{x_1, \cdots, x_n\}$. Denote by $\nu(\Phi)$ the n-tuple $(a_1, \cdots, a_n)$, where a_i is the number of occurrences of the unknown x_i in Φ. Let $\tilde{\Phi}$ denote the projection of Φ on the alphabet $\mathbf{И}$.

4.4. *Suppose* $\Phi_1\Phi_2 \to \Psi_1\Psi_2$ *is an equation in* n *unknowns, where* $\nu(\Phi_1) = \nu(\Psi_1) \neq 0$. *Let* $d = [\tilde{\Phi}_1^{\partial} - [\tilde{\Psi}_1^{\partial}$, *and suppose* $d \geqq 0$. *Then we have the equivalences*

(1) $$(\Phi_1\Phi_2 \to \Psi_1\Psi_2) \Leftrightarrow \bigvee_{[R^\partial=d} (\Phi_1 \to \Psi_1 R \,\&\, R\Phi_2 = \Psi_2),$$

(2) $$(\Phi_1\Phi_2 = \Psi_1\Psi_2) \Leftrightarrow \bigvee_{[R^\partial=d} (\Phi_1 = \Psi_1 R \,\&\, R\Phi_2 = \Psi_2),$$

(3) $$(\Phi_1\Phi_2 = \Psi_1\Psi_2) \Leftrightarrow \bigvee_{[R^\partial=d} (R\Phi_2\Phi_1 = \Psi_2\Psi_1 R),$$

where the summation $\bigvee$ *is taken over all words* R *in the alphabet* $\mathbf{И}$ *such that* $[R^\partial = d$.

PROOF. It follows from our assumptions that if $(X_1, \cdots, X_n)$ is a solution of the equation $\Phi_1\Phi_2 = \Psi_1\Psi_2$, then $\overline{\Phi}_1\overline{\Phi}_2 = \overline{\Psi}_1\overline{\Psi}_2$ and $[\overline{\Phi}_1^\partial = [\overline{\Psi}_1^\partial + d$, where $\overline{\Phi}_i$, $\overline{\Psi}_i$ are the results of substituting into Φ_i and Ψ_i the words $X_1, \cdots, X_n$ in place of the unknowns $x_1, \cdots, x_n$. But then there exists a word R in the alphabet $\mathbf{И}$ such that $[R^\partial = d$, $\overline{\Phi}_1 = \overline{\Psi}_1 R$ and $R\overline{\Phi}_2 = \overline{\Psi}_2$. We have proved that the left-hand side of (1), (2) implies the right. The converse is obvious, since if $\overline{\Phi}_1 = \overline{\Psi}_1 R$ and $R\overline{\Phi}_2 = \overline{\Psi}_2$, then $\overline{\Phi}_1\overline{\Phi}_2 = \overline{\Psi}_1\overline{\Psi}_2$. Thus the equivalences (1) and (2) are proved.

Equivalence (3) follows from (2), since we have $[\tilde{\Phi}_1^\partial = [\tilde{\Psi}_1 R^\partial$ and $\nu(\Phi_1) = \nu(\Psi_1 R)$. Then, obviously, if $R\overline{\Phi}_2\overline{\Phi}_1 = \overline{\Psi}_2\overline{\Psi}_1 R$, we have $\overline{\Phi}_1 = \overline{\Psi}_1 R$ and $R\overline{\Phi}_2 = \overline{\Psi}_2$, which implies the equivalence of the system $\Phi_1 = \Psi_1 R$, $R\Phi_2 = \Psi_2$ and the equation $R\Phi_2\Phi_1 = \Psi_2\Psi_1 R$. The proposition is proved.

Words A and B in the alphabet $\mathbf{И} \cup \mathbf{X}$ are called similar if their projections on the alphabet $\mathbf{X}$ are graphically equal.

4.5. *Suppose* $\mathfrak{A}$ *is the word equation* $A_1 \cdots A_n = B_1 \cdots B_n$, $n > 0$, *and* $\nu(A_i) = \nu(B_i) \neq 0$, $i = 1, \cdots, n$. *Then the equation* $\mathfrak{A}$ *is equivalent to a disjunction of* m $(m > 0)$ *systems of the form* $\&_1^n(A_i' = B_i')$, *where* A_i' *and* B_i' *are similar to* A_i *and* B_i *respectively and* $[A_i'^\partial = [B_i'^\partial$. *If* $\mathfrak{A}$ *is the one-sided equation* $A_1 \cdots A_n \to B_1 \cdots B_n$, *the assertion is valid if* $A_i' = B_i'$ *is replaced by* $A_i' \to B_i'$.

PROOF. We proceed by induction on n. If $n = 1$, the assertion is trivial. Suppose $n > 1$. We have $\nu(A_1) = \nu(B_1)$ and $\nu(A_2 \cdots A_n) = \nu(B_2 \cdots B_n)$. In view of 4.4 (2) we have the equivalence (for the case $[\widetilde{A}_1^\partial - [\widetilde{B}_1^\partial = d \geq 0$)

(1) $$\mathfrak{A} \Leftrightarrow \bigvee_{[R^\partial=d} (A_1 = B_1 R \,\&\, RA_2 \ldots A_n = B_2 \ldots B_n),$$

where the summation $\bigvee$ is taken over all words of length d in the alphabet $\mathbf{И}$. Since R is a word in the alphabet $\mathbf{И}$ and $\nu(A_2) = \nu(B_2)$, we have $\nu(RA_2) = \nu(B_2)$. The induction assumption is applicable to the factor $RA_2 \cdots A_n = B_2 \cdots B_n$. Denote this factor by $\mathfrak{B}_R$. By the induction assumption we have an equivalence

(2) $$\mathfrak{B}_R \Leftrightarrow \bigvee_{i=1}^{m} \omega_{R,i}.$$

Here $\omega_{R,i}$ is a system of equations of the form $\&_2^n (C_i = D_i)$, where C_i and D_i are similar to A_i and B_i $(i = 2, \cdots, n)$. From (1) we have

(3) $$\mathfrak{A} \Leftrightarrow \bigvee_{[R^\partial=d} (\mathfrak{A}_R \,\&\, \mathfrak{B}_R),$$

where $\mathfrak{A}_R$ is the equation $A_1 = B_1 R$. Using the distributive law of the propositional calculus

$$p \mathbin{\&} \bigvee_{i=1}^{N} q_i \Leftrightarrow \bigvee_{i=1}^{N} (p \mathbin{\&} q_i),$$

where p and $q_1, \cdots, q_N$ are propositions, we obtain from (2) and (3) the necessary equivalence for the equation $\mathfrak{A}$.

The validity of the assertion with $=$ replaced by $\rightarrow$ is obvious; instead of 4.4 (2) we use 4.4 (1). The proposition is proved.

4.5′. *Suppose there is given the equation $x\Phi_1\Phi_2\Phi_3 \rightarrow y\Psi_1\Psi_2\Psi_3$, and suppose $\nu(x\Phi_1) = \nu(y\Psi_1)$ and $\nu(\Phi_2) = \nu(\Psi_2)$. Then this equation is equivalent to a disjunction of equations of the form $x\Phi_1\Phi_3'\Phi_2' \rightarrow y\Psi_1\Psi_3'\Psi_2'$, where Φ_2', Φ_3', Ψ_2' and Ψ_3' are similar to Φ_2, Φ_3, Ψ_2 and Ψ_3 respectively.*

PROOF. Let $d_1 = [\tilde{\Phi}_1^{\partial} - [\tilde{\Psi}_1^{\partial}$, $d_2 = [\tilde{\Phi}_2^{\partial} - [\tilde{\Psi}_2^{\partial}$ and $d = d_1 + d_2$. We restrict ourselves to the case $d_1 \geqq 0$, $d \leqq 0$. The other cases are handled analogously. By 4.4 (1) we have the equivalence

$$x\Phi_1\Phi_2\Phi_3 \rightarrow y\Psi_1\Psi_2\Psi_3 \Leftrightarrow \bigvee_{[R^{\partial}=d_1} (x\Phi_1 \rightarrow y\Psi_1 R \mathbin{\&} R\Phi_2\Phi_3 = \Psi_2\Psi_3). \tag{1}$$

Obviously, $\nu(R\Phi_2) = \nu(\Psi_2)$. By hypothesis, $[R\tilde{\Phi}_2^{\partial} - [\tilde{\Psi}_2^{\partial} = d_1 + d_2 \leqq 0$. Then by 4.4 (3) we have

$$(R\Phi_2\Phi_3 = \Psi_2\Psi_3) \Leftrightarrow \bigvee_{[Q^{\partial}=d} (Q\Psi_3\Psi_2 = \Phi_3 R\Phi_2 Q). \tag{2}$$

From (1) and (2) we obtain

$$x\Phi_1\Phi_2\Phi_3 \rightarrow y\Psi_1\Psi_2\Psi_3 \Leftrightarrow \bigvee_{[R^{\partial}=d_1} \bigvee_{[Q^{\partial}=d} (x\Phi_1\Phi_3 R\Phi_2 Q \rightarrow y\Psi_1 R Q\Psi_3\Psi_2), \tag{3}$$

since $\nu(x\Phi_1) = \nu(y\Psi_1 R)$ and $[\widetilde{x\Phi_1}^{\partial} = [\widetilde{y\Psi_1}R^{\partial}$. The equivalence (3) is the desired one. The proposition is proved.

We will write $A(x,y,z)$, $A(x,y)$ or $A(x)$ if A is a word in the alphabet $\text{И} \cup \{x,y,z\}$, $\text{И} \cup \{x,y\}$ or $\text{И} \cup \{x\}$ respectively. We will write $\mathfrak{A}(x,y,z)$, $\mathfrak{A}(x,y)$ or $\mathfrak{A}(z)$ if $\mathfrak{A}$ is a word equation in the three unknowns x, y, z, the two unknowns x, y, or the one unknown z respectively. Let us consider a one-sided decomposable equation $\mathfrak{A}(x,y,z)$. It has the form $x\varphi(x,y)z \cdots \rightarrow yf(x,y)z \cdots$, where by the decomposability of the equation we have $\nu(x\varphi) = \nu(yf)$, and the dots denote the unwritten ends of the left- and right-hand sides of the equation. There are two possible cases.

The equation $\mathfrak{A}$ can have the form (possibly after multiplying on the right by z)

$$[\Phi_i(x,y)A_i(z)]_{i=1}^{n} \rightarrow [\Psi_i(x,y)B_i(z)]_{i=1}^{n}, \tag{1}$$

where $n > 0$, $\nu(\Phi_i) = \nu(\Psi_i)$ and $\nu(A_i) = \nu(B_i)$, $i = 1, \cdots, n$; each of the words Φ_i, Ψ_i, A_i and B_i begins with an unknown.

In the other case $\mathfrak{A}$ has the form (possibly after multiplying on the right by z)

$$[\Phi_i(x,y)A_i(z)]_{i=1}^{n}\Phi_{n+1}(x,y)z \cdots \rightarrow [\Psi_i(x,y)B_i(z)]_{i=1}^{n}\Psi_{n+1}(x,y)z \cdots, \tag{2}$$

where $n > 0$, and for $i = 1, \cdots, n$ the words Φ_i, Ψ_i, A_i and B_i are as in (1), but $\nu(\Phi_{n+1}) \neq \nu(\Psi_{n+1})$.

Thus any decomposable equation has the form (1) or (2). Equations of the form (1), in essence, are not equations in three unknowns. By 4.5 such an equation easily splits into pieces, each of which is an equation in the two unknowns x, y or in the one unknown z. Equations of the form (1) will be called degenerate, the remaining one-sided equations nondegenerate.

4.6. *Suppose $\mathfrak{A}(x,y,z)$ is a decomposable equation of the form $x\cdots\to y\cdots$.*

1) If $\mathfrak{A}$ is degenerate, it is equivalent to a disjunction of m $(m>0)$ systems of the form $\mathfrak{B}(x,y)\ \&\ \mathfrak{C}(z)$, where $\mathfrak{B}$ is a one-sided equation in two unknowns, $\mathfrak{C}$ an equation in one unknown.

2) If $\mathfrak{A}$ is nondegenerate and has the form $x\varphi(x,y)z\cdots\to yf(y)x\cdots$, then it is equivalent to a disjunction of indecomposable equations of the form $x\varphi\cdots\to yfx\cdots$ (with the same prefix $x\varphi\to yfx$, but with other "tails").

PROOF OF 1). By 4.5, equation (1) is equivalent to a disjunction of m $(m>0)$ systems, each of which is made up of a subsystem β consisting of equations in the two unknowns x, y and a subsystem γ consisting of equations in the one unknown z. Among the equations of the system β will be a one-sided equation $x\cdots\to y\cdots$. Returning each of the systems β and γ to a single equation,[10] we obtain a system $\mathfrak{B}(x,y)\ \&\ \mathfrak{C}(z)$, where $\mathfrak{B}$ is a one-sided equation. Assertion 1) is proved.

PROOF OF 2). Consider equation (2). Put $\Phi=A_1[\Phi_iA_i]_{i=2}^n$ and $\Psi=B_1[\Psi_iB_i]_{i=2}^n$. By our assumptions we have $\nu(\Phi)=\nu(\Psi)$. Equation (2) has the form $\Phi_1\Phi\Phi_{n+1}zA(x,y,z)\to\Psi_1\Psi\Psi_{n+1}zB(x,y,z)$. We have $\nu(\Phi_1)=\nu(\Psi_1)$ and $\nu(\Phi)=\nu(\Psi)$. By 4.5′ equation (2) is equivalent to a disjunction of equations of the form

$$\Phi_1\Phi'_{n+1}zA'\Phi'\to\Psi_1\Psi'_{n+1}zB'\Psi', \tag{3}$$

where Φ'_{n+1} and Ψ'_{n+1} are similar to Φ_{n+1} and Ψ_{n+1}. But $\nu(\Phi_{n+1})\neq\nu(\Psi_{n+1})$, whence $\nu(\Phi'_{n+1})\neq\nu(\Psi'_{n+1})$ and $\nu(\Phi_1\Phi'_{n+1})\neq\nu(\Psi_1\Psi'_{n+1})$. Hence (3) is indecomposable. Moreover, (3) has the form $x\varphi\cdots\to yfx\cdots$ (we have $x\varphi\angle\Phi_1$ and $yfx\angle\Psi_1$). Assertion 2) is proved. This completes the proof of the proposition.

We now turn to the elimination of the parameter of transformations.

4.7. *Suppose $y\angle f(y)$ and we have the equivalence*

$$[\Phi_i(x,\ y)\ A_i(z)]_{i=1}^n\to[\Psi_i(x,\ y)B_i(z)]_{i=1}^n, \tag{1}$$

where Φ, Ψ, Φ_i and Ψ_i are words in the alphabet $\mathbf{H}\cup\{x,y,z\}$. If the parameter is eliminable in some region $\mathscr{P}$ in transformations of each equation $x\Phi_i\to fx\Psi_i$, then it is eliminable in $\mathscr{P}$ in transformations of the equation $x\Phi\to fx\Psi$.

PROOF. Eliminability of the parameter in transformations of an equation means that if the equation has solutions of sufficiently large height, then it has solutions of small height (in a precise sense, §3).

Suppose (X,Y,Z) is a solution of the equation $x\Phi\to fx\Psi$ of height h. By (1) the triple (X,Y,Z) is a solution of some equation $x\Phi_i\to fx\Psi_i$, of the same height h, since the prefix f, which does not contain x or y, is the same for both equations. The height is uniquely determined by a solution of the equation $x\cdots\to fx\cdots$ and the prefix f, which does not contain x (see the

[10] Obviously, 1.28 remains valid if one of the equations of the system is one-sided. The system $\Phi_1\to\Psi_1$, $\Phi_2=\Psi_2$ is equivalent to the equation $\Phi_1a\Phi_2\Phi_1b\Phi_2\to\Psi_1a\Psi_2\Psi_1b\Psi_2$.

proof of Lemma 3.1). Now use the eliminability of the parameter in transformations of the equation $x\Phi_i \to fx\Psi_i$ and the equivalence (1). The proposition is proved.

4.8. *For any nondegenerate equation in three unknowns there is a number $L > 0$ such that the parameter is eliminable in transformations of this equation in the region $\neg\tau(x, y, z; L)$, and the elimination bound does not depend on the coefficients of the equation.*

PROOF. For indecomposable equations the assertion follows from 4.3, and for decomposable equations from 4.6.2), 4.7 and 4.3. There remain equations of the form $x \cdots \to D(y, z)x \cdots$, where y and z occur in D. The constant of which we spoke in 4.1 generally depends on the coefficients. However, in a region where y and z are of sufficient length compared to the sum of the lengths of the coefficients this constant will not depend on the coefficients. This is particularly true in the region $\neg\tau(x, y, z, L)$ for a suitable L. The proposition is proved.

4.8′. *For any degenerate equation in three unknowns of the form $x \cdots \to y \cdots$ there is a number $L > 0$ such that the parameter is eliminable in transformations of this equation if $\partial(y) > L$, and the elimination bound does not depend on the coefficients of the equation.*

PROOF. This follows from 4.6.1), 4.7 and 4.1. We need only return each of the systems $\mathfrak{B} \,\&\, \mathfrak{L}$ of 4.6.1) to a one-sided equation $x \cdots \to y \cdots$ and use the fact that the elimination bound of the parameter in transformations of the resulting equation is the same as in transformations of the equation $\mathfrak{B}$ and, for sufficiently large L, does not depend on the coefficients.

For any one-sided equation in three unknowns we define a complete system of images and a complete system of infinite images. Suppose $\mathfrak{A}(x, y, z)$ is a one-sided equation $x \cdots \to y \cdots$ in three unknowns. If $\mathfrak{A}$ is degenerate, we denote by N an elimination bound of the parameter in transformations of the equation $\mathfrak{A}$ under the condition $\partial(y) > L$, where L is the number in 4.8′. If $\mathfrak{A}$ is nondegenerate, we denote by N an elimination bound of the parameter in transformations of $\mathfrak{A}$ under the condition $\neg\tau(x, y, z, L)$, where L is the number in 4.8. A set of (infinite) images of $\mathfrak{A}$ obtained for the values $t = 0, 1, \cdots, N$ of the parameter of transformations is called a complete system of (infinite) images of the equation $\mathfrak{A}$.

4.9. THEOREM. *Suppose $\mathfrak{A}_1, \cdots, \mathfrak{A}_n$ is a complete system of infinite images of the equation $\mathfrak{A}$. Then we can construct a finite number of one-stage exponential equations $\pi_1, \cdots, \pi_k$ and word equations $\mathfrak{B}_1, \cdots, \mathfrak{B}_s$ in two unknowns such that $\mathfrak{A}$ has solutions if and only if at least one of the equations $\pi_1, \cdots, \pi_k$, $\mathfrak{B}_1, \cdots, \mathfrak{B}_s$, $\mathfrak{A}_1, \cdots, \mathfrak{A}_n$ has.*

PROOF. If $\mathfrak{A}$ is degenerate, then $k = 0$ and the list $\mathfrak{B}_1, \cdots, \mathfrak{B}_s$ consists of the finite images in a complete system of images of the equation $\mathfrak{A}$ and the equations $\mathfrak{A}(x, y, z)$, where y ranges over the words in the alphabet H of length $\leq L$ (see 4.8′). If $\mathfrak{A}$ is a nondegenerate equation $\Phi(x, y, z) \to \Psi(x, y, z)$, the list $\mathfrak{B}_1, \cdots, \mathfrak{B}_s$ consists of the finite images in a complete system of images of $\mathfrak{A}$, and the list $\pi_1, \cdots, \pi_k$ consists of the one-stage exponential equations

$$\Phi(A^\lambda A_1, B^\mu B_1, C^\nu C_1) = \Psi(A^\lambda A_1, B^\mu B_1, C^\nu C_1),$$

where λ, μ, ν are natural unknowns, A, B, C, A_1, B_1, C_1 are words in the alphabet $\mathbf{И}$, $[A^\partial, [B^\partial, [C^\partial \leq L$, $A_1 \angle A$, $B_1 \angle B$, $C_1 \angle C$, and L is the number in 4.8. To the equations π_i it is necessary to add the LDR corresponding to the arrow in $\mathfrak{A}$. The assertion follows from 4.8, 4.8′, 3.1 and the definitions of the predicate τ and a complete system of images.

4.9′. Theorem. *For any one-sided equation $\mathfrak{A}$ in three unknowns and any $k > 0$ we can construct a finite number of one-stage exponential equations $\pi_1, \cdots, \pi_m$, word equations $\mathfrak{B}_1, \cdots, \mathfrak{B}_s$ in two unknowns and infinite images $\mathfrak{A}_1, \cdots, \mathfrak{A}_n$ of $\mathfrak{A}$ of order k such that $\mathfrak{A}$ has solutions if and only if at least one of the equations $\pi_1, \cdots, \pi_m$, $\mathfrak{B}_1, \cdots, \mathfrak{B}_s$, $\mathfrak{A}_1, \cdots, \mathfrak{A}_n$ has.*

Proof (by induction on k). It suffices to construct the equations $\mathfrak{A}_1, \cdots, \mathfrak{A}_n$ of 4.9 and apply the induction assumption to them. The images of order $k-1$ of the equations $\mathfrak{A}_1, \cdots, \mathfrak{A}_n$ are images of order k of $\mathfrak{A}$. The theorem is proved.

Theorems 4.9 and 4.9′ provide the following program for investigating the decision problem for one-sided word equations in three unknowns. An algorithm for solving one-stage exponential equations follows from 2.3′, this being the algorithm for solving an LDR. An algorithm for solving equations in two unknowns is obtained in §7. It remains to consider infinite images. These are one-sided equations in three unknowns and we can apply the same argument to them. This leads to the study of chains of images generated by equations in three unknowns. Here (see 4.9′) it suffices to consider chains $\mathfrak{A}_0, \cdots, \mathfrak{A}_k$ for large k. This is the case for one-sided equations. As regards arbitrary equations in three unknowns, the passage from them to one-sided equations requires in addition only the ability to solve equations in two unknowns. This is a consequence of the following general assertion.

4.10. *For any nontrivial*[11] *word equation $\mathfrak{A}$ in n unknowns, $n > 1$, we can construct a finite number of word equations $\mathfrak{A}_1, \cdots, \mathfrak{A}_k$ in $n-1$ unknowns and one-sided equations $\mathfrak{B}_1, \cdots, \mathfrak{B}_s$ in n unknowns such that $\mathfrak{A}$ has solutions if and only if at least one of the equations $\mathfrak{A}_1, \cdots, \mathfrak{A}_k$, $\mathfrak{B}_1, \cdots, \mathfrak{B}_s$ has.*

Proof. Consider a nontrivial word equation in the n unknowns $x_1, \cdots, x_n$. We will first reduce it to an equation $x \cdots = y \cdots$, where x and y are unknowns, $x \neq y$. Cancel the equation on the left. We may assume that after cancellation unknowns remain in the left- and right-hand sides. The equation then takes the form $P\xi \cdots = \eta \cdots$, where P is a coefficient, and ξ and η are unknowns. If $P = \Lambda$, then $\xi \neq \eta$, and everything is in order. Suppose $P \neq \Lambda$. If ξ and η are distinct unknowns (for example, $\xi = x_1$ and $\eta = x_2$), then the change of variable $x_2 = Py$ reduces the equation $P\xi \cdots = \eta \cdots$ to an equation in the n unknowns $x_1, y, x_3, \cdots, x_n$ of the form $x_1 \cdots = y \cdots$. Here, however, we lose the solutions $(X_1, \cdots, X_n)$ where $[X_2^\partial < [P^\partial$. To determine them it suffices to consider a finite number of equations in $n-1$ unknowns.

There remains the case $\xi = \eta$. In this case we write the equation in the form

$$P[\xi_i P_i]_{i=1}^n \Phi = [\xi_i Q_i]_{i=1}^n \Psi, \tag{1}$$

[11] A word equation is called trivial (nontrivial) if its left-hand side is graphically equal (unequal) to its right-hand side.

where $n > 0$, P_i and Q_i are coefficients, the ξ_i are unknowns, Φ and Ψ are ends of the equation, and either one (possibly both) of the words Φ or Ψ is empty or Φ and Ψ begin with different unknowns. If one of Φ or Ψ is empty and the other is nonempty (for example, $\Psi \neq \Lambda$ and begins with x_1), then for any solution $(X_1, \cdots, X_n)$ of (1) we have $[X_1^\partial \leqq [PP_1 \cdots P_n^\partial$, and the equation reduces to equations in $n-1$ unknowns. There remain the case $\Phi = \Psi = \Lambda$ and the case $\Phi, \Psi \neq \Lambda$.

If $\Phi = \Psi = \Lambda$, then by 4.5 equation (1) is equivalent to a disjunction of systems ω_i consisting of equations in one unknown. Subsystems of ω_i containing the same unknown can be returned to a single equation (see 1.28). Testing the solvability of the resulting disjunction reduces to testing the solvability of equations in one unknown.

Consider the case $\Phi, \Psi \neq \Lambda$. In this case (1) has the form $\Phi_1 \xi \Phi_2 = \Psi_1 \eta \Psi_2$, where ξ and η are distinct unknowns, $\Phi = \xi \Phi_2$, $\Psi = \eta \Psi_2$ and $\nu(\Phi_1) = \nu(\Psi_1)$. By 4.4 the equation $\Phi_1 \xi \Phi_2 = \Psi_1 \eta \Psi_2$ is equivalent (if $[\tilde{\Phi}_1^\partial - [\tilde{\Psi}_1^\partial = d \geqq 0$) to a disjunction of equations of the form $R\xi\Phi_2\Phi_1 = \eta\Psi_2\Psi_1 R$, where R is a word in the alphabet $\mathbf{H}$ and $[R^\partial = d$. Each of them, as we have seen above, reduces to equations of the form $x \cdots = y \cdots$ or to equations in $n-1$ unknowns.

From this argument it follows that we can construct word equations $\mathfrak{A}_1, \cdots, \mathfrak{A}_k$ in $n-1$ unknowns and word equations $\mathfrak{B}_1, \cdots, \mathfrak{B}_s$ of the form $x \cdots = y \cdots$, where x and y are distinct unknowns, such that (1) has solutions if and only if at least one of the equations $\mathfrak{A}_i$, $\mathfrak{B}_j$ has. It remains to pass from $\mathfrak{B}_1, \cdots, \mathfrak{B}_s$ to one-sided equations. This passage requires in addition only the ability to solve equations in $n-1$ unknowns. The proposition is proved.

§5. Degenerate chains. The function F

We will begin the study of chains of images with degenerate chains. Let us define degenerate and nondegenerate transformations and images.

Suppose there is given a one-sided equation $x\Phi \to Dx\Psi$, where D is nonempty and does not contain x. Transformations of this equation of the form $x = D^t y$, where $t > 0$ and y is a new unknown, will be called degenerate. Moreover, if D contains only one unknown (which may occur many times), then any transformation of the equation $x\Phi \to Dx\Psi$ is by definition degenerate. Nondegenerate transformations are infinite transformations which are not degenerate. Images of equations under degenerate (nondegenerate) transformations will be called degenerate (nondegenerate) images. A chain $\mathfrak{A}_0, \cdots, \mathfrak{A}_k$ of word equations will be called a degenerate chain of $\mathfrak{A}_0$ if for any i, $1 \leqq i \leqq k$, the equation $\mathfrak{A}_i$ is a degenerate image of $\mathfrak{A}_{i-1}$. When $k = 0$ we have by definition a degenerate chain. Infinite chains of images which are not degenerate chains will be called nondegenerate. In this section we consider degenerate chains.

We make several definitions which are closely related to the notion of a degenerate chain. Consider a procession of the form

$$(*) \qquad [A, B, \lambda_1, \cdots, \lambda_k],$$

where A and B are nonempty words in some alphabet $\mathbf{Б}$, $k \geqq 0$, and $\lambda_1, \cdots, \lambda_k$ are positive integers. If $k = 0$, then the procession (*) is the pair $[A, B]$. It will also be convenient to consider a procession (*) with $k = -1$, which by definition is the pair $[B, A]$. We define on the processions (*) a function F. Put

$$F(A, B, \lambda_1, \cdots, \lambda_k) = B \quad \text{if } k = -1,$$
$$F(A, B, \lambda_1, \cdots, \lambda_k) = A \quad \text{if } k = 0,$$
$$F(A, B, \lambda_1, \cdots, \lambda_k) = (F(A, B, \lambda_1, \cdots, \lambda_{k-1}))^{\lambda_k} F(A, B, \lambda_1, \cdots, \lambda_{k-2}) \quad \text{if } k \geqq 1.$$

These equalities define on the processions (*) a function F whose values are nonempty words in the alphabet Б.

Let us observe several properties of F.

5.1. *If* $0 \leqq p \leqq k$, *then*

$$(1) \quad \begin{aligned} &F(A, B, \lambda_1, \cdots, \lambda_k) \\ &\quad = F(F(A, B, \lambda_1, \cdots, \lambda_p), F(A, B, \lambda_1, \cdots, \lambda_{p-1}), \lambda_{p+1}, \cdots, \lambda_k). \end{aligned}$$

Proof (by induction on k). First note that if $p = k$ or $p = k - 1$ the assertion is true. Indeed, if $p = k$, then (1) has the form $F(A, B, \lambda_1, \cdots, \lambda_k) = F(A, B, \lambda_1, \cdots, \lambda_k)$, and if $p = k - 1$, then $k > 0$ and (1) is true because

$$\begin{aligned} F(A, B, \lambda_1, \cdots, \lambda_k) &= (F(A, B, \lambda_1, \cdots, \lambda_{k-1}))^{\lambda_k} F(A, B, \lambda_1, \cdots, \lambda_{k-2}) \\ &= F(F(A, B, \lambda_1, \cdots, \lambda_{k-1}), F(A, B, \lambda_1, \cdots, \lambda_{k-2}), \lambda_k). \end{aligned}$$

We now carry out the induction. If $k = 0$, then $p = 0$, and the assertion is proved, since $k = p$. Suppose $m > 0$ and the assertion is true if $k < m$. Let us prove it for $k = m$. Suppose there is given a procession $[A, B, \lambda_1, \cdots, \lambda_m]$ and suppose $0 \leqq p \leqq m$. It suffices to consider the case $p < m - 1$. For brevity, put $F_i = F(A, B, \lambda_1, \cdots, \lambda_i)$, $i = p$, $p - 1$. Since $p \leqq m - 2$, by the induction assumption

$$\begin{aligned} F(A, B, \lambda_1, \cdots, \lambda_{m-1}) &= F(F_p, F_{p-1}, \lambda_{p+1}, \cdots, \lambda_{m-1}), \\ F(A, B, \lambda_1, \cdots, \lambda_{m-2}) &= F(F_p, F_{p-1}, \lambda_{p+1}, \cdots, \lambda_{m-2}). \end{aligned}$$

Therefore,

$$\begin{aligned} F(A, B, \lambda_1, \cdots, \lambda_m) &= (F(A, B, \lambda_1, \cdots, \lambda_{m-1}))^{\lambda_m} F(A, B, \lambda_1, \cdots, \lambda_{m-2}) \\ &= (F(F_p, F_{p-1}, \lambda_{p+1}, \cdots, \lambda_{m-1}))^{\lambda_m} F(F_p, F_{p-1}, \lambda_{p+1}, \cdots, \lambda_{m-2}) \\ &= F(F_p, F_{p-1}, \lambda_{p+1}, \cdots, \lambda_m), \end{aligned}$$

as required.

5.2. *Suppose* $k > 0$. *Then the word* $F(A, B, \lambda_1, \cdots, \lambda_k)$ *ends in* AB *if* k *is odd and in* BA *if* k *is even.*

This follows from the equalities $F(A, B, \lambda_1) = A^{\lambda_1} B$ and $F(A, B, \lambda_1, \lambda_2) = (A^{\lambda_1} B)^{\lambda_2} A$, and from the definition of the function F.

5.3. *Suppose* $k \geqq 0$, $X = F(A, B, \lambda_1, \cdots, \lambda_k)$ *and* $Y = F(A, B, \lambda_1, \cdots, \lambda_{k-1})$. *Then* $\mathscr{L}(X, Y)$ *if and only if* $\mathscr{L}(A, B)$.

Proof. If A and B are powers of some word S, then X and Y are also powers of S. On the other hand, if X and Y are powers of some word S, then starting from the definition of the function F it is easy to see that for any p, $-1 \leqq p \leqq k$, the word $F(A, B, \lambda_1, \cdots, \lambda_p)$ is a power of S. In particular, A and B are powers of S. The proposition is proved.

The following proposition explains the significance of the function F in the study of degenerate chains. Here $[A * B]$ denotes the beginning C of the word A such that $A = CB$.

5.4. *Consider a word equation in n unknowns*

$$x_1 A x_2 \Phi(x_1, \cdots, x_n) \to x_2 B x_1 \Psi(x_1, \cdots, x_n), \tag{1}$$

where the words A and B do not contain the unknowns x_1 and x_2. Suppose k is even and $\mathfrak{A}_0, \cdots, \mathfrak{A}_k$ is a degenerate chain of equation (1). *Then there exist $\lambda_1, \cdots, \lambda_k$ such that $\mathfrak{A}_k$ has the form* (*to within the designation of the unknowns x_1 and x_2*)

$$x_1 A x_2 \Phi(C_1, C_2, x_3, \cdots, x_n) \to x_2 B x_1 \Psi(C_1, C_2, x_3, \cdots, x_n), \tag{2}$$

where

$$C_1 = [F(x_1 A, x_2 B, \lambda_1, \cdots, \lambda_k) * A], \qquad C_2 = [F(x_1 A, x_2 B, \lambda_1, \cdots, \lambda_{k-1}) * B].$$

PROOF (BY INDUCTION ON k). When $k = 0$ the assertion is trivial, since when $k = 0$ the word C_1 is x_1 and C_2 is x_2.

Suppose m is an even positive integer and the assertion being proved is true for all k less than m. Suppose $\mathfrak{A}_0, \cdots, \mathfrak{A}_m$ is a degenerate chain of equation (1). By the induction assumption there exists a sequence $\lambda_1, \cdots, \lambda_{m-2}$ such that $\mathfrak{A}_{m-2}$ has the form

$$x_1 A x_2 \Phi(D_1, D_2, x_3, \cdots, x_n) \to x_2 B x_1 \Psi(D_1, D_2, x_3, \cdots, x_n), \tag{3}$$

where

$$D_1 = [F(x_1 A, x_2 B, \lambda_1, \cdots, \lambda_{m-2}) * A], \qquad D_2 = [F(x_1 A, x_2 B, \lambda_1, \cdots, \lambda_{m-3}) * B].$$

Since the unknowns x_1 and x_2 do not occur in B, any degenerate transformation of (3) has the form $x_1 = (x_2 B)^\lambda y$, where $\lambda > 0$ and y is a new unknown. Moreover, x_1 does not occur in A. Hence $\mathfrak{A}_{m-1}$, which is a degenerate image of (3), has the form (the symbol $S_{\cdots}^{\cdots}|$ denotes the result of the substitution (see [9], Chapter II, §4.14))

$$y A x_2 S_{(x_2 B)^\lambda y}^{x_1} \Phi(D_1, D_2, x_3, \ldots, x_n) | \leftarrow x_2 B y S_{(x_2 B)^\lambda y}^{x_1} \Psi(D_1, D_2, x_3, \ldots, x_n) |. \tag{4}$$

Similarly, a degenerate image of (4), i.e. the equation $\mathfrak{A}_m$, has the form

$$y A z S_{A_1 A_2}^{x_1 x_2} \Phi(D_1, D_2, x_3, \ldots, x_n) | \to z B y S_{A_1 A_2}^{x_1 x_2} \Psi(D_1, D_2, x_3, \ldots, x_n) |, \tag{5}$$

where $A_2 = (yA)^t z$, $A_1 = ((yA)^t zB)^\lambda y$, t is a positive integer and z is a new unknown. We now have

$$\begin{aligned} S_{A_1 A_2}^{x_1 x_2} D_1 | &= [F(((yA)^t zB)^\lambda yA, (yA)^t zB, \lambda_1, \ldots, \lambda_{m-2}) * A] \\ &= [F(F(yA, zB, t, \lambda), F(yA, zB, t), \lambda_1, \ldots, \lambda_{m-2}) * A] \\ &= [F(yA, zB, t, \lambda, \lambda_1, \ldots, \lambda_{m-2}) * A]. \end{aligned}$$

Similarly, $S_{A_1 A_2}^{x_1 x_2} D_2 | = [F(yA, zB, t, \lambda, \lambda_1, \cdots, \lambda_{m-3}) * B]$. But then (5) has the form

$$y A z \Phi(C_1, C_2, x_3, \cdots, x_n) \to z B y \Psi(C_1, C_2, x_3, \cdots, x_n),$$

where

$$C_1 = [F(yA, zB, t, \lambda, \lambda_1, \cdots, \lambda_{m-2}) * A],$$
$$C_2 = [F(yA, zB, t, \lambda, \lambda_1, \cdots, \lambda_{m-3}) * B],$$

so that the sequence $t, \lambda, \lambda_1, \cdots, \lambda_{m-2}$ is the desired one. The proposition is proved.

5.5. *Suppose A, B, X and Y are words in the alphabet $\mathbf{H}$, k and l are integers, $k \geqq l+6$, $|[X^\partial - [Y^\partial| l \leqq [AB^\partial$, $\mathscr{B}(A,B)$ and $XF(A,B,\lambda_1, \cdots, \lambda_{k-2}) V_1 = YF(A,B,\lambda_1, \cdots, \lambda_{k-2}) V_2$. Then $X = Y$.*

The proof is broken down into a series of lemmas.

5.5.1. *Suppose $\mathscr{B}(A,B)$, $[X^\partial \leqq [A^\partial$, $\lambda > 1$ and $XA^\lambda BV_1 = A^\lambda BA V_2$. Then $X = \Lambda$.*

PROOF. Since $[A^\partial \geqq [X^\partial$ and $\lambda > 1$, we have $A = XC$ for some C. But then $(XC)^\lambda BV_1 = C(XC)^{\lambda-1}BXCV_2$. Since $\lambda > 1$, the last equality implies that $XC = CX$ and $XCB = CBX$; hence $XB = BX$. If $X \neq \Lambda$, then from $XC = CX$ and $XB = BX$ we deduce, using 1.2 and 1.3, that X, C and B are powers of some word S. But then A is also a power of S. We have obtained a contradiction to the condition $\mathscr{B}(A,B)$. Consequently $X = \Lambda$, as required.

5.5.2. *Suppose $\mathscr{B}(A,B)$, $[X^\partial \leqq [A^\partial$, $\lambda \geqq 1$ and $X(AB)^\lambda A^2 V_1 = (AB)^\lambda A^2 V_1$. Then $X = \Lambda$.*

PROOF. Arguing as in 5.5.1, we obtain for some C that $A = XC$, $(XCB)^\lambda (XC)^2 V_1 = (CBX)^\lambda CXCV_2$, $XCB = CBX$ and $XC = CX$. If $X \neq \Lambda$, we obtain a contradiction from the last two equalities exactly as in 5.5.1. The proposition is proved.

5.5.3. *If $k \geqq 0$, then $k[AB^\partial \leqq [F(A,B,\lambda_1, \cdots, \lambda_{k+1})^\partial$.*

This follows from the definition of the function F.

5.5.4. *Suppose $\mathscr{B}(A,B)$, k and l are integers, $l[AB^\partial \geqq [X^\partial$, $k \geqq l+4$ and $XF(A,B,\lambda_1, \cdots, \lambda_k)V_1 = F(A,B,\lambda_1, \cdots, \lambda_k) V_2$. Then $X = \Lambda$.*

PROOF. Since $0 \leqq l \leqq k-4$, by 5.1 we have

$$F(A,B,\lambda_1, \cdots, \lambda_k) = F(P,Q,\lambda_{l+2}, \cdots, \lambda_k),$$

where $P = F(A,B,\lambda_1, \cdots, \lambda_{l+1})$ and $Q = F(A,B,\lambda_1, \cdots, \lambda_l)$. Since $\mathscr{B}(A,B)$, then by 5.3 we have $\mathscr{B}(P,Q)$. Since $[X^\partial \leqq l[AB^\partial$, by 5.5.3 we have $[X^\partial \leqq [P^\partial$. From the definition of F and the condition $k \geqq l+4$ it follows that the word $F(P,Q,\lambda_{l+2}, \cdots, \lambda_k)$ begins with $(P^{\lambda_{l+2}}Q)^{\lambda_{l+3}}PP$. If $\lambda_{l+2} > 1$, then our assertion follows from 5.5.1, and if $\lambda_{l+2} = 1$, from 5.5.2. The proposition is proved.

The validity of 5.5 obviously follows from 5.5.4.

The following proposition plays a fundamental role in the study of degenerate chains of equations in three unknowns.

5.6. *Consider an equation in three unknowns*

$$xA(y)zP(y)\xi\Phi(x,y,z) \to zB(y)xQ(y)\eta\Psi(x,y,z), \tag{1}$$

where A, B, P and Q are words in the alphabet $\mathbf{H} \cup \{y\}$, ξ and η are unknowns *different from y, and Φ and Ψ are words in the alphabet $\mathbf{H} \cup \{x, y, z\}$. Denote by a, b, p and q the number of occurrences of y in the words A, B, P and Q respectively. Suppose $a+b>0$, $a+p \neq b+q$, $k \geq 8+|p-q|$, k is even, and $\mathfrak{A}_0, \cdots, \mathfrak{A}_k$ is a degenerate chain of equation (1). Then $\mathfrak{A}_k$ is equivalent to a disjunction $(xAzBx\Phi_1 \rightarrow zBxAz\Phi_2) \vee \bigvee_{i=1}^n \mathfrak{B}_i$, where $x\Phi_1 \rightarrow z\Phi_2$ is the equation $\mathfrak{A}_k$ and $\mathfrak{B}_1, \cdots, \mathfrak{B}_n$ are word equations in two unknowns.*

Proof. Applying 5.4 to equation (1), we conclude that there exists a sequence $\lambda_1, \cdots, \lambda_k$ such that $\mathfrak{A}_k$ has the form

$$xA(y)\,zP(y)\,\mathrm{S}^{xz}_{C_1C_2}\xi\Phi(x, y, z) \mid \rightarrow zB(y)\,xQ(y)\,\mathrm{S}^{xz}_{C_1C_2}\eta\Psi(x, y, z) \mid, \tag{2}$$

where $C_1 = [F(xA, zB, \lambda_1, \cdots, \lambda_k) * A]$, $C_2 = [F(xA, zB, \lambda_1, \cdots, \lambda_{k-1}) * B]$. Since $k \geq 2$, the word $F(xA, zB, \lambda_1, \cdots, \lambda_{k-2})$ is a beginning of the words C_1 and C_2. If (X, Y, Z) is a solution of (2), then for some V_1 and V_2 we have

$$\begin{aligned} &XA\,(Y)\,ZP\,(Y)\,F\,(XA\,(Y), ZB\,(Y), \lambda_1, \ldots, \lambda_{k-2})\,V_1 \\ &= ZB\,(Y)\,XQ\,(Y)\,F\,(XA(Y), ZB\,(Y), \lambda_1, \ldots, \lambda_{k-2})\,V_2. \end{aligned} \tag{3}$$

Let C and D denote the projections on the alphabet $\mathbf{H}$ of the words AP and BQ respectively. Put $M = |[C^\partial - [D^\partial|$. We will consider the case $[Y^\partial > M$. We have $|[XA(Y)ZP(Y)^\partial - [ZB(Y)XQ(Y)^\partial| \leq M + |(a+p) - (b+q)|[Y^\partial \leq [Y^\partial + (a+b)[Y^\partial + |p-q|[Y^\partial \leq (2+|p-q|)[XA(Y)ZB(Y)^\partial$. Thus,

$$|[XA(Y)ZP(Y)^\partial - [ZB(Y)XQ(Y)^\partial| \leq (2+|p-q|)[XA(Y)ZB(Y)^\partial. \tag{4}$$

If $\mathscr{B}(XA(Y), ZB(Y))$, then, applying 5.5 to (3) and using (4) and the assumption $k \geq 8+|p-q|$, we conclude that $XA(Y)ZP(Y) = ZB(Y)XQ(Y)$. Since by hypothesis the letter y does not occur the same number of times in the words AP and BQ, from the last equality and the definition of M we obtain $[Y^\partial \leq M$, in spite of the assumption $[Y^\partial > M$. From this contradiction we conclude that $\neg\mathscr{B}(XA(Y), ZB(Y))$, which means that $XA(Y)ZB(Y) = ZB(Y)XA(Y)$. We have proved that if (X, Y, Z) is a solution of $\mathfrak{A}_k$ such that $[Y^\partial > M$, then $XA(Y)ZB(Y) = ZB(Y)XA(Y)$. Hence if (X, Y, Z) is a solution of $\mathfrak{A}_k$, then either $[Y^\partial \leq M$ or $XA(Y)ZB(Y) = ZB(Y)XA(Y)$. In other words, we have the equivalence

$$\mathfrak{A}_k\,(x, y, z) \Leftrightarrow (\mathfrak{A}_k\,(x, y, z)\,\&\,\partial\,(y) \leqslant M) \vee (xAzB\Psi_1 \rightarrow zBxA\Psi_2), \tag{5}$$

where Ψ_1 and Ψ_2 are the left- and right-hand sides, respectively, of $\mathfrak{A}_k$. It follows from (2) that $x \angle \Psi_1$ and $z \angle \Psi_2$. The proposition is proved.

As an application to coefficient-free equations we obtain the following result.

5.6′. *Consider a coefficient-free equation in three unknowns*

$$xy^azy^p\xi\Phi \rightarrow zy^bxy^q\eta\Psi, \tag{1}$$

where Φ and Ψ are words in the alphabet $\{x, y, z\}$, and ξ and η are unknowns different from y. Suppose $a+b>0$, $a+p \neq b+q$, $k \geq 8+|p-q|$, k is even, and $\mathfrak{A}_0, \cdots, \mathfrak{A}_k$ is a degenerate chain of equation (1). Let $\mathscr{B}(x, y, z)$ denote the following predicate: x, y and z are not powers of the same word. Then in the region $\mathscr{B}(x, y, z)$ the equation $\mathfrak{A}_k$ is equivalent to the equation $xy^azy^bx\Phi_1 \rightarrow zy^bxy^az\Psi_1$, where $x\Phi_1 \rightarrow z\Psi_1$ is $\mathfrak{A}_k$.

Proof. We will prove that the term $\mathfrak{A}_k(x, y, z)\,\&\,\partial(y) \leq M$ in the equivalence 5.6 (5) has no solutions in the region $\mathscr{B}(x, y, z)$. Indeed, the number

M in 5.6 is now equal to zero. The inequality $\partial(y) \leqq M$ is equivalent to the relation $y = \Lambda$. By 1.1, if (x, y, z) is a solution of $x\Phi_1 \to z\Psi_1$ such that $y = \Lambda$, then x and z are powers of the same word S. But then $y = S^0$. Hence in the region $\mathscr{L}(x, y, z)$ the equation $\mathfrak{A}_k$ has no solutions such that $y = \Lambda$. The term $\mathfrak{A}_k(x, y, z) \& \partial(y) \leqq M$ has no solutions in $\mathscr{L}(x, y, z)$. The equivalence 5.6 (5) takes the form $\mathfrak{A}_k \& \mathscr{L} \Leftrightarrow (xy^azy^b\Psi_1 \to zy^bxy^a\Psi_2) \& \mathscr{L}$, where $\Psi_1 \to \Psi_2$ is $\mathfrak{A}_k$. Here $x \angle \Psi_1$ and $z \angle \Psi_2$. The proposition is proved.

We will use the function F and Proposition 5.4 to study the structure of the solutions of an equation in two unknowns

$$(*) \qquad xAy \to yBx,$$

where A and B are coefficients, $[A^\partial = [B^\partial$ and $A \neq B$. We will establish a criterion for the solvability of (*) and develop formulas permitting us to find all solutions of this equation.

Let us begin with some definitions.

Suppose A, B are words in the alphabet И. Put

$$(5.1) \qquad \langle A, B\rangle_k = \begin{cases} AB & \text{if } k \text{ is odd,} \\ BA & \text{if } k \text{ is even.}\end{cases}$$

In view of 5.2, if $k > 0$, then the word $F(A, B, \lambda_1, \cdots, \lambda_k)$ ends with $\langle A, B\rangle_k$. Thus when $k > 0$ this allows us to define on the processions $[A, B, \lambda_1, \cdots, \lambda_k]$ a function G by

$$F(A, B, \lambda_1, \cdots, \lambda_k) = G(A, B, \lambda_1, \cdots, \lambda_k)\ \langle A, B\rangle_k.$$

We will use the following abbreviated notation.

Suppose there is given a procession $[A, B, \lambda_0, \cdots, \lambda_{k-1}]$, where $k \geqq -1$. Put

$$F_i = F(A, B, \lambda_0, \cdots, \lambda_{i-1}) \qquad (i = -1, 0, \cdots, k),$$
$$G_i = G(A, B, \lambda_0, \cdots, \lambda_{i-1}) \qquad (i = 1, 2, \cdots, k).$$

In particular, $F_{-1} = B$ and $F_0 = A$. From the definitions of F and G we have

$$F_i = F_{i-1}^{\lambda_{i-1}} F_{i-2},$$

if $1 \leqq i \leqq k$, and

$$(5.2) \qquad G_i = F_{i-1}^{\lambda_{i-1}} G_{i-2},$$

if $3 \leqq i \leqq k$.

5.7. *Suppose there is given a procession* $[A, B, \lambda_0, \cdots, \lambda_{k-1}]$, $k > 1$. *Then*

$$(1) \qquad F_k G_{k-1} = F_{k-1} G_k.$$

PROOF. We will deduce (1) from a lemma.

LEMMA. *Suppose there is given a procession* $[A, B, \lambda_0, \cdots, \lambda_{k-1}]$, $k \geqq 0$. *Then there exists a word V such that* $F_k F_{k-1} = V\langle A, B\rangle_{k-1}$ *and* $F_{k-1}F_k = V\langle A, B\rangle_k$.

PROOF OF THE LEMMA (BY INDUCTION ON k). If $k = 0$, then $F_0 = A$, $F_{-1} = B$, and the assertion is obvious ($V = \Lambda$). Suppose $m > 0$ and the assertion being proved is true for $k = m - 1$. Let us prove it for $k = m$. Consider a procession $[A, B, \lambda_0, \cdots, \lambda_{m-1}]$. By the induction assumption there exists V_1 such that

$$F_{m-1}F_{m-2} = V_1\langle A, B\rangle_{m-2}, \qquad F_{m-2}F_{m-1} = V_1\langle A, B\rangle_{m-1}.$$

Then

$$F_m F_{m-1} = F_{m-1}^{\lambda_{m-1}} F_{m-2} F_{m-1} = F_{m-1}^{\lambda_{m-1}} V_1 \langle A, B\rangle_{m-1},$$

$$F_{m-1} F_m = F_{m-1} F_{m-1}^{\lambda_{m-1}} F_{m-2} = F_{m-1}^{\lambda_{m-1}} F_{m-1} F_{m-2} = F_{m-1}^{\lambda_{m-1}} V_1 \langle A, B\rangle_{m-2} = F_{m-1}^{\lambda_{m-1}} V_1 \langle A, B\rangle_m,$$

and our assertion is true with $V = F_{m-1}^{\lambda_{m-1}} V_1$. The lemma is proved.

We obtain (1) if we cancel each equality of the lemma on the right by a word of length $[AB^\partial$ and take into account that $k > 1$. The proposition is proved.

If we multiply the equality $F_k G_{k-1} = F_{k-1} G_k$ on the right by $\langle A, B\rangle_k$, we obtain the following result.

5.8. *Suppose there is given a procession* $[A, B, \lambda_0, \cdots, \lambda_{k-1}]$, $k > 1$. *Then* $F_{k-1} F_k = F_k G_{k-1} \langle A, B\rangle_k$.

We will now use the word procession to denote a sequence (possibly empty) of positive integers.

The following proposition establishes the structure of the solutions of the equation $xAy \to yBx$.

5.9. a) *Suppose* $k > 1$, *and suppose the words* A, B, P, Q, R, X, Y *and the procession* $(\lambda_1, \cdots, \lambda_k)$ *are such that* $Q \neq \Lambda$ *and*

$$X = G(QP, QR, \lambda_1, \cdots, \lambda_k) Q, \tag{1}$$

$$Y = G(QP, QR, \lambda_1, \cdots, \lambda_{k-1}) Q, \tag{2}$$

$$QA = \langle QP, QR\rangle_k, \tag{3}$$

$$QB = \langle QP, QR\rangle_{k-1}. \tag{4}$$

Then the pair (X, Y) *satisfies the equation* $xAy \to yBx$.

b) *Suppose* $A \neq B$ *and the pair* (X, Y) *is a solution of the equation* $xAy \to yBx$. *Then there exist words* P, Q, R *and a procession* $(\lambda_1, \cdots, \lambda_k)$ *such that* $Q \neq \Lambda$, $k > 1$ *and* (1) – (4) *hold*.

PROOF OF a). Obviously, $[X^\partial > [Y^\partial > 0$. Using 5.7 (in passing from the second to the third line) and the relation between the functions F and G, we have from (1) – (4)

$$\begin{aligned} XAY &= G(QP, QR, \lambda_1, \ldots, \lambda_k) \langle QP, QR\rangle_k G(QP, QR, \lambda_1, \ldots, \lambda_{k-1}) Q \\ &= F(QP, QR, \lambda_1, \ldots, \lambda_k) G(QP, QR, \lambda_1, \ldots, \lambda_{k-1}) Q \\ &= F(QP, QR, \lambda_1, \ldots, \lambda_{k-1}) G(QP, QR, \lambda_1, \ldots, \lambda_k) Q \\ &= G(QP, QR, \lambda_1, \ldots, \lambda_{k-1}) \langle QP, QR\rangle_{k-1} X = YBX, \end{aligned}$$

as required.

PROOF OF b). Let $(\mathfrak{A}_0, \Sigma_0)$ denote the pair consisting of the equation $xAy \to yBx$ and the solution (X, Y) of this equation. Consider a complete chain of pairs generated by $(\mathfrak{A}_0, \Sigma_0)$. Suppose this is the chain $(\mathfrak{A}_0, \Sigma_0), \cdots, (\mathfrak{A}_k, \Sigma_k)$, where $k \geqq 0$. Any infinite transformation of $xAy \to yBx$ has the form $x = (yB)^t v$, where $t > 0$ and v is a new unknown, i.e. is a degenerate transformation. The image has the form $vAy \leftarrow yBv$. We have, to within a change of notation, the original equation $xAy \to yBx$. It follows that the chain $\mathfrak{A}_0, \cdots, \mathfrak{A}_k$ is a degenerate chain of the equation $xAy \to yBx$. Suppose k is even. Then by 5.4 (with empty Φ and Ψ) $\mathfrak{A}_k$ has the form $xAy \to yBx$. The solution Σ_k is a finite solution of $xAy \to yBx$. Denote it by (X_0, Y_0). Arguing as in 5.4, it is not difficult to conclude that (X, Y) is obtained from (X_0, Y_0) by means of the formulas

$$X = [F(X_0A, Y_0B, \lambda_1, \cdots, \lambda_k) * A], \tag{5}$$

$$Y = [F(X_0A, Y_0B, \lambda_1, \cdots, \lambda_{k-1}) * B]. \tag{6}$$

Consider the solution (X_0, Y_0). This is a finite solution of $xAy \to yBx$. Hence $X_0 = (Y_0B)^\lambda Y_0C$ for some $\lambda \geqq 0$ and $C \angle B$. Substituting X_0 and Y_0 into the equation, we obtain after cancellation $CAY_0 = BY_0C$. Since $C \angle B$, we have $B = CR$ for some R, and then $AY_0 = RY_0C$. This last relation implies that for some D we have $A = RD$ and $DY_0 = Y_0C$. If $D = \Lambda$, then $C = \Lambda$, $A = R$ and $B = R$, which contradicts the hypothesis $A \neq B$. Hence $D \neq \Lambda$. Furthermore, (X_0, Y_0) is a solution of the one-sided equation $xAy \to yBx$. Hence $Y_0 \neq \Lambda$. But then, in view of 1.16′, from $DY_0 - Y_0C$ we obtain $Y_0 = D^tQ$ and $D = QP$, where $t \geqq 0$ and $Q \neq \Lambda$. After cancellations we have $C = PQ$. Then $A = RQP$, $B = PQR$, and

$$Y_0B = (QP)^tQPQR = (QP)^{t+1}QR = F(QP, QR, t+1), \tag{7}$$

$$\begin{aligned} X_0A &= (Y_0B)^\lambda Y_0CA = ((QP)^{t+1}QR)^\lambda (QP)^tQPQRQP \\ &= ((QP)^{t+1}QR)^{\lambda+1}QP = F(QP, QR, t+1, \lambda+1). \end{aligned} \tag{8}$$

Using 5.1, we obtain from (5) – (8)

$$X = [F(QP, QR, t+1, \lambda+1, \lambda_1, \cdots, \lambda_k) * A], \tag{9}$$

$$Y = [F(QP, QR, t+1, \lambda+1, \lambda_1, \cdots, \lambda_{k-1}) * B]. \tag{10}$$

Furthermore, since k is even, we have $QA = QRQP = \langle QP, QR\rangle_{k+2}$, $QB = QPQR = \langle QP, QR\rangle_{k+1}$. Now from (9) and (10) it follows that

$$X = G(QP, QR, t+1, \lambda+1, \lambda_1, \cdots, \lambda_k)Q,$$
$$Y = G(QP, QR, t+1, \lambda+1, \lambda_1, \cdots, \lambda_{k-1})Q.$$

The assertion is proved for the case where k is even. If k is odd, then consider the pair $(\mathfrak{A}_{k-1}, \Sigma_{k-1})$. The equation $\mathfrak{A}_{k-1}$ has the form $xAy \to yBx$, and Σ_{k-1} is a solution (X_1, Y_0) of this equation. The equation $\mathfrak{A}_k$ has the form $yBx \to xAy$, and Σ_k is a finite solution (X_0, Y_0) of $\mathfrak{A}_k$, and $X_1 = (Y_0B)^\mu X_0$, where $\mu > 0$. A solution (X, Y) is obtained from (X_1, Y_0) by means of (5) and (6), with X_1 and $k-1$ instead of X_0 and k. But $X_1A = (Y_0B)^\mu X_0A = F(Y_0B, X_0A, \mu)$. Thus (X, Y) is obtained from the solution (X_0, Y_0) of the equation $yBx \to xAy$ by the formulas (an application of 5.1)

$$X = [F(Y_0B, X_0A, \mu, \lambda_1, \cdots, \lambda_{k-1}) * A],$$
$$Y = [F(Y_0B, X_0A, \mu, \lambda_1, \cdots, \lambda_{k-2}) * B].$$

The situation is the same as for even k, only X_0, Y_0, A, B and x, y have changed places. Arguing as in the case k even, we obtain

$$A = PQR, \qquad B = RQP, \qquad Q \neq \Lambda,$$
$$X_0A = F(QP, QR, t+1), \qquad Y_0B = F(QP, QR, t+1, \lambda+1),$$
$$X = [F(QP, QR, t+1, \lambda+1, \mu, \lambda_1, \cdots, \lambda_{k-1}) * A],$$
$$Y = [F(QP, QR, t+1, \lambda+1, \mu, \lambda_1, \cdots, \lambda_{k-2}) * B].$$

Since k is odd, we have $QA = QPQR = \langle QP, QR\rangle_{k+2}$ and $QB = \langle QP, QR\rangle_{k+1}$.

As in the case k even, we obtain expressions for X, Y in terms of the function G:

$$X = G(QP, QR, t+1, \lambda+1, \mu, \lambda_1, \cdots, \lambda_{k-1}),$$
$$Y = G(QP, QR, t+1, \lambda+1, \mu, \lambda_1, \cdots, \lambda_{k-2}).$$

The proposition is proved.

From 5.9 we obtain the following criterion for the solvability of the equation $xAy \to yBx$, first formulated by A. A. Markov.

5.10. *In order that the equation $xAy \to yBx$, where A and B are coefficients and $A \neq B$, have a solution it is necessary and sufficient that there exist words P, Q and R such that $A = PQR$, $B = RQP$ and $Q \neq \Lambda$.*

PROOF. If $xAy \to yBx$ has a solution, then by 5.9. b) we have $QA = \langle QP, QR \rangle_k$ and $QB = \langle QP, QR \rangle_{k-1}$, where $k > 1$ and $Q \neq \Lambda$. Hence, for k even, $A = RQP$ and $B = PQR$, and, for k odd, $A = PQR$ and $B = RQP$. On the other hand, if $A = PQR$, $B = RQP$ and $Q \neq \Lambda$, then the pair ($X = QRQ$, $Y = Q$) satisfies the equation $xAy \to yBx$. The proposition is proved.

In conclusion, we will prove several other assertions concerning the functions F and G.

5.11. *Suppose $k > 1$. Then*

$$G(A, B, \lambda_0, \cdots, \lambda_{k-1}, 1) = G(A, B, \lambda_0, \cdots, \lambda_{k-2}, \lambda_{k-1}+1).$$

PROOF. Since $k > 1$, by the lemma used to prove 5.7 we have

$$F_{k-1}F_{k-2} = V\langle A, B \rangle_{k-2}, \qquad F_{k-2}F_{k-1} = V\langle A, B \rangle_{k-1}$$

for some V. But then

$$F(A, B, \lambda_0, \cdots, \lambda_{k-1}, 1) = F_k F_{k-1} = F_{k-1}^{\lambda_{k-1}} F_{k-2} F_{k-1} = F_{k-1}^{\lambda_{k-1}} V\langle A, B \rangle_{k-1}$$

and

$$F(A, B, \lambda_0, \cdots, \lambda_{k-2}, \lambda_{k-1}+1) = F_{k-1}^{\lambda_{k-1}+1} F_{k-2} = F_{k-1}^{\lambda_{k-1}} V\langle A, B \rangle_{k-2}.$$

Hence

$$G(A, B, \lambda_0, \cdots, \lambda_{k-1}, 1) = F_{k-1}^{\lambda_{k-1}} V = G(A, B, \lambda_0, \cdots, \lambda_{k-2}, \lambda_{k-1}+1),$$

as required.

The following proposition establishes that if $\mathscr{B}(A, B)$, then sufficiently long values of F in comparison to $[A^\partial$ and $[B^\partial$ are simple words.

5.12. *Suppose $\mathscr{B}(A, B)$, and suppose $k > 2$ or $\max\{[A^\partial, [B^\partial\} \leq M$ and $[F(A, B, \lambda_0, \cdots, \lambda_{k-1})^\partial > 6M$. Then $F(A, B, \lambda_0, \cdots, \lambda_{k-1})$ is a simple word.*

PROOF. By our assumptions

$$[F(A, B, \lambda_0, \cdots, \lambda_{k-1})^\partial > \max\{2[AB^\partial, 2[ABA^\partial\},$$

so that $k > 0$. If $k = 1$, then $F(A, B, \lambda_0) = A^{\lambda_0}B$ and the assertion follows from 1.10. If $k = 2$, then $F(A, B, \lambda_0, \cdots, \lambda_{k-1}) = (A^{\lambda_0}B)^{\lambda_1}A$ and the assertion follows from 1.11. And if $k > 2$, then by 5.1 we have $F(A, B, \lambda_0, \cdots, \lambda_{k-1}) = F(F_{k-3}, F_{k-4}, \lambda_{k-3}, \lambda_{k-2}, \lambda_{k-1}) = ((F_{k-3}^{\lambda_{k-3}} F_{k-4})^{\lambda_{k-2}} F_{k-3})^{\lambda_{k-1}} F_{k-3}^{\lambda_{k-3}} F_{k-4}$. Since $\mathscr{B}(A, B)$, by 5.3 we have $\mathscr{B}(F_{k-3}, F_{k-4})$. We now apply 1.12. The proposition is proved.

5.12′. *Suppose $\mathscr{B}(A, B)$, $\max\{[A^\partial, [B^\partial\} \leq M$ and $F(A, B, \lambda_0, \cdots, \lambda_{k-1}) = S^a$. Then $[F(A, B, \lambda_0, \cdots, \lambda_{k-1})^\partial \leq \max\{6M, [S^\partial\}$.*

PROOF. If $a>1$, then the inequality follows from 5.12. If $a=1$, the inequality is trivial. The case $a=0$ is impossible.

5.13. *Suppose there is given a procession* $[QP, QR, \lambda_0, \cdots, \lambda_{k-1}]$, *and suppose* $k>1$ *and* $0 \leqq p < k$. *Then there exists a word* A *such that for any* m, $p < m \leqq k$, *we have*

(1) $$G(QP, QR, \lambda_0, \cdots, \lambda_{m-1})Q = G(F_p, F_{p-1}, \lambda_p, \cdots, \lambda_{m-1})A,$$

(2) $$A \angle F_pF_{p-1}, \quad A \angle F_{p-1}F_p, \quad A \angle F_p^2.$$

PROOF. It follows from the lemma used to prove 5.7 that there exists a word V such that for any $m \geqq p$

(3) $$V\langle QP, QR\rangle_m = \langle F_p, F_{p-1}\rangle_{m-p}.$$

Put $A = VQ$. We will prove that A is the desired word. Let us first prove (1). Suppose $m>p$. We have

(4) $$F(QP, QR, \lambda_0, \ldots, \lambda_{m-1}) = F(F_p, F_{p-1}, \lambda_p, \ldots, \lambda_{m-1}),$$

(5) $$F(QP, QR, \lambda_0, \ldots, \lambda_{m-1}) = G(QP, QR, \lambda_0, \ldots, \lambda_{m-1})\langle QP, QR\rangle_m,$$

(6) $$F(F_p, F_{p-1}, \lambda_p, \ldots, \lambda_{m-1}) = G(F_p, F_{p-1}, \lambda_p, \ldots, \lambda_{m-1})\langle F_p, F_{p-1}\rangle_{m-p}.$$

From (3) – (6) we conclude that

$$G(QP, QR, \lambda_0, \cdots, \lambda_{m-1}) = G(F_p, F_{p-1}, \lambda_p, \cdots, \lambda_{m-1})V,$$

from which (1) follows, if we take into account that $A = VQ$. Let us now prove (2). If $p=0$, then $V=\Lambda$ and (2) is obvious. If $p=1$, then $F_p = (QP)^{\lambda_0}QR$, $F_{p-1}=QP$, $V=(QP)^{\lambda_0}$, $A=(QP)^{\lambda_0}Q$ and (2) is obvious. And if $p>1$, then from (3) we have $VQ=F_pG_{p-1}Q$ or $VQ=F_{p-1}G_pQ$, depending on the parity of $m-p$. By 5.7, $F_pG_{p-1}=F_{p-1}G_p$. Obviously $G_{p-1}Q \angle F_{p-1}$ and $G_pQ \angle F_p$. Now the relations (2) are obvious, since $F_{p-1} \angle F_p$ when $p>1$. The proposition is proved.

§6. Trees of equations

In this section we prove that the solution of arbitrary word equations in three unknowns reduces to the solution of equations of a special type, the so-called basic equations.

The main tool in the reduction is the concept of chains of images. We also use inversions, admissible permutations and sets of substitutions. Together these lead to the important notion of a tree of an equation.

It is more convenient to construct the trees of interest to us for coefficient-free equations and then transfer the results obtained to equations with coefficients. Accordingly, this section consists of two parts: 1) trees of coefficient-free equations; 2) trees of equations with coefficients.

Trees of coefficient-free equations

We assume that all words considered below are words in a three-letter alphabet of unknowns. The term "equation" refers to a coefficient-free word equation in three unknowns. Words in the alphabet of unknowns are denoted by the letters A, B, C, D, P, Q, R, Φ, Ψ, and the unknowns themselves by the letters x, y, z, u, v, w, ξ, η, ζ. We write $A(x,y,z)$, $A(x,y)$ or $A(x)$ if A is a word in the alphabet $\{x,y,z\}$, $\{x,y\}$ or $\{x\}$ respectively.

Equations are considered the same if they agree after a suitable renaming of the unknowns. This is true, for example, of the equations $xyz \to zxy$ and $zxy \to yzx$.

By transformations and images we mean infinite transformations and images. Namely, if there is given a coefficient-free equation in three unknowns

$$(1) \qquad x\Phi(x,y,z) \to Dx\Psi(x,y,z),$$

where D is nonempty and does not contain x, then a transformation of (1) is any equality of the form

$$(2) \qquad x = D^t D_1 v,$$

where $t \geqq 0$, $D_1 \angle D$, $D^t D_1 \neq \Lambda$, and v is a new unknown.

The image of (1) under the transformation (2) is the equation

$$v\Phi(D^t D_1 v, y, z) \leftarrow D_2 D_1 v \Psi(D^t D_1 v, y, z),$$

where the word D_2 is determined by the (graphical) equality $D_1 D_2 = D$. Degenerate and nondegenerate transformations, images and chains are defined as in §5.

In the consideration of equations the information of interest to us is usually contained in certain beginnings of the left- and right-hand sides of an equation. In such cases we replace the remaining ends by three dots. For example, the sentence "Consider the equation $xyz \cdots \to zy^2x \cdots$" means that we are considering a one-sided equation whose left-hand side begins with the word xyz and whose right-hand side begins with the word zy^2x.

We will now prove several assertions concerning images of equations.

6.1. *Consider the equation*

$$(*) \qquad x^a y^b \eta \cdots \to zyA(y,z)x \cdots,$$

where $a, b \geqq 1$ and η is an unknown different from y. If $a = 1$, then any image of (*) has the form

$$(1) \qquad vy^b z \cdots \leftarrow Dv \cdots,$$

where v is a new unknown and D is a cyclic permutation of the word zyA. If $a > 1$, then the image of (*) under the transformation $x = zv$ has the form

$$(2) \qquad v(zv)^{a-1} y \cdots \leftarrow yAzv \cdots,$$

and images under all other transformations have the form

$$(3) \qquad vzy \cdots \leftarrow Dv \cdots,$$

where v is a new unknown and D is a cyclic permutation of the word zyA.

Proof. Any transformation of (*) has the form $x = zRv$, where v is a new unknown, R is a word in the alphabet $\{y, z\}$, and either $R = \Lambda$ or R begins with the letter y. Since $\eta \neq y$, the image of (*) under the transformation $x = zRv$ has the form

$$(4) \qquad v(zRv)^{a-1} y^b z \cdots \leftarrow Dv \cdots,$$

where D is a cyclic permutation of the word zyA. It is obvious that if $a = 1$, then (4) has the form (1). If $a > 1$, then when $R \neq \Lambda$ equation (4) has the form (3), and if $R=\Lambda$, then the transformation has the form $x=zv$, D has the form yAz, and (4) has the form (2). The proposition is proved.

We define a special class of equations, the class of supporting equations. These are of two types.

1. *σ-supporting equations.* These are equations of the following forms:

$$x^a y^b \eta \cdots \to zyx \cdots, \qquad x^a y^b \eta \cdots \to zxy \cdots,$$

where $1 \leqq a \leqq 2$, $1 \leqq b \leqq 2$, and η is an unknown different from y.

2. *τ-supporting equations.* These equations have the form

$$x^a y^b \eta \cdots \to z(yz)^d x \cdots,$$

where $1 \leqq a \leqq 2$, $1 \leqq b \leqq 2$, $d \geqq 1$, and η is an unknown different from y.

6.2. *Consider the equation*

$$(*) \qquad x^a y \eta \cdots \to zyA(y,z)x \cdots,$$

where $a \geqq 1$ and η is an unknown different from y. If each cyclic permutation of the word zyA contains no occurrences of y^3 or z^3, then the images of (*) are supporting equations. Equation (*) has a τ-supporting image only for $a > 1$ and the transformation $x = zv$; in all other cases the images are σ-supporting equations.

PROOF. Equation (*) is equation 6.1 (*) with $b=1$. The assertion obviously follows from 6.1 and the definition of supporting equations.

We now define basic equations. Basic equations are equations of the following seven types (parentheses are used for clarifications):

$$(6.1) \qquad x^a y \cdots = y^b x \cdots \qquad (a, b \geqq 1),$$

$$(6.2) \qquad x^2 \cdots \to y^c x \cdots \qquad (c \geqq 1),$$

$$(6.3) \qquad xy\eta \cdots \to zxy \cdots \qquad (\eta \neq y),$$

$$(6.4) \qquad xy\eta \cdots \to zyx \cdots \qquad (\eta \neq y),$$

$$(6.5) \qquad xyz \cdots = zxy \cdots,$$

$$(6.6) \qquad xyz \cdots = zyx \cdots,$$

$$(6.7) \qquad xy^a z \cdots = xy^a x \cdots \qquad (a \geqq 2).$$

We distinguish α- and β-basic equations. The α-basic equations are equations of types (6.1) – (6.6), and the β-basic equations are all α-basic equations and (6.7) with $a=2$.

Our aim is to study the form of images in chains of equations. We are interested in the image itself, not the image in relation to the original equation. Accordingly, so as not to introduce new letters, we use the same x, y, z for the equation and for the images. The phrase "the equation has the form" means that it has this form after a suitable renaming of the unknowns.

Simple calculations (done in 6.17) show that any chain $\mathfrak{A}_0$, $\mathfrak{A}_1$, $\mathfrak{A}_2$ of any equation in three unknowns contains a basic equation, an equation of the form 6.1 (*) or an equation of the form $xy \cdots \to z^c x \cdots$. Images of equation 6.1 (*) were considered in 6.1. Let us consider images of the equation $xy \cdots \to z^c x \cdots$. Multiplying this equation on the right by suitable unknowns, we bring it to the form 6.3 (*).

6.3. *Consider an equation*

$$(*) \qquad xy^a\eta A(x,y,z) \to z^c xB(x,z)yC(x,y,z),$$

where $a, c \geq 1$, $A, C \neq \Lambda$, *and* η *is an unknown different from* y.

Suppose $\mathfrak{A}_0, \mathfrak{A}_1, \mathfrak{A}_2, \mathfrak{A}_3$ *is a chain of* (*). *Consider two cases:* a) $\mathfrak{A}_2$ *is a degenerate image of* $\mathfrak{A}_1$; b) $\mathfrak{A}_2$ *is a nondegenerate image of* $\mathfrak{A}_1$.

1) *In case* a), $\mathfrak{A}_2$ *has the form* $xy^az\cdots \to zx\cdots$, *and in case* b), $\mathfrak{A}_3$ *is a supporting equation.*

2) *If* $a = 2$, $c = 1$, $B = \Lambda$ *and* $\urcorner(y \angle C)$, *then in case* a), $\mathfrak{A}_2$ *has the form* $xy^2z\cdots \to zxyx\cdots$, *and in case* b), $\mathfrak{A}_3$ *either is* α*-basic or has the form* $yxzy\cdots \to zxzy\cdots$.

3) *If* $a = 2$, $c = 1$ *and* $B = x$, *then in case* a), $\mathfrak{A}_2$ *has the form* $xy^2z\cdots \to zx^2y\cdots$, *and in case* b), $\mathfrak{A}_3$ *either is* σ*-supporting or has the form* $x^2y\xi\cdots \to zyzxy\cdots$ $(\xi \neq y)$.

4) *If* $a = 1$, *then in case* a), $\mathfrak{A}_2$ *either is* α*-basic or has the form* $xyz\xi\cdots \to zx^2y\cdots$ $(\xi \neq z)$, *and in case* b), $\mathfrak{A}_3$ *is a* σ*-supporting equation.*

PROOF. Any transformation of (*) can be written in the form $x = z^\lambda\xi$, where $\lambda > 0$ and ξ is a new unknown. By hypothesis, $\eta \neq y$. Hence $\mathfrak{A}_1$, being an image of (*), will have the form

$$(1) \qquad \xi y^a zA_1(\xi,y,z) \leftarrow z^c\xi B(z^\lambda\xi,z)yC(z^\lambda\xi,y,z),$$

where $A_1 \neq \Lambda$. Any transformation of (1) has the form $z = (\xi y^a)^t D_1 v$, where $t \geq 0$, $D_1 \not\angle \xi y^a$, $(\xi y^a)^t D_1 \neq \Lambda$, and v is a new unknown. Hence any image of (1) has the form

$$(2) \qquad D_2D_1vA_2(\xi,y,v) \to vS^z_{(\xi y^a)^t D_1 v} z^{c-1}\xi B(z^\lambda\xi,z)yC(z^\lambda\xi,y,z)\,|,$$

where $D_1D_2 = \xi y^a$, $A_2 \neq \Lambda$ and A_2 begins with either y or ξ. Equation (2) is the $\mathfrak{A}_2$ of our chain. There are four possible cases.

Case 1. $D_1 = \Lambda$. Then a) holds. Since $D_1 = \Lambda$, the equality $D_1D_2 = \xi y^a$ implies that $D_2 = \xi y^a$, and $(\xi y^a)^t D_1 \neq \Lambda$ implies that $t > 0$. But then (2) has the form

$$(3) \qquad \xi y^a vA_2 \to vS^z_{(\xi y^a)^t v} z^{c-1}\xi B(z^\lambda\xi,z)yC(z^\lambda\xi,y,z)\,|.$$

Since $t > 0$, (3) has the form $\xi y^a v\cdots \to v\xi\cdots$, from which assertion 1) follows. Furthermore, if $c = 1$, $B = \Lambda$ and $\urcorner(y \angle C)$, then the right-hand side of (3) begins with the word $v\xi y\xi$, and if $c = 1$ and $B \neq \Lambda$, then the right-hand side of (3) begins with the word $v\xi^2y$. The validity of 2) and 3) is now obvious. Consider the case $a = 1$. In this case the left-hand side of (3) has the form $\xi yvu\cdots$, where $u \neq v$ by what was said above about the word A_2. The right-hand side of (3) either has the form $v\xi y\cdots$ (and then (3) is an α-basic equation of the form (6.3)) or has the form $v\xi^2y\cdots$; the validity of 4) follows from this argument.

In the remaining three cases $D_1 \neq \Lambda$, so that b) holds.

Case 2. $t = 0$, $D_1 = \xi$. Then $D_2 = y^a$ and (2) has the form

$$(4) \qquad y^a\xi vA_2 \to (v\xi)^cB((\xi v)^\lambda\xi,\xi v)yC((\xi v)^\lambda\xi,y,\xi v).$$

Equation (4), up to the designation of the unknowns, is equation 6.2 (*). It is not difficult to see that cyclic permutations of the word $(v\xi)^cB((\xi v)^\lambda\xi,\xi v)$ do not contain occurrences of the words ξ^3 or v^3. Hence by 6.2 the images of

(4) are supporting equations, and if $a=1$, they are σ-supporting. The validity of 1) and 4) is now obvious. If $a=2$, $c=1$, $B=\Lambda$ and $\neg(y \angle C)$, then (4) has the form $y^2\xi v\cdots \to v\xi y\xi\cdots$. The image of this last equation under the transformation $y=vu$ has the form $uvu\xi\cdots \leftarrow \xi vu\xi\cdots$, and under the remaining transformations we obtain α-basic equations of the forms (6.3) and (6.4). Thus 2) is proved. It remains to consider the case $a=2$, $c=1$, $B=x$. In this case (4) has the form $y^2\xi v\cdots \to v\xi(\xi v)^\lambda \xi y\cdots$, where $\lambda>0$. The image of this equation under the transformation $y=vu$ has the form $uvu\xi v\cdots \leftarrow \xi^2 v\xi\cdots$, but by 6.2 under all remaining transformations we obtain σ-supporting images. This implies 3).

Case 3. $t>0$, $D_1=\xi$. In this case $D_2=y^a$. If $c>1$ or $c=1$, $B=\Lambda$, then (2) has the form $y^a\xi v\cdots \to v\xi y\cdots$. If $c=1$, $B\neq\Lambda$, then (2) has the form $y^a\xi v\cdots \to v\xi^2 y\cdots$. The validity of 1), 3) and 4) becomes obvious if we apply Proposition 6.2 to the equations $y^a\xi v\cdots \to v\xi y\cdots$ and $y^a\xi v\cdots \to v\xi^2 y\cdots$ and observe that with $a=2$ and the transformation $y=vu$ the image of $y^a\xi v\cdots \to v\xi^2 y\cdots$ has the form $uvu\xi v\cdots \leftarrow \xi^2 vu\cdots$. The validity of 2) is established exactly as in Case 2, since when $c=1$, $B=\Lambda$, $a=2$, $\neg(y\angle C)$ and $C\neq\Lambda$, then (2) has the form $y^2\xi v\cdots \to v\xi y\xi\cdots$.

Case 4. $t\geqq 0$, $D_1=\xi y^n$, where $1\leqq n<a$. In this case $D_2=y^m$, where $m=a-n$. Since we have $1\leqq n<a$, then $a>1$ and 4) is trivially true. Furthermore, (2) has the form $y^m\xi y\cdots \to v\xi^b y\cdots$, where $b=1$ if $c=1$, $B=\Lambda$ or $c>1$, and $b=2$ if $c=1$, $B\neq\Lambda$. Assertions 1) and 3) are obtained by applying 6.2 to the equation $y^m\xi y\cdots \to v\xi^b y\cdots$. Furthermore, if $a=2$, $c=1$, $B=\Lambda$ and $\neg(y\angle C)$, then (2) has the form $y\xi yv\cdots \to v\xi y\xi\cdots$. It is not difficult to compute that the images of this last equation are α-basic equations of one of the forms (6.3) or (6.4). Thus 2) is true. The proposition is proved.

We now turn to trees of coefficient-free equations.

Let us begin with the notions of an inversion and an admissible permutation. Suppose there is given an equation $\xi_1\xi_2\cdots\xi_k=\eta_1\eta_2\cdots\eta_l$, where $k,l\geqq 1$ and ξ_i and η_j are unknowns. By the inversion of this equation we mean the equation $\xi_k\cdots\xi_2\xi_1=\eta_l\cdots\eta_2\eta_1$. We define admissible permutations of an equation. Suppose there is given an equation $AC=BD$, where $\nu(A)=\nu(B)$ (see the definition preceding 4.4) and the words C and D begin with different unknowns. Then the equation $CA=DB$ is an admissible permutation of the equation $AC=BD$.

We now define "reduction by an n-tuple of substitutions". Suppose an equation $\mathfrak{A}(x,y,z)$ has the form

$$xA(x,y,z)\to \xi_1\cdots\xi_k xB(x,y,z),$$

where $k\geqq 1$ and $\xi_1,\cdots,\xi_k$ are unknowns different from x. Suppose $1\leqq n\leqq k$, $1\leqq i<n$ and v is a new unknown. Let $\mathfrak{A}_i$ denote the equation

$$vA\ (\xi_1\ldots\xi_i v,\ y,\ z)\leftarrow \xi_{i+1}\ldots\xi_k\xi_1\ldots\xi_i vB\ (\xi_1\ldots\ \xi_i v,\ y,\ z),$$

and $\mathfrak{A}_n$ the equation

$$vA\ (\xi_1\ \ldots\ \xi_n v,\ y,\ z)=\xi_{n+1}\ldots\ \xi_k\xi_1\ \ldots\ \xi_n vB(\xi_1\ \ldots\ \xi_n v,\ y,\ z).$$

Having thus defined the equations $\mathfrak{A}_1,\cdots,\mathfrak{A}_n$, we say that the n-tuple of substitutions $x=\xi_1 v,\cdots,x=\xi_1\cdots\xi_n v$ reduces $\mathfrak{A}$ to the equations $\mathfrak{A}_1,\cdots,\mathfrak{A}_n$.

For example, ($n=1$) the substitution $x=yv$ reduces the equation $xyx\to yzxz$ to the equation $vy^2v=zyvz$, and ($n=2$) the pair of substitutions $x=yv$, $x=yzv$ reduces the same equation to the equations $vy^2v\leftarrow zyvz$, $vy^2zv=yzvz$.

We now define a neighborhood of an equation. A set of equations $\mathfrak{A}_1,\cdots,\mathfrak{A}_n$ is called a neighborhood of an equation $\mathfrak{A}$ in the following cases.

1) $\mathfrak{A}_1,\cdots,\mathfrak{A}_n$ form a complete system of infinite images of $\mathfrak{A}$ (see the definition preceding 4.9).

2) $\mathfrak{A}$ reduces to the equations $\mathfrak{A}_1,\cdots,\mathfrak{A}_n$ by some n-tuple of substitutions.

3) If $\mathfrak{A}$ is "two-sided" of the form $A=B$, where the words A and B begin with different unknowns, then $n=2$ and $\mathfrak{A}_1$ and $\mathfrak{A}_2$ are the equations $A\to B$ and $B\to A$.

4) $n=1$ and $\mathfrak{A}_1$ is the inversion of $\mathfrak{A}$.

5) $n=1$ and $\mathfrak{A}_1$ is an admissible permutation of $\mathfrak{A}$.

6) $n=1$ and $\mathfrak{A}_1$ is obtained from $\mathfrak{A}$ by multiplication on the right by a word in the alphabet $\{x,y,z\}$.

We define trees of equations. A tree of an equation $\mathfrak{A}$ is a graph without loops whose vertices are word equations and whose edges are arrows. No vertex receives two arrows. There is only one vertex which receives no arrows; this vertex corresponds to $\mathfrak{A}$. The following condition is satisfied: if a vertex $\mathfrak{B}$ transmits exactly n arrows and these arrows are received at the vertices $\mathfrak{B}_1,\cdots,\mathfrak{B}_n$, then the equations $\mathfrak{B}_1,\cdots,\mathfrak{B}_n$ form a neighborhood of $\mathfrak{B}$. A sequence $\mathfrak{A}_1,\cdots,\mathfrak{A}_k$ of vertices of a tree is called a path if for any i, $1\leqq i<k-1$, an arrow is transmitted from $\mathfrak{A}_i$ and received at $\mathfrak{A}_{i+1}$. In connection with Theorem 5.6 we introduce the following "replacement rule". Suppose a vertex $\mathfrak{A}_0$ corresponds to the equation $xy^azy^p\xi\cdots\to zy^bxy^q\eta\cdots$, where $a+p\neq b+q$, and ξ and η are unknowns different from y; suppose k is even, $k\geqq 8+|p-q|$, and $\mathfrak{A}_0,\cdots,\mathfrak{A}_k$ is a degenerate chain of $\mathfrak{A}_0$. We replace a subgraph which is a tree of $\mathfrak{A}_k$ by a subgraph which is a tree of the equation $xy^azy^bxP\to zy^bxy^azQ$, where $xP\to zQ$ is the equation $\mathfrak{A}_k$. The graph thus obtained is considered a tree of $\mathfrak{A}_0$.

Vertices which do not transmit arrows are called final vertices. We will distinguish trees according to the type of their final vertices. A tree is called a basic tree (α-, β-tree) if all of its final vertices are basic (α-, β-basic) equations. We say that an equation $\mathfrak{A}$ possesses a basic tree (α-, β-tree) if it is possible to construct a tree of $\mathfrak{A}$ which is a basic tree (α-, β-tree). Obviously, any α-tree is a β-tree, and any β-tree is a basic tree. The aim of this section is to prove that any equation in three unknowns possesses a basic tree. Using this result, it is not difficult to obtain a reduction which is of interest to us.

We now prove a number of assertions to the effect that an equation of a certain type possesses a basic tree (α-, β-tree). Which form these trees have is not now essential for us (this knowledge is of value in the investigation of the structure of solutions). For the purpose of illustrating the arguments, we will sketch the trees in the next three propositions.

We will use the following obvious property: if we have a tree of an equation $\mathfrak{A}$ and the final vertices of this tree possess basic trees, then $\mathfrak{A}$ also possesses a basic tree. In particular, if any image of $\mathfrak{A}$ possesses a basic tree, then $\mathfrak{A}$ possesses a basic tree [see item 1) in the definition of a neighborhood].

6.4. *Any equation of the form* $xyxz\cdots\to zx^2y\cdots$ *possesses an* α*-tree.*

PROOF. Denote the given equation by $\mathfrak{A}$. Consider the images of $\mathfrak{A}$. Any image of $\mathfrak{A}$ is obtained under a transformation $x = z^t v$, where $t > 0$ and v is a new unknown, and has the form

$$(1) \qquad vyz^t vz \cdots \leftarrow zvz^t vy \cdots.$$

We will prove that any equation of the form (1) possesses a basic tree. Consider an equation of the form (1). Denote it by $\mathfrak{A}_1$. The pair of substitutions $z = vx$, $z = vyx$ reduces $\mathfrak{A}_1$ to the equations $y(vx)^t v^2 x \cdots \rightarrow xv(vx)^t vy \cdots$, $(vyx)^t v^2 yxP = xv(vyx)^t vyQ$, where P and Q are words in the alphabet $\{x, y, v\}$. Denote these equations by $\mathfrak{A}_{11}$ and $\mathfrak{A}_{12}$ respectively. Consider $\mathfrak{A}_{12}$. An admissible permutation of it is the equation $vyxP(vyx)^t v = xvyQxv^2 y(xvy)^{t-1}$, which is α-basic of the form (6.5). Consider $\mathfrak{A}_{11}$. It is not difficult to see that any image of $\mathfrak{A}_{11}$ is an equation of one of the following forms: $z(vx)^t v^2 x \cdots \leftarrow xv(vx)^t vz \cdots$ (degenerate images), $v^2 \cdots \rightarrow zv \cdots$, $vxv \cdots \rightarrow zvx \cdots$, $xvx \cdots \rightarrow zvx \cdots$ (nondegenerate images), where z is a new unknown. The last three equations are α-basic equations of one of the forms (6.2), (6.3) or (6.4). The first equation is reduced by the substitution $x = zu$ to the equation $(vzu)^t v^2 zuA = uv(vzu)^t vzB$, where A and B are words in the alphabet $\{u, v, z\}$. An admissible permutation of the last equation is the α-basic equation $vzuA(vzu)^t v = uvzBuv^2 z(uvz)^{t-1}$. The existence of an α-tree of the equation $xyxz \cdots \rightarrow zx^2 y \cdots$ is proved. This tree is represented in Figure 1. In this figure $\mathfrak{A}_1, \cdots, \mathfrak{A}_n$ is a complete system of images of $\mathfrak{A}$; $\mathfrak{B}_1, \cdots, \mathfrak{B}_q$ is a complete system of images of $\mathfrak{A}_{11}$, and $\mathfrak{B}_1, \cdots, \mathfrak{B}_s$ are degenerate images, $\mathfrak{B}_{s+1}, \cdots, \mathfrak{B}_q$ are nondegenerate. For the equations $\mathfrak{A}_2, \cdots, \mathfrak{A}_n$ trees have been constructed analogous to the tree of $\mathfrak{A}_1$. These have been denoted by Д$_2, \cdots,$ Д$_n$. The black circles denote final vertices. The arrows have been omitted.

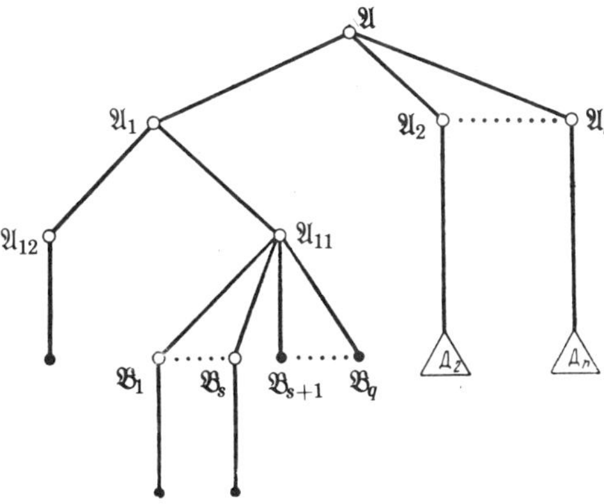

FIGURE 1

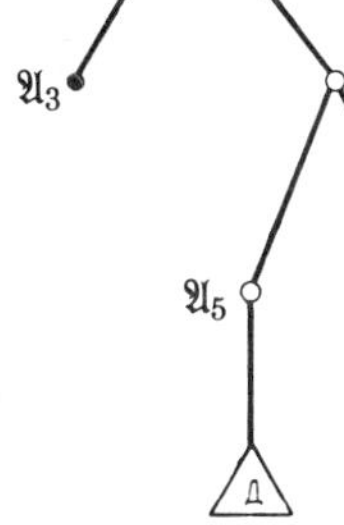

FIGURE 2

FIGURE 3

6.5. *Any equation of the form* $xyz^2 A(x, y, z) = yz^2 xB(x, y, z)$ *possesses a* β*-tree.*

PROOF. Denote the given equation by $\mathfrak{A}$. The equation $Axyz^2 = Byz^2 x$ is an admissible permutation of $\mathfrak{A}$. Denote it by $\mathfrak{A}_1$. Consider the inversion $\mathfrak{A}_2$ of $\mathfrak{A}_1$. It has the form $z^2 yxP = xz^2 yQ$, where P and Q are words in the alphabet $\{x, y, z\}$. Consider the one-sided equations $z^2yxP \to xz^2yQ$ and $xz^2yQ \to z^2yxP$. Denote them by $\mathfrak{A}_3$ and $\mathfrak{A}_4$. Equation $\mathfrak{A}_3$ is β-basic of the form (6.2). Consider $\mathfrak{A}_4$. The pair of substitutions $x = zv$, $x = z^2v$ reduces $\mathfrak{A}_4$ to the equations $vz^2y \cdots \leftarrow zyzv \cdots$ and $vz^2y \cdots = yz^2v \cdots$. Denote them by $\mathfrak{A}_5$ and $\mathfrak{A}_6$. Equation $\mathfrak{A}_6$ is β-basic of the form (6.7), and, by 6.4, $\mathfrak{A}_5$ possesses an α-tree Д. Connecting this tree to $\mathfrak{A}_5$, we obtain the β-tree of $\mathfrak{A}$ represented in Figure 2. The proposition is proved.

6.6. *Any equation of the form $x^2yz \cdots \to zyxy \cdots$ possesses a β-tree.*

PROOF. Denote the given equation by $\mathfrak{A}$. The pair of substitutions $x = zv$, $x = zyv$ reduces this equation to the equations $vzvy \cdots \leftarrow yzvy \cdots$ and $vzy \cdots = zyv \cdots$. Denote them by $\mathfrak{A}_1$ and $\mathfrak{A}_2$. Equation $\mathfrak{A}_2$ is β-basic of the form (6.5). Consider $\mathfrak{A}_1$. The substitution $y = vx$ reduces $\mathfrak{A}_1$ to an equation $\mathfrak{A}_3$ of the form $zv^2x \cdots = xzv^2 \cdots$, which by 6.5 possesses a β-tree Д. Connecting this tree to $\mathfrak{A}_3$, we obtain the β-tree of $\mathfrak{A}$ represented in Figure 3. The proposition is proved.

6.7. *Any nondegenerate image of an equation $xy^2z \cdots \to zy^2x \cdots$ possesses an α-tree.*

PROOF. Nondegenerate images of an equation $xy^2z \cdots \to zy^2x \cdots$ have the form $vy^2z \cdots \leftarrow y^nzy^mv \cdots$, where $n = 1$ or 2 and $m = 2 - n$. If $n = 2$, we have an α-basic equation of the form (6.2). If $n = 1$, we have an equation $yzyv \cdots \to vy^2z \cdots$, which by 6.4 possesses an α-tree. Thus any nondegenerate image of $xy^2z \cdots \to zy^2x \cdots$ possesses an α-tree, as required.

Our next aim is to prove that supporting equations possess β-trees. For this we need the following two assertions.

6.8. *Consider an equation $xyz\eta A(x, y, z) \to zx^2yB(x, y, z)$, where η is an unknown different from z. Suppose $\mathfrak{A}_0$, $\mathfrak{A}_1$, $\mathfrak{A}_2$ is a chain of this equation and $\mathfrak{A}_2$ is a nondegenerate image of $\mathfrak{A}_1$. Then $\mathfrak{A}_2$ possesses an α-tree.*

PROOF. Any image $\mathfrak{A}_1$ of $\mathfrak{A}_0$ is obtained under a transformation $x = z^\lambda v$, where $\lambda > 0$ and v is a new unknown, and has the form

$$vyzC(z^\lambda v, y, z) \leftarrow zvz^\lambda vyB(z^\lambda v, y, z), \tag{1}$$

where $C = \eta A$. Look at the nondegenerate images of (1). These are obtained under transformations $z = (vy)^t vu$, where $t \geqq 0$ and u is a new unknown, and have the form

$$yvu\mathrm{S}^z_{(vy)^t vu} C(z^\lambda v, y, z) | \to uv((vy)^t vu)^\lambda vy\mathrm{S}^z_{(vy)^t vu} B(z^\lambda v, y, z) |. \tag{2}$$

Equation (2) is the $\mathfrak{A}_2$ of our chain. We will prove that $\mathfrak{A}_2$ possesses an α-tree. We consider two cases.

Case 1. $t > 0$. By hypothesis, $\eta \neq z$. This means that the word C begins with either x or y. We consider two subcases.

1) $x \angle C$. Then the left-hand side of (2) begins with the word $yvu(vy)^t vuv$. Since t, $\lambda > 0$, the right-hand side of (2) begins with the word $uv(vy)^t vuvy$. Equation (2) has the form

$$yvuv(yv)^{t-1}yvuv \cdots \to uv^2y(vy)^{t-1}vuvy \cdots.$$

The substitution $y = ux$ reduces this last equation to the equation

(3) $$xvuv(uxv)^{t-1}uxvuvP = v^2ux(vux)^{t-1}vuvuxQ.$$

If $t > 1$, then the equation $(uxv)^{t-1}uxvuvPxvuv = (vux)^{t-1}vuvuxQv^2ux$ is an admissible permutation of (3) and, moreover, is α-basic of the form (6.5). Consider the case $t = 1$. In this case (3) has the form $xvuvuxvuvP = v^2uxvuvuxQ$. Denote this last equation by $\mathfrak{A}$. Consider the one-sided equations corresponding to $\mathfrak{A}$. One of them has the form $v^2 \cdots \to xv \cdots$ and is α-basic of the form (6.2). The other has the form $xvuvux \cdots \to v^2uxvu \cdots$ and is reduced by the substitution $x = vz$ to an equation $zvuvuvP_1 = vuvzvuQ_1$. An admissible permutation of this last equation is the α-basic equation $uvP_1zvuv = vuQ_1vuvz$. Our assertion is proved for the case $x \angle C$.

2) $y \angle C$. Then (2) can be written in the form

(4) $$yvuy \cdots \to uv^2y(vy)^{t-1}vu \cdots.$$

The triple of substitutions $y = ux$, $y = uvx$, $y = uv^2x$ reduces (4) to the equations $xv \cdots \leftarrow v^2 \cdots$, $xvu \cdots \leftarrow vuv \cdots$ and $xvu^2v^2P = uv^2xvuQ$, where P and Q are words in the alphabet $\{u, v, x\}$. The first two equations are α-basic of the forms (6.2) and (6.3). The equation $xvu^2v^2P = uv^2xvuQ$ has an admissible permutation $Pxvu^2v^2 = Quv^2xvu$, whose inversion is an α-basic equation $v^2u \cdots = uv \cdots$. The proposition is proved for the case $y \angle C$, hence also for Case 1.

Case 2. $t = 0$. In this case (2) has the form

(5) $$yvuC((vu)^\lambda v, y, vu) \to uv(vu)^\lambda vyB((vu)^\lambda v, y, vu).$$

Consider the images of (5). Put $D = uv(vu)^\lambda v$. Then any transformation of (5) has the form $y = D^\mu D_1 x$, where $\mu \geqq 0$, $D_1 \underset{\cdot}{\angle} D$, $D^\mu D_1 \neq \Lambda$, and x is a new unknown. The images of (5) have the form

(6) $$xvuC\left((vu)^\lambda v,\ D^\mu D_1x,\ vu\right) \leftarrow D_2D_1xB\left((vu)^\lambda v,\ D^\mu D_1x,\ vu\right),$$

where D_2 is determined by the relation $D_1D_2 = uv(vu)^\lambda v$. It is not difficult to verify that D_2D_1 begins with one of the words v^2, vuv, uvu or uv^2, and also $uv^2 \angle D_2D_1$ only when $D_1 = \Lambda$. But each of the equations $v^2 \cdots \to xv \cdots$, $vuv \cdots \to xvu \cdots$ and $uvu \cdots \to xvu \cdots$ is α-basic. There remains the case $uv^2 \angle D_2D_1$. Then we must have $D_1 = \Lambda$, $\mu > 0$ and $D_2 = uv(vu)^\lambda v$, and (6) has the form

(7) $$xvuC\left((vu)^\lambda v,\ D^\mu x,\ vu\right) \leftarrow uv\,(vu)^\lambda\, vx\,B\left((vu)^\lambda v,\ D^\mu x,\ vu\right).$$

1) $x \angle C$. Then (7) has the form $uv^2(uv)^\lambda x \cdots \to xvuv(uv)^\lambda \cdots$. This last equation is reduced by the substitution $u = xz$ to an equation $zv^2(xzv)^\lambda xP = vxzv(xzv)^\lambda Q$, an admissible permutation of which is the α-basic equation $(zvx)^\lambda Pzv^2x = (xzv)^\lambda Qvxzv$. The proposition is proved for the case $x \angle C$.

2) $y \angle C$. Then (7) has the form

(8) $$xvu\,(uv(vu)^\lambda\, v)^\mu\, x\,A\left((vu)^\lambda v,\ D^\mu x,\ vu\right) \leftarrow uv\,(vu)^\lambda vx\,B\left((vu)^\lambda v, D^\mu x,\ vu\right).$$

Consider the case $\lambda > 1$. Since $\mu > 0$, the substitution $u = xz$ reduces (8) to an equation $vxzxzv(vxz)^{\lambda-1}vP = zv(vxz)^\lambda vxQ$. An admissible permutation of this last equation is the α-basic equation $(xzv)^{\lambda-1}Pvxzxzv^2 = (zvx)^{\lambda-1}Qzv^2xzvx$.

Consider the case $\lambda = 1$. In this case the substitution $u = xz$ reduces (8) to the equation

(9) $$vxz(xzv^2xzv)^{\mu}xA(X,Y,Z) = zv^2xzvxB(X,Y,Z),$$

where $X = vxzv$, $Y = (xzv^2xzv)^{\mu}x$ and $Z = vxz$. The equation

(10) $$xzv(xzv^2xzv)^{\mu-1}xA(X,Y,Z)vxzxzv^2 = B(X,Y,Z)zv^2xzvx$$

is an admissible permutation of (9). In view of item 6) in the definition of a neighborhood of an equation we may assume that the words A and B are nonempty. If B begins with one of the letters x or z, then (10) has the form $xzv\cdots = vxz\cdots$, i.e. (10) is α-basic. If B begins with y, then (10) has an admissible permutation

(11) $$(xzv^2xzv)^{\mu-1}xA_1(X,Y,Z) = vxzvD^{\mu-1}xB_1(X,Y,Z).$$

If $\mu > 1$, then (11) is α-basic of the form (6.5). If $\mu = 1$, then (11) has the form

(12) $$xA_1(X,Y,Z) = vxzvxB_1(X,Y,Z).$$

If A_1 begins with one of the letters x, z, then (12) has the form $xv\cdots = vx\cdots$, i.e. (12) is α-basic. If A_1 begins with y, then (12) has the form $x^2zv^2\cdots = vxzvx\cdots$. The corresponding one-sided equations have the form $x^2\cdots \to vx\cdots$ and $vxzvx\cdots \to x^2zv^2\cdots$. The first is α-basic of the form (6.2); the second is reduced by the substitution $v = xu$ to an equation $uxzxuxP = xzxuxuQ$. An admissible permutation of this last equation is the α-basic equation $uxPuxzx = xuQxzxu$. The proposition is proved.

A tree whose final vertices consist of basic equations, σ-supporting equations or equations of the form $x^2y\xi\cdots \to zyzxy\cdots$ $(\xi \neq y)$ is called an h-tree.

6.9. *Any equation of the form*

(1) $$xy^2z\cdots \to zx^2y\cdots$$

possesses an h-tree. Each equation of the form

(2) $$xyz\eta\cdots \to zx^2y\cdots \qquad (\eta \neq x),$$

(3) $$xy^2z\cdots \to zxy\eta\cdots \qquad (\eta \neq y),$$

(4) $$xyz\eta\cdots \to zy^2x\cdots \qquad (\eta \neq y),$$

(5) $$xyz\cdots \to zy^2x\cdots,$$

possesses a β-tree.

PROOF. Write each of the equations (1)–(4) in the form $xy^azy^p\xi\cdots \to zy^bxy^q\zeta\cdots$, where $\xi, \zeta \neq y$ (this can always be done by multiplying the equations by suitable unknowns). For equation (1) $a = 2$ and $b = q = 0$, for equation (2) $a = 1$ and $b = q = 0$, for equation (3) $a = 2$, $b = 0$ and $q = 1$, and for equation (4) $a = 1$, $p = 0$ and $b = 2$. Hence $a + p \neq b + q$ for each of (1)–(4). Choose $l \geqq 8 + |p - q|$, l even. Let $\mathfrak{A}_0$ be any of the equations (1)–(4). We construct a tree of $\mathfrak{A}_0$ in the following manner. Take a complete system of images of $\mathfrak{A}_0$ and connect $\mathfrak{A}_0$ to these images by means of arrows. Connect each of the vertices thus obtained to the equations of a complete system of images of that vertex by means of arrows, and so on. Repeat this procedure l times. The tree Д thus constructed is called a tree of images of $\mathfrak{A}_0$ of height l.

We will show how, starting from the tree Д, to construct the desired tree of $\mathfrak{A}_0$. Consider a path connecting $\mathfrak{A}_0$ to a final vertex of Д. Such a path is a chain of images $\mathfrak{A}_0, \cdots, \mathfrak{A}_l$. Consider first paths for which the chain $\mathfrak{A}_0, \cdots, \mathfrak{A}_l$ is degenerate. For such paths, according to the "replacement rule", $\mathfrak{A}_l$ can be replaced by an equation of the form $xy^azy^b \cdots \to zy^bxy^a \cdots$. For equations (1) – (4) the "replacing" equations will have, respectively, the forms

$$(1')\qquad xy^2z \cdots \to zxy^2 \cdots,$$

$$(2')\qquad xyz \cdots \to zxy \cdots,$$

$$(3')\qquad xy^2z \cdots \to zxy^2 \cdots,$$

$$(4')\qquad xyzy \cdots \to zy^2x \cdots.$$

Making this replacement, we obtain a tree $\text{Д}'$ of $\mathfrak{A}_0$. We will prove that each of the equations (1′) – (4′) possesses a β-tree. Indeed, (2′) is β-basic of the form (6.3). Equations (1′) and (3′) are reduced by the substitution $x = zv$ to an equation $vy^2z \cdots = zvy^2 \cdots$, which by 6.5 possesses a β-tree. Consider (4′). The substitution $x = zv$ reduces it to an equation $vyzyP = y^2zvQ$, an admissible permutation of which is the equation $Pvyzy = Qy^2zv$. The inversion of this last equation has the form $yzyv \cdots = vzy^2 \cdots$. Consider the corresponding one-sided equations $yzyv \cdots \to vzy^2 \cdots$ and $vzy^2 \cdots \to yzyv \cdots$. The first is β-basic of the form (6.4); the second is reduced by the substitution $v = yx$ to an equation $xzy^2 \cdots = zy^2x \cdots$, which by 6.5 possesses a β-tree. Hence each of the equations (1) – (4) possesses a β-tree. Connecting these trees to the corresponding vertices, we obtain a tree Д_1 of $\mathfrak{A}_0$ all of whose final vertices are β-basic, with the possible exception of those corresponding to nondegenerate chains $\mathfrak{A}_0, \cdots, \mathfrak{A}_l$.

Now consider a nondegenerate chain $\mathfrak{A}_0, \cdots, \mathfrak{A}_l$. There exists q, $1 \leqq q \leqq l$, such that $\mathfrak{A}_0, \cdots, \mathfrak{A}_{q-1}$ is a degenerate chain but $\mathfrak{A}_q$ is a nondegenerate image of $\mathfrak{A}_{q-1}$. We consider two subcases.

1) $\mathfrak{A}_0$ is an equation of one of the forms (1), (2) or (3). Then $\mathfrak{A}_0$ is an equation of the form 6.3 (*) with $c = 1$. Any image of such an equation, as is easily verified, is degenerate. Applying Propositions 6.3.2) – 4) in case a), we deduce by induction on q that q is even and that $\mathfrak{A}_{q-2}$ has the form (1), (2) or (3) respectively if $\mathfrak{A}_0$ has the form (1), (2) or (3), or $\mathfrak{A}_{q-2}$ is α-basic (in the case of (2)). But $\mathfrak{A}_q$ is a nondegenerate image of $\mathfrak{A}_{q-1}$. Hence if $\mathfrak{A}_{q-2}$ has the form (2), then $\mathfrak{A}_q$ possesses an α-tree by 6.8. If $\mathfrak{A}_{q-2}$ has the form (3), then by 6.3.2b) any image of $\mathfrak{A}_q$ is either α-basic or has the form $yxzy \cdots \to zxzy \cdots$. The substitution $y = zv$ reduces this last equation to an equation $vxz^2 \cdots = xz^2v \cdots$, which by 6.5 possesses a β-tree. Finally, if $\mathfrak{A}_{q-2}$ has the form (1), then by 6.3.3b) the equation $\mathfrak{A}_q$ possesses an h tree. Hence in any nondegenerate chain $\mathfrak{A}_0, \cdots, \mathfrak{A}_l$ there is an equation possessing a β-tree if $\mathfrak{A}_0$ has the form (2) or (3), or an h-tree if $\mathfrak{A}_0$ has the form (1). Now to construct the desired tree of $\mathfrak{A}_0$ it suffices to "cut off" any path $\mathfrak{A}_0, \cdots, \mathfrak{A}_l$ of the tree Д_1 corresponding to a nondegenerate chain $\mathfrak{A}_0, \cdots, \mathfrak{A}_l$ at a vertex $\mathfrak{A}_j$ possessing a β-tree (h-tree) and connect this tree to $\mathfrak{A}_j$. Thus the proposition is proved for equations (1), (2) and (3).

2) Now suppose $\mathfrak{A}_0$ is an equation of the form (4). It is not difficult to see that if $\mathfrak{A}_0, \cdots, \mathfrak{A}_k$ is a degenerate chain of (4), then for even k the equation $\mathfrak{A}_k$ has the form $xyz \cdots \to xy^2x \cdots$, and for odd k the form $zy^2x \cdots \to xyz \cdots$.

Accordingly, we consider two cases. 1) q is even. Then $\mathfrak{A}_{q-1}$ has the form $zy^2x\cdots\to xyzQ$, where Q can be assumed to be nonempty in view of item 6) in the definition of a neighborhood. But then $\mathfrak{A}_q$, being a nondegenerate image of $\mathfrak{A}_{q-1}$, is obtained under a transformation $z=(xy)^t xv$, where $t\geqq 0$ and v is a new unknown, and has the form $vy^2x\cdots\leftarrow yxv\eta\cdots$, where $\eta\neq v$. In other words, $\mathfrak{A}_{q-1}$ is an equation of the form (2) and, by what has been proved, possesses a β-tree. 2) q is odd. Then $\mathfrak{A}_{q-1}$ has the form $xyz\cdots\to zy^2x\cdots$, and its nondegenerate images have one of the forms $y^2\cdots\to vy\cdots$ or $yzy\cdots\to vyz\cdots$, where v is a new unknown. The last two equations are β-basic of the forms (6.2), (6.3), so that $\mathfrak{A}_q$ is β-basic. Hence in any nondegenerate chain $\mathfrak{A}_0,\cdots,\mathfrak{A}_l$ of (4) there is an equation possessing a β-tree. Thus our assertion is proved for an equation (4). It remains to observe that an equation (5) has the form (4) or (4′). The proposition is proved.

6.10. *Any σ-supporting equation possesses a β-tree.*

PROOF. We must prove that each equation of the form

$$x^ay^b\eta\cdots\to zyx\cdots\qquad(1\leqq a\leqq 2,\ 1\leqq b\leqq 2,\ \eta\neq y),\tag{1}$$

$$x^ay^b\eta\cdots\to zxy\cdots\qquad(1\leqq a\leqq 2,\ 1\leqq b\leqq 2,\ \eta\neq y)\tag{2}$$

possesses a β-tree.

Consider (1). If $a=b=1$, we have a β-basic equation of the form (6.4). If $a=1$ and $b=2$, we have an equation $xy^2\eta\cdots\to zyxQ$, where in view of item 6) in the definition of a neighborhood we may assume that $Q\neq\Lambda$. Since $\eta\neq y$ and $Q\neq\Lambda$, the images of this equation are equations of one of the forms $zyv\cdots\to vy^2z\cdots$ and $yzv\xi\cdots\to vy^2z\cdots$, where $\xi\neq v$ and v is a new unknown, i.e. equations of the forms (5) and (2) of 6.9, and the assertion being proved follows from 6.9. Consider the case $a=2$. Then (1) has the form $x^2y\cdots\to zyxQ$, where we may assume that $Q\neq\Lambda$. The pair of substitutions $x=zv$, $x=zyv$ reduces this equation to the equations $vzvy\cdots\to yzv\xi\cdots$ $(\xi\neq v)$ and $vzy\cdots=zyv\cdots$. The last equation is β-basic of the form (6.5). Consider the first. If $\xi=y$, we have an equation $yzvy\cdots\to vzvy\cdots$, which is reduced by the substitution $y=vx$ to an equation $xzv^2\cdots=zv^2x\cdots$. This last equation, in view of 6.5, possesses a β-tree. Consider the case $\xi=z$. Then we have an equation $yzvz\cdots\to vzvy\cdots$. Any image of this equation has the form $xzvz\cdots\leftarrow Dx\cdots$, where x is a new unknown and D is a cyclic permutation of the word vzv. If $D=vzv$, we have a β-basic equation of the form (6.4). If $D=zv^2$, we have an equation $zv^2x\cdots\to xzvz\cdots$, which by 6.9 (3) possesses a β-tree. And if $D=v^2z$, we have an equation $v^2zx\cdots\to xzvz\cdots$, which by 6.6 possesses a β-tree. We have proved that when $a=2$, equation (1) possesses a β-tree. Our assertion is proved for equations (1).

Now consider an equation (2). If $a=2$ or $a=b=1$, then we have a β-basic equation of one of the forms (6.2) or (6.3). There remains the case $a=1$, $b=2$. Then (2) has the form $xy^2\eta\cdots\to zxy\xi\cdots$, where $\eta\neq y$ and ξ is x, y or z. If $\xi=y$, then the substitution $x=zv$ reduces (2) to an equation $vy^2z\cdots=zvy^2\cdots$, which by 6.5 possesses a β-tree. And if $\xi\neq y$, then by 6.3.2) any chain $\mathfrak{A}_0,\cdots,\mathfrak{A}_3$ of (2) contains an α-basic equation or equations of the form $yxzy\cdots\to zxzy\cdots$ or $xy^2z\cdots\to zxyx\cdots$. The next to last equation, as we have seen in 6.9, possesses a β-tree. The last possesses a β-tree by 6.9

(3). Thus when $a=1$ and $b=2$, equation (2) possesses a β-tree. The proposition is proved.

Before proving that τ-supporting equations possess β-trees, we will prove the following assertion.

6.11. *Any equation of the form* $x^2y\eta\cdots\to zyzxy\cdots$, *where* $\eta\neq y$, *possesses a* β-*tree.*

PROOF. In view of 6.2, any image of the equation either is σ-supporting or is obtained under the transformation $x=zv$ and has the form $vzvyz\cdots\leftarrow yz^2vy\cdots$. The substitution $y=vx$ reduces this last equation to the equation $zv^2xz\cdots=xz^2v^2\cdots$. In view of 6.10, it remains to prove that $zv^2xz\cdots=xz^2v^2\cdots$ possesses a β-tree. Consider the corresponding one-sided equations:

1) $zv^2xz\cdots\to xz^2v^2\cdots$,

2) $xz^2v^2\cdots\to zv^2xz\cdots$.

1) Consider the equation

$$zv^2xz\cdots\to xz^2v^2\cdots. \tag{1}$$

The transformations of (1) have the form $z=x^\lambda y$, $\lambda>0$, and hence the images of (1) have the form

$$yv^2x^{\lambda+1}y\cdots\leftarrow xyx^\lambda yv^2\cdots. \tag{2}$$

The triple of substitutions $x=yz$, $x=yvz$, $x=yv^2z$ reduces (2) to the equations

$$v^2(yz)^{\lambda+1}y\cdots\to zy(yz)^\lambda yv^2\cdots, \tag{3}$$

$$vyvz\cdots\to zy(yvz)^\lambda\cdots, \tag{4}$$

$$(yv^2z)^{\lambda+1}y\cdots=zy(yv^2z)^\lambda yv^2\cdots. \tag{5}$$

In view of 6.2, (4) has only σ-supporting images. By 6.10, (4) possesses a β-tree.

Consider (3). By 6.2, any image of (3) is either σ-supporting or is obtained under the transformation $v=zx$ and has the form $xzx(yz)^{\lambda+1}y\cdots\leftarrow y^2(zy)^\lambda(zx)^2\cdots$. By 6.2, any image of this last equation is either σ-supporting or is obtained under the transformation $y=xv$ and has the form

$$zx(xvz)^{\lambda+1}xv\cdots\to vxv(zxv)^\lambda(zx)^2\cdots.$$

This last equation is reduced by the substitution $z=vy$ to an equation $yx(xv^2y)^{\lambda+1}xvP=xv(vyxv)^\lambda(vyx)^2Q$. An admissible permutation of this is the β-basic equation $yxvPyx^2v^2(yxv^2)^\lambda=vyxQxv(vyxv)^\lambda vyx$. We have proved that (3) possesses a β-tree.

Now consider (5). It is not difficult to verify that it has an admissible permutation of the form $v^2zy\cdots=zyv^2\cdots$. We will prove that this last equation possesses a β-tree. Consider the one-sided equations $v^2zy\cdots\to zyv^2\cdots$ and $zyv^2\cdots\to v^2zy\cdots$. Denote them by $\mathfrak{A}_1$, $\mathfrak{A}_2$. Any image of $\mathfrak{A}_2$ has the form $xyv^2\cdots\leftarrow v^2xy\cdots$, i.e. the form $\mathfrak{A}_1$. It remains to consider $\mathfrak{A}_1$. The pair of substitutions $v=zx$, $v=zyx$ reduces $\mathfrak{A}_1$ to the equations $xzxzy\cdots\leftarrow yzx\cdots$ and $xzy\cdots=zyx\cdots$. The first of these, in view of 6.2, has only σ-supporting images; the second is β-basic of the form (6.5). We have proved that (5) possesses a β-tree. Thus we have proved that (1) possesses a β-tree.

2) Consider the equation

(6) $$xz^2v^2 \cdots \to zv^2xz \cdots.$$

Any image of (6) has one of the following three forms (y is a new unknown):

(7) $$yz^2v^2 \cdots \leftarrow v^2zyz \cdots,$$

(8) $$yz^2v^2 \cdots \leftarrow vzvyz \cdots,$$

(9) $$yz^2v^2 \cdots \leftarrow zv^2yz \cdots.$$

By 6.2, (8) has only σ-supporting images and therefore, by 6.10, possesses a β-tree. Consider (7). By 6.2, any image of (7) either is σ-supporting or is obtained under the transformation $v = yx$ and has the form $z^2(yx)^2 \cdots \to xyxzyx \cdots$. By 6.2, any image of this last equation either is σ-supporting or is obtained under the transformation $z = xv$ and has the form $vxv(yx)^2 \cdots \leftarrow yx^2vyxv \cdots$. This last equation is reduced by the substitution $y = vx$ to an equation $xv(vzx)^2P = zx^2v^2zxvQ$. An admissible permutation of this is the β-basic equation $vzxPxv^2zx = zxvQzx^2v^2$. We have proved that (7) and (8) possess β-trees. Equation (9) has the same form as (1) and, by what has been proved, possesses a β-tree. We have thus shown that (6) possesses a β-tree. The proposition is proved.

6.12. *Any supporting equation possesses a β-tree.*

Proof. In view of 6.10, it suffices to prove that any τ-supporting equation possesses a β-tree. Let us prove this. Consider a τ-supporting equation. It has the form

(1) $$x^ay^b\eta \cdots \to z(yz)^dx \cdots \qquad (\eta \neq y, d \geqq 1),$$

where $1 \leqq a \leqq 2$ and $1 \leqq b \leqq 2$. There are three possible cases.

Case 1. $a = b = 1$. Then by 6.2 any image of (1) is σ-supporting and the assertion being proved follows from 6.10.

Case 2. $a = 1$, $b = 2$. Then any image of (1) has one of the following four forms (v is a new unknown):

(2) $$vy^2z \cdots \leftarrow zyz \cdots,$$

(3) $$vy^2z \cdots \leftarrow yzy \cdots$$

(4) $$vy^2z \cdots \leftarrow yz^2\xi \cdots \qquad (\xi \neq z),$$

(5) $$vy^2z \cdots \leftarrow z^2y\xi \cdots \qquad (\xi \neq y).$$

We will prove that each of the equations (2) – (5) possesses a β-tree. For equation (2) this follows from the fact that, in view of 6.2, any image of (2) is σ-supporting. Consider (3). This is equation 6.3 (*) with $a = 1$. By 6.3.4), for any chain $\mathfrak{A}_0, \cdots, \mathfrak{A}_3$ of (3), either $\mathfrak{A}_3$ is σ-supporting, $\mathfrak{A}_2$ is α-basic or $\mathfrak{A}_2$ is equation 6.9 (2). Thus in any chain $\mathfrak{A}_0, \cdots, \mathfrak{A}_3$ of (3) there is an equation possessing a β-tree. But then (3) itself possesses a β-tree.

Equation (4) is the equation described in 6.3.3). It possesses a β-tree by 6.3.3), 6.10, 6.11 and 6.9 (1).

Consider, finally, equation (5). By 6.2, any image of this equation either is σ-supporting or is obtained under the transformation $z = vx$ and has the form $y^2vx \cdots \to xvxyv \cdots$. Applying Proposition 6.11 to this last equation, we conclude that (5) possesses a β-tree. We have proved that when $a=1$ and $b=2$, equation (1) possesses a β-tree.

Case 3. $a=2$. In this case, using 6.2 and the fact that any cyclic permutation of the word $z(yz)^d$ contains no occurrences of y^3 or z^3, we deduce that any image of (1) either is σ-supporting or is obtained under the transformation $x=zv$ and has the form $vzvy\cdots \leftarrow (yz)^d zv\cdots$. Consider this last equation. If $d>1$, we have an equation of the form $yzy\cdots \rightarrow vzvy\cdots$, all of whose images are σ-supporting by 6.2. If $d=1$, we have an equation $yz^2v\cdots \rightarrow vzvy\cdots$, i.e. to within the designation of unknowns, equation (1) with $a=1$, $b=2$ and $d=1$. As was shown in Case 2, such an equation possesses a β-tree. The proposition is proved.

6.13. *Any equation of the form* $xy^azy^p\xi\cdots \rightarrow zy^bxy^q\eta\cdots$, *where* $a>0$, $a+p=b+q$, *and* ξ *and* η *are unknowns different from* y, *possesses a basic tree.*

PROOF. The system of substitutions $x=zy^cv$ $(c=0,1,\cdots,b)$ reduces the given equation to the equations $vy^az\cdots \leftarrow y^{b-c}zy^cv\cdots$ $(c=0,1,\cdots,b-1)$, $vy^azy^p\xi P=zy^bvy^q\eta Q$. If $b-c>1$, we have a basic equation of the form (6.2). If $b-c=1$, we have an equation of the form $yzw\cdots \rightarrow vy^az\cdots$, where $w\neq z$, i.e. equation 6.3 (*) with $a=1$. This last equation possesses a basic tree by 6.3.4), 6.9 (2) and 6.10. There remains the equation $vy^azy^p\xi P=zy^bvy^q\eta Q$. If $a=b$, we have a basic equation of the form (6.6) or (6.7). Consider the case $a\neq b$.

We will not use the assumption that $a>0$. This allows us to assume without loss of generality that $a<b$. Then an admissible permutation of the equation $vy^azy^p\xi P=zy^bvy^q\eta Q$ is the equation $y^q\xi Pvy^azy^{b-a}=y^q\eta Qzy^bv$. The inversion of this last equation has the form $y^dzy^av\cdots \rightarrow vy^bz\cdots$, where $d=b-a\geqq 1$. Consider the one-sided equations $y^dzy^av\cdots \rightarrow vy^bz\cdots$, $vy^bz\cdots \rightarrow y^dzy^av\cdots$. Denote them by $\mathfrak{A}_1$ and $\mathfrak{A}_2$. If $d>1$, then $\mathfrak{A}_1$ is basic of the form (6.2). If $d=1$, then $\mathfrak{A}_1$ has the form $yzw\cdots \rightarrow vy^bz\cdots$, where $w\neq z$. By 6.3.4), 6.9 (2) and 6.10, this equation possesses a basic tree. Hence $\mathfrak{A}_1$ possesses a basic tree. Consider $\mathfrak{A}_2$. The system of substitutions $v=y^cx$, where $c=1,\cdots,d$, reduces $\mathfrak{A}_2$ to the equations $xy^bz\cdots \leftarrow y^{d-c}zy^{a+c}x\cdots$ $(c=1,\cdots,d-1)$, $xy^bz\cdots =zy^{a+d}x\cdots$. This last equation is basic of the form (6.7), (6.6). The remaining equations are handled just like $\mathfrak{A}_1$. The proposition is proved.

6.14. *Any equation of the form* $xy^az\cdots \rightarrow zy^bx\cdots$, *where* $a>0$, *possesses a basic tree.*

PROOF. Our argument will be close to that of 6.9. Denote the given equation by $\mathfrak{A}_0$ and write it in the form $xy^azy^p\xi\cdots \rightarrow zy^bxy^q\eta\cdots$, where ξ and η are unknowns different from y. If $a+p-b+q$, the assertion follows from 6.13. Consider the case $a+p\neq b+q$. Choose $l\geqq 8+|p-q|$, l even. By analogy with 6.9, it suffices to prove two assertions:

I. A "replacing" equation $xy^azy^bxP\rightarrow zy^bxy^azQ$ possesses a basic tree.

II. In any nondegenerate chain $\mathfrak{A}_0,\cdots,\mathfrak{A}_l$ of $\mathfrak{A}_0$ there is an equation possessing a basic tree.

Assertion I follows from 6.13. Let us prove II. Consider a nondegenerate chain $\mathfrak{A}_0,\cdots,\mathfrak{A}_l$ of the equation $xy^az\cdots \rightarrow zy^bx\cdots$. There exists j, $1\leqq j\leqq l$, such that $\mathfrak{A}_0,\cdots,\mathfrak{A}_{j-1}$ is a degenerate chain but $\mathfrak{A}_j$ is a nondegenerate image of $\mathfrak{A}_{j-1}$. We consider two cases.

1) $b=0$. Then $\mathfrak{A}_0$ has the form $xy^az\cdots\to zx\cdots$, i.e. $\mathfrak{A}_0$ is equation 6.3 (*) with $c=1$. Arguing as in subcase 1) of 6.9 and using 6.3.1), case a), we deduce that j is even and $\mathfrak{A}_{j-2}$ has the form $xy^az\cdots\to zx\cdots$. Since $\mathfrak{A}_j$ is a nondegenerate image of $\mathfrak{A}_{j-1}$, by 6.3.1) any image of $\mathfrak{A}_j$ is supporting. We now conclude from 6.12 that $\mathfrak{A}_j$ possesses a basic tree.

2) $b>0$. Then it is not difficult to see that the equations of the chain $\mathfrak{A}_0,\cdots,\mathfrak{A}_{j-1}$ with even indices have the form $xy^az\cdots\to zy^bx\cdots$, and those with odd indices the form $zy^bx\cdots\to xy^az\cdots$. Hence $\mathfrak{A}_{j-1}$ has the form $xy^nz\cdots\to zy^mx\cdots$, where $n,m\geqq 1$. Equation $\mathfrak{A}_j$ is a nondegenerate image of $\mathfrak{A}_{j-1}$ and, as is easily seen, has the form $y^czy^dv\cdots\to vy^nz\cdots$, where $c+d=m$ and $c\geqq 1$. If $c>1$, we have a basic equation of the form (6.2). If $c=1$, we have an equation of the form $yzw\cdots\dot{\to} vy^nz\cdots$, where $w\neq z$. This last equation, in view of 6.3.4), 6.9 (2) and 6.10, possesses a basic tree. We have shown that $\mathfrak{A}_j$ possesses a basic tree. Thus II, and hence also Proposition 6.14, is proved.

We can now prove that any equation in three unknowns possesses a basic tree (6.15–6.17).

6.15. *Any equation of the form*

(*) $$xy^a\eta\cdots\to z^cxB(x,z)y\cdots\qquad (a,c\geqq 1,\eta\neq y)$$

possesses a basic tree.

PROOF. We have an equation of the form 6.3 (*). By 6.3.1), for any chain $\mathfrak{A}_0,\cdots,\mathfrak{A}_3$ of (*), either $\mathfrak{A}_2$ has the form $xy^az\cdots\to zx\cdots$ or $\mathfrak{A}_3$ is supporting. Hence, by 6.14 and 6.12, in any chain $\mathfrak{A}_0,\cdots,\mathfrak{A}_3$ of (*) there is an equation possessing a basic tree. But then (*) itself possesses a basic tree. The proposition is proved.

6.16. *Any equation of the form*

(*) $$x^ny^m\xi\cdots\to zyA(y,z)x\cdots\qquad (n,m\geqq 1,\ \xi\neq y)$$

possesses a basic tree.

PROOF. We will first prove that any image of (*) either possesses a basic tree or has one of the forms

(1) $$x^ay^b\eta\cdots\to zy^dx\cdots\qquad (a,b,d\geqq 1,\ \eta\neq y),$$

(2) $$x^ay^b\eta\cdots\to z(yz)^dx\cdots\qquad (a,b,d\geqq 1,\ \eta\neq y).$$

Indeed, by 6.1 the images of (*) have one of the forms 6.1 (1)–(3), where the word D contains both y and z. Consider equation 6.1 (1). If $y^2\angle D$, then we have a basic equation of the form (6.2); if $yz\angle D$, then we have equation 6.15 (*), which by 6.15 possesses a basic tree; and if $z\angle D$, then we have an equation of the form (1). Equation 6.1 (3) is handled analogously (with an interchange of the y and z places). Finally, 6.1 (2) is an equation of the form (2). We have proved that any image of (*) either has the form (1) or (2) or possesses a basic tree.

It now suffices to prove that each of the equations (1) and (2) possesses a basic tree. Let us prove this.

Consider (1). We begin with the case $a=1$. By 6.1, any image of (1) has the form $vy^bz\cdots\leftarrow Dv\cdots$, where D is a cyclic permutation of the word

zy^d and v is a new unknown. If $z \angle D$, we have an equation $zy^d v \cdots \to vy^b z \cdots$, which by 6.14 possesses a basic tree. If $y^2 \angle D$, we have a basic equation of the form (6.2), and if $yz \angle D$, we have equation 6.15 (*). Thus the assertion is proved for $a = 1$. Consider the case $a > 1$. By 6.1, we obtain under the transformation $x = zv$ the image $v(zv)^{a-1}y \cdots \leftarrow y^d zv \cdots$, and all other images have the form $vzy \cdots \leftarrow Dv \cdots$, where D is a cyclic permutation of the word zy^d. The first equation, in view of 6.2, has only supporting images and, by 6.12, possesses a basic tree. Consider the equation $Dv \cdots \to vzy \cdots$. If $zy \angle D$, then we have equation 6.15 (*). Otherwise, the equation has the form $y^c zw \cdots \to vzy \cdots$, where $w \neq z$, $c \geqq 1$, and by 6.2 all of its images are supporting equations. Thus, the equation $Dv \cdots \to vzy \cdots$, hence also (1), possesses a basic tree.

Consider (2). We begin with the case $a = 1$. The images of (2) have the form $vy^b z \cdots \leftarrow Dv \cdots$, where v is a new unknown and D is a cyclic permutation of the word $z(yz)^d$. If $yz \angle D$, we have equation 6.15 (*). Otherwise, the word Dv has the form $z^c yw \cdots$, where $c = 1$ or 2 and $w \neq y$. But then the equation $Dv \cdots \to vy^b z \cdots$ is an equation of the form (1) and, by what has been proved, possesses a basic tree. Hence if $a = 1$, then (2) possesses a basic tree. Let us now assume that in (2) we have $a > 1$. Then by 6.1 the image of (2) under the transformation $x = zv$ has the form $v(zv)^{a-1}y \cdots \leftarrow (yz)^d zv \cdots$, and under all other transformations the images have the form $vzy \cdots \leftarrow Dv \cdots$, where D is a cyclic permutation of the word $z(yz)^d$. The first equation is (2) with $a = 1$ and, by what has been proved, possesses a basic tree. The second, as is easily seen, is σ-supporting. Applying 6.10, we conclude that (2) also possesses a basic tree when $a > 1$. The proposition is proved.

6.17. THEOREM. *Any coefficient-free equation in three unknowns possesses a basic tree.*

PROOF. It suffices to prove the assertion for one-sided equations. Suppose $\mathfrak{A}$ is a one-sided coefficient-free equation in three unknowns. It is not difficult to see that $\mathfrak{A}$ has one of the following six forms (after multiplication on the right by suitable unknowns):

(1) $\quad x^2 \cdots \to y^c x \cdots \quad (c \geqq 1),$

(2) $\quad xy \cdots \to y^c x \cdots \quad (c \geqq 1),$

(3) $\quad xz^a \eta \cdots \to y^c x B(x,y) z \cdots \quad (a, c \geqq 1,\ \eta \neq z),$

(4) $\quad x^a y^b \eta \cdots \to y^c z B(y,z) x \cdots \quad (a, b, c \geqq 1,\ \eta \neq y),$

(5) $\quad x^a z^b \eta \cdots \to yz B(y,z) x \cdots \quad (a, b \geqq 1,\ \eta \neq z),$

(6) $\quad x^a z^b \eta \cdots \to y^c z B(y,z) x \cdots \quad (c > 1;\ a, b \geqq 1,\ \eta \neq z).$

We will prove that each of the equations (1)–(6) possesses a basic tree. Equation (1) is basic of the form (6.2). Equation (2) is reduced by the substitution $x = yv$ to an equation $vy \cdots = y^c v \cdots$, which is basic of the form (6.1). Equation (3) is equation 6.15 (*) and, by 6.15, possesses a basic tree. Equation (5) is equation 6.16 (*) and, by 6.16, possesses a basic tree. Consider (4). Look at the images of (4). It is easy to see that any image of (4) has the form $vy \cdots \leftarrow Dv \cdots$, where v is a new unknown and D is a cyclic permutation of the word $y^c zB$. The word D contains both of the unknowns y and z.

If $y^2 \angle D$, we have an equation of the form (1); if $yz \angle D$, the form (3); and if $z \angle D$, the form (5). Hence any image of (4) possesses a basic tree. But then (4) itself possesses a basic tree. It remains to consider (6). Any image $\mathfrak{B}$ of (6) has the form $v(y\cdots)^{a-1}z^b y\cdots \leftarrow Dv\cdots$, where v is a new unknown and D is a cyclic permutation of the word $y^c zB$. When $a>1$ we have an equation $Dv\cdots \to vy\cdots$, and when $a=1$ we have $Dv\cdots \to vz\cdots$. The word D contains both y and z. It is now easy to verify that $\mathfrak{B}$ is an equation of one of the forms (1), (3), (5) and, by what has been proved, possesses a basic tree. Hence any image of (6) possesses a basic tree, so (6) itself possesses a basic tree. The theorem is proved.

6.18. Theorem. *Any coefficient-free equation in three unknowns possesses a tree in which each final vertex has the form $xy^a z\cdots = zy^a x\cdots$, where $a \geqq 0$ need not have the same value at each vertex.*

Proof. By 6.17, it suffices to prove that any basic equation possesses such a tree. Let us prove this. Consider the basic equations (6.1) – (6.7).

First consider (6.1). It suffices to prove the assertion for the one-sided equation $x^a y\cdots \to y^b x\cdots$. Consider the images of this last equation. They are obtained under a transformation $x = y^t v$, where $t>0$ and v is a new unknown, and have the form $v(y^t v)^{a-1} y\cdots \leftarrow y^b v\cdots$, i.e. the form $y^b v\cdots \to vy\cdots$. This last equation is reduced by the substitution $y = vz$ to an equation $zv\cdots = vz\cdots$, which has the form $xy^a z\cdots = zy^a x\cdots$ with $a=0$. Thus our assertion is proved for (6.1).

Consider (6.2). The substitution $x = yv$ reduces this equation to an equation $vy\cdots = y^c v\cdots$, which is an equation of the form (6.1). Our assertion is proved for (6.2).

Consider (6.5). The corresponding one-sided equations have the forms $xyz\cdots \to zxy\cdots$ and $zxy\cdots \to xyz\cdots$. Denote them by $\mathfrak{A}_1$ and $\mathfrak{A}_2$. The images of $\mathfrak{A}_1$ have the form $vyz\cdots \leftarrow zvy\cdots$, i.e. the form $\mathfrak{A}_2$. But $\mathfrak{A}_2$ is reduced by the substitution $z = xv$ to an equation $vxy\cdots = yxv\cdots$, which has the form $xy^a z\cdots = zy^a x\cdots$ with $a=1$. Our assertion is proved for (6.5).

Equations (6.3) and (6.4) are reduced by the substitution $x = zv$ to the equations $vyz\cdots = zvy\cdots$ and $vyz\cdots = yzv\cdots$ respectively. Each of these is, to within designation of the unknowns, an equation of the form (6.5). Thus our assertion is proved for (6.3) and (6.4).

There remain equations (6.6) and (6.7). Our assertion is trivial for these. The theorem is proved.

The following two assertions are used in the reduction of certain basic equations to others.

6.19. *If $a>2$, then any nondegenerate image of an equation $xy^a z\cdots \to zy^a x\cdots$ possesses a β-tree.*

Proof. Nondegenerate images of an equation $xy^a z\cdots \to zy^a x\cdots$ have the form $vy^a z\cdots \leftarrow y^n zy^m v\cdots$, where $1 \leqq n \leqq a$ and $m = a-n$. If $n>1$, we have a β-basic equation of the form (6.2). If $n=1$, we have an equation $yzy\cdots \to vy^a z\cdots$, which by 6.3.4), 6.10 and 6.9 (2) possesses a β-tree. The proposition is proved.

Basic equations of the forms (6.1) and (6.2) are called ω-equations. From the argument in 6.18 (the second and third paragraphs of the proof) it follows that any ω-equation possesses a tree with final vertices of the form $xy\cdots\to yx\cdots$.

6.20. *Suppose $a \geqq 1$. Consider an equation*

$$(*)\qquad xy^azA(x,y,z)\to zy^axB(x,y,z).$$

Suppose the words A and B contain each of the letters x, y and z, and suppose one of A or B begins with y, the other with x or z. Then it is possible to construct a tree of $(*)$ whose final vertices have the form $xy\cdots\to yx\cdots$.

PROOF. Trees with final vertices of the form $xy\cdots\to yx\cdots$ are called ω-trees. We must prove that $(*)$ possesses an ω-tree. The pair of substitutions $x=zv$, $x=zyv$ reduces $(*)$ to the equations

$$(1)\qquad vy^azA(zv,y,z)\leftarrow y^azvB(zv,y,z),$$

$$(2)\qquad vy^azA(zyv,y,z)=y^{a-1}zyvB(zyv,y,z).$$

Consider (2) first. An admissible permutation of (2) is the equation

$$(3)\qquad A(zyv,y,z)vy^az=B(zyv,y,z)y^{a-1}zyv.$$

From our assumptions concerning the occurrences of unknowns in A and B it follows that one side of (3) begins with the word y^cz, where $c\geqq 1$, and the other with the word z^dy, where $d\geqq 1$. Equation (3) is an ω-equation of the form (6.1). It possesses an ω-tree.

Now consider (1). If $a>1$, then we have an ω-equation of the form (6.2). Consider the case $a=1$. Then (1) has the form $vyzA(zv,y,z)\leftarrow yzvB(zv,y,z)$. Look at the images of this last equation. They have the form

$$(4)\qquad vuzA(zv,v^tu,z)\to uzvB(zv,v^tu,z),$$

where $t>0$ and u is a new unknown. The substitution $v=ux$ reduces (4) to the equation

$$xuzA(zux,(ux)^tu,z)=zuxB(zux,(ux)^tu,z).$$

Consider the corresponding one-sided equations

$$(5)\qquad xuzA(zux,(ux)^tu,z)\to zuxB(zux,(ux)^tu,z),$$

$$(6)\qquad xuzA(zux,(ux)^tu,z)\leftarrow zuxB(zux,(ux)^tu,z).$$

Consider (5). The cases $y\angle A$ and $y\angle B$ are handled analogously. Let us restrict ourselves to the case $y\angle A$. Then (5) has the form $xuzux\cdots\to zuxz^cu\cdots$, $c\geqq 1$. This last equation is reduced by the substitution $x=zv$ to an equation $vuzuzP=uzvz^cuQ$, an admissible permutation of which is the ω-equation $uzPvuz=z^cuQuzv$. From the above argument it follows that if $a>1$, then 6.20 is true, and if $a=1$, then it is possible to construct a tree of $(*)$ whose final vertices are ω-equations or have the form (6). But (6) is an equation of the form $(*)$ with $a=1$ (the role of y in $(*)$ is now played by u, and x and z have interchanged places). Applying to (6) the above deduction concerning $(*)$ (which is possible if we consider instead of y a word beginning with u, and instead of x and z words beginning with z), we conclude that it is possible to construct a tree of (6) whose final vertices either are ω-equations or have the form

$$xyzA(\bar{z}\bar{u}x, (\bar{u}x)^t\bar{u}, \bar{z}) \to zyxB(\bar{z}\bar{u}x, (\bar{u}x)^t\bar{u}, \bar{z}), \tag{7}$$

where $\bar{z} = xyz$, $\bar{u} = (yz)^\lambda y$, $\lambda > 0$, and y and z are new unknowns. Let us consider the case $y \angle A$. Then B begins with x or z. Hence the right-hand side of (7) begins with the word $zyxxyz$, the left-hand side with the word $xyzyzy$. Equation (7) has the form

$$xyzyzy \cdots \to zyxxyz \cdots. \tag{8}$$

Applying our argument once more, we conclude that it is possible to construct a tree of (8) whose final vertices are ω-equations or have the form (cf. (6))

$$xuz\bar{y}z\bar{y} \cdots \leftarrow zux\bar{x}\bar{y}z \cdots, \tag{9}$$

where $\bar{x} = zux$ and $\bar{y} = (ux)^\mu u$, $\mu > 0$. Equation (9) has the form

$$xuz(ux)^\mu uzux \cdots \leftarrow zuxzux(ux)^\mu u \cdots.$$

The substitution $z = xv$ reduces this last equation to an equation

$$uxv(ux)^\mu uxvuxP = vuxxvux(ux)^\mu uQ.$$

An admissible permutation of this is the ω-equation

$$uxPuxv(ux)^\mu uxv = xuQvuxxvu(xu)^\mu.$$

Thus 6.20 is proved when $a = 1$. The proposition is proved.

Trees of equations with coefficients

We will apply our results to equations with coefficients. The term "equation" means an arbitrary word equation in three unknowns (one-sided or two-sided).

Let us define trees of equations. We begin with the notions of an inversion and an admissible permutation.

By the inversion of the word $P = \xi_1\xi_2 \cdots \xi_k$, where $\xi_1, \cdots, \xi_k$ are letters, we mean the word $\overline{P} = \xi_k \cdots \xi_2\xi_1$. By the inversion of the equation $\Phi = \Psi$ we mean the equation $\overline{\Phi} = \overline{\Psi}$. If $(X_1, \cdots, X_n)$ is a solution of $\Phi = \Psi$, then $(\overline{X}_1, \cdots, \overline{X}_n)$ is a solution of $\overline{\Phi} = \overline{\Psi}$. Hence an equation and its inversion either both have solutions or both do not have solutions. We state one other property of an inversion: the projection on the alphabet $\mathbf{X}$ of the inversion of an equation agrees with the inversion of the projection of this equation on $\mathbf{X}$ (this is the invariance of the inversion with respect to projection on the alphabet of unknowns).

We now turn to admissible permutations. Consider an equation $A\Phi = B\Psi$. Denote it by $\mathfrak{A}(x, y, z)$. Suppose each unknown occurs in A as many times as in B ($\nu(A) = \nu(B)$ in the notation of §4), and suppose Φ and Ψ begin with different unknowns. Let a and b denote the lengths of the projections of A and B on the alphabet **И**. Put $d = a - b$. Consider the case $d \geqq 0$. Let $\mathfrak{A}_R(x, y, z)$ denote the equation $R\Phi A = \Psi BR$, where R is a word in the alphabet **И** of length d. We say that the set of equations $\mathfrak{A}_R$ forms an admissible permutation of $\mathfrak{A}$. Note that the projection of $\mathfrak{A}_R$ on the alphabet $\mathbf{X}$ is an admissible permutation of the projection of $\mathfrak{A}$ on $\mathbf{X}$ (this is the invariance of an admissible permutation with respect to projection on the alphabet of unknowns). Our utilization of admissible permutations is based on the following fact, which follows from 4.4 (3). The equation $A\Phi = B\Psi$ is equivalent to a disjunction of equations $R\Phi A = \Psi BR$, extending over all words R in the alphabet И such that $[R^\partial = d$. According to this equivalence, $\mathfrak{A}$ has solutions if and only if at least one of the equations $\mathfrak{A}_R$ has.

We now consider reduction by an n-tuple of substitutions. Suppose an equation $\mathfrak{A}(x,y,z)$ has the form

$$x\Phi(x,y,z) \to [\xi_i A_i]_{i=1}^k x\Psi(x,y,z),$$

where $k \geqq 1$, $\xi_1, \cdots, \xi_k$ are unknowns different from x, and $A_1, \cdots, A_k$ are coefficients. Suppose $1 \leqq n \leqq k$, $1 \leqq q < n$, and v is a new unknown. Let $\mathfrak{A}_q$ denote the equation

$$v\Phi([\xi_i A_i]_{i=1}^q v, y, z) \leftarrow [\xi_i A_i]_{i=q+1}^k [\xi_i A_i]_{i=1}^q v\Psi([\xi_i A_i]_{i=1}^q v, y, z),$$

and $\mathfrak{A}_n$ the equation

$$v\Phi([\xi_i A_i]_{i=1}^n v, y, z) = [\xi_i A_i]_{i=n+1}^k [\xi_i A_i]_{i=1}^n v\Psi([\xi_i A_i]_{i=1}^n v, y, z).$$

Having thus defined the equations $\mathfrak{A}_1, \cdots, \mathfrak{A}_n$, we say that the n-tuple of substitutions $x = \xi_1 A_1 v, \cdots, x = [\xi_i A_i]_{i=1}^n v$ reduces $\mathfrak{A}$ to the equations $\mathfrak{A}_1, \cdots, \mathfrak{A}_n$. Our utilization of "reduction by an n-tuple of substitutions" is based on the following equivalence:

$$\mathfrak{A}(x,y,z) \Leftrightarrow \bigvee_{q=1}^{n} \bigvee_{R \angle A_q} (x = [\xi_i A_i]_{i=1}^{q-1} \xi_q R \,\&\, \mathfrak{A}_{q,R}) \tag{1}$$
$$\vee \bigvee_{q=1}^{n} \exists v\, (x = [\xi_i A_i]_{i=1}^q v \,\&\, \mathfrak{A}_q(v,y,z)),$$

where $\mathfrak{A}_{q,R}$ is the equation in two unknowns which is obtained from $\mathfrak{A}$ by replacing x by the word $[\xi_i A_i]_{i=1}^{q-1} \xi_q R$.

The equivalence (1) is almost obvious. Indeed, if (X,Y,Z) is a solution of $\mathfrak{A}$, then $[X^\partial > [\bar{\xi}_1^\partial$ ($\bar{\xi}_i$ denotes the value of the unknown ξ_i). Two cases are possible: 1) $[X^\partial \leqq [[\bar{\xi}_i A_i]_{i=1}^{n\partial}$; 2) $[X^\partial > [[\bar{\xi}_i A_i]_{i=1}^{n\partial}$. In the second case $X = [\bar{\xi}_i A_i]_{i=1}^n V$ for some V, and (V,Y,Z) is a solution of $\mathfrak{A}_n$. In the first case we have $\bar{\xi}_1 \not\angle X \angle [\bar{\xi}_i A_i]_{i=1}^n$, i.e. X "cuts" the word $[\bar{\xi}_i A_i]_{i=1}^n$. If a coefficient is cut, i.e. $X = [\bar{\xi}_i A_i]_{i=1}^{q-1} \bar{\xi}_q R$, where $1 \leqq q \leqq n$ and $R \angle A_q$, then (Y,Z) satisfies the equation $\mathfrak{A}_{q,R}$. If the "unknown" ξ_{q+1} is cut, then $X = [\bar{\xi}_i A_i]_{i=1}^q V$, where $1 \leqq q < n$ and $\Lambda \neq V \not\angle \bar{\xi}_{q+1}$, so that $0 < [V^\partial < [\bar{\xi}_{q+1}^\partial$ and the triple (V,Y,Z) satisfies the one-sided equation $\mathfrak{A}_q$. We have proved that the left-hand side of the equivalence (1) implies the right. But the right-hand side of (1) does not always imply the left. Namely, if (v,y,z) is a solution of $\mathfrak{A}_n$ and one of the components y or z is empty, then by the nonemptiness of the components of solutions of one-sided equations the right-hand side of (1) does not imply the left. In order to preserve the equivalence (1), we henceforth agree, if nothing is said to the contrary, that by solutions of two-sided word equations we understand nonempty solutions, i.e. solutions all of whose components are nonempty words. Then the equivalence (1) will be true.

This convention, of course, does not limit the generality of our arguments. If there exists an algorithm α recognizing the existence of solutions with nonempty components for equations with n unknowns, then there also exists an algorithm β recognizing the existence of solutions (in the sense of section 1 of the Introduction) for equations in n unknowns. Indeed, the algorithm β is constructed from the algorithm α in the following manner. Suppose there is given an equation $\mathfrak{A}(x_1, \cdots, x_n)$. Adjoin to it the equations $\mathfrak{A}_1, \cdots, \mathfrak{A}_m$, which are obtained from $\mathfrak{A}$ by all possible substitutions of empty words in place of the unknowns $x_1, \cdots, x_n$. We obtain a finite set $\mathfrak{M}$ of word equations in n unknowns. The equation $\mathfrak{A}$ has a solution (in the sense of section 1 of the Introduction) if and only if at least one equation in the set $\mathfrak{M}$ has a nonempty solution. It remains to check by means of the algorithm α whether at least one of the equations in $\mathfrak{M}$ has a (nonempty) solution. This completes the construction of the algorithm β. It is easy to see that all assertions proved earlier concerning solutions of two-sided word equations remain valid if by solutions we understand nonempty solutions. We have in mind Propositions 1.20, 3.1 and 4.10.

According to the equivalence (1), $\mathfrak{A}$ has solutions if and only if at least one of $\mathfrak{A}_1, \cdots, \mathfrak{A}_n$ or the equations in two unknowns $\mathfrak{A}_{q,R}$ has. We also note that if $\mathfrak{A}$ is reduced to the equations $\mathfrak{A}_1, \cdots, \mathfrak{A}_n$ by an n-tuple of substitutions, then the projection of $\mathfrak{A}$ on the alphabet $\mathbf{X}$ is reduced (in the sense of the definition following 6.3) by the corresponding n-tuple of substitutions to the projections of $\mathfrak{A}_1, \cdots, \mathfrak{A}_n$ on $\mathbf{X}$ (this is the invariance of substitutions with respect to projection on the alphabet of unknowns).

We also define a rule for the transformation of equations, the "rule of removing the first coefficient". Suppose an equation $\mathfrak{A}$ has the form $Px\Phi(x,y,z) = y\Psi(x,y,z)$, where P is a coefficient and x and y are distinct unknowns. Then we say that the equation $x\Phi(x,Py,z) = y\Psi(x,Py,z)$ is obtained from $Px\Phi = y\Psi$ by the rule of removing the first coefficient. It is obvious that if an equation $\mathfrak{B}$ is obtained from an equation $\mathfrak{A}$ by the rule of removing the first coefficient, then the projections of $\mathfrak{A}$ and $\mathfrak{B}$ on the alphabet of unknowns agree (this is the invariance of the rule with respect to projection on the alphabet of unknowns). It is also obvious that $\mathfrak{A}$ has solutions if and only if $\mathfrak{B}$ or one of the equations in two unknowns $Px\Phi(x, Y, z) = Y\Psi(x, Y, z)$ has, Y being a word in the alphabet $\mathbf{H}$ of length $\leqq [P^\partial$.

We can now define a neighborhood of an equation.

A set of equations $\mathfrak{A}_1, \cdots, \mathfrak{A}_n$ is called a neighborhood of $\mathfrak{A}$ in the following cases (cf. the definition of neighborhood following 6.3).

1) The equations $\mathfrak{A}_1, \cdots, \mathfrak{A}_n$ form a complete system of infinite images of $\mathfrak{A}$.

2) $\mathfrak{A}$ is reduced to the equations $\mathfrak{A}_1, \cdots, \mathfrak{A}_n$ by some n-tuple of substitutions.

3) If $\mathfrak{A}$ is a two-sided equation of the form $x\Phi = y\Psi$, where x and y are distinct unknowns, then $n = 2$, and $\mathfrak{A}_1$ and $\mathfrak{A}_2$ have the forms $x\Phi \to y\Psi$ and $y\Psi \to x\Phi$.

4) $n = 1$ and $\mathfrak{A}_1$ is the inversion of $\mathfrak{A}$.

5) The equations $\mathfrak{A}_1, \cdots, \mathfrak{A}_n$ form an admissible permutation of $\mathfrak{A}$.

6) $n = 1$ and $\mathfrak{A}_1$ is obtained from $\mathfrak{A}$ by multiplication on the right or cancellation on the left by a word in the alphabet $\mathbf{H} \cup \mathbf{X}$.

7) $n = 1$ and $\mathfrak{A}_1$ is obtained from $\mathfrak{A}$ by the rule of removing the first coefficient.

6.21. *Suppose the equations $\mathfrak{A}_1, \cdots, \mathfrak{A}_n$ form a neighborhood of $\mathfrak{A}$. Then it is possible to construct a finite number of one-stage exponential equations $\pi_1, \cdots, \pi_k$ and word equations $\mathfrak{B}_1, \cdots, \mathfrak{B}_s$ in two unknowns such that $\mathfrak{A}$ has solutions if and only if at least one of the equations $\mathfrak{A}_1, \cdots, \mathfrak{A}_n$, $\mathfrak{B}_1, \cdots, \mathfrak{B}_s$, $\pi_1, \cdots, \pi_k$ has.*

PROOF. In accordance with the definition of a neighborhood, we consider seven cases. In case 1), our assertion follows from 4.9. Only in this case is it possible that $k > 0$. In the remaining cases, $k = 0$. In case 2), our assertion follows from the equivalence (1). In case 3), our assertion follows from the equivalence $(x\Phi = y\Psi) \Leftrightarrow (x\Phi \to y\Psi) \vee (y\Psi \to x\Phi) \vee (x = y, \Phi = \Psi)$. The validity of our assertion in case 4) was noted in the definition of an inversion, in case 5) in the definition of admissible permutations, and in case 7) in the definition of the rule of removing the first coefficient. In case 6), our assertion is trivial ($k = s = 0$). The proposition is proved.

If $\mathfrak{A}$ is an equation in two unknowns, we have

6.21′. *Suppose $\mathfrak{A}$ is a word equation in two unknowns and the equations $\mathfrak{A}_1, \cdots, \mathfrak{A}_n$ form a neighborhood of $\mathfrak{A}$. Then it is possible to construct a finite number of word equations $\mathfrak{B}_1, \cdots, \mathfrak{B}_s$ in one unknown such that $\mathfrak{A}$ has solutions if and only if at least one of the equations $\mathfrak{A}_1, \cdots, \mathfrak{A}_n$, $\mathfrak{B}_1, \cdots, \mathfrak{B}_s$ has.*

The proof is the same as that of 6.21, only now the equations $\mathfrak{B}_1, \cdots, \mathfrak{B}_s$ of 6.21 are equations in one unknown, and when considering case 1) we refer to 4.2 instead of 4.9.

Using the notion of a neighborhood, trees of word equations are defined in exactly the same way as trees of coefficient-free equations. The replacement rule is as follows. Suppose $\mathfrak{A}_0$ has the form $xA(y)zP(y)\xi \cdots \to zB(y)xQ(y)\eta \cdots$, where ξ and η are unknowns different from y, and let a, b, p and q be the number of occurrences of y in the words A, B, P and Q, where $a + p \neq b + q$. Suppose l is even, $l \geqq 8 + |p - q|$, and suppose $\mathfrak{A}_0, \cdots, \mathfrak{A}_l$ is a degenerate chain of $\mathfrak{A}_0$. Then a tree of the equation $\mathfrak{A}_l$ can be replaced by a tree of the equation $xAzBx\Phi \to zBxAz\Psi$, where $x\Phi \to z\Psi$ is the equation $\mathfrak{A}_l$.

The following assertion explains the role played by the trees of equations.

6.22. *Suppose $\mathfrak{A}_1, \cdots, \mathfrak{A}_n$ are the final vertices of a tree Д of an equation $\mathfrak{A}$. Then it is possible to construct a finite set $\mathfrak{B}_1, \cdots, \mathfrak{B}_s$ of one-stage exponential equations and word equations in two unknowns such that $\mathfrak{A}$ has solutions if and only if at least one of the equations $\mathfrak{A}_1, \cdots, \mathfrak{A}_n$, $\mathfrak{B}_1, \cdots, \mathfrak{B}_s$ has.*

PROOF. If $\mathfrak{B}_1, \cdots, \mathfrak{B}_k$ is a path of the tree, then the number $k - 1$ is called the length of the path. A number h is called the height of the tree if the tree contains at least one path of length h but no path of length $> h$. Obviously, the height of a tree is uniquely determined by the tree.

We will prove our assertion by induction on the height $h(Д)$ of the tree Д. If $h(Д) = 0$, the assertion is trivial. Suppose there is given a tree Д of height $h(Д) = h \geqq 1$. The tree Д contains paths of length h. Discard from Д the final vertices of these paths. We obtain a tree Д′ of $\mathfrak{A}$ of height $h - 1$. Divide the final vertices of Д′ into two groups. In the first put the equations $\mathfrak{A}_1, \cdots, \mathfrak{A}_k$ which are also final in the tree Д. In Figure 4 the vertices in this group are denoted by light circles. The second group $\pi_1, \cdots, \pi_s$ contains all remaining final vertices of Д′. By the induction assumption it is possible to construct a finite set $\mathfrak{B}_1, \cdots, \mathfrak{B}_q$ of one-stage exponential equations and word equations in two unknowns such that $\mathfrak{A}$ has solutions if and only if at least one of the equations $\mathfrak{A}_1, \cdots, \mathfrak{A}_k$, $\mathfrak{B}_1, \cdots, \mathfrak{B}_q$, $\pi_1, \cdots, \pi_s$ has. Consider the equations $\pi_1, \cdots, \pi_s$. Let i be one of the numbers $1, \cdots, s$. By definition of a tree there are two possible cases.

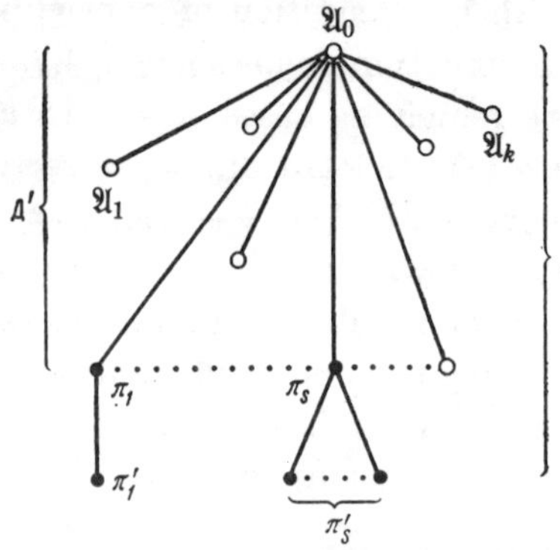

FIGURE 4

1) In the construction of the tree Д there occurs a transition from the vertex π_i to a neighborhood of the equation π_i. Denote this neighborhood by π_i'. Then by 6.21 it is possible to construct a finite set π_i'' of one-stage exponential equations and word equations in two unknowns such that the equation π_i has solutions if and only if at least one equation in the set $\pi_i' \cup \pi_i''$ has.

2) In the construction of the tree Д the equation π_i is replaced in accordance with the replacement rule. Denote the replacing equation by π_i'. By 5.6 it is possible to construct a finite set π_i'' of word equations in two unknowns such that π_i has solutions if and only if at least one equation in the set $\pi_i' \cup \pi_i''$ has.

It follows from the above argument that the final vertices of the tree Д comprise the set $\bigcup_{i=1}^k \mathfrak{A}_i \cup \bigcup_{i=1}^s \pi_i'$, and the equations $\mathfrak{B}_1, \cdots, \mathfrak{B}_s$, whose existence the theorem asserts, comprise the set $\bigcup_{i=1}^q \mathfrak{B}_i \cup \bigcup_{i=1}^s \pi_i''$. The proposition is proved.

If $\mathfrak{A}$ is an equation in two unknowns, we have

6.22′. *Suppose $\mathfrak{A}$ is a word equation in the two unknowns x and y. Suppose Д is a tree of $\mathfrak{A}$ whose construction does not involve the rule of multiplication by an unknown different from x and y (item 7), and suppose $\mathfrak{A}_1, \cdots, \mathfrak{A}_n$ are the final vertices of Д. Then it is possible to construct a finite number of word equations $\mathfrak{B}_1, \cdots, \mathfrak{B}_s$ in one unknown such that $\mathfrak{A}$ has solutions if and only if at least one of the equations $\mathfrak{A}_1, \cdots, \mathfrak{A}_n$, $\mathfrak{B}_1, \cdots, \mathfrak{B}_s$ has.*

The proof is the same as that of 6.22, only instead of 6.21 we use 6.21′.

By the projection of an equation $\Phi = \Psi$ ($\Phi \to \Psi$) on the alphabet **X** we mean the equation $\Phi^* = \Psi^*$ ($\Phi^* \to \Psi^*$), where Φ^* and Ψ^* are the projections of the words Φ and Ψ on **X**. The application of our results on trees of coefficient-free equations to equations with coefficients is based on the following assertion.

6.23. *Let $\mathfrak{A}^*$ denote the projection of an equation $\mathfrak{A}$ on the alphabet* **X**. *Then for any tree Д* of $\mathfrak{A}^*$ it is possible to construct a tree Д of $\mathfrak{A}$ such that the set of projections on* **X** *of the final vertices of Д coincides with the set of final vertices of Д*.*

PROOF. A tree Д which is constructed with respect to a given tree Д* will be said to correspond to Д*. We will construct Д by induction on the height $h(\text{Д}^*)$ of Д^*. For each final vertex $\mathfrak{B}$ of Д we will determine a corresponding final vertex $\mathfrak{B}^*$ of Д^* such that $\mathfrak{B}^*$ is the projection of $\mathfrak{B}$ on the alphabet of unknowns.

If $h(\text{Д}^*) = 0$, then the tree Д^* consists of one vertex, the equation $\mathfrak{A}^*$. The tree Д also consists of one vertex, the equation $\mathfrak{A}$. To the vertex $\mathfrak{A}$ corresponds the vertex $\mathfrak{A}^*$.

Now suppose there is given a tree Д^* of $\mathfrak{A}^*$ of height $h > 0$. Discard from Д^* the final vertices of the paths of length h (cf. 6.22). We obtain a tree Д_1^* of $\mathfrak{A}^*$ of height $h-1$. Let Д_1 be the tree of $\mathfrak{A}$ corresponding to Д_1^* (Д_1 is constructed by the induction assumption). We construct the desired tree Д in the following way. Take an arbitrary final vertex $\mathfrak{B}$ of Д_1, and its corresponding final vertex $\mathfrak{B}^*$ of Д_1^*. The equation $\mathfrak{B}^*$ is the projection of $\mathfrak{B}$ on the alphabet $\mathbf{X}$. Consider the vertex $\mathfrak{B}^*$. In accordance with the definition of a tree following 6.3 there are three possible cases.

1. $\mathfrak{B}^*$ is a final vertex of Д^*. Then $\mathfrak{B}$ is a final vertex of Д. The vertex $\mathfrak{B}^*$ corresponds to the vertex $\mathfrak{B}$ in Д.

2. In the construction of Д^* there occurs a transition from the vertex $\mathfrak{B}^*$ to a neighborhood of the equation $\mathfrak{B}^*$ in accordance with one of the items 1) –6) following 6.3.

The vertices $\mathfrak{B}_1^*, \cdots, \mathfrak{B}_k^*$ thus obtained are the final vertices of Д^*. Then in the construction of Д there occurs a transition from the vertex $\mathfrak{B}$ to a neighborhood of the equation $\mathfrak{B}$ in accordance with the matching item from among 1) –6) preceding 6.21. In this regard, before applying 3) to the equation $\mathfrak{B}$, remove the first coefficient in accordance with 7). The vertices $\mathfrak{A}_1, \cdots, \mathfrak{A}_m$ thus obtained are the final vertices of Д. In view of the invariance with respect to projection on the alphabet $\mathbf{X}$ of the operations of inversion, an admissible permutation, an n-tuple of substitutions, removing the first coefficient, passing from two-sided equations to one-sided equations, and taking a complete system of images, we conclude that to each vertex $\mathfrak{A}_i$ there corresponds a completely determined vertex $\mathfrak{B}_j^*$ which is the projection of $\mathfrak{A}_i$ on the alphabet $\mathbf{X}$. In the case of 1) we also note that the elimination bound of the parameter in transformations of an equation does not depend on the coefficients of the equation (see 4.8 and 4.8′).

3. In the construction of Д^* the vertex $\mathfrak{B}^*$ is replaced in accordance with the replacement rule. The replacing vertex is $\mathfrak{B}_1^*$. Then in the construction of Д the equation $\mathfrak{B}$ is also replaced in accordance with the replacement rule. We obtain an equation $\mathfrak{B}_1$. Obviously, $\mathfrak{B}_1^*$ is the projection of $\mathfrak{B}_1$ on the alphabet $\mathbf{X}$. The vertex $\mathfrak{B}_1^*$ corresponds to the vertex $\mathfrak{B}_1$. The proposition is proved.

6.24. Theorem. *For any one-sided equation in three unknowns it is possible to construct a tree of this equation whose final vertices have the form* $xA(y)z\cdots = zB(y)x\cdots$, *where* A *and* B *are words in the alphabet* $\mathbf{H}\cup\{y\}$ *containing the same number* a ($a \geq 0$) *of occurrences of* y (*a need not have the same value at each vertex*).

This follows from 6.18 and 6.23 plus, possibly, the rule of removing the first coefficient [before applying item 3), preceding 6.21, to equations with projections $xy^az\cdots = zy^ax\cdots$].

We formulate one more rule, which we will use in the construction of trees. We will call it the extended replacement rule. This rule consists of the following.

To the definition of a neighborhood preceding 6.21 we add item 8). If we have an equivalence

$$\exists xyz\mathfrak{A}(x, y, z) \Leftrightarrow \bigvee_{i=1}^{n} \exists xyz\mathfrak{A}_i(x, y, z) \vee \bigvee_{i=1}^{m} \exists uv\mathfrak{B}_i(u, v),$$

where $n \geqq 1$, $m \geqq 0$ and $\mathfrak{B}_1, \cdots, \mathfrak{B}_m$ are word equations in two unknowns, then the equations $\mathfrak{A}_1, \cdots, \mathfrak{A}_n$ form a neighborhood of $\mathfrak{A}$. In particular, any equation which is equivalent to $\mathfrak{A}$ forms a neighborhood of $\mathfrak{A}$. Moreover, this permits us to discard in disjunctions word equations in two unknowns.

The reduction theorem 6.22 remains valid, of course, for trees with the extended replacement rule. But Theorem 6.23, in general, is not true: the equivalence of projections does not imply the equivalence of the equations. Accordingly, 6.23 will be applied in the sequel if the coefficient-free tree $Д^*$ is constructed without the extended replacement rule.

We can now prove Theorem 6.25, which is the goal of §6.

Let us henceforth designate as *basic* any word equations of the form

$$(*)\qquad xA(y)z\Phi(x,y,z) \to zB(y)x\Psi(x,y,z),$$

where A and B are words in the alphabet $\mathbf{И} \cup \{y\}$ containing the same number ($\geqq 0$) of occurrences of y, and where $[A^\partial = [B^\partial$.

6.25. Theorem. *Suppose $\mathfrak{A}$ is a word equation in three unknowns. Then it is possible to construct a finite set $\mathfrak{A}_1, \cdots, \mathfrak{A}_n$ of one-stage exponential equations, word equations in two unknowns and basic equations such that $\mathfrak{A}$ has solutions if and only if at least one of the equations $\mathfrak{A}_1, \cdots, \mathfrak{A}_n$ has.*

Proof. By 4.10, it suffices to prove the assertion for one-sided equations. By 6.24 and 6.22 it suffices to obtain that in 6.24 the relation $[A^\partial = [B^\partial$ is satisfied. This is done in the following way. Consider a final vertex $xA(y)z\Phi = zB(y)x\Psi$ of 6.24. Put $d = [A^\partial - [B^\partial$. We restrict ourselves to the case $d > 0$. By 4.4(2), the equation $xAz\Phi = zBx\Psi$ is equivalent to the disjunction

$$\bigvee_{[R^\partial = d} (xAz = zBxR \,\&\, R\Phi = \Psi),$$

where R is a word in the alphabet $\mathbf{И}$ of length d. In the region $\partial(z) > d$ we have the equivalence

$$xAz = zBxR \Leftrightarrow \exists u(z = uR,\ xAu = uRBx),$$

where $[A^\partial = [BR^\partial$. Thus the equation $xAz\Phi = zBx\Psi$ has solutions in the region $\partial(z) > d$ if and only if the same is true of at least one of the equations

$$(1)\qquad xAuR\Phi(x, y, uR) = uRBx\Psi(x, y, uR).$$

But in the domain $\partial(z) \leq d$ the equation $xAz\Phi = zBx\Psi$ reduces in the obvious way to word equations in two unknowns. It remains to pass from the equations (1) to one-sided equations and use the extended replacement rule. The theorem is proved.

We state an analogous result for equations in two unknowns.

6.25′. *For any word equation $\mathfrak{A}$ in two unknowns it is possible to construct a finite set $\mathfrak{A}_1, \cdots, \mathfrak{A}_n$ of word equations in one unknown and word equations in two unknowns of the form $xAy\Phi(x,y) \to yBx\Psi(x,y)$, where A and B are coefficients, $[A^\partial = [B^\partial$, such that $\mathfrak{A}$ has solutions if and only if at least one of the equations $\mathfrak{A}_1, \cdots, \mathfrak{A}_n$ has.*

PROOF. By 4.10, it suffices to prove the assertion for one-sided equations. Suppose $\mathfrak{A}$ is a one-sided equation in two unknowns of the form $x \cdots \to y \cdots$. Let $\mathfrak{A}^*$ denote the projection of $\mathfrak{A}$ on the alphabet $\{x,y\}$. The equation $\mathfrak{A}$ has the form $x^a y \cdots \to y^b x \cdots$, where $a, b > 0$. From the argument in 6.18 (the second paragraph of the proof) it follows that the equation $x^a y \cdots \to y^b x \cdots$ possesses a tree with final vertices of the form $xy\Phi = yx\Psi$, where Φ and Ψ are words in the alphabet $\{x,y\}$. By 6.23, $\mathfrak{A}$ possesses a tree with final vertices of the form $xAy\Phi(x,y) = yBx\Psi(x,y)$, where A and B are coefficients, and Φ and Ψ are words in the alphabet $\mathbf{H} \cup \{x,y\}$. Using 6.22′, we obtain that 6.25′ is true if we do not consider the requirement $[A^\partial = [B^\partial$. It remains to obtain the fulfilment of the condition $[A^\partial = [B^\partial$. This is done as in 6.25, only now the "redundant" equations (in the region $\partial(y) \leq d$) will involve one unknown. The proposition is proved.

The number $\nu(A)$ of occurrences of y in the word A is called the characteristic of the basic equation $xAz\Phi \to zBx\Psi$. The following two propositions are used in the reduction of arbitrary basic equations to basic equations of characteristic ≤ 1.

6.26. *Any nondegenerate image of a basic equation of characteristic > 2 possesses a tree whose final vertices are basic equations of characteristic ≤ 2.*

PROOF. Suppose $\mathfrak{A}$ is a basic equation of characteristic > 2 and $\mathfrak{A}^*$ is its projection on the alphabet of unknowns. By 6.19, $\mathfrak{A}^*$ possesses a β-tree. As shown in 6.18 (the second through fifth paragraphs of the proof), any α-basic equation possesses a tree whose final vertices have the form $xz \cdots = zx \cdots$ or the form $xyz \cdots = zyx \cdots$. Thus $\mathfrak{A}^*$ possesses a tree whose final vertices have the form $xy^a z \cdots = zy^a x \cdots$, where $a \leq 2$. We now use 6.23, obtaining the condition $[A^\partial = [B^\partial$ as in 6.25.

6.27. *Any nondegenerate image of a basic equation of characteristic two possesses a tree whose final vertices are basic equations of characteristic ≤ 1.*

PROOF. This follows from 6.7 in the same way as 6.26 follows from 6.19.

In view of 6.25, the solution[12)] of arbitrary word equations in three unknowns is reduced to the solution of one-stage exponential equations, word equations in two unknowns and basic equations. The first we know how to solve (see 2.3′). By 6.25 and 1.19 the solution of the second is reduced to the solution of basic equations in two unknowns. There remain, consequently, basic equations.

In view of the conditions $\nu(A) = \nu(B)$ and $[A^\partial = [B^\partial$, the basic equation

[12)] Here and henceforth, if a certain class Ω of equations is distinguished, the words "to solve equations of the class Ω" mean "to construct an algorithm recognizing whether or not a given equation of the class Ω has solutions".

(*) is equivalent to the system composed of the equations $xAz \to zBx$ and $\Phi = \Psi$. One of the important questions arising in the study of basic equations is: under what conditions is this system equivalent to the equation $xAz \to zBx$? The elucidation of these conditions (more precisely, conditions of non-equivalence) comprises an essential part of our algorithm. The notion of a tree continues to play a fundamental role.

Of course, to solve our problem we need some information concerning the structure of the solutions of the equations $xAz \to zBx$. We extract this information from chains of transformations of $xAz \to zBx$. We distinguish two cases: $\nu(A) = 0$ and $\nu(A) > 0$ ($\nu(A)$ is the characteristic of the basic equation, the number of occurrences of y in the word A). When $\nu(A) > 0$ we can extract from the chains information involving all three components x, y and z. This permits us to investigate the equivalence. When $\nu(A) = 0$ the unknown y is not affected in chains of transformations of (*). Here we use the following direct method. Express x and z from the equation $xAz \to zBx$ by means of 5.9 if $A \neq B$, and by means of 1.23 if $A = B$. Substitute the expressions obtained for x and z in place of x and z in the equation $\Phi = \Psi$ and analyze the resulting equality. This has the form $P(\bar{x}, \bar{z})y \cdots = Q(\bar{x}, \bar{z})y \cdots$, where $P(x, z)$ and $Q(x, z)$ are prefixes of the equation $\Phi = \Psi$ not containing y, and $\bar{x}$ and $\bar{z}$ are the formulas for x and z obtained from 5.9 and 1.23. In order to express y it is necessary to annihilate the prefix on one side of the equation: divide $P(\bar{x}, \bar{z})$ by $Q(\bar{x}, \bar{z})$ (or Q by P). Proceeding thus to y (by induction if $P(\bar{x}, \bar{z})$ and $Q(\bar{x}, \bar{z})$ cancel entirely), we express y in terms of $\bar{x}$ and $\bar{z}$ (more precisely, in terms of the remainder from the division) by means of 1.16. We obtain an expression $y = \bar{y}$ as a function of certain variables. The final step is the analysis of the equality $\Phi(\bar{x}, \bar{y}, \bar{z}) = \Psi(\bar{x}, \bar{y}, \bar{z})$. When $A \neq B$ the main technique is related to the function F. When $A = B$ the main tool is the notion of exponential equations.

The cases where $\nu(A) > 0$ are handled in §§7 and 8. The main result of these sections is the reduction of arbitrary basic equations to basic equations of characteristic zero, basic equations without the tail $\Phi = \Psi$ (fundamental basic equations) and basic equations with special restrictions allowing us to reduce a basic equation to exponential equations (γ-equations). Equations of characteristic zero are considered in §9; γ-equations and fundamental basic equations, in §10.

§7. μ-images of basic equations

This section is devoted to the reduction of arbitrary basic equations to a narrower class of basic equations. Moreover, we construct an algorithm recognizing the solvability of word equations in two unknowns.

The main object of our study is the equation in $k + 2$ unknowns

$$xAz\Phi \to zBx\Psi, \tag{*}$$

where A and B are words in the alphabet $\text{И} \cup \{y_1, \cdots, y_k\}$, and Φ and Ψ are words in the alphabet $\text{И} \cup \{x, z, y_1, \cdots, y_k\}$. We will be interested in the cases $k = 0$ and $k = 1$. However, it is convenient to state and prove the main lemmas in the general case, without restrictions on k.

With each equation (*) we associate special equations of the form (*), the so-called μ-images of (*). They are obtained from (*) by substituting expressions of a special form in place of x and z in the words Φ and Ψ.

Suppose that $\mu \geqq 1$ is an integer. Put $\bar{x} = (xAzB)^\mu x$ and $\bar{z} = xAz$. Let $\overline{C}$ denote the word which is obtained from C by substituting $\bar{x}$ and $\bar{z}$ in place of the unknowns x and z. By definition, the μ-images of (*) are the equations $xAz\overline{\Phi} \to zBx\overline{\Psi}$ and $zBx\overline{\Psi} \to xAz\overline{\Phi}$ (μ is concealed in the bar on top).

We will indicate the role played by μ-images in the study of basic equations. Let us define nondegenerate images of order k of an equation $\mathfrak{A}_0$. Suppose $\mathfrak{A}_0, \cdots, \mathfrak{A}_k$ is a chain of images. We say that $\mathfrak{A}_k$ is a nondegenerate image of order k of $\mathfrak{A}_0$ if $\mathfrak{A}_0, \cdots, \mathfrak{A}_{k-1}$ is a degenerate chain but $\mathfrak{A}_k$ is a nondegenerate image of $\mathfrak{A}_{k-1}$.

7.1. *For any basic equation $\mathfrak{A}$ it is possible to construct a tree whose final vertices are either nondegenerate images of the first or second order of $\mathfrak{A}$ or μ-images of $\mathfrak{A}$.*

PROOF. Any basic equation has the form $xA(y)z\Phi(x,y,z) \to zB(y)x\Psi(x,y,z)$. We construct a tree of this equation as follows. Take a complete system of infinite images of the equation $xAz\Phi \to zBx\Psi$. The nondegenerate images are declared to be final vertices. The degenerate images have the form $vAz\Phi_1 \leftarrow zBv\Psi_1$, where Φ_1 and Ψ_1 are obtained from Φ and Ψ by substituting $(zB)^\mu v$ for x (μ is the parameter of the transformation; v is a new unknown). The word vA has the form $vA_1[yA_i]_{i=2}^n$, where $n \geqq 1$ and $A_1, \cdots, A_n$ are words in the alphabet $\mathbf{H}$. The n-tuple of substitutions $z = vA_1[yA_i]_{i=2}^q u$ $(q = 1, \cdots, n)$ reduces the equation $zBv\Psi_1 \to vAz\Phi_1$ to the equations $\mathfrak{A}_1, \cdots, \mathfrak{A}_n$. It is not difficult to see that $\mathfrak{A}_1, \cdots, \mathfrak{A}_{n-1}$ are nondegenerate images of the equation $zBv\Psi_1 \to vAz\Phi_1$ and, consequently, nondegenerate images of the second order of the equation $xAz\Phi \to zBx\Psi$. We declare them to be final vertices. The equation $\mathfrak{A}_n$ is obtained by the substitution $z = vAu$ and has the form $uBv\overline{\Psi} = vAu\overline{\Phi}$, where $\bar{x} = (vAuB)^\mu v$ and $z = vAu$. The one-sided equations corresponding to $\mathfrak{A}_n$ are, to within designation of the unknowns, μ-images of the equation $xAz\Phi \to zBx\Psi$. It remains to declare them to be final vertices. The proposition is proved.

When $\mathfrak{A}$ is a basic equation of characteristic $\geqq 2$ we obtain the following result.

7.2. *For any basic equation $\mathfrak{A}$ of characteristic > 2 (of characteristic two) it is possible to construct a tree whose final vertices either are μ-images of $\mathfrak{A}$ or are basic equations of characteristic $\leqq 2$ (of characteristic $\leqq 1$).*

PROOF. Indeed, in the tree constructed in 7.1 the final vertices which were not μ-images were nondegenerate images of the equation $xAz\Phi \to zBx\Psi$ or the equation $zBv\Psi_1 \to vAz\Phi_1$. Our assertion now follows from 6.26 and 6.27.

The tree constructed in 7.1 and 7.2 will be called a supporting tree of $\mathfrak{A}$.

Proposition 7.2 allows us to reduce an arbitrary basic equation to basic equations of characteristic $\leqq 1$ and μ-images of basic equations. Equations of characteristic $\leqq 1$ are considered in §§8 and 9. We now turn our attention to μ-images.

We will first prove that identically named μ-images of equivalent equations (*) are equivalent. More precisely, if $\mathfrak{A}$ and $\mathfrak{B}$ are equivalent equations of the form (*) with the same prefix $xAz \to zBx$, and $\mathfrak{A}'$ and $\mathfrak{B}'$ are μ-images of $\mathfrak{A}$ and $\mathfrak{B}$ obtained under the same μ and with the same direction of the arrow, then $\mathfrak{A}'$ and $\mathfrak{B}'$ are equivalent. To prove this we consider the sets $\Sigma = (\overline{X}, \overline{Z}, Y_1, \cdots, Y_k)$, $\Sigma' = (X, Z, Y_1, \cdots, Y_k)$ of words in the alphabet $\mathbf{H}$, where $\overline{X} = (XAZB)^\mu X$, $\overline{Z} = XAZ$, $[X^\partial \neq [Z^\partial$ and $X, Z \neq \Lambda$. We have

$$\overline{X}A\overline{Z} = (XAZB)^\mu XAXAZ, \quad \overline{Z}B\overline{X} = XAZB(XAZB)^\mu X = (XAZB)^\mu XAZBX.$$

We see that $\overline{X}A\overline{Z} = \overline{Z}B\overline{X}$ if and only if $XAZ = ZBX$. It is now evident that the set Σ is a solution of (*) if and only if the set Σ' is a solution of a μ-image (with same μ as in $\overline{X}$) of (*). The direction of the arrow in the μ-image depends on whether $[X^\partial > [Z^\partial$ or $[X^\partial < [Z^\partial$. If now Σ' is a solution of a μ-image $\mathfrak{A}'$, then, by what has been proved, Σ is a solution of $\mathfrak{A}$. In view of the equivalence of $\mathfrak{A}$ and $\mathfrak{B}$, the set Σ satisfies $\mathfrak{B}$. But then by what has been proved, taking into account that the μ-images $\mathfrak{A}'$ and $\mathfrak{B}'$ are identically named, the set Σ' is a solution of the μ-image $\mathfrak{B}'$. Thus the equivalence of the μ-images $\mathfrak{A}'$ and $\mathfrak{B}'$ has been proved. Therefore, identically named μ-images of equivalent equations (*) are equivalent.

In the study of μ-images the unknowns x and z, on one hand, and $y_1, \cdots, y_k$, on the other, play unequal roles. The unknowns x and z are active: new expressions are substituted for them. The unknowns $y_1, \cdots, y_k$ are not affected under the taking of a μ-image. This is what is meant by the "passive" role of $y_1, \cdots, y_k$ in the study of μ-images. The coefficients of (*) interest us in this section only when $k = 0$. Therefore, and only in this section, we alter the notion of a coefficient.

Let us define the coefficients of the equation (*).

The tail $\Phi = \Psi$ of (*) has the form

$$A_0[\xi_i A_i]_{i=1}^n = B_0[\eta_i B_i]_{i=1}^m. \tag{7.1}$$

where $n, m \geqq 0$; A_i and B_j are words in the alphabet $\mathbf{H} \cup \{y_1, \cdots, y_k\}$; and ξ_i and η_j are letters from the list $\{x, z\}$. The words $A, B, A_0, \cdots, A_n, B_0, \cdots, B_m$ are called the coefficients of (*). When $k = 0$ these are the usual coefficients, words in the alphabet $\mathbf{H}$. Throughout this section, words in the alphabet $\mathbf{H} \cup \{y_1, \cdots, y_k\}$ will be denoted by the letters A, B, C, D, P, Q, R (with or without indices). In another sense we will not use these letters.

We define a relation $\equiv$ (congruence) between words in the alphabet $\mathbf{H} \cup \{y_1, \cdots, y_k\}$. Suppose $P(y_1, \cdots, y_k)$ and $Q(y_1, \cdots, y_k)$ are words in the alphabet $\mathbf{H} \cup \{y_1, \cdots, y_k\}$. By definition, $P \equiv Q$ if $[P(Y_1, \cdots, Y_k)^\partial = [Q(Y_1, \cdots, Y_k)^\partial$ for any values $Y_1, \cdots, Y_k$ of the unknowns $y_1, \cdots, y_k$. It is easy to verify that $P \equiv Q$ if and only if (in the notation of §4) $\nu(P) = \nu(Q)$ and $[P^\partial = [Q^\partial$. Thus if (*) is a basic equation, then $A \equiv B$. It is also evident that if $P \equiv Q$, then the equations $P\Phi = Q\Psi$, $\Phi P = \Psi Q$ and the system $P = Q \,\&\, \Phi = \Psi$ are equivalent to each other.

Systems composed of equations $P_i = Q_i$, where P_i and Q_i are words in the alphabet $\mathbf{H} \cup \{y_1, \cdots, y_k\}$, and where $P_i \equiv Q_i$ for all i, are called τ-systems. Obviously, any τ-system is equivalent to a τ-system consisting of a single τ-equation.

We have already stated that if (*) is a basic equation, then $A \equiv B$. Henceforth, when considering (*), we always assume that $A \equiv B$. Under this assumption (*) is equivalent to the system consisting of the equations $xAz \rightarrow zBx$ and $\Phi = \Psi$. It follows that if in the tail $\Phi = \Psi$ of (*) we replace an occurrence of xAz by zBx, or conversely, then the resulting equation will be equivalent to the original. We make use of this property below.

We define for equation (*) the property α. We say that (*) possesses property α if it is equivalent to a system $xAzP\xi\cdots\rightarrow zBxQ\cdots \,\&\, \mathfrak{B}$, where ξ and η are letters from the list $\{x,z\}$; P and Q are words in the alphabet $\mathbf{H}\cup\{y_1,\cdots,y_k\}$, $P\not\equiv Q$, and $\mathfrak{B}$ is a τ-system. We remark that in this definition thc factor $\mathfrak{B}$ can be discarded, since a τ-system $\mathfrak{B}$ is equivalent to a τ-equation $C=D$, where $C\equiv D$. Now multiply the equation $xAzP\xi\cdots\rightarrow zBxQ\eta\cdots$ on the right by the equation $C=D$ (i.e. to the left-hand side of the equation attach C on the right, and to the right-hand side D, or the other way around). The resulting equation has the form $xAzP\xi\cdots\rightarrow zBxQ\eta\cdots$ and is equivalent to the original system.

By fundamental basic equations we mean equations of the form $xAzP\rightarrow zBxQ$, where $A\equiv B$, $P\equiv Q$, and the words A, B, P and Q do not contain the letters x and z.

We will now prove two lemmas concerning μ-images.

7.3. *Suppose $A\equiv B$ and $\mathfrak{A}$ is a μ-image of the equation* (*). *Then one of the following four assertions is true*:

1) *$\mathfrak{A}$ is equivalent to a fundamental basic equation with the same A and B as in* (*).

2) *$\mathfrak{A}$ has no solutions.*

3) *$\mathfrak{A}$ possesses property α.*

4) *Any solution of $\mathfrak{A}$ satisfies the equation $A=B$.*

PROOF. The μ-images of (*) have the form $xAz\bar\Phi\rightarrow zBx\bar\Psi$, $xAz\bar\Phi\leftarrow zBx\bar\Psi$, where $\bar x=(xAzB)^\mu x$, $\bar z=xAz$, and the bar over a word denotes the result of substituting $\bar x$ and $\bar z$ into this word in place of x and z. Consider the tail $\bar\Phi=\bar\Psi$.

The equation $\bar\Phi=\bar\Psi$ has the form (see (71))

$$A_0\,[\xi_iA_i]_{i=1}^n = B_0\,[\eta_iB_i]_{i=1}^m, \tag{1}$$

where $n,m\geqq 0$; A_i and B_j are the coefficients, and ξ_i and η_j are x or z.

The number of occurrences of x and z in the word $\Phi\Psi$ is called the height of equation (*). We apply induction on the height $h=n+m$ of (*).

If $h=0$, then $n=m=0$ and $\mathfrak{A}$ has the form $xAzA_0\rightleftarrows zBxB_0$ (i.e. the form $xAzA_0\rightarrow zBxB_0$ or the form $zBxB_0\rightarrow xAzA_0$). If $A_0\equiv B_0$, then 1) holds. If $A_0\not\equiv B_0$, then 3) holds, since equation (1) is equivalent to the equation $A_0x=B_0x$. The lemma is proved for $h=0$.

Consider the case $h>0$. Suppose one of n or m is zero. Then the other is greater than zero. If $A_0\not\equiv B_0$, then 3) holds, since (1) has the form (after multiplication on the right by x or z) $A_0u\cdots=B_0v\cdots$, where u and v are x or z. If $A_0\equiv B_0$, then one of the components x or z of any solution of $\mathfrak{A}$ is empty. The equation $\mathfrak{A}$ has no solutions (because of the arrow), so that 2) holds.

Consider the case $n,m>0$. Compare the unknowns ξ_1 and η_1.

Suppose $\xi_1=\eta_1$. If $A_0\not\equiv B_0$, then obviously 3) holds. Suppose $A_0\equiv B_0$. Then any solution of (*) satisfies the equation $A_0=B_0$. But then (*) is equivalent to the equation $xAz\Phi_1\rightarrow zBx\Psi_1$, where $\Phi_1=\Psi_1$ is the equation

$A_1[\xi_i A_i]_{i=2}^n A_0 = B_1[\eta_i B_i]_{i=2}^m B_0$. The equation $xAz\Phi_1 \to zBx\Psi_1$ is an equation of the form (*) of height $h-2$, so that the induction assumption is applicable to it. In view of the equivalence of μ-images of equivalent equations, $\mathfrak{A}$ is equivalent to an identically named μ-image $\mathfrak{A}_1$ of the equation $xAz\Phi_1 \to zBx\Psi_1$. By the induction assumption, we have for $\mathfrak{A}_1$ an assertion i), where $1 \leq i \leq 4$. But then i) also holds for $\mathfrak{A}$ (this is obvious if we take into account that $\mathfrak{A}_1$ and $\mathfrak{A}$ are equivalent). Hence if $\xi_1 = \eta_1$, the lemma is true.

Suppose $\xi_1 \neq \eta_1$. Without loss of generality, we may assume that $\xi_1 = x$ and $\eta_1 = z$. Then the equation $\overline{\Phi} = \overline{\Psi}$ has the form

$$A_0\,(\bar{z}B)^{\mu} x A_1\,[\bar{\xi}_i A_i]_{i=2}^n = B_0\,[\bar{z}B_i]_{i=1}^q\,[\bar{\eta}_i B_i]_{i=q+1}^m, \tag{2}$$

where $1 \leq q \leq m$, and either $q = m$ or $\eta_{q+1} = x$.

There are two possible cases.

Case 1. $\mu \geq q$. We have $x \angle \bar{z}$. Using the obvious induction on q, we deduce that either 3) holds or $A_0 \equiv B_0$ and $B_i \equiv B$ $(i = 1, \cdots, q-1)$, and in the second alternative any solution of (2) satisfies the τ-system $A_0 = B_0$, $B_i = B$ $(i = 1, \cdots, q-1)$. The interest lies in the second alternative. In that case (2) is equivalent to the system

$$B\,(\bar{z}B)^{\mu-q} x A_1\,[\bar{\xi}_i A_i]_{i=2}^n = B_q\,[\bar{\eta}_i B_i]_{i=q+1}^m \;\&\; \mathfrak{B}_0, \tag{3}$$

where $\mathfrak{B}_0$ is the τ-system $A_0 = B_0$, $B_i = B$ $(i = 1, \cdots, q-1)$.

We have either $q = m$ or $\eta_{q+1} = x$.

Consider the case $q = m$. Then $[\bar{\eta}_i B_i]_{i=q+1}^m$ is empty. If $B \equiv B_q$, then the component x of any solution of the system (3) is empty and $\mathfrak{A}$ has no solutions (because of the arrow): 2) holds. If $B \not\equiv B_q$, then we write the equation in (3) in the form $Bx \cdots = B_q x \cdots$ (after multiplication on the right by x) and we see that 3) holds.

Consider the case $\eta_{q+1} = x$. We have $x \angle \bar{x}$. Hence the left-hand side of the equation in (3) begins with the word Bx, and the right-hand side with $B_q x$. If now $B \not\equiv B_q$, then 3) holds. Suppose $B \equiv B_q$. Then any solution of the system (3) satisfies the equation $B = B_q$. We obtain that any solution of $\mathfrak{A}$ satisfies the τ-system $B_i = B$ $(i = 1, \cdots, q)$, $A_0 = B_0$. Denote it by $\mathfrak{B}$. Since $\eta_{q+1} = x$, $\mathfrak{A}$ has the form

$$xAzA_0\bar{x}A_1\,[\bar{\xi}_i A_i]_{i=2}^n \rightleftarrows zBxB_0\,[\bar{z}B_i]_{i=1}^q \bar{x} B_{q+1}\,[\bar{\eta}_i B_i]_{i=q+2}^m.$$

Since any solution of $\mathfrak{A}$ satisfies the system $\mathfrak{B}$, $\mathfrak{A}$ is equivalent to the equation

$$xAz\bar{x}A_1\,[\bar{\xi}_i A_i]_{i=2}^n P \rightleftarrows zBx(\bar{z}B)^q \bar{x} B_{q+1}\,[\bar{\eta}_i B_i]_{i=q+2}^m Q, \tag{4}$$

where $P = Q$ is the τ-equation equivalent to the τ-system $\mathfrak{B}$. The word $(\bar{z}B)^q\bar{x}$ has the form $(xAzB)^{q+\mu}x = (xAzB)^{\mu}x(AzBx)^q$. Replacing in the factor $(AzBx)^q$ the word zBx by xAz, we obtain the word $(xAzB)^{\mu}x(AxAz)^q$, which is equal to $\bar{x}(A\bar{z})^q$. It is obvious that if in (4) we replace zBx by xAz, we obtain an equation equivalent to the original one (since $A \equiv B$, any solution of (4) satisfies the equation $xAz = zBx$). Hence if in (4) we replace $(\bar{z}B)^q\bar{x}$ by $\bar{x}(A\bar{z})^q$, then the resulting equation $\mathfrak{A}'$ is equivalent to (4). The equation $\mathfrak{A}'$ has the form $xAz\bar{x}\Phi' \rightleftarrows zBx\bar{x}\Psi'$. Throwing out $\bar{x}$, we obtain an equation $\mathfrak{A}''$ equivalent to $\mathfrak{A}'$. The equation $\mathfrak{A}''$ has the form

$$xAzA_1\,[\bar{\xi}_i A_i]_{i=2}^n P \rightleftarrows zBx\,(A\bar{z})^q B_{q+1}\,[\bar{\eta}_i B_i]_{i=q+2}^m Q$$

and is a μ-image of the equation

$$xAzA_1\ [\xi_i A_i]_{i=2}^n P \rightarrow zBx\ (Az)^q B_{q+1}\ [\eta_i B_i]_{i=q+2}^m Q.$$

The latter is an equation (*) of height $h-2$. By the induction assumption, we have for $\mathfrak{A}''$ an assertion i), where $1 \leqq i \leqq 4$. The same is true of the equation $\mathfrak{A}$, which is equivalent to $\mathfrak{A}''$. Our assertion is proved when $\mu \geqq q$.

Case 2. $\mu < q$. Using the obvious induction on μ, we see that either 3) holds or $A_0 \equiv B_0$ and $B_i \equiv B$ $(i = 1, \cdots, \mu)$, and in the second alternative any solution of $\mathfrak{A}$ satisfies the τ-system $A_0 = B_0$, $B_i = B$ $(i = 1, \cdots, \mu)$. Consider the second alternative. In that case (2) is equivalent to the system (cancel (2) on the left by $A_0(\bar{z}B)^\mu x$)

$$A_1\ [\xi_i A_i]_{i=2}^n = AzB_{\mu+1}\ [\eta_i B_i]_{i=\mu+2}^m\ \&\ \mathfrak{B}_1, \tag{5}$$

where $\mathfrak{B}_1$ is the τ-system $A_0 = B_0$, $B_i = B$ $(i = 1, \cdots, \mu)$.

If $n = 1$, then when $A \equiv A_1$ we have 2), and if $A \not\equiv A_1$ we have 3).

Suppose $n > 1$. Then either $A \equiv A_1$ or 3) holds.

Consider the case $A \equiv A_1$. Then any solution of the system (5) satisfies the equation $A = A_1$ and (5) is equivalent to the system

$$[\xi_i A_i]_{i=2}^n = zB_{\mu+1}\ [\eta_i B_i]_{i=\mu+2}^m\ \&\ \mathfrak{B}_2, \tag{6}$$

where $\mathfrak{B}_2$ is the τ-system $\mathfrak{B}_1 \& (A = A_1)$.

There are two possible cases: $\xi_2 = x$ or $\xi_2 = z$.

Consider the case $\xi_2 = x$. The tail $\overline{\Phi} = \overline{\Psi}$ of $\mathfrak{A}$ is equivalent to the system (6). Since $\xi_2 = x$, the equation in (6) has the form $xAzBx \cdots = zB_{\mu+1} \cdots$. Replacing here xAz by zBx and cancelling the resulting equation on the left by z, we deduce that the equation $xAz\overline{\Phi} \rightleftarrows zBx\overline{\Psi}$, i.e. $\mathfrak{A}$, is equivalent to a system $xAz\Phi_1 \rightleftarrows zBx\Psi_1 \& \mathfrak{B}_2$, where the equation $\Phi_1 = \Psi_1$ has the form $BxBx \cdots = B_{\mu+1}[\eta_i B_i]_{i=\mu+2}^m$. If here $m = \mu+1$, then when $B \equiv B_{\mu+1}$ we have 2), and when $B \not\equiv B_{\mu+1}$ we have 3). If $m > \mu+1$, then the equation $\Phi_1 = \Psi_1$ has the form $BxBx \cdots = B_{\mu+1}xAz \cdots$. If $B \not\equiv B_{\mu+1}$, then 3) holds. If $B \equiv B_{\mu+1}$, then since $A \equiv B$ any solution of the equation $BxBx \cdots = B_{\mu+1}xAz \cdots$ (and, consequently, the equation $\mathfrak{A}$) satisfies the equation $A = B$. But then 4) holds. Hence if $\xi_2 = x$, then 4) holds.

Consider the case $\xi_2 = z$. Then $\mathfrak{A}$ has the form

$$xAzA_0\bar{x}A_1\bar{z}A_2\ [\xi_i A_i]_{i=3}^n \rightleftarrows zBxB_0\bar{z}B_1\ [\eta_i B_i]_{i=2}^m.$$

In this case any solution of $\mathfrak{A}$ satisfies the τ-system $A_0 = B_0$, $A_1 = A$. Hence $\mathfrak{A}$ is equivalent to the equation

$$xAz\bar{x}A\bar{z}A_2\ [\xi_i A_i]_{i=3}^n A_0 A_1 \rightleftarrows zBx\bar{z}B_1\ [\eta_i B_i]_{i=2}^m B_0 A. \tag{7}$$

We have $\bar{x}A\bar{z} = (xAzB)^\mu xAxAz$. Replacing the last occurrence of xAz by zBx, we obtain the word $\bar{z}B\bar{x}$. Hence replacing in the left-hand side of (7) the word $\bar{x}A\bar{z}$ by $\bar{z}B\bar{x}$, we obtain an equation $\mathfrak{A}'$ which is equivalent to $\mathfrak{A}$. The equation $\mathfrak{A}'$ has the form $xAz\bar{z}\Phi' \rightleftarrows zBx\bar{z}\Psi'$. Throwing out $\bar{z}$, we obtain an equation $\mathfrak{A}''$ which is equivalent to $\mathfrak{A}$. The equation $\mathfrak{A}''$ has the form

$$xAzB\bar{x}A_2\ [\xi_i A_i]_{i=3}^n A_0 A_1 \rightleftarrows zBxB_1\ [\eta_i B_i]_{i=2}^m B_0 A$$

and is a μ-image of the equation

$$xAzBxA_2\ [\xi_i A_i]_{i=3}^n A_0 A_1 \rightarrow zBxB_2\ [\eta_i B_i]_{i=2}^m B_0 A.$$

The latter is an equation (*) of height $h-2$. By the induction assumption, we obtain for $\mathfrak{A}''$ an assertion i), where $1 \leqq i \leqq 4$. The same is true of $\mathfrak{A}$, which is equivalent to $\mathfrak{A}''$. Our assertion is proved when $\mu < q$. The proposition is proved.

Along with the one-sided equation (*) it is convenient to consider the two-sided equation obtained from it by replacing the symbol $\rightarrow$ by the symbol $=$.

We will prove one more lemma concerning μ-images of (*). It has the same character as 7.3, only the congruence relation between the coefficients is replaced by graphical equality, and instead of one-sided μ-images we consider two-sided ones.

We define for the equation (*) a property α'. We say that an equation (*) possesses property α' if it is equivalent to an equation $xAzP\xi\cdots \rightarrow zBxQ\eta\cdots$, where ξ and η are letters from the list $\{x, z\}$, and P and Q are coefficients of (*) such that $P \neq Q$. Analogously, by replacing $\rightarrow$ by $=$ we define the property α' for two-sided equations (*). The property α' differs from the property α in that in place of $\not\equiv$ stands the symbol $\neq$. Obviously, if an equation possesses property α, it also possesses property α'. But the equation $xyzay \rightarrow zyxyz$, where a is a letter of the alphabet $\mathbf{И}$, possesses property α' but not property α. Note that if $A \neq B$ and (*) has no solutions, then it possesses property α', since the equation $xAzAx \rightarrow zBxBx^2$ also has no solutions.

We say that an equation (*) is an ϵ-equation if after multiplication on the left the equation $\Phi = \Psi$ has the form $x\cdots = z\cdots$, but does not have the form $xA\xi\cdots = zB\eta\cdots$, where ξ and η are x or z.

7.4. *Suppose $A \equiv B$ but $A \neq B$. The following assertions are true.*

a) *Either* (*) *is equivalent to the equation $xAz \rightarrow zBx$ or any two-sided μ-image of* (*) *possesses property α'.*

b) *If* (*) *is an ϵ-equation, then any two-sided μ-image of* (*) *possesses property α'.*

Proof. The idea behind the argument is the same as in 7.3. We prove a) by induction on the height h of (*). We obtain b) in the course of the argument.

We will assume that the tail $\Phi = \Psi$ has the form (7.1). It is obvious that if $A_0 \neq B_0$, then any μ-image of (*) possesses property α', and a) and b) are true. We will assume that $A_0 = B_0$.

We proceed by induction on h. If $h = 0$, then $n = m = 0$. Since $A_0 = B_0$, the tail $\Phi = \Psi$ is trivial and (*) is equivalent to the equation $xAz \rightarrow zBx$. Suppose $h > 0$. If one of n or m is zero, then the other is greater than zero. Since $A_0 = B_0$, the components x and z of solutions of μ-images of (*) are empty. In view of the convention in the section on trees of equations with coefficients, the μ-images have no solutions and, consequently, possess property α'. Suppose $n, m > 0$. If $\xi_1 = \eta_1$, then a) obviously follows from the induction assumption. Consider the case $\xi_1 \neq \eta_1$. Without loss of generality, we will assume that $A_0 = B_0 = \Lambda$, $\xi_1 = x$ and $\eta_1 = z$. The equation $\overline{\Phi} = \overline{\Psi}$ can be written in the form (we know $\Phi = \Psi$ has the form (7.1))

$$(\bar{z}B)^{\mu} x A_1 \, [\bar{\xi}_i A_i]_{i=2}^{n} = (\bar{z}B)^{q} [\bar{z}B_i]_{i=q+1}^{k} \, [\bar{\eta}_i B_i]_{i=k+1}^{m}, \tag{1}$$

where $k > 0$, $0 \leqq q \leqq k \leqq m$, and where if $k > q$, then $B_{q+1} \neq B$, and if $m > k$, then $\eta_{k+1} = x$. We consider two cases as preliminary.

1) $k = q$, $m > k$ and $\eta_{k+1} = x$. Then the equation $\Phi = \Psi$ has the form $xA_1[\xi_i A_i]_{i=2}^n = (zB)^q xB_{q+1}[\eta_i B_i]_{i=q+2}^m$. Replacing zBx in the word $(zB)^q x$ by xAz q times, we obtain the word $x(Az)^q$. Now substituting $x(Az)^q$ for $(zB)^q x$ in the equation $\Phi = \Psi$ and cancelling the resulting equation on the left by x, we obtain the equation $A_1[\xi_i A_i]_{i=2}^n = (Az)^q B_{q+1}[\eta_i B_i]_{i=q+2}^m$. Denote it by $\Phi_1 = \Psi_1$. The equation (*), taking into account the nature of our transformations, is equivalent to the equation $xAz\Phi_1 \to zBx\Psi_1$. The induction assumption is applicable to the latter. Applying it and using the equivalence of two-sided μ-images of equivalent equations (*), we see that a) is true. (The equivalence of two-sided μ-images is stated and proved exactly as for one-sided ones.)

2) $A_1 = A$ and $\xi_2 = z$. Then the equation $\Phi = \Psi$ has the form $xAzA_2[\xi_i A_i]_{i=3}^n = zB_1[\eta_i B_i]_{i=2}^m$. Replacing xAz in the left-hand side by zBx and cancelling on the left by z, we obtain the equation $BxA_2[\xi_i A_i]_{i=3}^n = B_1[\eta_i B_i]_{i=2}^m$. Denote it by $\Phi_1 = \Psi_1$ and continue the argument exactly as in case 1) following the formula $\Phi_1 = \Psi_1$.

We will write $\alpha'(\mu)$ if a two-sided μ-image of the (given) equation (*) possesses property α'. We will prove that if 1) and 2) do not hold, then for any $\mu \geqq 1$ we have $\alpha'(\mu)$. Since $q \leqq k$, there are three possible cases.

I) $\mu > q$ and $k > q$. Then after cancelling on the left by $(\bar{z}B)^q\bar{z}$, equation (1) assumes the form (since $x \angle \bar{z}$) $Bx \cdots = B_{q+1}[\bar{\eta}_i B_i]_{i=q+2}^m$. The latter has the form (possibly after multiplication on the right by x) $Bx \cdots = B_{q+1}x \cdots$, and since $B \neq B_{q+1}$ we have $\alpha'(\mu)$.

II) $\mu \geqq q$ and $k = q$. After multiplication on the left by $(\bar{z}B)^q$, equation (1) assumes the form $x \cdots = [\bar{\eta}_i B_i]_{i=k+1}^m$. If $m > k$, then by definition of k we have $\eta_{k+1} = x$, which is case 1). Hence $m = q$. But then $\alpha'(\mu)$, and the component x of any solution of the equation $\bar{\Phi} = \bar{\Psi}$ is empty and the μ-images have no solutions.

III) $\mu \leqq q$ and $\mu < k$. Cancelling (1) on the left by $(\bar{z}B)^\mu x$, we obtain the equation $A_1[\bar{\xi}_i A_i]_{i=2}^n = AzB_{\mu+1}[\bar{\eta}_i B_i]_{i=\mu+2}^m$. If $n = 1$, then the component z of any solution of the equation $\bar{\Phi} = \bar{\Psi}$ is empty and the μ-images have no solutions; hence $\alpha'(\mu)$. Suppose $n > 1$. If $A \neq A_1$, then since $x \angle \bar{\xi}_2$ we have $\alpha'(\mu)$. Consider the case $A = A_1$. If $\xi_2 = z$, we have 2). Hence $\xi_2 = x$. After multiplications on the left by $(\bar{z}B)^\mu xA$, equation (1) has the form (since $xAzBx \angle \bar{x}$) $xAzBx \cdots = zB_{\mu+1}[\bar{\eta}_i B_i]_{i=\mu+1}^m$. Replacing xAz in the left-hand side by zBx and cancelling on the left by z, we obtain $BxBx \cdots = B_{\mu+1}[\bar{\eta}_i B_i]_{i=\mu+2}^m$. Denoting this by $\bar{\Phi}_1 = \bar{\Psi}_1$, we see that the μ-image is equivalent to the equation $xAz\bar{\Phi}_1 = zBx\bar{\Psi}_1$. If now $m = \mu + 1$, then x is empty for any solution of the μ-image, and hence $\alpha'(\mu)$. If $m > \mu + 1$, then the equation $\bar{\Phi}_1 = \bar{\Psi}_1$ has the form $BxBx \cdots = B_{\mu+1}xAz \cdots$. If now $B \neq B_{\mu+1}$, then $\alpha'(\mu)$, and if $B = B_{\mu+1}$, then since $A \neq B$ we again have $\alpha'(\mu)$.

Hence in all cases except 1) and 2) we have $\alpha'(\mu)$ for any $\mu \geqq 1$. Thus a) is proved.

It is easy to prove that in cases 1) and 2), if (*) is an ϵ-equation, we also have $\alpha'(\mu)$ for any $\mu \geqq 1$. Indeed, in case 1) the equation $\Phi = \Psi$ has the form $xA_1[\xi_i A_i]_{i=2}^n = (zB)^q xB_{q+1} \cdots$. The equation (*) can be an ϵ-equation only in two cases: when $A_1 \neq A$ and when $n = 1$. As was shown in case 1), the equation (*) is equivalent to the equation $xAz\Phi_1 \to zBx\Psi_1$, where $\Phi_1 = \Psi_1$ is the equation $A_1[\xi_i A_i]_{i=2}^n = (Az)^q B_{q+1} \cdots$. In view of the equivalence of μ-images of equivalent equations, any μ-image of (*) is equivalent to $xAzA_1[\xi_i A_i]_{i=2}^n = zBx(Az)^q B_{q+1} \cdots$. If $n > 1$ and $A_1 \neq A$, then obviously $\alpha'(\mu)$ for any $\mu \geqq 1$. If $n = 1$, then, considering the cases $A_1 = A$ and $A_1 \neq A$, we see that $\alpha'(\mu)$ for any $\mu \geqq 1$.

Consider the case 2). The equation $\Phi = \Psi$ has the form $xAzA_2 \cdots = zB_1[\eta_i B_i]_{i=2}^m$. The equation (*) can be an ϵ-equation only in two cases: when $B_1 \neq B$ and when $m = 1$. Replacing the word xAz in the equation $\Phi = \Psi$ by zBx and arguing as in case 1), we see that if $m = 1$ or $m > 1$, $B_1 \neq B$, then for any $\mu \geqq 1$ we have $\alpha'(\mu)$. Thus b) is proved. The proposition is proved.

We will apply Lemmas 7.3 and 7.4 in the cases $k = 0$ and $k = 1$. This will permit us to solve word equations in two unknowns (see 7.7) and also to reduce the solution of arbitrary basic equations to the solution of basic equations of a special form (see 7.12). We begin with the case $k = 0$.

The case $k = 0$. Equations in two unknowns

Let us first explain what property α' means when $k = 0$. When $k = 0$ the coefficients of (*) are words in the alphabet $\mathbf{H}$. It follows from the definition of property α' that if $k = 0$ and (*) possesses property α', then any solution (X, Z) of (*) satisfies the equality $PX_1V_1 = QX_2V_2$, where X_1 and X_2 are X or Z, and P and Q are coefficients of (*), not depending on X and Z, such that $P \neq Q$. Cancelling $PX_1V_1 = QX_2V_2$ on the left, we obtain $RX_1V_1 = X_2V_2$, where $R \neq \Lambda$ since $P \neq Q$. Because of the arrow, we have $Z \angle X$ for any solution of (*). Thus the component Z of any solution (X, Z) of (*) satisfies $RZV_1 = ZV_2$ for some V_1 and V_2 and some fixed R. Here $[R^\partial \leq M$, where M is the maximum coefficient length of (*). Since $R \neq \Lambda$, it follows from $RZV_1 = ZV_2$, in view of 1.16, that $Z = R^t R_1$, where $t \geqq 0$ and $R_1 \angle R$, i.e. we have $\tau(Z, [R^\partial)$ (the definition of the predicate τ follows 4.2). Hence if (*) with $k = 0$ possesses property α', then for any solution (X, Z) we have $\tau(Z, M)$, where M is the maximum coefficient length of (*).

7.5. *Consider the word equation in two unknowns*

$$(*) \qquad xAz\Phi(x, z) \to zBx\Psi(x, z),$$

where A and B are words in the alphabet $\mathbf{H}$, $A \neq B$, $[A^\partial = [B^\partial$; and Φ and Ψ are words in the alphabet $\mathbf{H} \cup \{x, y\}$. Let M denote the maximum coefficient length of (*). Then one and only one of the following two assertions holds:

1) *The equation (*) is equivalent to the equation $xAz \to zBx$.*

2) *If $\mathfrak{A}(x, z)$ is a μ-image of (*) (one-sided or two-sided), then any solution of $\mathfrak{A}$ satisfies the condition $\tau(x, M) \vee \tau(z, M)$.*

PROOF. Note that the coefficients of μ-images of (*) are coefficients of (*). Applying 7.4 a) with $k = 0$ and taking into account the meaning of property α' when $k = 0$, we see that 1) or 2) holds.

Let us prove that 1) and 2) are mutually exclusive. Indeed, any two-sided μ-image of the equation $xAz \to zBx$ has the form $xAz = zBx$. In view of the equivalence of identically named μ-images of equivalent equations, we obtain that in case 1) any two-sided μ-image of (*) is equivalent to the equation $xAz = zBx$. We will prove that if $A \neq B$ and the equation $xAz = zBx$

has solutions, then it has solutions (x,z) such that $\urcorner\, \tau(x,M)\,\&\,\urcorner\, \tau(z,M)$. By 5.9 it suffices to prove that if $\mathscr{L}(QP,QR)$ (this condition is equivalent in 5.9 to $A \neq B$), then for sufficiently large k and for $X = G(QP,QR,\lambda_0,\cdots,\lambda_k)Q$ we have $\urcorner\, \tau(X,M)$. Let us prove this.

The word X is representable in the form $X = T^{\lambda}T_1$, where $\lambda > 1$, T is a simple word, and $T_1 \angle T$. Indeed (formula (5.2)), $X = F_k^{\lambda_k}G_{k-1}Q$. By 5.12, we may assume that F_k is simple. Obviously, $G_{k-1}Q \angle F_k$. If $\lambda_k > 1$, then everything is in order. If $\lambda_k = 1$, then by 5.11, (5.2) we have

$$X = G(QP,QR,\lambda_0,\cdots,\lambda_{k-2},\lambda_{k-1}+1)Q = F_{k-1}^{\lambda_{k-1}+1}G_{k-2}Q,$$

where again F_{k-1} is simple and $G_{k-2}Q \angle F_{k-1}$. The representation $X = T^{\lambda}T_1$, $\lambda > 1$, is proved. If we now assume $\tau(X,M)$, then $X = S^tS_1$, where $t \geqq 0$, $S_1 \angle S$, S is simple, and $[S^{\partial} \leqq M$. Since $[X^{\partial}$ is sufficiently large, we have $t > 1$, and so $X = T^{\lambda}T_1 = S^tS_1$. By 1.7 we must have $S = T$, which is impossible, since $[T^{\partial}$ is sufficiently large for large k. We have proved that if $\mathscr{L}(QP,QR)$ and k is sufficiently large, then $\urcorner\, \tau(X,M)$. Thus our assertion concerning the equation $xAz = zBx$ is proved.

The proposition is proved.

In §11 we use the following form of Theorem 7.5.

7.6. *Consider the word equation in two unknowns*

$$(*) \qquad xAz\Phi(x,z) \to zBx\Psi(x,z),$$

where A and B are words in the alphabet $\mathbf{И}$, $A \neq B$, and $[A^{\partial} = [B^{\partial}$. Let M be the maximum coefficient length of the equation, and suppose the equation has a solution $(\overline{X},\overline{Z})$ such that $\overline{X} = (\overline{Z}B)^{\mu}X$, $\overline{Z} = XAZ$, where $\mu \geqq 1$, and the words X and Z satisfy the condition $\urcorner\, \tau(X,M)\,\&\,\urcorner\, \tau(Z,M)$. Then (*) is equivalent to the equation $xAz \to zBx$.

PROOF. The pair (X,Z), as is easily verified, satisfies the equation $xAz\Phi(\bar{x},\bar{z}) = zBx\Psi(\bar{x},\bar{z})$, where $\bar{x} = (xAzB)^{\mu}x$ and $\bar{z} = xAz$, i.e. a two-sided μ-image of (*). Since $\urcorner\, \tau(X,M)\,\&\,\urcorner\, \tau(Z,M)$, we see that 7.5.2) is impossible. There remains only the possibility 7.5.1), as required.

The proposition is proved.

Using 7.5 we will prove the following theorem, first established by A. A. Markov.

7.7. THEOREM. *There exists an algorithm recognizing the solvability of word equations in two unknowns.*

PROOF. In view of 6.25′ and 1.20 it suffices to construct an algorithm recognizing the solvability of equations of the form

$$(1) \qquad xAy\Phi(x,y) \to yBx\Psi(x,y),$$

where A and B are coefficients, $[A^{\partial} = [B^{\partial}$, and Φ and Ψ are words in the alphabet $\mathbf{И} \cup \{x,y\}$.

Consider equation (1). There are two possible cases: $A = B$ or $A \neq B$.

First suppose $A \neq B$. Apply Proposition 7.5 to (1). According to 7.5, either 7.5.1) or 7.5.2) holds. Moreover, the proof of 7.5 (like any other existence statement in this paper) allows us to effectively determine for a given equation 7.5 (*) which of 1) or 2) is actually true. If 7.5.1) holds, we take the desired algorithm from 5.10. Suppose 7.5.2) holds. Then we construct a supporting tree Д of (1) in accordance with 7.1. Any infinite transformation of (1), as is easily verified, is degenerate. Hence all final vertices of Д are μ-images of (1). According to 6.22′ and 1.20, it remains to solve these μ-images. Any μ-image of (1) has the form (1) to within designation of the unknowns x and y, the others Φ and Ψ, and, possibly, an interchange of the places A and B. In the case 7.5.2) any solution (X, Y) of a μ-image (1) satisfies the condition $\tau(Y, M)$. Hence it remains for us to construct an algorithm recognizing the solvability of those equations of the form (1), each of whose solutions satisfies $\tau(Y, M)$. Such equations are called γ-equations.

Consider a γ-equation (1). We construct a supporting tree Д of this equation in accordance with 7.1. The final vertices of Д, as we observed above, are μ-images of (1). From the definition of μ-images it follows that any solution (X, Y) of a μ-image of a γ-equation (1) satisfies $\tau(XAY, M)$. This condition means that X and Y are representable in the forms $X = S^\lambda S_1$ and $Y = S_2 S^\mu S_3$, where $[S^\partial \leqq M$; S_1 and S_3 are beginnings and S_2 is an end of S; and λ and μ are natural numbers. To determine such solutions of (1) it suffices to solve the one-stage exponential equations $\varphi(S^\lambda, S^\mu) = f(S^\lambda, S^\mu)$, where f and φ are words in the alphabet $\text{И} \cup \{x, y\}$ (the S_i have been excluded from φ and f). By 2.3, the solution of such equations is reduced to the solution of LDR's in two unknowns. The solution of these LDR's completes the algorithm recognizing the solvability of equation (1) when $A \neq B$.

We have described an algorithm for solving one-sided equations in two unknowns of the form (1) in the case $A \neq B$. Now we consider the case $A = B$. We have the equation

$$xAy\Phi(x, y) \to yAx\Psi(x, y). \tag{2}$$

Any solution of it is also a solution of the equation $xAy \to yAx$. We will express x and y from the prefix $xAy \to yAx$. If (X, Y) is a solution of the equation $xAy \to yAx$, then $XAYA = YAXA$, and $[X^\partial > [Y^\partial > 0$. By 1.2 we have

$$XA = Z^n, \qquad YA = Z^m \tag{3}$$

for some $n > m > 0$ and simple word Z. If $[Z^\partial \leqq [A^\partial$, then X and Y are periodic words with period Z (the definition precedes Proposition 2.8). The solution of (2) in the region $[Z^\partial \leqq [A^\partial$ is reduced to the solution of one-stage exponential equations in two unknowns. Consider the region $[Z^\partial > [A^\partial$. In this region $Z = VA$ for some V, and we have from (3) that $X = (VA)^\lambda V$ and $Y = (VA)^\mu V$, where $\lambda = n - 1$ and $\mu = m - 1$, so that $\lambda > \mu \geqq 0$. Thus the determination of the solutions of (2) in the region $[Z^\partial > M$ is reduced to the determination of the solutions of the equation in the three unknowns v, λ and μ

$$\Phi((vA)^\lambda v, (vA)^\mu v) = \Psi((vA)^\lambda v, (vA)^\mu v) \tag{4}$$

in the region $\mathscr{S}(vA)$, where $\mathscr{S}(x)$ is the predicate: x is a simple word.

Consider equation (4). Let M denote the maximum coefficient length of (2). We may, of course, assume that $M > 0$, by multiplying (2) on the right by a letter. We will solve (4) separately in the region $\tau(v, M)$ and in the region $\urcorner\, \tau(v, M)$.

In the region $\tau(v, M)$ we have for v the representations $v = S^\sigma S_1$, where S is a simple word, $[S^\partial \leqq M$, $S_1 \angle S$ and σ is a natural unknown (these representations depend on S_1 and S). We have $vA = S^\sigma S_1 A$. If $\sigma \leqq 1$, then vA is short, and (4) is an exponential equation in two unknowns. If $\sigma > 1$, then $\mathscr{L}(vA)$ implies $\mathscr{L}(S, S_1A)$. We arrive at a two-stage exponential equation

(5) $$\varphi((S^\sigma C)^\lambda S^\sigma, (S^\sigma C)^\mu S^\sigma) = \psi((S^\sigma C)^\lambda S^\sigma, (S^\sigma C)^\mu S^\sigma),$$

where $C = S_1 A$ and $\mathscr{L}(S, C)$; φ and ψ are words in the alphabet $\mathbf{H} \cup \{x, y\}$. By 2.10, equation (5) is equivalent to some LDR. Thus the solution of (5) in the region $\tau(v, M)$ is reduced to the solution of a LDR.

Consider the region $\urcorner\, \tau(v, M)$. Note that if in the alphabet $\mathbf{H}$ we have an equality $A_0[VA_i]_{i=1}^n = B_0[VB_i]_{i=1}^m$, where $\urcorner\, \tau(V, M)$ and $M \geqq \max\{[A_i^\partial, [B_i^\partial\}$, then necessarily $n = m$ and $A_i = B_i$, i.e. in the alphabet $\mathbf{H} \cup \{v\}$ we have the (graphical) equality $A_0[vA_i]_{i=1}^n = B_0[vB_i]_{i=1}^m$. This is easily deduced from 1.16. Hence if $(V, \bar\lambda, \bar\mu)$ is a solution of (4), then in the alphabet $\mathbf{H} \cup \{v\}$ we have $\Phi((vA)^{\bar\lambda} v, (vA)^{\bar\mu} v) = \Psi((vA)^{\bar\lambda} v, (vA)^{\bar\mu} v)$. It follows that $(\bar\lambda, \bar\mu)$ satisfies the exponential equation (4) with $\mathbf{H} \cup \{v\}$ in the role of $\mathbf{H}$. By 2.3, this equation is equivalent to the LDR $\mathscr{D}(\lambda, \mu)$. Hence in the region $\mathscr{L}(vA)\,\&$ $\urcorner\, \tau(v, M)$ equation (4) is equivalent to the predicate $\mathscr{L}(vA)\,\&\urcorner\, \tau(v, M)\,\&$ $\mathscr{D}(\lambda, \mu)$. It remains to solve the LDR $\mathscr{D}$, since the conjunction $\mathscr{L}(vA)\,\&$ $\urcorner\, \tau(v, M)$ has the solutions v (this is obvious). An algorithm for solving equations in two unknowns of the form (1) has been constructed. The theorem is proved.

Suppose Д is a tree of an equation $\mathfrak{A}$ with final vertices $\mathfrak{A}_1, \cdots, \mathfrak{A}_n$, and suppose for each of the equations $\mathfrak{A}_1, \cdots, \mathfrak{A}_n$ it is known whether or not it has solutions. Then it is possible to find out whether $\mathfrak{A}$ has solutions. Indeed, construct equations $\mathfrak{B}_1, \cdots, \mathfrak{B}_s$ in accordance with 6.22. By 6.22, if at least one of the equations $\mathfrak{A}_1, \cdots, \mathfrak{A}_n$ has solutions, then $\mathfrak{A}$ also has solutions. Suppose none of $\mathfrak{A}_1, \cdots, \mathfrak{A}_n$ has solutions. Then we check, using 2.3′ and 7.7, whether $\mathfrak{B}_1, \cdots, \mathfrak{B}_s$ have solutions. Equation $\mathfrak{A}$ has solutions if and only if at least one of $\mathfrak{B}_1, \cdots, \mathfrak{B}_s$ has solutions. We obtain the following result.

7.8. *The decision problem for equations in three unknowns reduces to the decision problem for the final vertices of the trees of these equations.*

The case $k = 1$. A-equations

We now turn to the case $k = 1$. In this case the equation (*) is an equation in the three unknowns x, y, z.

Before applying Lemma 7.3 to the case $k = 1$, consider (*) with $A = B$ and $k = 1$. In other words, consider the equation

(**) $$xAz\Phi(x, y, z) \to zAx\Psi(x, y, z),$$

where A is a word in the alphabet $\mathbf{H} \cup \{y\}$, and Φ and Ψ are words in the alphabet $\mathbf{H} \cup \{x, y, z\}$.

It would be possible (in the case of (*)) to proceed directly: express x, y and z from the equation $xAz \to zBx$ and, substituting the resulting expressions into the equation $\Phi = \Psi$, analyze the result of the substitution. The difficulties arising here when $A \neq B$ can be estimated by means of 5.9. The case $A = B$ is simpler. In this case, by 1.23, x, A and z are expressed from the equation $xAz \to zAx$ as follows:

$$x = (uv)^{\lambda}u, \qquad A = v(uv)^{\mu}, \qquad z = (uv)^{\nu}u,$$

where u and v are word unknowns, and λ, μ and ν are natural unknowns. However, even when $k = 1$, if the number of occurrences of y in the word A is greater than one, the solution of the equation $A(y) = v(uv)^{\mu}$ with respect to y, u and v turns out to be cumbersome. For our purposes it is not necessary to know the solutions of $xAz \rightarrow zAx$ in such a strong form. We prefer trees of the equation to explicit expressions for x, y and z. In our case $k = 1$ these are the supporting trees of 7.2.

Hence we will consider the equation (**). The coefficients of (**) are to be understood in the sense of the definition following the proof of 7.2. According to that definition the coefficients of (**) are words in the alphabet $\mathbf{И} \cup \{y\}$. The usual coefficients, words in the alphabet $\mathbf{И}$, are not encountered in this section.

Obviously, any solution of (**) satisfies the equation $xAz \rightarrow zAx$ and therefore the equation $xAzA = zAxA$. Hence if the equation $\Phi_1 = \Psi_1$ is obtained from $\Phi = \Psi$ by commutation of the words xA and zA, then the equation $xAz\Phi \rightarrow zAx\Psi$ is equivalent to $xAz\Phi_1 \rightarrow zAx\Psi_1$. In other words, commutation of xA and zA in the tail $\Phi = \Psi$ of (**) changes (**) into an equivalent equation. Furthermore, suppose the tail $\Phi = \Psi$ has the form $(xA)^a(zA)^b\xi \cdots = (xA)^c(zA)^d\eta \cdots$, where ξ and η are x or z. Then, replacing the tail by a new one obtained by cancelling the equation $\Phi = \Psi$ on the left by $(xA)^p(zA)^q$, where $p = \min\{a, c\}$ and $q = \min\{b, d\}$, we obtain an equation equivalent to the original one (**). The new tail has the form $(xA)^{a-p}(zA)^{b-q}\xi \cdots = (xA)^{c-p}(zA)^{d-q}\eta \cdots$.

Equations of the form (**) are also called A-equations. The procedure described above for transforming A-equations, commutation and cancellation, changes an A-equation into an equivalent A-equation. This permits us to consider A-equations to within the indicated transformations, i.e. to identify A-equations, one of whose tails is obtained from the other by commutation of xA, zA and cancellation by the common beginning.

We define several characteristics of A-equations. The tail $\Phi = \Psi$ of the A-equation (**) has the form (after commutation of xA and zA)

$$(7.2) \quad A_0[(xA)^{a_i}(zA)^{b_i}\xi_i A_i]_{i=1}^{n}(xA)^a(zA)^b = B_0[(xA)^{c_i}(zA)^{d_i}\eta_i B_i]_{i=1}^{m}(xA)^c(zA)^d$$

where $n, m, a, b, c, d, a_i, b_i, c_i, d_i \geqq 0$; ξ_i and η_i are x or z; $A_0, \cdots, A_n$, $B_0, \cdots, B_m$ are the coefficients of the A-equation, and $A_1, \cdots, A_n$, $B_1, \cdots, B_m$ are different from A. The number $h = n + m$ is called the height of the A-equation. In other words, the height of an A-equation is the number of coefficients of the tail, not counting those on the extreme left, which are different from A. We also define the weight of an A-equation. If one of n or m is zero, then the weight of the A-equation is zero. If $n, m > 0$, then the weight of the A-equation is the sum $|a_1 - c_1| + |b_1 - d_1|$.

Let us say a few words about μ-images of A-equations. These μ-images have the form $xAz\bar{\Phi} \rightleftarrows zAx\bar{\Psi}$, where $\bar{x} = (xAzA)^{\mu}x$ and $\bar{z} = xAz$. It is obvious that any μ-image of an A-equation $\mathfrak{A}$ has the same height as $\mathfrak{A}$.

Before proving the main result concerning the equation (**), Theorem 7.10, we consider A-equations of a special form, the so-called σ_A-equations. These are A-equations in which the tail $\Phi = \Psi$ has the form

$$(7.3) \qquad (xA)^a xP\xi\Phi_1 = (zA)^b zQ\eta\Psi_1,$$

where $a, b \geqq 0$; ξ and η are x or z; P and Q are words in the alphabet $\mathbf{H} \cup \{y\}$, and $P \neq A$, $Q \neq A$. The numbers a and b are called the exponents of the σ_A-equation. Note that the weight of a σ_A-equation with tail (7.3) is $a+b$. The following lemma concerns μ-images of σ_A-equations.

7.9. *Suppose $\mathfrak{A}$ is a σ_A-equation of weight g and height h, and $\overline{\mathfrak{A}}$ is a μ-image of $\mathfrak{A}$. Then one of the following three assertions holds.*

1) *$\overline{\mathfrak{A}}$ is equivalent to a σ_A-equation of weight $<g$ and height h.*

2) *$\overline{\mathfrak{A}}$ is equivalent to a σ_A-equation of height h with a zero exponent.*

3) *$\overline{\mathfrak{A}}$ is equivalent to an A-equation of height $<h$.*

If one exponent of $\overline{\mathfrak{A}}$ is zero, then 3) *holds.*

PROOF. Suppose there is given a σ_A-equation with tail $\Phi = \Psi$ of the form (7.3). The equation $\overline{\Phi} = \overline{\Psi}$ (i.e. the tail of $\overline{\mathfrak{A}}$) has the form

$$(1) \qquad ((xAzA)^\mu xA)(xAzA)^\mu xP\bar{\xi}\overline{\Phi}_1 = (xAzA)^b xAzQ\bar{\eta}\overline{\Psi}_1.$$

Commuting xA and zA, we transform (1) to the form

$$(2) \qquad (xA)^{a(\mu+1)+\mu}(zA)^{a\mu+\mu}xP\bar{\xi}\overline{\Phi}_1 = (xA)^{b+1}(zA)^b zQ\bar{\eta}\overline{\Psi}_1.$$

Put $d_1 = b+1-[a(\mu+1)+\mu]$ and $d_2 = a\mu+\mu-b$. We have $d_1+d_2 = 1-a$. We consider three cases.

1) $d_1d_2 > 0$. If $d_2 > 0$, then $d_1 = 1-a-d_2 \leqq 0$, and then $d_1d_2 \leqq 0$. Hence in this case $d_2 < 0$, and then $d_1 < 0$ also. Thus, $d_1, d_2 < 0$. But then (2) may be cancelled on the left (taking into account the commutation of xA and zA) by $(xA)^{b+1}(zA)^{a\mu+\mu}$. After cancellation we obtain the equation $(xA)^{-d_1}xP\bar{\xi}\overline{\Phi}_1 = (zA)^{-d_2}zQ\bar{\eta}\overline{\Psi}_1$. Denote this by $\mathfrak{A}_1$. It follows from our reasoning that $\overline{\mathfrak{A}}$ is equivalent to an A-equation $\mathfrak{B}$ with tail $\mathfrak{A}_1$. The equation $\mathfrak{B}$ is a σ_A-equation with exponents $-d_1$, $-d_2$ and with weight $-d_1-d_2$ and height h. Estimating the weight, we obtain $-d_1-d_2 = a-1 < a+b$, so that $\mathfrak{B}$ has a smaller weight. Thus 1) holds for $\overline{\mathfrak{A}}$.

2) $d_1d_2 < 0$. The cases $d_1 < 0$ and $d_1 > 0$ are handled analogously. Consider the case $d_1 < 0$. Then $d_2 > 0$. Equation (2) can be cancelled on the left by $(xA)^{b+1}(zA)^b$. After cancellation we obtain the equation $(xA)^{-d_1}(zA)^{d_2}xP\bar{\xi}\overline{\Phi}_1 = zQ\bar{\eta}\overline{\Psi}_1$. Cancelling (after commutation of xA and zA) the last equation on the left by z, we obtain the equation $A(xA)^{-d_1}(zA)^{d_2-1}xP\bar{\xi}\overline{\Phi}_1 = Q\bar{\eta}\overline{\Psi}_1$. Denote it by $\mathfrak{A}_1$. It follows from our reasoning that $\overline{\mathfrak{A}}$ is equivalent to an A-equation $\mathfrak{B}$ with tail $\mathfrak{A}_1$. The equation $\mathfrak{B}$ has height $<h$, since the coefficient Q, which is different from A, is on the extreme left. Thus 3) holds for $\overline{\mathfrak{A}}$.

3) $d_1d_2 = 0$. To the left of P and Q either all powers of xA or all powers of zA completely cancel. After cancellation, (2) has the form

$$(3) \qquad (zA)^{d_2}xP\bar{\xi}\overline{\Phi}_1 = zQ\bar{\eta}\overline{\Psi}_1,$$

if $d_1 = 0$, $d_2 > 0$; the form

$$xP\bar{\xi}\overline{\Phi}_1 = (zA)^{-d_2}zQ\bar{\eta}\overline{\Psi}_1,$$

if $d_1 = 0$, $d_2 \leqq 0$; the form

$$(xA)^{-d_1}xP\bar{\xi}\overline{\Phi}_1 = zQ\bar{\eta}\overline{\Psi}_1,$$

if $d_2 = 0$, $d_1 \leqq 0$; or the form

$$xP\bar{\xi}\bar{\Phi}_1 = (xA)^{d_1} zQ\bar{\eta}\bar{\Psi}_1, \tag{4}$$

if $d_2 = 0$, $d_1 > 0$.

It is now clear that if $d_1 = 0$, $d_2 \leqq 0$ or if $d_2 = 0$, $d_1 \leqq 0$, then 2) holds. If $d_1 = 0$, $d_2 > 0$ or if $d_2 = 0$, $d_1 > 0$, then 3) holds (we need only cancel (3) on the left by z, and (4) by x).

To complete the proof, it remains to show that if one (or both) of the exponents a or b is zero, then 3) holds. Indeed, if $b = 0$, then (2) can be cancelled on the left by xAz, since $\mu \geqq 1$. After cancellation, (2) assumes the form $A\gamma \cdots = Q\bar{\eta}\bar{\Psi}_1$, where γ is x or z, which implies 3). If $a = 0$, then (2) has the form

$$(xA)^{\mu} (zA)^{\mu} xP\bar{\xi}\bar{\Phi}_1 = (xA)^{b+1} (zA)^{b} zQ\bar{\eta}\bar{\Psi}_1. \tag{5}$$

Considering the cases $\mu \geqq b$ and $\mu < b$, we see that after all possible left cancellations, (5) assumes the form $P\bar{\xi}\bar{\Phi}_1 = A\gamma \cdots$ if $\mu \leqq b$, and the form $A\gamma \cdots = Q\bar{\eta}\bar{\Psi}_1$ if $\mu > b$, where γ is x or z. This easily implies 3). The proposition is proved.

7.10. *For any A-equation $\mathfrak{A}$ it is possible to construct a tree whose final vertices are of the following four kinds:*

a) *they have no solutions,*

b) *they have the form $xAzP(xA)^n \to zAxQ(zA)^m$, where $m, n \geqq 0$ and P and Q are words in the alphabet $\mathbf{H} \cup \{y\}$, $P \equiv Q$,*

c) *they possess property α,*

d) *they are basic equations of characteristic $\leqq 2$.*

If $\mathfrak{A}$ has characteristic $\leqq 2$, then in d) *$\leqq 2$ can be replaced by $\leqq 1$.*

PROOF. We introduce an ordering on pairs of numbers. Suppose (a, b) and (c, d) are (ordered) pairs of natural numbers. The pair (a, b) precedes the pair (c, d) in exactly two cases: 1) $a < c$; 2) $a = c$, $b < d$. We prove the theorem by induction on the pair (h, g), where h is the height and g the weight of $\mathfrak{A}$.

The tail $\Phi = \Psi$ of $\mathfrak{A}$ has the form (7.2). If $A_0 \not\equiv B_0$, then $\mathfrak{A}$ possesses property α. This is obvious if the unknowns x and z occur on both sides of the equation $\Phi = \Psi$. Otherwise, it is necessary to multiply the equation on the right by x or z. Hence if $A_0 \not\equiv B_0$, then $\mathfrak{A}$ possesses property α. The desired tree consists of one vertex of the kind c), the same kind as $\mathfrak{A}$ itself. There remain equations $\mathfrak{A}$ with $A_0 \equiv B_0$. Any solution of such an equation satisfies the equation $A_0 = B_0$. But then $\mathfrak{A}$ is equivalent to the system $xAzA_0 \to zAxB_0 \,\&\, \Phi_1 = \Psi_1$, where the equation $\Phi_1 = \Psi_1$ is obtained from $\Phi = \Psi$ by dropping on the left the coefficients A_0 and B_0. Writing for brevity

$$u_i = (xA)^{a_i} (zA)^{b_i} \xi_i A_i, \quad v_i = (xA)^{c_i} (zA)^{d_i} \eta_i B_i, \quad u = (xA)^a (zA)^b,$$

$v = (xA)^c (zA)^d$, we can write the equation $\Phi_1 = \Psi_1$ in the form

$$[u_i]_{i=1}^n u = [v_i]_{i=1}^m v. \tag{1}$$

We henceforth assume that the tail $\Phi = \Psi$ of $\mathfrak{A}$ has the form (1). We also assume that the words u and v in (1) begin with x or z. For this it suffices to multiply the equation on the right by xA. The resulting equation has the same height and weight as the original.

We proceed by induction. Consider an equation $\mathfrak{A}$ of height zero. Then $n = m = 0$ and the tail (1) has the form $u = v$, i.e. $(xA)^a(zA)^b = (xA)^c(zA)^d$. After all left and right cancellations, (1) assumes the form $(xA)^p(zA)^q = \Lambda$, where $p, q \geqq 0$, or the form $(xA)^p = (zA)^q$, where $p, q > 0$. In the first case, $\mathfrak{A}$ either has no solutions $(p + q > 0)$ or is equivalent to the equation $xAzA_0 \to zBxB_0$ $(p = q = 0)$. The desired tree consists of one vertex of the kind a) or b), the same kind as $\mathfrak{A}$ itself. In the second case, $\mathfrak{A}$ is equivalent to the equation $xAzA_0(xA)^p \to zAxB_0(zA)^q$. The desired tree consists of one vertex of the kind b), the same kind as $\mathfrak{A}$. The theorem is proved for A-equations of height $h = 0$.

Consider an A-equation $\mathfrak{A}$ with pair (h,g), $h > 0$, where h is the height of the equation and g its weight. Suppose first that one of n or m is zero. Then the other is greater than zero. The equation $\Phi = \Psi$ has the form (for the case $n = 0$, $m > 0$)

$$(2) \qquad (xA)^a(zA)^b = (xA)^{c_1}(zA)^{d_1}\eta_1 B_1[v_i]_{i=2}^m v.$$

Multiplying (2) on the right by suitable powers of xA and zA and commuting xA and zA if necessary (h and g remain unchanged), we may assume that $a > c_1$, $b > d_1$ and to the right of B_1 stands either x or z. After all left cancellations, (2) assumes the form $(xA)^p(zA)^q = \eta_1 B_1[v_i]_{i=1}^m v'$, where $p, q > 0$. By commuting xA and zA we may cancel the last equation on the left by η_1. After cancellation we obtain the equation $A(xA)^{p-1}(zA)^q = B_1[v_i]_{i=2}^m v'$ (for the case $\eta_1 = x$). Denote it by $\Phi_1 = \Psi_1$. It follows from our reasoning that $\mathfrak{A}$ is equivalent to the equation $xAzA_0\Phi_1 \to zAxB_0\Psi_1$. This is an A-equation of height $< h$ (the coefficient B_1 is on the extreme left). By the induction assumption, the equation $xAzA_0\Phi_1 \to zAxB_0\Psi_1$ possesses a tree with final vertices of the kinds a) – d). In view of the extended replacement rule (see the definition following the proof of Theorem 6.24), $\mathfrak{A}$ possesses the desired tree.

Consider the case $m, n > 0$. Then (1) has the form

$$(3) \qquad (xA)^{a_1}(zA)^{b_1}\xi_1 A_1\ [u_i]_{i=2}^n u = (xA)^{c_1}(zA)^{d_1}\eta_1 B_1\ [v_i]_{i=2}^m v.$$

There are two cases to consider.

1) $(a_1 - c_1)(b_1 - d_1) > 0$. After all left cancellations, (3) assumes the form

$$(xA)^p(zA)^q\xi_1 A_1 \cdots = \eta_1 B_1 \cdots$$

if $a_1 > c_1$, and the form

$$\xi_1 A_1 \cdots = (xA)^p(zA)^q\eta_1 B_1 \cdots$$

if $a_1 < c_1$, where $p, q > 0$. Both cases are handled analogously. Consider the case $a_1 > c_1$. Since $p, q > 0$, after commuting, if necessary, xA and zA we can cancel the equation on the left by η_1. After cancellation we obtain the equation $A(xA)^{p-1}(zA)^q\xi_1 A_1[u_i]_{i=2}^n u = B_1[v_i]_{i=2}^m v$ (for the case $\eta_1 = x$). The argument continues as in the sixth paragraph of the proof, following the formula

$$A(xA)^{p-1}(zA)^q = B_1[v_i]_{i=2}^m v'.$$

2) $(a_1 - c_1)(b_1 - d_1) \leq 0$. After all left cancellations, (3) assumes the form

$$(4) \qquad (\xi A)^p\xi_1 A_1[u_i]_{i=2}^n u = (\eta A)^2\eta_1 B_1[v_i]_{i=2}^m v,$$

where $p, q \geqq 0$; ξ and η are x or z, and $\xi \neq \eta$. Let us consider the possible cases.

2.1) $p = q = 0$, $\xi_1 = \eta_1$. After left cancellation by ξ_1 we obtain the equation $A_1[u_i]_{i=2}^n u = B_1[v_i]_{i=2}^m v$. Denote it by $\Phi_1 = \Psi_1$. The equation $\mathfrak{A}$ is equivalent to the equation $xAzA_0\Phi_1 \to zAxB_0\Psi_1$. This is an A-equation of height $h - 2$. By the induction assumption, the equation $xAzA_0\Phi_1 \to zAxB_0\Psi_1$ possesses a tree with final vertices a) – d). But then, by the extended replacement rule, $\mathfrak{A}$ possesses such a tree.

2.2) $p = q = 0$, $\xi_1 \neq \eta_1$. Here we consider the equation

$$xAz\xi_1 A_1[u_i]_{i=2}^n u \to zAx\eta_1 B_1[v_i]_{i=2}^m v. \tag{5}$$

The equation (5) is a σ_A-equation with zero exponents (since A_1 and B_1 are different from A). It is equivalent to $\mathfrak{A}$. We construct in accordance with 7.2 a supporting tree of (5). The final vertices of this tree are either basic equations of characteristic $\leqq 2$ (the first group) or μ-images of (5) (the second group). If (5) has characteristic two, the equations of the first group have characteristic $\leqq 1$. The vertices of the first group, consequently, are of the kind d). Each vertex of the second group, in view of the second assertion of 7.9, is equivalent to an A-equation (with the same A as in (5)) of smaller height. By the induction assumption (and the extended replacement rule), each vertex of the second group possesses a tree with final vertices of the kinds a) – d). Connecting these trees to the vertices of the second group, we obtain the desired tree of $\mathfrak{A}$.

2.3) $q = 0$, $p > 0$. If $\xi = \eta_1$, then after left cancellation by ξ we obtain the equation $A(\xi A)^{p-1}\xi_1 A_1[u_i]_{i=2}^n u = B_1[v_i]_{i=2}^m v$. The argument continues as in the sixth paragraph, following the equation $A(xA)^{p-1}(zA)^q = B_1[v_i]_{i=2}^m v$.

Consider the case $\xi \neq \eta_1$. Let $\mathfrak{A}_1$ denote the equation $xAzA_0\Phi_1 \to zAxB_0\Psi_1$, where $\Phi_1 = \Psi_1$ is equation (4). The equation $\mathfrak{A}_1$ is a σ_A-equation with zero exponent q. The argument continues as in 2.2), after (5).

2.4) $p = 0$, $q > 0$. This case is handled analogously to 2.3).

2.5) $p > 0$, $q > 0$. Without loss of generality, we will assume that $\xi = x$ and $\eta = z$. Equation (4) has the form

$$(xA)^p \xi_1 A_1[u_i]_{i=2}^n u = (zA)^q \eta_1 B_1[v_i]_{i=2}^m v. \tag{6}$$

There are four possible cases.

2.5.1) $\xi_1 = x$, $\eta_1 = z$. Let $\mathfrak{A}_1$ denote the equation $xAzA_0\Phi_1 \to zAxB_0\Psi_1$, where $\Phi_1 = \Psi_1$ is equation (6). The equation $\mathfrak{A}_1$ is equivalent to $\mathfrak{A}$. In view of the extended replacement rule, it suffices to construct a tree of $\mathfrak{A}_1$ with final vertices of the kinds a) – d). Let us construct such a tree.

Consider $\mathfrak{A}_1$. It is a σ_A-equation with positive exponents p and q. It has height h. We construct in accordance with 7.2 a supporting tree of $\mathfrak{A}_1$. Its final vertices are either equations of the kind d) or μ-images of $\mathfrak{A}_1$. We are interested in the latter. Let $\overline{\mathfrak{A}}_1$ be any of these μ-images. By 7.9 and the extended replacement rule, we may assume that $\overline{\mathfrak{A}}_1$ is a σ_A-equation of weight $< g$ and height h, a σ_A-equation with zero exponent of height h, or an A-equation of height $< h$. Consider each of the three cases which can occur for $\overline{\mathfrak{A}}_1$.

In the first case (smaller weight) and the third case (smaller height) the induction assumption is applicable to $\overline{\mathfrak{A}}_1$; hence it has a tree with final vertices of the kinds a) – d). Connect this tree to $\overline{\mathfrak{A}}_1$. In the second case (zero exponent) connect to $\overline{\mathfrak{A}}_1$ the supporting tree of $\overline{\mathfrak{A}}_1$ from 7.2. From among its final vertices we are interested only in the μ-images of $\overline{\mathfrak{A}}_1$. The other final vertices are of the kind d). By 7.9, each of these μ-images is equivalent to an A-equation of smaller height and, by the induction assumption, possesses a tree with final vertices of the kinds a) – d). It remains to connect these trees to the corresponding vertices. This completes the construction of the desired tree of $\mathfrak{A}_1$.

2.5.2) $\xi_1 = x$, $\eta_1 = x$.

2.5.3) $\xi_1 = z$, $\eta_1 = x$.

2.5.4) $\xi_2 = z$, $\eta_1 = z$.

The treatment of these three cases is similar and based on the fact that the replacement of xAz by zAx in the tail of an A-equation changes the equation into an equivalent one. Consider, for example, the case $\xi_1 = x$, $\eta_1 = x$. Replacing in the right-hand side of (6) the word $(zA)^q\eta_1$ by $x(Az)^q$ (we replace xAz by zAx q times) and cancelling the resulting equation on the left by xA, we obtain

$$(xA)^{p-1}xA_1u_2\cdots u_nu = (zA)^{q-1}zB_1v_2\cdots v_mv. \tag{7}$$

Let $\mathfrak{A}_1$ be an A-equation with tail (7). The equation $\mathfrak{A}_1$ is equivalent to $\mathfrak{A}$ and is an A-equation of height h and weight $g_1 = p-1+q-1 = g-2$. It remains to apply the induction assumption to $\mathfrak{A}_1$ and use the extended replacement rule. The theorem is proved.

We will now explain what property α means for basic equations. A basic equation $x\cdots \to z\cdots$ is called a γ-equation if there exists an integer $M > 0$ such that for any solution (X, Y, Z) of the equation the condition $\tau(Z, M)$ is satisfied.

7.11. *If a basic equation possesses property α, then either it is a γ-equation or it possesses a tree whose final vertices are basic equations of characteristic zero.*

Proof. If the equation possesses property α, then it is equivalent to an equation $xAzP\xi\cdots \to zBxQ\eta\cdots$, where P and Q are words in the alphabet $\mathbf{H}\cup\{y\}$, $P\not\equiv Q$, and ξ and η are x or z. Since $P\not\equiv Q$, there are two possible cases.

1) $\nu(P) = \nu(Q)$ (see the definition preceding 4.4). Let p and q denote the lengths of the projections of P and Q on the alphabet $\mathbf{H}$. Put $d = p - q$ and consider the case $d \geqq 0$. Since $\nu(P) = \nu(Q)$, the difference $[P(Y)^\partial - [Q(Y)^\partial$ does not depend on Y and is equal to d. Since $P\not\equiv Q$, we have $d > 0$. Hence any solution (X, Y, Z) of the equation $xAzP\xi\cdots \to zBxQ\eta\cdots$ satisfies $RX_1\cdots = X_2\cdots$, where $[R^\partial = d$ and X_1 and X_2 are X or Z. But since (*) is one-sided we have $Z \angle X$ for any solution of (*). Hence the component Z satisfies $RZV_1 = ZV_2$. Since $[R^\partial = d > 0$, by 1.16 we obtain $Z = R^tR_1$, where $R_1 \angle R$ and $t \geqq 0$, i.e. $\tau(Z, M)$ holds with $M = d$. But this means that the equation is a γ-equation.

2) $\nu(P) \neq \nu(Q)$. Putting $p = \nu(P)$ and $q = \nu(Q)$, we consider only the case $p > q$. Represent P in the form $P = CD$, where $\nu(C) = q$ and $\nu(D) = p - q$. By 4.5′ the equation $xAzCD\xi\Phi \to zBxQ\eta\Psi$ is equivalent to a disjunction of equations similar (in the sense of the definition preceding 4.5) to the equation $xAzD\xi\Phi C \to zBx\eta\Psi Q$. The projection of this last equation on the

alphabet $\{x, y, z\}$ has the form $xy^azy^{p-q}\cdots \to zy^ax\cdots$, where $a = \nu(A)$ and $p > q$. In accordance with the extended replacement rule, it suffices to prove that any equation with projection $xy^azy\cdots \to zy^ax\cdots$, where $\eta \neq y$, possesses a tree with final vertices of characteristic zero. By 6.20, the equation $xy^azy\cdots \to zy^ax\eta\cdots$ possesses a (coefficient-free) tree with final vertices of the form $xy\cdots \to yx\cdots$. Applying 6.23, we obtain that any equation with projection $xy^azy\cdots \to zy^ax\eta\cdots$, $\eta \neq y$, possesses a tree with final vertices of characteristic zero. The proposition is proved.

On p. 85 we defined fundamental basic equations. We will broaden this class somewhat. By fundamental basic equations we will mean equations of the forms

$$xAzP \to zBxQ,$$
$$xAzP(xA)^n \to zAxQ(zA)^m,$$

where A, B, P and Q are words in the alphabet $\mathit{И} \cup \{y\}$, $A \equiv B$, $P \equiv Q$, and $n, m > 0$.

We now prove the following theorem on the reduction of basic equations.

7.12. **Theorem.** *For any basic equation $\mathfrak{A}$ of characteristic $\nu \geqq 2$ it is possible to construct a tree with final vertices of the following three kinds:*

1) *fundamental basic equations of characteristic* $\leqq \nu$,
2) *γ-equations of characteristic* $\leqq \nu$,
3) *basic equations of characteristic* $\leqq 1$.

Proof. We construct the desired tree of $\mathfrak{A}$ in the following manner.

Construct a supporting tree of $\mathfrak{A}$ in accordance with 7.2. Its final vertices are of two kinds: they are either μ-images of $\mathfrak{A}$ or basic equations of characteristic $\leqq 2$. If $\mathfrak{A}$ itself has characteristic two, then the vertices which are not μ-images have characteristic $\leqq 1$.

Consider the final vertices which are μ-images. We apply Lemma 7.3 to $\mathfrak{A}$. By 7.3, for any μ-image of $\mathfrak{A}$ one of 7.3.1)–7.3.4) holds. The μ-images for which 7.3.1) holds are (by the extended replacement rule) vertices of the kind 1). The μ-images for which 7.3.2) holds are also vertices of the kind 1), since it is always possible to select a fundamental basic equation having no solutions, for example, $xAza \to zBxb$, where a and b are letters. Furthermore, for μ-images possessing property α (see 7.3.3)) it is possible, by 7.11, to construct a tree with final vertices of the kinds 2) and 3). Finally, the μ-images of 7.3.4) are equivalent to A-equations. For these μ-images it is possible, by 7.10 and 7.11, to construct a tree with final vertices of the kinds 1), 2) and 3) plus, possibly, vertices of the kind d) from 7.10.

It follows that it is possible to construct a tree Д of $\mathfrak{A}$ whose final vertices are of the kinds 1)–3) plus, possibly, basic equations of characteristic two. If $\mathfrak{A}$ is itself of characteristic two, the final vertices of Д are of the kinds 1)–3). Thus for equations of characteristic two, hence also for equations of characteristic > 2, it is possible to construct the desired tree. The theorem is proved.

§8. Basic equations of characteristic one

In this section we consider basic equations of characteristic one. These have the form

$$(*) \qquad xAyBz\Phi(x,y,z) \to zCyDx\Psi(x,y,z),$$

where A, B, C and D are coefficients, $[AB^\partial = [CD^\partial$, and Φ and Ψ are words in the alphabet $\mathbf{H} \cup \{x,y,z\}$. Along with the one-sided equation (*) we consider the two-sided equation (*), where the arrow is replaced by the symbol of equality.

It follows from the assumption $[AB^\partial = [CD^\partial$ that the equation is equivalent to the system consisting of the equations $xAyBz \to zCyDx$ and $\Phi = \Psi$. Hence if the equation $\Phi = \Psi$ is equivalent to an equation $\Phi' = \Psi'$, then (*) is equivalent to $xAyBz\Phi' \to zCyDx\Psi'$. A similar argument holds, of course, for the two-sided equation (*).

With regard to (*), we are principally interested in how the prefix $xAyBz \to zCyDx$ and the tail $\Phi = \Psi$ are related to each other. In particular, we would like to know the conditions under which (*) is equivalent (or not equivalent) to its prefix $xAyBz \to zCyDx$. We distinguish two cases: $AyB \neq CyD$ and $AyB = CyD$. In the first case we consider trees of (*). This allows us to reduce (*) to basic equations of characteristic zero. In the second case we express x, y and z from the prefix $xAyBz \to zAyBx$ and reduce (*) to one-stage exponential equations with variable bases.

The equation (*). The case $AyB \neq CyD$

A convenient tool for investigating (*) in the case $AyB \neq CyD$ is the notion of a θ-image of (*). We give the definition. Consider three triples $(\bar{x}, \bar{y}, \bar{z})$ of parametric words:

1) $\bar{x} = (\bar{z}C\bar{y}D)^\lambda zCyDx, \quad \bar{y} = (yDxA)^\mu y, \quad \bar{z} = z, \quad (\lambda \geqslant 0,\ \mu > 0),$

2) $\bar{x} = (\bar{z}C\bar{y}D)^\lambda x, \quad \bar{y} = (yBzC)^\mu y, \quad \bar{z} = xAyBz \quad (\lambda,\ \mu > 0),$

3) $\bar{x} = (\bar{z}C\bar{y}D)^\lambda x, \quad \bar{y} = y, \quad \bar{z} = xAyBz \quad (\lambda > 0),$

where λ and μ are natural variables. Consider the equation

$$(8.1) \qquad xAyBz\Phi(\bar{x},\bar{y},\bar{z}) = zCyDx\Psi(\bar{x},\bar{y},\bar{z}),$$

where $(\bar{x},\bar{y},\bar{z})$ is one of the triples 1) – 3). The θ-images of (*) are the equations which are obtained from (8.1) by substituting numbers in place of the variables λ and μ. If $(\bar{x},\bar{y},\bar{z})$ is the triple i), then an image (8.1) is called a θ_i-image.

Along with θ-images of the form (8.1) (two-sided) we consider one-sided θ-images. These are obtained from (8.1) by replacing the symbol $=$ by the symbols $\to$, $\leftarrow$ (in both directions). Unless otherwise stated, by θ-images we mean two-sided θ-images.

REMARK 1. It is not difficult to prove that identically named θ-images of equivalent equations (*) are equivalent. More precisely, if $\mathfrak{A}$ and $\mathfrak{B}$ are equivalent equations (*) with the same prefix $xAyBz \to zCyDx$, then their θ_i-images with identical i, λ and μ are also equivalent. It suffices to observe that for any fixed i, λ and μ, if (X, Y, Z) is a solution of a θ_i-image of (*), then the triple $(\overline{X}, \overline{Y}, \overline{Z})$ obtained from the formulas i) with fixed λ and μ is a solution of (*). Conversely, if $(\overline{X}, \overline{Y}, \overline{Z})$ is a solution of (*), then (X, Y, Z) is a solution of the θ_i-image with fixed i, λ and μ. This can be verified by a direct computation. Now suppose that $\mathfrak{A}'$ and $\mathfrak{B}'$ are identically named θ-images of $\mathfrak{A}$ and $\mathfrak{B}$. Let (X, Y, Z) be a solution of $\mathfrak{A}'$. Then $(\overline{X}, \overline{Y}, \overline{Z})$ is a solution of $\mathfrak{A}$ and, consequently, of $\mathfrak{B}$, in view of the equivalence of $\mathfrak{A}$ and

$\mathfrak{B}$. Since $(\overline{X}, \overline{Y}, \overline{Z})$ is a solution of $\mathfrak{B}$, (X, Y, Z) is a solution of $\mathfrak{B}'$. Thus the equivalence of $\mathfrak{A}'$ and $\mathfrak{B}'$ is established.

8.1. *It is possible to construct a tree of* (*) *whose final vertices are* θ-*images of* (*).

PROOF. We construct the desired tree in the following manner. Take a complete system of infinite images of (*). The equations of this complete system are of the following two forms:

$$vAyBz\Phi\ ((zCyD)^{\lambda}\, v,\ y,\ z) \leftarrow zCyDv\Psi\ ((zCyD)^{\lambda}v,\ y,\ z), \tag{1}$$

$$vAyBz\Phi\ ((zCyD)^{\lambda}zCv,\ y,\ z) \leftarrow yDzCv\Psi\ ((zCyD)^{\lambda}zCv,\ y,\ z). \tag{2}$$

Here λ is the parameter of the transformation of (*), and v is a new unknown. For equation (1) we have $\lambda > 0$, and for (2) we have $\lambda \geqq 0$. We consider separately each of the equations (1) and (2).

Let us begin with (2). A complete system of infinite images of (2) consists of equations of the form

$$vAuBz\Phi\ ((zC\bar{y}D)^{\lambda}zCv,\ \bar{y},\ z) \rightarrow uDzCv\Psi\ ((zC\bar{y}D)^{\lambda}zCv,\ \bar{y},\ z), \tag{3}$$

where $\bar{y} = (vA)^{\mu}u$, $\mu > 0$, and u is a new unknown. Equation (3) is reduced by the substitution $v = uD\xi$ to the equation

$$\xi AuBz\Phi\ (T_1,\ T_2,\ z) = zCuD\xi\Psi\ (T_1,\ T_2,\ z), \tag{4}$$

where $T_2 = (uD\xi A)^{\mu}u$ and $T_1 = (zCT_2D)^{\lambda}zCuD\xi$. Equation (4), to within designation of the unknowns x, y and z, is a θ_1-image of (*). Thus we have proved that (2) possesses a tree whose final vertices are θ_1-images of (*).

We now turn to (1). The pair of substitutions $z = vAu$, $z = vAyBu$ reduces (1) to the equations

$$yBvAu\Phi\ ((\bar{z}CyD)^{\lambda}v,\ y,\ \bar{z}) \rightarrow uCyDv\Psi\ ((\bar{z}CyD)^{\lambda}v,\ y,\ \bar{z}), \tag{5}$$

$$vAyBu\Phi\ (T_1,\ y,\ T_3) = uCyDv\Psi\ (T_1,\ y,\ T_3), \tag{6}$$

where $\bar{z} = vAu$, $T_3 = vAyBu$ and $T_1 = (T_3CyD)^{\lambda}v$. Equation (6), to within designation of the unknowns x and z, is a θ_3-image of (*). There remains equation (5). A complete system of infinite images of (5) consists of equations of the form $\eta BvAu\Phi((\bar{z}C\bar{y}D)^{\lambda}v, \bar{y}, \bar{z}) \rightarrow uC\eta Dv\Psi((\bar{z}C\bar{y}D)^{\lambda}v, \bar{y}, \bar{z})$, where $\bar{y} = (uC)^{\mu}\eta$, $\mu > 0$, and η is a new unknown. The substitution $u = \eta B\xi$ reduces this equation to

$$vA\eta B\xi\Phi\ (T_1,\ T_2,\ T_3) = \xi\ C\eta Dv\Psi\ (T_1,\ T_2,\ T_3), \tag{7}$$

where $T_3 = vA\eta B\xi$, $T_2 = (\eta B\xi C)^{\mu}\eta$ and $T_1 = (T_3CT_2D)^{\lambda}v$. Equation (7), to within designation of the unknowns x, y and z, is a θ_2-image of (*).

We have proved that (1) possesses a tree whose final vertices are θ-images of (*).

Thus, each of the equations (1) and (2) possesses a tree whose final vertices are θ-images of (*). Hence (*) itself possesses such a tree. The proposition is proved.

In view of 7.8 and 8.1, the decision problem for (*) is reduced to the decision problem for θ-images of this equation. These equations have the form $xAyBz\overline{\Phi} = zCyDx\overline{\Psi}$. We will prove that any θ-image of (*) either is equivalent to the prefix $xAyBz = zCyDx$ or has a specific nature, which is described by means of the so-called ϵ-equations. These are two-sided equations of the form (*) in which the tail $\Phi=\Psi$ has a special form. The solution of these ϵ-equations is reduced to the solution of basic equations of characteristic zero.

Let us now define ϵ-equations. We first define ϵ_0-, ϵ_1-, and ϵ_2-equations. An equation (*) (one-sided or two-sided) is called:

an ϵ_0-equation if the equation $\Phi=\Psi$ has one of the following forms: $P=Q$, $P_1\alpha\cdots=Q_1$, or $P\xi\cdots=Q\eta\cdots$, where P, Q, P_1 and Q_1 are coefficients, $P\neq Q$, and α, ξ and η are unknowns and either $\xi=\eta$ or $\xi,\eta\neq y$.

an ϵ_1-equation if the equation $\Phi=\Psi$ has the form $P\xi\cdots=Qy\cdots$, where P and Q are coefficients and ξ is x or z.

an ϵ_2-equation if after left cancellation the equation $\Phi=\Psi$ has the form $x\cdots=z\cdots$, but does not have the form $xAy\cdots=zCy\cdots$. Here it is permitted to multiply the left- and right-hand sides of $\Phi=\Psi$ on the right by a common factor.

An equation (*) (one-sided or two-sided) is called an ϵ-equation if it is an ϵ_i-equation for some i, $0\leqq i\leqq 2$. It follows from the definitions that if any of the equations $x\varphi(x,y,z)\to zf(x,y,z)$, $x\varphi\leftarrow zf$ or $x\varphi=zf$ is an ϵ_i-equation, so are the other two.

Suppose $\mathfrak{A}$ is an equation in three unknowns. We will write $\epsilon(\mathfrak{A})$ if $\mathfrak{A}$ is equivalent to an ϵ-equation, and $\epsilon_i(\mathfrak{A})$ if $\mathfrak{A}$ is equivalent to an ϵ_i-equation.

The following lemma plays a fundamental role in the investigation of (*) in the case $AyB\neq CyD$.

8.2. *Suppose* $[AB^\partial=[CD^\partial$, $AyB\neq CyD$, *and* $\overline{\mathfrak{A}}$ *is a* θ*-image of* (*). *Then the following assertions are true.*

a) *Either* $\overline{\mathfrak{A}}$ *is equivalent to the equation* $xAyBz=zCyDx$, *or* $\epsilon(\overline{\mathfrak{A}})$.

b) *If* (*) *is an* ϵ_2*-equation, then* $\epsilon_i(\overline{\mathfrak{A}})$, *where* $0\leqq i\leqq 1$.

PROOF. The tail $\Phi=\Psi$ of (*) has the form

(1) $$A_0\,[\xi_iA_i]_{i=1}^n=B_0\,[\eta_iB_i]_{i=1}^m,$$

where $n,m\geqq 0$, ξ_i and η_j are unknowns, and A_i and B_j are coefficients. The number $h=n+m$ is called the height of (*). We prove a) by induction on the height of the equation. We obtain b) in the course of the argument.

The equation $\overline{\mathfrak{A}}$ is (8.1) for some i, λ and μ. We will assume that λ and μ in the expressions for $\bar{x}$, $\bar{y}$ and $\bar{z}$ are fixed numbers. Put $\overline{\Phi}=\Phi(\bar{x},\bar{y},\bar{z})$ and $\overline{\Psi}=\Psi(\bar{x},\bar{y},\bar{z})$. Then $\overline{\mathfrak{A}}$ has the form $xAyBz\overline{\Phi}=zCyDx\overline{\Psi}$. The tail $\overline{\Phi}=\overline{\Psi}$ of $\overline{\mathfrak{A}}$ has the form

(2) $$A_0\,[\bar{\xi}_iA_i]_{i=1}^n=B_0\,[\bar{\eta}_iB_i]_{i=1}^m.$$

We begin the induction. If one of the numbers h, n or m is zero, then a) and b) are trivial: if $h=0$ and $A_0=B_0$, then $\overline{\mathfrak{A}}$ is equivalent to the equation $xAyBz=zCyDx$, and in the other cases we have $\epsilon_0(\overline{\mathfrak{A}})$. Note that if we have $\epsilon_i(\overline{\mathfrak{A}})$, where $0\leqq i\leqq 1$, then a) and b) are trivially true.

Consider the case $n,m>0$. Assume that $\xi_1=\eta_1$. If $A_0\neq B_0$, then we have $\epsilon_0(\overline{\mathfrak{A}})$. Suppose $A_0=B_0$. Then (*) is equivalent to $\mathfrak{A}$, which has the form $xAyBzA_1[\xi_iA_i]_{i=2}^n\to zCyDxB_1[\eta_iB_i]_{i=2}^m$. The equation $\mathfrak{A}$ is an equation of the form (*) of height $h-2$. It is an ϵ_2-equation if and only if the original equation (*) is. The induction assumption is applicable to $\mathfrak{A}$. Applying it and noting Remark 1, we see that a) and b) are true.

The main case is $\xi_1 \neq \eta_1$. Here we consider three cases, depending on whether $\overline{\mathfrak{A}}$ is a θ_1-, θ_2- or θ_3-image.

Case 1. $\overline{\mathfrak{A}}$ is a θ_1-image. In this case we have

$$\bar{x} = (zC\bar{y}D)^{\lambda}\, zCyDx, \qquad \bar{y} = (yDxA)^{\mu}\, y, \qquad \bar{z} = z, \qquad \lambda \geqslant 0, \qquad \mu > 0.$$

If one of the letters ξ_1 or η_1 is y and the other is different from y, then obviously $\epsilon_1(\overline{\mathfrak{A}})$, and a) and b) are true.

Suppose one of the unknowns ξ_1 or η_1 is x and the other is z. Without loss of generality, suppose $\xi_1 = x$ and $\eta_1 = z$. Then the equation $\overline{\Phi} = \overline{\Psi}$ has the form

$$A_0\,(zC\bar{y}D)^{\lambda}\, zCyDxA_1\,[\bar{\xi}_i A_i]_{i=2}^{n} = B_0 zB_1\,[\bar{\eta}_i B_i]_{i=2}^{m}, \tag{3}$$

where $\lambda \geqq 0$. We list the possible cases and the consequences for $\overline{\mathfrak{A}}$.

1) $A_0 \neq B_0$. Then $\epsilon_0(\overline{\mathfrak{A}})$, and a) and b) are true.

2) $A_0 = B_0$, $m = 1$. Then $\epsilon_0(\overline{\mathfrak{A}})$, and a) and b) are true.

3) $A_0 = B_0$, $m > 1$, $\eta_2 \neq y$. Then $\epsilon_1(\overline{\mathfrak{A}})$, and a) and b) are true.

4) $A_0 = B_0$, $\eta_2 = y$. Then (3) can be written in the form

$$A_0\,(zC\bar{y}D)^{\lambda}\, zCyDxA_1\,\,[\bar{\xi}_i A_i]_{i=2}^{n} = A_0\,[zP_i\bar{y}Q_i]_{i=1}^{k}\,[\bar{\zeta}_i R_i]_{i=1}^{t}, \tag{4}$$

where $k \geqq 1$, $t = m - 2k$, the ζ_i are unknowns, P_i, Q_i and R_i are coefficients, and $t = 0$, $t = 1$ or $\zeta_1\zeta_2 \neq zy$. The argument now splits into two subcases.

Subcase 4.1. $\lambda \geqq k$. Using induction on k, it is not difficult to see that either $P_1 = \cdots = P_k = C$ and $Q_1 = \cdots = Q_{k-1} = D$, or $\epsilon_0(\overline{\mathfrak{A}})$. The interest lies in the first alternative. Let us consider it. Equation (4) can be cancelled on the left by $A_0(zC\bar{y}D)^{k-1}zC\bar{y}$. After cancellation we obtain

$$D\,(zC\bar{y}D)^{\lambda-k}\, zCyDxA_1\,[\bar{\xi}_i A_i]_{i=2}^{n} = Q_k\,[\bar{\zeta}_i R_i]_{i=1}^{t}. \tag{5}$$

It is obvious that $\overline{\mathfrak{A}}$ is equivalent to the equation $xAyBz\Phi_1 = zCyDx\Psi_1$, where $\Phi_1 = \Psi_1$ is equation (5). We list below the possible cases and the consequences for $\overline{\mathfrak{A}}$.

1) $t = 0$. Then $\epsilon_0(\overline{\mathfrak{A}})$.

2) $t > 0$, $\zeta_1 = y$. Then $\epsilon_1(\overline{\mathfrak{A}})$.

3) $\zeta_1 = z$. Then (5) differs from (3) only in that $\lambda - k$ replaces λ. Arguing as above, following (3), and taking into account that $\zeta_2 \neq y$ (so that case 4) is now impossible), we see that $\epsilon_i(\overline{\mathfrak{A}})$, where $0 \leqq i \leqq 1$.

4) $\zeta_1 = x$, $Q_k \neq D$. Then $\epsilon_0(\overline{\mathfrak{A}})$.

5) $\zeta_1 = x$, $Q_k = D$. From the above argument it is evident that either $\epsilon_i(\overline{\mathfrak{A}})$, where $0 \leqq i \leqq 1$, or we have the relations $A_0 = B_0$, $\xi_1 = x$, $P_1 = \cdots = P_k = C$ and $Q_1 = \cdots = Q_k = D$. In the second alternative the equation $\Phi = \Psi$ has the form $A_0xA_1[\xi_i A_i]_{i=2}^{n} = A_0(zCyD)^k xR_1[\zeta_i R_i]_{i=2}^{t}$. Let $\amalg$ denote the relation of equality in the semigroup generated by the letters of the alphabet $\mathbf{H} \cup \{x, y, z\}$ with defining relation $xAyBz \amalg zCyDx$. It is obvious that if $\Psi \amalg \Psi_1$, then the equation $xAyBz\Phi \rightarrow zCyDx\Psi$ is equivalent to the equation $xAyBz\Phi \rightarrow zCyDx\Psi_1$. This follows from the fact that, in view of the assumption $[AB^{\partial} = [CD^{\partial}$, any solution of $xAyBz \cdots \rightarrow zCyDx \cdots$ satisfies the equation $xAyBz \rightarrow zCyDx$. The analogous statement is true if $\rightarrow$ is replaced by $=$. Obviously, $x(AyBz)^k \amalg (zCyD)^k x$. Hence (*) is equivalent to the equation $\mathfrak{A}$ having the form $xAyBz\Phi_1 \rightarrow zCyDx\Psi_1$, where $\Phi_1 = \Psi_1$ is obtained from $\Phi = \Psi$ by replacing $(zCyD)^k x$ by $x(AyBz)^k$ and then cancelling

on the left by A_0x. The equation $\mathfrak{A}$ is an equation of the form (*) of height $h-2$. Hence the induction assumption is applicable to $\mathfrak{A}$. Applying it and noting Remark 1, we see that a) is true.

We continue the argument to prove b). Put $\overline{\Phi}_1 = \Phi_1(\bar{x},\bar{y},\bar{z})$ and $\overline{\Psi}_1 = \Psi_1(\bar{x},\bar{y},\bar{z})$. By Remark 1, $\overline{\mathfrak{A}}$ is equivalent to the equation $xAyBz\overline{\Phi}_1 = zCyDx\overline{\Psi}_1$. The equation $\overline{\Phi}_1 = \overline{\Psi}_1$ itself has the form

$$(6) \qquad A_1\,[\bar{\xi}_iA_i]_{i=2}^n = (A\bar{y}Bz)^k\,R_1\,[\bar{\zeta}_iR_i]_{i=2}^t.$$

It is now easy to see that either $n>1$, $\xi_2 = y$ and $A_1 = A$, or $\epsilon_i(\overline{\mathfrak{A}})$, where $0 \leq i \leq 1$. It follows from the above argument that either $\Phi = \Psi$ has the form $A_0xAy\cdots = A_0zCy\cdots$ or $\epsilon_i(\overline{\mathfrak{A}})$, where $0\leq i\leq 1$. In particular, if (*) is an ϵ_2-equation, then $\epsilon_i(\overline{\mathfrak{A}})$, where $0\leq i\leq 1$.

Subcase 4.2. $\lambda<k$. Using induction on λ, we see that either $P_1=\cdots=P_{\lambda+1}=C$ and $Q_1=\cdots=Q_\lambda=D$, or $\epsilon_0(\overline{\mathfrak{A}})$. It suffices to consider the first alternative. Let us do so. Equation (4) can be cancelled on the left by $A_0(zC\bar{y}D)^\lambda zCyDx$. After cancellation we obtain

$$(7) \qquad A_1\,[\bar{\xi}_iA_i]_{i=2}^n = Ay\,(DxAy)^{\mu-1}\,Q_{\lambda+1}\,[\bar{\zeta}_iR_i]_{i=1}^t,$$

where $t=m-2(\lambda+1)$, the ζ_i are unknowns and the R_i are coefficients. The equation $\overline{\mathfrak{A}}$ is equivalent to $xAyBz\Phi_1 = zCyDx\Psi_1$, where $\Phi_1=\Psi_1$ is equation (7). Hence if $n=1$, then $\epsilon_0(\overline{\mathfrak{A}})$; if $n>1$, $\xi_2\neq y$, then $\epsilon_1(\overline{\mathfrak{A}})$; and if $\xi_2=y$, $A_1\neq A$, then $\epsilon_0(\overline{\mathfrak{A}})$. There remains the case $A_1=A$, $\xi_2=y$.

It follows from the above argument that either $\Phi=\Psi$ has the form $A_0xAy\cdots=A_0xCy\cdots$ or $\epsilon_i(\overline{\mathfrak{A}})$, where $0\leq i\leq 1$. In particular, if (*) is an ϵ_2-equation, then $\epsilon_i(\overline{\mathfrak{A}})$, where $0\leq i\leq 1$. Hence b) is completely proved for θ_1-images.

We now turn to the case $A_1=A$, $\xi_2=y$. Equation (7) can be cancelled on the left by $Ay(DxAy)^{\mu-1}$. We obtain

$$(8) \qquad DxAyA_2\,[\bar{\xi}_iA_i]_{i=3}^n = Q_{\lambda+1}\,[\bar{\zeta}_iR_i]_{i=1}^t.$$

The equation $\overline{\mathfrak{A}}$ is equivalent to the equation $xAyBz\Phi_1 = zCyDx\Psi_1$, where $\Phi_1=\Psi_1$ is equation (8). We list the possible cases and the consequences for $\overline{\mathfrak{A}}$.

1) $t=0$. Then $\epsilon_0(\overline{\mathfrak{A}})$.

2) $t>0$, $\zeta_1=y$. Then $\epsilon_1(\overline{\mathfrak{A}})$.

3) $\zeta_1\neq y$, $Q_{\lambda+1}\neq D$. Then $\epsilon_0(\overline{\mathfrak{A}})$.

4) $Q_{\lambda+1}=D$, $\zeta_1=x$. Then the equation $\Phi=\Psi$ has the form $A_0xA_1[\xi_iA_i]_{i=2}^n = (zCyD)^{\lambda+1}xR_1[\zeta_iR_i]_{i=2}^t$. This case is handled just like subcase 4.1.5), with $\lambda+1$ in place of k. Thus a) and b) are true.

5) $Q_{\lambda+1}=D$, $\zeta_1=z$. Then (8) has the form

$$(9) \qquad DxAyA_2\,[\bar{\xi}_iA_i]_{i=3}^n = DzR_1\,[\bar{\zeta}_iR_i]_{i=2}^t.$$

Recalling the definition of an ϵ_2-equation, we see that if $t=1$, then $\epsilon_2(\overline{\mathfrak{A}})$, and if $t>1$, $R_1\neq C$, then $\epsilon_2(\overline{\mathfrak{A}})$, and also if $t>1$, $\zeta_2\neq y$, then $\epsilon_2(\overline{\mathfrak{A}})$. There remains the case $R_1=C$, $\zeta_2=y$. In this case (9) has the form

$$(10) \qquad DxAyA_2\,[\bar{\xi}_iA_i]_{i=3}^n = DzCyDxAy\,(DxAy)^{\mu-1}\,R_2\,[\bar{\zeta}_iR_i]_{i=3}^t.$$

This is the equation $\Phi_1=\Psi_1$. The equation $\overline{\mathfrak{A}}$, as we have observed, is equivalent to the equation $xAyBz\Phi_1=zCyDx\Psi_1$. Denote this last equation by $\overline{\mathfrak{B}}$. The equation $\overline{\mathfrak{B}}$ is obviously equivalent to the equation $xAyBz\Phi_2=$

$zCyDx\Psi_2$, where $\Phi_2 = \Psi_2$ is obtained from $\Phi_1 = \Psi_1$ by replacing the first occurrence of $zCyDx$ on the right-hand side by $xAyBz$ and then cancelling on the left by $DxAy$. The equation $\Phi_2 = \Psi_2$ has the form

$$(11) \qquad A_2\,[\xi_i A_i]_{i=3}^n = BzAy\,(DxAy)^{\mu-1}\,R_2\,[\zeta_i R_i]_{i=3}^t.$$

It follows that $\overline{\mathfrak{A}}$ is equivalent to the equation $xAyBz\Phi_2 = zCyDx\Psi_2$. If now $n = 2$, then $\epsilon_0(\overline{\mathfrak{A}})$; if $n > 2$, $\xi_3 = y$, then $\epsilon_1(\overline{\mathfrak{A}})$, and if $\xi_3 \neq y$, $A_2 \neq B$, then $\epsilon_0(\overline{\mathfrak{A}})$. There remains the case $A_2 = B$, $\xi_3 \neq y$. Here there are two possible subcases.

1) $\xi_3 = z$. Then the equation $\Phi = \Psi$ has the form $A_0xAyBzA_3[\xi_iA_i]_{i=4}^n = A_0zCyD[\eta_iB_i]_{i=3}^m$. Now a) is established in the same way as in subcase 4.1.5), using $xAyBz \perp\!\!\!\perp zCyDx$ and the induction assumption.

2) $\xi_3 = x$. Then (11) has the form

$$(12) \qquad B\,(zC\bar{y}D)^\lambda\, zCyDxA_3\,[\xi_iA_i]_{i=4}^n = BzAy\,(DxAy)^{\mu-1}\,R_2\,[\bar{\zeta}_iR_i]_{i=3}^t.$$

If $A \neq C$, then $\epsilon_0(\overline{\mathfrak{A}})$. Suppose $A = C$. Then, cancelling (12) on the left by $BxAy$, we obtain

$$(13) \qquad ((DxAy)^\mu DzCy)^\lambda\, DxA_3\,[\bar{\xi}_iA_i]_{i=4}^n = (DxAy)^{\mu-1}\,R_2[\bar{\zeta}_iR_i]_{i=3}^t.$$

We have $A = C$. Hence $B \neq D$, inasmuch as $AyB \neq CyD$. This allows us to use the following.

Remark. Suppose $B \neq D$ and $\overline{\mathfrak{A}}$ is equivalent to the equation $xAyBz\varphi = zCyDx\psi$, where the equation $\varphi = \psi$ has the form $PxAyD\xi \cdots = Q[\bar{\alpha}_iC_i]_{i=1}^d$, where P, Q and C_i are coefficients, ξ and α_i are unknowns, and $d \geqq 0$. Then $\epsilon(\overline{\mathfrak{A}})$.

Indeed, if $d = 0$, then $\epsilon_0(\overline{\mathfrak{A}})$; if $d > 0$, $\alpha_1 = y$, then $\epsilon_1(\overline{\mathfrak{A}})$, and if $\alpha_1 \neq y$, $P \neq Q$, then $\epsilon_0(\overline{\mathfrak{A}})$. If $\alpha_1 = x$, $P = Q$, then $Q\bar{\alpha}_1$ begins with the word $PzCyDx$, and since $xAyBz \perp\!\!\!\perp zCyDx$, $B \neq D$, we obtain $\epsilon_i(\overline{\mathfrak{A}})$, $0 \leq i \leq 1$. Finally, if $P = Q$, $\alpha_1 = z$, then (by considering cases) we have either $\epsilon_2(\overline{\mathfrak{A}})$ or $C_1 = C$, $\alpha_2 = y$. In the latter case $zCyDx \angle \bar{\alpha}_1C_1\bar{\alpha}_2$, and since $B \neq D$, $xAyBz \perp\!\!\!\perp zCyDx$, we obtain $\epsilon_i(\overline{\mathfrak{A}})$, $0 \leq i \leq 1$. The remark is proved.

Thus, we have that $\overline{\mathfrak{A}}$ is equivalent to the equation $xAyBz\varphi = zCyDx\psi$, where $\varphi = \psi$ is equation (13). We consider three cases.

1) $\lambda > 0$. Then, cancelling (13) on the left by $(DxAy)^{\mu-1}$, we obtain an equation $DxAyDz \cdots = R_2[\bar{\zeta}_iR_i]_{i=3}^t$. By the remark we have $\epsilon(\overline{\mathfrak{A}})$, and a) is true.

2) $\lambda = 0$, $\mu > 1$. Cancelling (13) on the left by Dx, we obtain

$$(14) \qquad A_3\,[\bar{\xi}_iA_i]_{i=4}^n = Ay\,(DxAy)^{\mu-2}\,R_2[\bar{\zeta}_iR_i]_{i=3}^t.$$

We list the subcases and the consequences for $\overline{\mathfrak{A}}$.

2.1) $n = 3$. Then $\epsilon_0(\overline{\mathfrak{A}})$.

2.2) $n > 3$, $\xi_4 \neq y$. Then $\epsilon_1(\overline{\mathfrak{A}})$.

2.3) $\xi_4 = y$, $A_3 \neq A$. Then $\epsilon_0(\overline{\mathfrak{A}})$.

2.4) $\xi_4 = y$, $A_3 = A$. Cancelling (14) on the left by $Ay(DxAy)^{\mu-2}$, we obtain an equation $DxAyDx \cdots = R_2[\bar{\zeta}_iR_i]_{i=3}^t$; hence by the remark we must have $\epsilon(\overline{\mathfrak{A}})$.

3) $\lambda = 0$, $\mu = 1$. We have the equation

$$(15) \qquad DxA_3\,[\bar{\xi}_iA_i]_{i=4}^n = R_2\,[\bar{\zeta}_iR_i]_{i=3}^t.$$

We list the subcases and the consequences for $\overline{\mathfrak{A}}$.

3.1) $t=2$. Then $\epsilon_0(\overline{\mathfrak{A}})$.

3.2) $t>2$, $\zeta_3=y$. Then $\epsilon_1(\overline{\mathfrak{A}})$.

3.3) $\zeta_3 \neq y$, $R_2 \neq D$. Then $\epsilon_0(\overline{\mathfrak{A}})$.

3.4) $\zeta_3=z$, $R_2=D$. Here either $\epsilon_2(\overline{\mathfrak{A}})$ or we have the relations $t>3$, $A_3=A$ and $\xi_4=y$. In the latter case (15) has the form $DxAyDx\cdots = R_2[\overline{\zeta}_iR_i]_{i=3}^t$, and by the remark we must have $\epsilon(\overline{\mathfrak{A}})$.

3.5) $\zeta_3=x$, $R_2=D$. Then the equation $\Phi=\Psi$ has the form $A_0xA_1[\xi_iA_i]_{i=2}^n = A_0(zCyD)^{\lambda+2}xR_3[\zeta_iR_i]_{i=4}^t$. Arguing as in subcase 4.1.5) with $\lambda+2$ in place of k, we see that a) is true. Hence the assertions 8.2 are completely proved for θ_1-images.

Case 2. $\overline{\mathfrak{A}}$ is a θ_2-image. Then

$$\bar{x}=(\bar{z}C\bar{y}D)^{\lambda}x,\quad \bar{y}=(yBzC)^{\mu}y,\quad \bar{z}=xAyBz,\quad \lambda>0,\quad \mu>0.$$

The idea of the proof is the same as in Case 1. We sketch the proof, omitting those parts analogous to the corresponding parts in Case 1, and indicating the new points.

If one of the letters ξ_1 or η_1 is y and the other is different from y, then $\epsilon_1(\overline{\mathfrak{A}})$.

Consider the case $\xi_1=x$, $\eta_1=z$. The equation $\overline{\Phi}=\overline{\Psi}$ has the form

$$A_0\,(\bar{z}C\bar{y}D)^{\lambda}\,xA_1\,[\bar{\xi}_iA_i]_{i=2}^{n}=B_0\bar{z}B_1\,[\bar{\eta}_iB_i]_{i=2}^{m}. \tag{16}$$

Here (by considering cases) either $A_0=B_0$, $m>1$, $\eta_2=y$, or else $\epsilon_i(\overline{\mathfrak{A}})$, where $0\leq i\leq 1$. Suppose the first alternative holds. Equation (16) can be written in the form (see (4))

$$A_0\,(\bar{z}C\bar{y}D)^{\lambda}\,xA_1\,[\bar{\xi}_iA_i]_{i=2}^{n}=A_0\,[\bar{z}P_i\bar{y}Q_i]_{i=1}^{k}\,[\bar{\zeta}_iR_i]_{i=1}^{t}, \tag{17}$$

where $k\geq 1$, $t=m-2k$, $\zeta_1,\cdots,\zeta_t$ are unknowns, P_i, Q_i and R_i are coefficients, and $t=0$, $t=1$ or $\zeta_1\zeta_2\neq zy$. There are two subcases.

Subcase 2.1. $\lambda\geq k$. Here, arguing as in Subcase 4.1, p. 104, we see that either $\epsilon_i(\overline{\mathfrak{A}})$, where $0\leq i\leq 1$, or we have the relations $P_1=\cdots=P_k=C$, $Q_1=\cdots=Q_k=D$, $t>0$ and $\zeta_1\neq y$. Consider the second alternative. The case $\zeta_1=z$, $\lambda>k$ is handled just like Case 1 , Subcase 4.1.3); here we obtain $\epsilon_i(\overline{\mathfrak{A}})$, where $0\leq i\leq 1$.

Suppose $\zeta_1=z$ and $\lambda=k$. Cancelling (17) on the left by $A_0(\bar{z}C\bar{y}D)^{\lambda}x$, we obtain

$$A_1\,[\bar{\xi}_iA_i[_{i=2}^{n}=AyBzR_1\,[\bar{\zeta}_iR_i]_{i=2}^{t}. \tag{17'}$$

If now $n=1$, then $\epsilon_0(\overline{\mathfrak{A}})$; if $n>1$, $\xi_2\neq y$, then $\epsilon_1(\overline{\mathfrak{A}})$, and if $\xi_2=y$, $A_1\neq A$, then $\epsilon_0(\overline{\mathfrak{A}})$. There remains the case $A_1=A$, $\xi_2=y$. Then the left-hand side of (17′) begins with the word $AyBzCy$. Cancelling (17′) on the left by $AyBz$, we obtain an equation of the form $Cy\cdots=R_1[\overline{\zeta}_iR_i]_{i=2}^t$. If now $t=1$, then we have $\epsilon_0(\overline{\mathfrak{A}})$, and if $t>1$, then from the condition $\zeta_1\zeta_2\neq zy$ we obtain $\zeta_2\neq y$, and then, obviously, $\epsilon_1(\overline{\mathfrak{A}})$. Hence if $\zeta_1=z$ and $\lambda=k$, we have $\epsilon_i(\overline{\mathfrak{A}})$, $0\leq i\leq 1$.

There remains the case $\zeta_1=x$. Here, to prove a), we argue as in Subcase 4.1.5) of Case 1, using $xAyBz \perp\!\!\!\perp zCyDx$ and the induction assumption. To prove b), we argue as in Subcase 4.1.5), following (6), only now instead of z in (6) there stands $\bar{z}$.

Subcase 2.2. $\lambda<k$. We obtain in the usual way that either $\epsilon_i(\overline{\mathfrak{A}})$, where

$0 \leqq i \leqq 1$, or we have the relations $P_1 = \cdots = P_\lambda = C$ and $Q_1 = \cdots = Q_\lambda = D$. Note that we have $\lambda > 0$, so that, in particular, $P_1 = C$ and $Q_1 = D$. Consider the second alternative. Cancelling (17) on the left by $A_0(\bar{z}C\bar{y}D)^\lambda x$, we obtain

$$A_1 [\bar{\xi}_i A_i]_{i=2}^n = AyBzP_{\lambda+1}\bar{y}Q_{\lambda+1}[\bar{\eta}_i B_i]_{i=2\lambda+3}^m. \tag{18}$$

Now either $\epsilon_i(\overline{\mathfrak{A}})$, where $0 \leqq i \leqq 1$, or we have the relations $n > 1$, $\xi_2 = y$ and $A_1 = A$. Consider the second alternative. Cancelling (18) on the left by $AyBz$, we obtain

$$Cy(BzCy)^{\mu-1} A_2 [\bar{\xi}_i A_i]_{i=3}^n = P_{\lambda+1}\bar{y}Q_{\lambda+1}[\bar{\eta}_i B_i]_{i=2\lambda+3}^m. \tag{19}$$

If $P_{\lambda+1} \neq C$, then $\epsilon_0(\overline{\mathfrak{A}})$. Suppose $P_{\lambda+1} = C$. Then (19) can be cancelled on the left by $Cy(BzCy)^{\mu-1}$. After cancellation we obtain

$$A_2 [\bar{\xi}_i A_i]_{i=3}^n = BzCyQ_{\lambda+1}[\bar{\eta}_i B_i]_{i=2\lambda+3}^m. \tag{20}$$

Now either $\epsilon_i(\overline{\mathfrak{A}})$, where $0 \leqq i \leqq 1$, or we have the relations $n > 2$, $\xi_3 \neq y$ and $A_2 = B$.

It follows that either $\epsilon_i(\overline{\mathfrak{A}})$, where $0 \leqq i \leqq 1$, or the equation $\Phi = \Psi$ has the form $A_0xAyB\xi \cdots = A_0zCyD\eta \cdots$, where ξ and η are unknowns. In particular, if (*) is an ϵ-equation, then we must have $\epsilon_i(\overline{\mathfrak{A}})$, where $0 \leqq i \leqq 1$. Hence b) is proved for θ_2-images.

We continue with the case $\xi_3 \neq y$, $A_2 = B$. If $\xi_3 = z$, then the equation $\Phi = \Psi$ has the form $A_0xAyBz \cdots = A_0zCyD \cdots$. The rest of the argument is like Subcase 4.1.5), using $xAyBz \perp\!\!\!\perp zCyDx$ and the induction assumption. Thus a) is true.

There remains the case $\xi_3 = x$. We have

$$\bar{x} = (xAyBzC\bar{y}D)^\lambda x \perp\!\!\!\perp zCyDxC\bar{y}D(\bar{z}C\bar{y}D)^{\lambda-1}x.$$

Substituting the latter expression for $\bar{\xi}_3$ in (20) and cancelling the resulting equation on the left by $BzCy$, we obtain

$$DxC\bar{y}D(\bar{z}C\bar{y}D)^{\lambda-1}xA_3[\bar{\xi}_i A_i]_{i=4}^n = Q_{\lambda+1}[\bar{\eta}_i B_i]_{i=2\lambda+3}^m. \tag{21}$$

Obviously, $\overline{\mathfrak{A}}$ is equivalent to the equation $xAyBz\Phi_1 = zCyDx\Psi_1$, where $\Phi_1 = \Psi_1$ is equation (21). Consider (21). Either $\epsilon_i(\overline{\mathfrak{A}})$, where $0 \leqq i \leqq 1$, or we have the relations $m \geqq 2\lambda + 3$, $\eta_{2\lambda+3} \neq y$ and $Q_{\lambda+1} = D$. Consider the second alternative. If $\eta_{2\lambda+3} = x$, then the equation $\Phi = \Psi$ has the form $A_0xA_1 \cdots = A_0(zCyD)^{\lambda+1}x \cdots$, and we can argue as in Subcase 4.1.5).

Suppose $\eta_{2\lambda+3} = a$. Then (21) can be cancelled on the left by Dx. After cancellation we obtain

$$C\bar{y}D(\bar{z}C\bar{y}D)^{\lambda-1}x \ldots = AyBzB_{2\lambda+3}[\bar{\eta}_i B_i]_{i=2\lambda+4}^m. \tag{22}$$

Now either $A = C$ or $\epsilon_0(\overline{\mathfrak{A}})$. Suppose $A = C$. Then $B \neq D$. Cancelling (22) on the left by $AyBz$, we obtain

$$Cy(BzCy)^{\mu-1}Dx \ldots = B_{2\lambda+3}[\bar{\eta}_i B_i]_{i=2\lambda+4}^m. \tag{23}$$

Here either $\epsilon_i(\overline{\mathfrak{A}})$, where $0 \leqq i \leqq 1$, or we have the relations $m \geqq 2\lambda + 4$, $\eta_{2\lambda+1} = y$ and $B_{2\lambda+3} = C$. Consider the second alternative. Cancelling (23) on the left by $Cy(BzCy)^{\mu-1}$, we obtain an equation $Dx \cdots = Bz \cdots$, and hence $\epsilon_0(\overline{\mathfrak{A}})$, since $B \neq D$. Thus 8.2 is proved for θ_2-images.

Case 3. $\overline{\mathfrak{A}}$ is a θ_3-image. The θ_3-images of (*) are its two-sided μ-images in the sense of §7. By hypothesis, $[AB^\partial = [CD^\partial$ and $AyB \neq CyD$. Therefore, applying Proposition 7.4.a) to (*) (the roles of A and B in 7.4 are now played by AyB and CyD), we conclude that one of two things is true: 1) (*) is equivalent to the equation $xAyBz \to zCyDx$; 2) $\overline{\mathfrak{A}}$ possesses property α'. Furthermore, if (*) is an ϵ_2-equation, then it is an ϵ-equation in the sense of §7, and, in view of 7.4.b), $\overline{\mathfrak{A}}$ possesses property α'. Consider case 1). The θ-images of the equation $xAyBz \to zCyDx$ have the form $xAyBz = zCyDx$. In view of the equivalence of θ-images of equivalent equations, we see that in case 1) the equation $\overline{\mathfrak{A}}$ is equivalent to the equation $xAyBz = zCyDx$, i.e. we have 8.2.a). It remains to consider the case where $\overline{\mathfrak{A}}$ possesses property α'. We will prove that in this case we have $\epsilon_i(\overline{\mathfrak{A}})$, where $0 \leq i \leq 1$. Indeed, if $\overline{\mathfrak{A}}$ possesses property α', then it is equivalent to an equation

$$xAyBzP(y)\xi \cdots = zCyDxQ(y)\eta \cdots,$$

where ξ and η are x or z; P and Q are words in the alphabet $\mathbf{H} \cup \{y\}$ and $P \neq Q$. If P and Q contain the same number of occurrences of y ($\nu(P) = \nu(Q)$ in the notation preceding 4.4), then since $P \neq Q$ the equation $P = Q$ has (after left cancellation) one of the following three forms: 1) $Ry \cdots = y \cdots$, where R is a nonempty coefficient; 2) $a \cdots = b \cdots$, where a and b are distinct letters of the alphabet $\mathbf{H}$; 3) $R = \Lambda$, where R is a nonempty coefficient. In case 1) we have $\epsilon_0(\overline{\mathfrak{A}})$; in case 2) we also have $\epsilon_0(\overline{\mathfrak{A}})$, since in this case $\overline{\mathfrak{A}}$ has no solutions; in case 3) we again have $\epsilon_0(\overline{\mathfrak{A}})$, since $P\xi \cdots = Q\eta \cdots$ is equivalent to $R\xi \cdots = \eta \cdots$. Hence if $\nu(P) = \nu(Q)$, then $\epsilon_0(\overline{\mathfrak{A}})$. Suppose $\nu(P) \neq \nu(Q)$. Here either $\epsilon_0(\overline{\mathfrak{A}})$ for the same reasons as in the case $\nu(P) = \nu(Q)$, or the equation $P\xi \cdots = Q\eta \cdots$ has the form (if $\nu(P) > \nu(Q)$) $Sy \cdots = T\eta \cdots$, where S and T are coefficients, i.e. we have $\epsilon_1(\overline{\mathfrak{A}})$. Hence if $\overline{\mathfrak{A}}$ possesses property α', then $\epsilon_i(\overline{\mathfrak{A}})$, where $0 \leq i \leq 1$. Thus a) and b) are proved for θ_3-images. The proposition is proved.

Remark. If we group the arguments in 8.2 somewhat differently, we can prove that either (*) is equivalent to the equation $xAyBz \to zCyDx$ or any θ-image of (*) is equivalent to an ϵ-equation.

8.3. *Suppose* $[AB^\partial = [CD^\partial$ *and* $AyB \neq CyD$ *in* (*). *Then it is possible to construct a tree of* (*) *with final vertices of the following three kinds*: 1) *the equations* $xAyBz \rightleftarrows cZyDx$; 2) *basic equations of characteristic zero*; 3) ϵ_0-*equations*.

Proof. In the tree of 8.1 the final vertices are θ-images, and we may assume them to be one-sided equations. By 8.1, 8.2.a) and the extended replacement rule (following 6.24), it is possible to construct a tree with final vertices of the kinds 1) and 3) plus vertices which are one-sided ϵ_i-equations (ϵ_i-vertices), $1 \leq i \leq 2$. By 8.1 and 8.2.b), for each ϵ_2-vertex it is possible to construct a tree with final ϵ_i-vertices, $0 \leq i \leq 1$. Hence it is possible to construct a tree of (*) with final vertices of the kinds 1) and 3) plus, possibly, ϵ_i-vertices. But any ϵ_1-vertex possesses a tree with final vertices of the kind 2). It suffices to observe that the projection of any ϵ_1-vertex (one-sided) on the alphabet $\{x, y, z\}$ has the form $xyz\alpha \cdots \rightleftarrows zyx\beta \cdots$, where one of the unknowns α or β is y and the other is x or z, and use 6.20 and 6.23. The proposition is proved.

Now that 8.3 has been proved, for the complete solution of basic equations of characteristic ≤ 1 it remains to solve:

1) the equation $xAyBz\Phi(x,y,z) \to zAyBx\Psi(x,y,z)$;
2) the equation $xAyBz \to zCyDx$;
3) ϵ_0-equations;
4) basic equations of characteristic zero.

The phrase "solve the equation" means "construct an algorithm recognizing (in the first case), for given A, B, Φ and Ψ, whether or not the equation $xAyBz\Phi \to zAyBx\Psi$ has solutions". Analogous statements can be made for problems 2, 3 and 4.

We consider problem 4 in §9. Here we solve problems 1, 2 and 3. Let us begin with problem 1.

The equation $xAyBz\Phi(x,y,z) \to zAyBx\Psi(x,y,z)$

We now consider the case where $AyB = CyD$ in the equation (*). In other words, we consider the equation

$$(*) \qquad xAyBz\Phi(x,y,z) \to zAyBx\Psi(x,y,z),$$

where A and B are coefficients, and Φ and Ψ are words in the alphabet $\mathsf{H} \cup \{x,y,z\}$.

We choose the following method for solving (*). We express x, y and z from the prefix $xAyBz \to zAyBx$ and substitute the resulting expressions into the equation $\Phi = \Psi$. We obtain a one-stage exponential equation with variable bases: the bases contain word unknowns. The number of word unknowns does not exceed two, a fact which allows us to reduce the resulting equation to word equations in two unknowns.

We prove an assertion pertaining to exponential equations whose bases contain word unknowns.

8.4. *Suppose A is a word in the alphabet $\mathsf{H} \cup \{x_1, \cdots, x_n\}$ and each of the letters $x_1, \cdots, x_n$ occurs in A. Suppose*

$$\Phi(z_1, \cdots, z_m, x_1, \cdots, x_n), \qquad \Psi(z_1, \cdots, z_m, x_1, \cdots, x_n)$$

are words in the alphabet $\mathsf{H} \cup \{z_1, \cdots, z_m, x_1, \cdots, x_n\}$. Consider the equation in $m+n$ unknowns

$$(1) \qquad \Phi(A^{\lambda_1}, \ldots, A^{\lambda_m}, x_1, \ldots, x_n) = \Psi(A^{\lambda_1}, \ldots, A^{\lambda_m}, x_1, \ldots, x_n),$$

where $\lambda_1, \cdots, \lambda_m$ are natural unknowns and $x_1, \cdots, x_n$ are word unknowns. It is possible to construct a finite set $\mathfrak{M}$ of word equations in n unknowns such that (1) *has solutions if and only if at least one equation in the set $\mathfrak{M}$ has.*

PROOF. We proceed by induction on m. If $m=0$, the assertion is trivial. Suppose $m>0$ and the assertion is true for equations in less than m natural unknowns. Put $M = \max\{[\Phi^\partial, [\Psi^\partial, 8\}$. Let $\mathfrak{A}(\lambda_1, \cdots, \lambda_m, x_1, \cdots, x_n)$ denote the predicate defined by (1). Divide the domain of the variables $\lambda_1, \cdots, \lambda_m$ into two parts: the region $\Omega(\lambda_1, \cdots, \lambda_m)$ defined by the predicate $\bigvee_{i=1}^m (\lambda_i \leq 5M)$, and the region $\neg\Omega$ (defined by the inequalities $\lambda_1, \cdots, \lambda_m > 5M$). In Ω we have the equivalence

$$\mathfrak{A} \Leftrightarrow \bigvee_{i=1}^{m} \bigvee_{a=0}^{5M} [\mathfrak{A}(\lambda_1, \ldots, \lambda_{i-1}, a, \lambda_{i+1}, \ldots, \lambda_m, x_1, \ldots, x_n), \lambda_i = a].$$

Applying the induction assumption to each of the equations $\mathfrak{A}(\cdots, a, \cdots)$, we see that there exists a finite set $\mathfrak{M}_1$ of word equations in n unknowns such that $\mathfrak{A}$ has solutions in Ω if and only if at least one equation in $\mathfrak{M}_1$ is solvable.

Consider the region $\neg\Omega$. We will show that it is possible to construct a finite set $\mathfrak{N}$ of number m-tuples $(\lambda_{1i}, \cdots, \lambda_{mi})$, $i = 1, \cdots, N$, such that we have the equivalence

$$(2)\qquad \exists\lambda_1 \ldots \lambda_m x_1 \ldots x_n (\mathfrak{A} \,\&\, \neg\Omega) \leftrightarrow \bigvee_{i=1}^{N} \exists x_1 \ldots x_n \mathfrak{A}(\lambda_{1i}, \ldots, \lambda_{mi}, x_1, \ldots, x_n).$$

We construct the set $\mathfrak{N}$ as follows. Let Σ be the set of all possible systems of linear Diophantine equations in m unknowns $\lambda_1, \cdots, \lambda_m$ whose coefficients and free terms are in absolute value $\leqq 2M^2$. The set Σ is finite. Let Σ' be the subset consisting of those systems in Σ which have solutions in the region $\neg\Omega$ (the selection of Σ' from Σ is possible using the decidability of the theory of addition of natural numbers [7]). For each system of Σ' choose some solution in $\neg\Omega$. We obtain a set $\mathfrak{N}$ of natural m-tuples. We will prove that $\mathfrak{N}$ is the desired set. It is obvious that the right-hand side of the equivalence (2) implies the left. Let us prove that the left-hand side of (2) implies the right. Indeed, suppose $(\lambda_1, \cdots, \lambda_m, X_1, \cdots, X_n)$ is a solution of (1) in $\neg\Omega$. Put $T = A(X_1, \cdots, X_n)$. The case $T = \Lambda$ is trivial. Suppose $T \neq \Lambda$. Then the m-tuple $(\lambda_1, \cdots, \lambda_m)$ satisfies the exponential equation

$$(3)\qquad \Phi(T^{\lambda_1}, \ldots, T^{\lambda_m}, X_1, \ldots, X_n) = \Psi(T^{\lambda_1}, \ldots, T^{\lambda_m}, X_1, \ldots, X_n).$$

Let H, h and N denote the characteristic, height and maximum coefficient length of (3). Obviously, $h \leqq M$. Furthermore, $N \leqq (M-1)[X_1 \cdots X_n^\partial + M - 1$. Since $x_1, \cdots, x_n$ occur in A, we have $[T^\partial \geqq [X_1 \cdots X_n^\partial$. Hence for the characteristic H we have

$$H = \frac{\max\{N, [T^\partial\}}{[T^\partial} \leqslant M.$$

In the region $\neg\Omega$ we have $\lambda_1, \cdots, \lambda_m > 5M \geqq 5H$. Hence by 2.5 the set Σ' contains a system of equations $\mathscr{D}(\lambda_1, \cdots, \lambda_m)$ which is satisfied by $(\lambda_1, \cdots, \lambda_m)$ and any positive solution of which satisfies (3). Consequently, $\mathfrak{N}$ contains an m-tuple $(\tilde{\lambda}_1, \cdots, \tilde{\lambda}_m)$ such that $\mathfrak{A}(\tilde{\lambda}_1, \cdots, \tilde{\lambda}_m, X_1, \cdots, X_n)$, so that the right-hand side of (2) is true. The equivalence (2) is proved. Now let $\mathfrak{M}_2$ denote the set of word equations (1), where $(\lambda_1, \cdots, \lambda_m)$ ranges over all possible m-tuples in $\mathfrak{N}$. The set $\mathfrak{M}_1 \cup \mathfrak{M}_2$ is the desired one. The proposition is proved.

8.5. *There exists an algorithm recognizing the solvability of word equations of the form $xAyBz\Phi(x,y,z) \to zAyBx\Psi(x,y,z)$, where A and B are coefficients, and Φ and Ψ are words in the alphabet $\mathbf{H} \cup \{x,y,z\}$.*

PROOF. Suppose an equation (*) is given. Any of its solutions satisfies the equation $xAyBz \to zAyBx$. We will express x, y and z from the equation $xAyBz \to zAyBx$. By 1.23, we have expressions for x, AyB and z:

$$(1)\qquad x = (uw)^\lambda u, \quad AyB = w(uw)^\nu, \quad z = (uw)^\mu u, \quad \lambda > \mu,$$

where u and w are word unknowns, and λ, μ and ν are natural unknowns.

We will now express y, u and w from the equation $AyB = w(uw)^\nu$. It is convenient to divide the domain of the variables u and w into three parts.

1) *The region* $\partial(u), \partial(w) \leqq [AB^\partial$. In this region x, y and z are periodic

words with period of length $2[AB^\partial$. Determination of the solutions of (*) in the region $\partial(u), \partial(w) \leqq [AB^\partial$ reduces to the determination of the solutions of one-stage exponential equations.

2) *The region* $\partial(w) \geqq [AB^\partial$. In this region we have from the equation $AyB = w(uw)^\nu$ this expression for w: $w = AvB$. Then x, y and z are expressed by means of the formulas

$$x = (uAvB)^\lambda u, \quad y = (vBuA)^\nu v, \quad z = (uAvB)^\mu u, \quad \lambda > \mu.$$

If $\nu = 0$, then $y = v$. If $\nu > 0$, then $y = vB(uAvB)^{\nu-1}uAv$. It is now evident that one can construct a finite set $\mathfrak{M}$ of equations in the five unknowns u, v, λ, μ, ν of the form

$$\Phi(T^\lambda, T^\nu, T^\mu, u, v) = \Psi(T^\lambda, T^\nu, T^\mu, u, v),$$

where $T = uAvB$, and $\Phi(x, y, z, u, v)$ and $\Psi(x, y, z, u, v)$ are words in the alphabet $\mathbf{H} \cup \{x, y, z, u, v\}$, such that (*) has solutions in region 2) if and only if at least one of the equations in $\mathfrak{M}$ has. But, by 8.4, the solution of the equations of $\mathfrak{M}$ reduces to the solution of word equations in two unknowns. It remains to apply 7.7.

3) *The region* $\partial(w) < [AB^\partial$, $\partial(u) > [AB^\partial$. Then $\nu > 0$. Considering different cases depending on relations among the lengths of the words w, A and B, it is not difficult to see that in each case x, y and z are expressed in the form

(2) $$x = P_1(vQ)^\lambda vP_2, \; y = Q_1(vQ)^\nu vQ_2, \; z = P_1(vQ)^\mu vP_2, \; \lambda > \mu,$$

where v is a word unknown, and P_i, Q_i and Q are words in the alphabet $\mathbf{H}$ of length $\leqq 2[AB^\partial$. Consider, for example, the case $\partial(w) \geqq [A^\partial$, $\partial(w) \leqq [B^\partial$. In this case we have from the equation $AyB = w(uw)^\nu$, assuming $\nu > 0$: $w = AR$, $B = Sw$, $yS = R(uAR)^{\nu-1}u$. Also, $[S^\partial \leqq [B^\partial < \partial(u)$. Hence $u = vS$, $y = R(vSAR)^{\nu-1}v$ and, also (from (1)), $x = (vSAR)^\lambda vS$, $z = (vSAR)^\mu vS$, and we obtain for x, y and z the representation (2). The other cases: 1) $\partial(w) \leqq [A^\partial$, $\partial(w) > [B^\partial$; 2) $\partial(w) \leqq [A^\partial$, $\partial(w) \leqq [B^\partial$ are treated analogously.

Thus, it is possible to construct a finite set $\mathfrak{M}$ of equations in the four unknowns v, λ, μ and ν of the form

$$\Phi_1((vQ)^\lambda v, (vQ)^\nu v, (vQ)^\mu v) = \Psi_1((vQ)^\lambda v, (vQ)^\mu v, (vQ)^\mu v),$$

where Q is a word in the alphabet $\mathbf{H}$, and $\Phi_1(x, y, z)$ and $\Psi_1(x, y, z)$ are words in the alphabet $\mathbf{H} \cup \{x, y, z\}$, such that (*) has solutions in region 3) if and only if at least one equation in $\mathfrak{M}$ has. By 8.4, the solution of the equations of $\mathfrak{M}$ reduces to the solution of word equations in one unknown. It remains to apply 1.20. The proposition is proved.

Elementary equations

We now consider problem 2. It consists of finding an algorithm recognizing for given coefficients A, B, C and D whether the equation $xAyBz \to zCyDx$ has a solution. We will show that the decision problem of the equation $xAyBz \to zCyDx$ reduces to the decision problem of word equations in two unknowns. It is convenient to prove this in a more general form.

By an elementary equation in n unknowns we mean any word equation of the form

(1) $$[x_iA_i]_{i=1}^{n-1} x_n \to [y_iB_i]_{i=1}^{n-1} y_n,$$

where A_i and B_j are coefficients, $x_1, \cdots, x_n$ are distinct unknowns, and $y_1, \cdots, y_n$ is some permutation (without repetitions) of the letters $x_1, \cdots, x_n$. Instead of the symbol $\rightarrow$ in (1) there may also appear the symbol $=$. Two elementary equations in n unknowns are called similar if they have the same (unordered) set of coefficients. It is evident that the number of elementary equations similar to a given one is finite (equations which agree after a suitable renaming of unknowns are considered to be the same).

Suppose M is a natural number. A set $(X_1, \cdots, X_n)$ of words is called an M-set if there exist numbers i and j such that $i \neq j$, $X_i = X_j R$ and $[R^\partial \leq M$.

Lemma. *Suppose $(X_1, \cdots, X_n)$ is a solution of a one-sided elementary equation $\mathfrak{A}$ in n unknowns, and suppose the maximum coefficient length of $\mathfrak{A}$ is equal to M. Then there exists an elementary equation $\mathfrak{B}$ which is similar to $\mathfrak{A}$ and has an M-solution. If $\mathfrak{B}$ has a solution, then $\mathfrak{A}$ also has a solution.*

Proof. We proceed by induction on $[X_1 \cdots X_n^\partial$. Suppose $(X_1, \cdots, X_n)$ is a solution of (1). Then $X_1, \cdots, X_n \neq \Lambda$ and $[X_1^\partial > [Y_1^\partial$. If $[X_1^\partial \leq [Y_1 B_1^\partial$, then the assertion is trivial: as $\mathfrak{B}$ we can take $\mathfrak{A}$. Suppose $[X_1^\partial > [Y_1 B_1^\partial$. Then $X_1 = Y_1 B_1 V$, where $V \neq \Lambda$ and $(V, X_2, \cdots, X_n)$ is a solution of the equation $vA_1[x_i A_i]_{i=1}^{n-1} x_n = [z_i C_i]_{i=1}^{n-1} z_n$, where the right-hand side is obtained from the right-hand side of (1) by discarding $y_1 B_1$ and substituting $y_1 B_1 v$ for x_1; $z_1, \cdots, z_n$ are unknowns, and $C_1, \cdots, C_n$ are coefficients. The resulting equation (call if $\mathfrak{C}$), as is easily verified, is elementary and similar to (1). If $(v, x_2, \cdots, x_n)$ is a solution of $\mathfrak{C}$, then the set $(x_1 = y_1 B_1 v, x_2, \cdots, x_n)$ satisfies (1). If $[V^\partial = [Z_1^\partial$, then the lemma is true: as $\mathfrak{B}$ we take $\mathfrak{C}$. If $[V^\partial \neq [Z_1^\partial$, then either $[V^\partial > [Z_1^\partial$ or $[V^\partial < [Z_1^\partial$, so that the set $(V, X_2, \cdots, X_n)$ satisfies a one-sided elementary equation $\mathfrak{C}$. We have $Y_1 \neq \Lambda$ (because of the arrow). Hence $[VX_2 \cdots X_n^\partial < [X_1 X_2 \cdots X_n^\partial$. By the induction assumption, there exists an elementary equation $\mathfrak{B}$ which is similar to $\mathfrak{C}$ and has an M-solution. If $\mathfrak{B}$ has a solution, so does $\mathfrak{C}$. The equation $\mathfrak{B}$ is the desired one. The lemma is proved.

Since the number of elementary equations similar to a given one is finite, we see by the lemma that the test for the solvability of elementary equations in n unknowns reduces to the determination of the M-solutions of a finite number of elementary equations in n unknowns. The latter problem deals with equations in $n-1$ unknowns.

In particular, the equation $xAyBz \rightarrow zCyDx$ is an elementary equation in three unknowns, and for its solution it suffices to look at a finite number of word equations in two unknowns. Hence we have

8.6. *There exists an algorithm recognizing, for given coefficients A, B, C and D, whether the equation $xAyBz \rightarrow zCyDx$ has a solution.*

Remark. It is easy to verify that the equations in $n-1$ unknowns obtained in determining M-solutions in accordance with the lemma are, after removing the first (and last) coefficient (see the definition preceding 6.21), elementary equations in $n-1$ unknowns. Thus (by induction on n) we obtain an algorithm recognizing the solvability of elementary equations.

ϵ_0-equations

These are equations of the form

$$(*) \qquad xAyBz\Phi(x, y, z) \rightarrow zCyDx\Psi(x, y, z),$$

where $[AB^\partial = [CD^\partial$ and the tail $\Phi = \Psi$ has one of the following forms: 1) $P = Q$, where P and Q are coefficients, $P \neq Q$; 2) $P\xi \cdots = Q$, where P and Q are coefficients and ξ is an unknown; 3) $P\xi \cdots = Q\eta \cdots$, where P and Q are coefficients, $P \neq Q$, and either $\xi = \eta$ or ξ and η are unknowns different from y. In case 1) the equation has no solutions. In case 2) the length of the component ξ of solutions of the equation does not exceed $[Q^\partial$, and the problem reduces to the solution of word equations in two unknowns. There remains case 3).

In considering case 3) we use the notion of an M-periodic word. Recall that a word P is called periodic with period A if it has the form $P = A^t A_1$, where $t \geqq 0$ and $A_1 \angle A$. Obviously, any beginning, any end and, in general, any subword of a periodic word with period A is a periodic word with period of length $[A^\partial$ (the corresponding periods are cyclic permutations of A). Suppose M is a positive integer. A word P is called M-periodic if it is a periodic word with period of length $\leqq M$. A set $(X_1, \cdots, X_n)$ of words is called M-periodic if each of the words $X_1, \cdots, X_n$ is M-periodic. The determination of the M-periodic solutions of word equations obviously reduces to the solution of one-stage exponential equations. In other terms (see 2.3′), there exists an algorithm recognizing the existence of M-periodic solutions of word equations.

If in case 3) we have $\xi = \eta = y$, then for any solution (X, Y, Z) of (*), since $P \neq Q$, we have $\tau(Y, M)$, where $M = |[P^\partial - [Q^\partial|$. If $\xi, \eta \neq y$, then, for the same reason, we at least have $\tau(Z, M)$, since Z is a beginning of X (see the arrow in (*)). Hence one of two situations arises:

a) For any solution (X, Y, Z) of (*) we have $\tau(Y, M)$.

b) For any solution (X, Y, Z) of (*) we have $\tau(Z, M)$.

Consider first case a). Equation (*) is called a β-equation if Φ and Ψ are words in the alphabet $\mathbf{H} \cup \{y\}$, and $\Phi \neq \Psi$. In view of the condition $[AB^\partial = [CD^\partial$, any solution of (*) satisfies the equation $\Phi = \Psi$. Hence if (*) is a β-equation, then the component Y of any solution is an M-periodic word, where M is the maximum coefficient length of the equation $\Phi = \Psi$ (see 1.19).

8.7. *There exists an algorithm recognizing the solvability of β-equations.*

Proof. We define degenerate and nondegenerate solutions of one-sided equations. Suppose there is given a pair $(\mathfrak{A}_0, \Sigma_0)$, where $\mathfrak{A}_0$ is a one-sided equation and Σ_0 a solution of this equation. Suppose $(\mathfrak{A}_0, \Sigma_0), \cdots, (\mathfrak{A}_k, \Sigma_k)$ is a complete chain (see §3) of the pair $(\mathfrak{A}_0, \Sigma_0)$. The solution Σ_0 is called degenerate if the chain $\mathfrak{A}_0, \cdots, \mathfrak{A}_k$ is degenerate, and nondegenerate if the chain $\mathfrak{A}_0, \cdots, \mathfrak{A}_k$ is nondegenerate. We construct separate algorithms recognizing the existence of degenerate and nondegenerate solutions.

Let us begin with degenerate solutions. It is easy to compute (by induction on k) that if $\mathfrak{A}_0, \cdots, \mathfrak{A}_k$ is a degenerate chain of the equation $xAyBz\Phi(y) \to zCyDx\Psi(y)$, then the equations of the chain with even indices have the form $xAyBz\Phi \to zCyDx\Psi$, and with odd indices the form $zCyDx\Psi \to xAyBz\Phi$. From the definition of a degenerate solution, we see that if a β-equation (*) has a degenerate solution, then at least one of the equations $xAyBz\Phi \rightleftarrows zCyDx\Psi$ has a finite solution. The converse is obvious. Hence the existence of degenerate solutions of a β-equation (*) is equivalent to the existence of finite solutions of one of the equations $xAyBz\Phi \rightleftarrows zCyDx\Psi$. But the determination of finite solutions is a problem involving equations in two unknowns. Thus an algorithm for determining degenerate solutions has been constructed.

We now turn to nondegenerate solutions. We will prove that any nondegenerate image of a β-equation possesses a tree whose final vertices have only periodic solutions (with periods of bounded length). Any nondegenerate image of a β-equation (*) has the form $vAyBz\Phi \leftarrow yDzCv\Psi$, where v is a new unknown. Any image of the latter equation is obtained under a transformation $y=(vA)^{\lambda}u$, $\lambda > 0$, and has the form $vAuBz\Phi_1 \to uDzCv\Psi_1$. Since (*) is a β-equation, the component y is an M-periodic word. Then the equality $y=(vA)^{\lambda}u$ implies that the components v and u of any solution of the equation $vAuBz\Phi_1 \to uDzCv\Psi_1$ are also M-periodic words. The pair of substitutions $v=uDx$, $v=uDzCx$ reduces the equation $vAuBz\Phi_1 \to uDzCv\Psi_1$ to the equations $xAuBz\Phi_2 \leftarrow zCuDx\Psi_2$, $xAuBz\Phi_3 = uDzCx\Psi_3$. The latter equation has only M-periodic solutions, a fact which follows from the M-periodicity of the component v and the equality $v=uDzCx$. Regarding the first equation, its images have only M-periodic solutions. Indeed, any image of this equation has the form $\eta \cdots \to y \cdots$, where y is a new unknown and η is either x or u. Because of the arrow, we have $y \angle \eta$ for any solution (x,y,u) of the image. It remains to observe that the components x and u are M-periodic, a fact which follows from the equality $v=uDx$ and the M-periodicity of the component v. We have proved that any nondegenerate image of a β-equation possesses a tree whose final vertices have only M-periodic solutions. In view of 2.3′ and 7.8, this gives an algorithm recognizing the solvability of nondegenerate images of β-equations. We will now prove that if a β-equation (*) has a nondegenerate solution, then a nondegenerate image of one of the equations $xAyBz\Phi \rightleftarrows zCyDx\Psi$ has a solution. Indeed, suppose $\mathfrak{A}_0$ is a β-equation (*) and Σ_0 is a nondegenerate solution of $\mathfrak{A}_0$. Suppose $(\mathfrak{A}_0,\Sigma_0),\cdots,(\mathfrak{A}_k,\Sigma_k)$ is a complete chain of the pair $(\mathfrak{A}_0,\Sigma_0)$. The chain $\mathfrak{A}_0,\cdots,\mathfrak{A}_k$ is nondegenerate, i.e. there exists q, $1\le q\le k$, such that $\mathfrak{A}_0,\cdots,\mathfrak{A}_{q-1}$ is a degenerate chain and $\mathfrak{A}_q$ is a nondegenerate image of $\mathfrak{A}_{q-1}$. The equation $\mathfrak{A}_{q-1}$ has the form $xAyBz\Phi \rightleftarrows zCyDx\Psi$. The equation $\mathfrak{A}_q$ is the desired one. Thus we have proved that if a β-equation (*) has a nondegenerate solution, then a nondegenerate image of one of the equations $xAyBz\Phi \rightleftarrows zCyDx\Psi$ has a solution. The converse is obvious. Hence the existence of nondegenerate solutions of a β-equation (*) is equivalent to the existence of solutions of at least one nondegenerate image of one of the equations $xAyBz\Phi \rightleftarrows zCyDx\Psi$. We obtained above an algorithm recognizing the solvability of nondegenerate images of β-equations. It remains to observe that each of the equations $xAyBz\Phi \rightleftarrows zCyDx\Psi$ has a finite number of distinct nondegenerate images, since Φ and Ψ do not contain x or z. Thus we have constructed an algorithm recognizing the existence of nondegenerate solutions of β-equations. The proposition is proved.

We prove the following theorem on the reduction of ϵ_0-equations.

8.8. *If there exists an algorithm recognizing the solvability of basic equations of characteristic zero, then there exists an algorithm recognizing the solvability of* ϵ_0*-equations.*

PROOF. It suffices to prove the assertion for the cases a) and b).

Case a). The component y of any solution of (*) is an M-periodic word. In view of 7.8 and 8.1, it suffices to construct an algorithm for solving θ-

images of ϵ_0-equations. Let $(\bar{x},\bar{y},\bar{z})$ denote solutions of an ϵ_0-equation (*), and (x,y,z) solutions of its θ-images. In what follows, unless otherwise stated, by θ-images we will mean one-sided θ-images. Of course, 8.1 remains true. From the expression of $\bar{y}$ in terms of x, y and z [see formulas 1) $-$3) immediately preceding equation (8.1)] it follows that the components y and x of the solutions of θ_1-images and the components y and z of the solutions of θ_2-images are M-periodic words (we have by hypothesis that $\bar{y}$ is an M-periodic word; the components y and x of solutions of θ_1-images and the components y and z of solutions of θ_2-images are subwords of $\bar{y}$). But the θ_1- and θ_2-images of this equation have the form $xAyBz\cdots \rightleftarrows zCyDx\cdots$. From the expressions for $\bar{x}$, $\bar{y}$ and $\bar{z}$ it follows that if the components $\bar{x}$ and $\bar{y}$ (or the components $\bar{y}$ and $\bar{z}$) are M-periodic, then for θ-images all three components x, y and z are M-periodic. Thus we obtain an algorithm recognizing the solvability of θ_1- and θ_2-images of ϵ_0-equations.

There remain the θ_3-images. In taking θ_3-images the unknown y is not affected. Hence we do not obtain conditions on the components x, y and z of the solutions of θ_3-images from the conditions on $\bar{y}$. However, the θ_3-images (we have agreed that these are one-sided) are μ-images in the sense of §7. Hence by 7.3 (the roles of A and B in 7.3 are now played by AyB and CyD), for any θ_3-image $\mathfrak{A}$ of (*) one of four situations can occur:

1) $\mathfrak{A}$ is equivalent to an equation $xAyBzP(y) \rightleftarrows zCyDxQ(y)$, where P and Q are words in the alphabet $\mathbf{H} \cup \{y\}$.

2) $\mathfrak{A}$ has no solutions.

3) $\mathfrak{A}$ possesses property α.

4) Any solution of $\mathfrak{A}$ satisfies the equation $AyB = CyD$.

In case 2) the algorithm for solving $\mathfrak{A}$ is trivial. In case 4) it follows from 8.5, since in this case $\mathfrak{A}$ is equivalent to an equation of the form $xAyBz\cdots \to zAyBx\cdots$. In case 1), if $P=Q$, then the algorithm for solving $\mathfrak{A}$ follows from 8.6, and if $P\neq Q$, then $\mathfrak{A}$ is a β-equation and the algorithm for solving $\mathfrak{A}$ follows from 8.7. Finally, in case 3), by 7.11 the equation $\mathfrak{A}$ either is a γ-equation or possesses a tree whose final vertices are basic equations of characteristic zero. In the second alternative the algorithm for solving $\mathfrak{A}$ follows from 7.8 and the assumption of the existence of a solvability algorithm for equations of characteristic zero. Consider the first alternative. It means that the component z (if the θ_3-image has the form (*)) is an M-periodic word. Hence the components y and z are M-periodic and so, as we have seen above, the solution of θ-images of θ_3-images reduces to the solution of one-stage exponential equations. We have constructed an algorithm for solving θ_3-images of an ϵ_0-equation (*). Thus we have constructed an algorithm recognizing the solvability of an ϵ_0-equation (*) in Case a).

Note that it is exactly for Case a) that the alternative 7.4.a) is insufficient and the stronger 7.3 is necessary. Indeed, by 7.11 the property α reveals the periodicity of the component z, while the property α' reveals (see Case 3 of 8.2) the periodicity of one of the components x, y and z. If this happens to be y, then progress in Case a) is impossible. We would have to consider chains of θ_3-images (i.e. μ-images) and bring in the function F, which would be very cumbersome (see §9 below).

Case b). With reference to the formulas for $\bar{x}$, $\bar{y}$ and $\bar{z}$, we see that all three components of solutions of θ_2- and θ_3-images of (*) are M-periodic words.

There remain the θ_1-images. In taking a θ_1-image the unknown z is not affected. Hence if the arrow in a θ_1-image is directed from x to z, then in a new passage to θ_1-images z is again not affected. We do not obtain conditions on x and y from the condition on $\bar{z}$. This forces us to seek another means of reduction.

Consider the equation $xAyBz\Phi \rightarrow zCyDx\Psi$ in Case b). The substitution $x = zCv$ reduces this to the equation $vAyBz\Phi_1(v,y,z) = yDzCv\Psi_1(v,y,z)$. Look at the latter equation. Since $[AB^\partial = [CD^\partial$, any of its solutions satisfies the equation $vAyBz = yDzCv$. We consider separately two regions: 1) $\partial(v) > \partial(Bz)$; 2) $\partial(v) \leqq \partial(Bz)$.

1) In the region $\partial(v) > \partial(Bz)$ we have from the equation $vAyBz = yDzCv$ (see the ends of the equation) that $v = uBz$, where u is a new unknown. The triple (u,y,z) satisfies the equation $uBzAy\Phi_2(u,y,z) = yDzCu\Psi_2(u,y,z)$, where Φ_2 and Ψ_2 are obtained from Φ_1 and Ψ_1 by substituting uBz for v. The one-sided equations corresponding to this last equation have the form $uBzAy\Phi_2 \rightleftarrows yDzCu\Psi_2$. This is an equation of the form (*), and the middle component z has M-periodic solutions; we are back in Case a). Thus we have an algorithm recognizing the solvability of the equation $vAyBz\Phi_1 = yDzCv\Psi_1$ in the region $\partial(v) > \partial(Bz)$.

2) The region $\partial(v) \leqq \partial(Bz)$. In the region $\partial(z) < \partial(v) \leqq \partial(Bz)$ we arrive at equations in two unknowns. In the region $\partial(v) \leqq \partial(z)$ the word v is an end of z, i.e. v is an M-periodic word. Thus it remains to solve the equation $vAyBz\Phi_1 = yDzCv\Psi_1$, where v and z are M-periodic words. Consider the corresponding one-sided equations $vAyBz\Phi_1 \rightarrow yDzCv\Psi_1$, $vAyBz\Phi_1 \leftarrow yDzCv\Psi_1$. In solutions of the first equation all three components are M-periodic, since $y \angle v$ for any solution of this equation. For the second equation it suffices to take a complete system of images and observe that for each image all three components of the solutions are M-periodic words. Indeed, any image of the second equation has the form $v \cdots \rightarrow u \cdots$, where u is a new unknown, and, because of the arrow, we have $u \angle v$ for any solution (u,v,z) of the image. Thus we have an algorithm recognizing the solvability of the equation $vAyBz\Phi_1 = yDzCv\Psi_1$ in the region $\partial(v) \leqq \partial(Bz)$. The construction of the algorithm for solving ϵ_0-equations in Case b) is complete. The proposition is proved.

Combining 7.8, 8.3, 8.5, 8.6 and 8.8, we obtain the following theorem on the reduction of basic equations of characteristic one.

8.9. Theorem. *If there exists an algorithm recognizing the solvability of basic equations of characteristic zero, then there also exists an algorithm recognizing the solvability of basic equations of characteristic one.*

§9. Basic equations of characteristic zero

Basic equations of characteristic zero are equations of the form $xAy\Phi(x,y,z) \rightarrow yBx\Psi(x,y,z)$, where A and B are coefficients, $[A^\partial = [B^\partial$, and Φ and Ψ are words in the alphabet $\mathbf{H} \cup \{x,y,z\}$. The cases $A = B$ and $A \neq B$ are considered separately. In both cases we express x and y from the equation $xAy \rightarrow yBx$ and substitute the resulting expressions into the equation $\Phi = \Psi$. The main case is $A \neq B$. Here x and y are expressed from the equation $xAy \rightarrow yBx$ by means of the function G. The expression for z is obtained in terms of "cuts" of the functions F and G. These two functions play a fundamental role in this section.

This section consists of three parts. In 9.1 we establish an important δ-representation of F. In 9.2 we consider the equality $F(A,B,\lambda_1,\cdots,\lambda_k) = UV_1UV_2$, where $[U^\partial$ is sufficiently large in comparison with $[AB^\partial$. The problem is to prove that in this equality the word U cannot cut the word F arbitrarily "deeply". This allows us to reduce the equation $xAyx\Phi(x,y,z) \to yBxz\Psi(x,y,z)$, $A \neq B$, to equations in two unknowns. In 9.3 we solve basic equations of characteristic zero.

9.1. Words in the two-letter alphabet $\{0,1\}$ will be called δ-processions. Any δ-procession has the form $\delta_1 \cdots \delta_n$, where $n \geqq 0$ and each of $\delta_1, \cdots, \delta_n$ is either 0 or 1. When $n=0$ we have the empty δ-procession. Sequences of positive integers are called processions. Any procession has the form $(\lambda_1,\cdots,\lambda_k)$, where $\lambda_1,\cdots,\lambda_k$ are positive integers. When $k=0$ we have the empty procession. The processions $(\lambda_1,\cdots,\lambda_k)$ and $(\mu_1,\cdots,\mu_s)$ are considered to be equal if $k=s$ and $\lambda_i=\mu_i$ $(i=1,\cdots,k)$.

We fix a two-letter alphabet $\{\alpha,\beta\}$.

DEFINITION 1. Suppose $\delta_1\cdots\delta_n$ is a δ-procession, $(\lambda_1,\cdots,\lambda_k)$ a procession. We say that $\delta_1\cdots\delta_n$ is the δ-procession of $(\lambda_1,\cdots,\lambda_k)$ if for any $\lambda>0$ we have

$$F(\alpha,\beta,\lambda,\lambda_1,\cdots,\lambda_k) = [\alpha^{\lambda+\delta_i}\beta]_{i=1}^n\langle\alpha\rangle_k,$$

where $\langle\alpha\rangle_k=\alpha$ if k is odd and $\langle\alpha\rangle_k=\Lambda$ if k is even.

It is obvious that no procession can have two different δ-processions. We write $\delta_1\cdots\delta_n=\delta(\lambda_1,\cdots,\lambda_k)$ if $\delta_1\cdots\delta_n$ is the δ-procession of $(\lambda_1,\cdots,\lambda_k)$.

9.1.1. *The following equalities hold:*

(1) $$\delta(\lambda_1) = 0^{\lambda_1},$$

(2) $$\delta(\lambda_1,\lambda_2) = 0^{\lambda_1}(1\,0^{\lambda_1-1})^{\lambda_2-1}1,$$

(3) $$\delta(1,\lambda_2,\lambda_3) = (0\,1^{\lambda_2})^{\lambda_3}0.$$

PROOF. From the definition of the function F (§5) we have

(1′) $$F(\alpha,\beta,\lambda,\lambda_1) = (\alpha^\lambda\beta)^{\lambda_1}\alpha,$$

(2′) $$\begin{aligned} F(\alpha,\beta,\lambda,\lambda_1,\lambda_2) &= ((\alpha^\lambda\beta)^{\lambda_1}\alpha)^{\lambda_2}\alpha^\lambda\beta = (\alpha^\lambda\beta)^{\lambda_1}\alpha((\alpha^\lambda\beta)^{\lambda_1}\alpha)^{\lambda_2-1}\alpha^\lambda\beta \\ &= (\alpha^\lambda\beta)^{\lambda_1}(\alpha^{\lambda+1}\beta(\alpha^\lambda\beta)^{\lambda_1-1})^{\lambda_2-1}\alpha^{\lambda+1}\beta, \end{aligned}$$

(3′) $$F(\alpha,\beta,\lambda,1,\lambda_2,\lambda_3) = ((\alpha^\lambda\beta\alpha)^{\lambda_2}\alpha^\lambda\beta)^{\lambda_3}\alpha^\lambda\beta\alpha = (\alpha^\lambda\beta(\alpha^{\lambda+1}\beta)^{\lambda_2})^{\lambda_3}\alpha^\lambda\beta\alpha$$

for any $\lambda>0$. Then (1) – (3) follow from (1′) – (3′) and from Definition 1.

9.1.2. *For any nonempty procession $(\lambda_1,\cdots,\lambda_k)$ there exists a δ-procession $\delta_1\cdots\delta_n$ such that $\delta_1\cdots\delta_n=\delta(\lambda_1,\cdots,\lambda_k)$. The δ-procession $\delta_1\cdots\delta_n$ possesses the following properties:*

a) $0^{\lambda_1} \angle \delta_1\cdots\delta_n$.

b) $\lambda_1+\cdots+\lambda_k \leq n$.

c) $\delta_n=0$ *if k is odd,* $\delta_n=1$ *if k is even.*

d) *The word 0^{λ_1+1} does not occur in the word $\delta_1\cdots\delta_n$.*

e) *If $\lambda_1>1$, then the word 11 does not occur in the word $\delta_1\cdots\delta_n$.*

f) *If $\lambda_1=1$ and $k>1$, then the word 1^{λ_2+2} does not occur in the word $\delta_1\cdots\delta_n$.*

PROOF (BY INDUCTION ON k). Since the procession $(\lambda_1,\cdots,\lambda_k)$ is nonempty, we have $k\geqq 1$. For $k=1,2$, our assertion follows from formulas 9.1.1 (1) and (2) (the δ-processions comprising the right-hand sides of these formulas satisfy conditions a) – f)). Suppose $m\geqq 3$ and our assertion is true whenever

$k < m$. We will prove it for $k = m$. Suppose there is given a procession $(\lambda_1, \cdots, \lambda_m)$. We consider two cases, according to the parity of m.

Case 1. m is odd. The induction assumption is applicable to the processions $(\lambda_1, \cdots, \lambda_{m-1})$ and $(\lambda_1, \cdots, \lambda_{m-2})$; hence there exist δ-processions $\epsilon_1 \cdots \epsilon_p$ and $\gamma_1 \cdots \gamma_q$ such that $\epsilon_1 \cdots \epsilon_p = \delta(\lambda_1, \cdots, \lambda_{m-1})$ and $\gamma_1 \cdots \gamma_q = \delta(\lambda_1, \cdots, \lambda_{m-2})$; the δ-processions $\epsilon_1 \cdots \epsilon_p$ and $\gamma_1 \cdots \gamma_q$ satisfy conditions a) – f). By Definition 1, we have for any $\lambda > 0$

$$F(\alpha, \beta, \lambda, \lambda_1, \cdots, \lambda_{m-1}) = [\alpha^{\lambda+\epsilon_i}\beta]_{i=1}^{p},$$
$$F(\alpha, \beta, \lambda, \lambda_1, \cdots, \lambda_{m-2}) = [\alpha^{\lambda+\gamma_i}\beta]_{i=1}^{q}\alpha.$$

From the definition of the function F we have

$$\begin{aligned} F(\alpha, \beta, \lambda, \lambda_1, \ldots, \lambda_m) &= (F(\alpha, \beta, \lambda, \lambda_1, \ldots, \lambda_{m-1}))^{\lambda_m} F(\alpha, \beta, \lambda, \lambda_1, \ldots, \lambda_{m-2}) \\ &= ([\alpha^{\lambda+\varepsilon_i}\beta]_{i=1}^{p})^{\lambda_m} [\alpha^{\lambda+\gamma_i}\beta]_{i=1}^{q}\alpha \end{aligned} \tag{1}$$

for any $\lambda > 0$. Put

$$\delta_1 \cdots \delta_n = (\epsilon_1 \cdots \epsilon_p)^{\lambda_m} \gamma_1 \cdots \gamma_q. \tag{2}$$

Obviously, $\delta_1 \cdots \delta_n$ is a δ-procession. We will prove that it is the desired one. It follows from (1), (2) that $\delta_1 \cdots \delta_n = \delta(\lambda_1, \cdots, \lambda_m)$. From (2) we obtain $n = p\lambda_m + q$. By the induction assumption b), $\lambda_1 + \cdots + \lambda_{m-1} \leqq p$, and by the induction assumption a), $q \geqq \lambda_1 \geqq 1$. Hence $n \geqq (\lambda_1 + \cdots + \lambda_{m-1})\lambda_m + 1 \geqq \lambda_1 + \cdots + \lambda_m$, i.e. $\delta_1 \cdots \delta_n$ satisfies b). Also, by the induction assumptions a) and c) we have $\epsilon_1 = \gamma_1 = 0$, $\epsilon_p = 1$ and $\gamma_q = 0$. Using these relations, the equality (2) and the induction assumptions a), c), d), e) and f), it is easy to see that a), c), d), e) and f) are satisfied for $\delta_1 \cdots \delta_n$.

Case 2. m is even. By the induction assumption, there exist δ-processions $\epsilon_1 \cdots \epsilon_p$ and $\gamma_1 \cdots \gamma_q$ such that $\epsilon_1 \cdots \epsilon_p = \delta(\lambda_1, \cdots, \lambda_{m-1})$ and $\gamma_1 \cdots \gamma_q = \delta(\lambda_1, \cdots, \lambda_{m-2})$, and $\epsilon_1 \cdots \epsilon_p$ and $\gamma_1 \cdots \gamma_q$ satisfy a) – f). From Definition 1, we have for any $\lambda > 0$

$$F(\alpha, \beta, \lambda, \lambda_1, \ldots, \lambda_{m-1}) = [\alpha^{\lambda+\varepsilon_i}\beta]_{i=1}^{p}\alpha,$$

$$F(\alpha, \beta, \lambda, \lambda_1, \ldots, \lambda_{m-2}) = [\alpha^{\lambda+\gamma_i}\beta]_{i=1}^{q}.$$

Furthermore,

$$\begin{aligned} F(\alpha, \beta, \lambda, \lambda_1, \ldots, \lambda_m) &= ([\alpha^{\lambda+\varepsilon_i}\beta]_{i=1}^{p}\alpha)^{\lambda_m} [\alpha^{\lambda+\gamma_i}\beta]_{i=1}^{q} \\ &= [\alpha^{\lambda+\varepsilon_i}\beta]_{i=1}^{p}\alpha([\alpha^{\lambda+\varepsilon_i}\beta]_{i=1}^{p}\alpha)^{\lambda_m-1}[\alpha^{\lambda+\gamma_i}\beta]_{i=1}^{q} \\ &= [\alpha^{\lambda+\varepsilon_i}\beta]_{i=1}^{p}(\alpha^{\lambda+1+\varepsilon_1}\beta[\alpha^{\lambda+\varepsilon_i}\beta]_{i=2}^{p})^{\lambda_m-1}\alpha^{\lambda+1+\gamma_1}\beta[\alpha^{\lambda+\gamma_i}\beta]_{i=1}^{q} \\ &= [\alpha^{\lambda+\varepsilon_i}\beta]_{i=1}^{p}(\alpha^{\lambda+1}\beta[\alpha^{\lambda+\varepsilon_i}\beta]_{i=2}^{q})^{\lambda_m-1}\alpha^{\lambda+1}\beta[\alpha^{\lambda+\gamma_i}\beta]_{i=1}^{q}, \end{aligned} \tag{3}$$

since $\epsilon_1 = \gamma_1 = 0$ in view of the induction assumption a). Put

$$\delta_1 \cdots \delta_n = \epsilon_1 \cdots \epsilon_p(1\epsilon_2 \cdots \epsilon_p)^{\lambda_m-1}1\gamma_2 \cdots \gamma_q. \tag{4}$$

Obviously, $\delta_1 \cdots \delta_n$ is a δ-procession. It follows from (3) and (4) that $\delta_1 \cdots \delta_n = \delta(\lambda_1, \cdots, \lambda_m)$. It is easy to see, starting from (4) and the induction assumptions a) – d), that $\delta_1 \cdots \delta_n$ satisfies a) – d). In the proof of c) we must consider $q \geqq 2$, since $m - 2 \geqq 2$. Let us verify e) and f). If $\lambda_1 > 1$, then by the induction assumption a) we have $\epsilon_2 = \gamma_2 = 0$. By the induction assumption c) we have $\epsilon_p = 0$. Now property e) for $\delta_1 \cdots \delta_n$ follows from (4) and the induction assumption e). Suppose $\lambda_1 = 1$. Then from 9.1.1 (2) and (3) we have $\delta(1,\lambda_2)=01^{\lambda_2}$ and $01^{\lambda_2}0 \angle \delta(1,\lambda_2,\lambda_3)$. The definition of F implies that if $t \geqq 3$, then $F(\alpha,\beta,\lambda,\lambda_1,\lambda_2,\lambda_3) \angle F(\alpha,\beta,\lambda,\lambda_1,\cdots,\lambda_t)$. Hence $\delta(\lambda_1,\lambda_2,\lambda_3) \angle \delta(\lambda_1,\cdots,\lambda_t)$ when $t \geqq 3$. From these relations and 9.1.1 (3) we conclude that $01^{\lambda_2}0 \angle \epsilon_1 \cdots \epsilon_p$ and, furthermore, that $01^{\lambda_2}0 \angle \gamma_1 \cdots \gamma_q$ if $m > 4$. If $m=4$, then by 9.1.1 (2) we have $01^{\lambda_2}=\gamma_1 \cdots \gamma_q$. Now property f) for $\delta_1 \cdots \delta_n$ follows from (4), the relation $\epsilon_p=0$ and the induction assumption f). The proposition is proved.

The following proposition uses the symbol $\langle A,B\rangle_k$ (see (5.1)).

9.1.3. *If $k>0$ and $\delta_1 \cdots \delta_n=\delta(\lambda_1,\cdots,\lambda_k)$, then for any $\lambda>0$ we have*

(1) $F(A,B,\lambda,\lambda_1,\ldots,\lambda_k)=[A^{\lambda+\delta_i}B]_{i=1}^{n}\langle A\rangle_k=[A^{\lambda+\delta_i}B]_{i=1}^{n-1}A^{\lambda}\langle A,B\rangle_{k+1}.$

PROOF. From the definitions of the functions $\langle A,B\rangle_k$ and $\langle A\rangle_k$ and from 9.1.2 c) it follows that $A^{\lambda+\delta_n}B\langle A\rangle_k=A^{\lambda}\langle A,B\rangle_{k+1}$. Now (1) obviously follows from Definition 1.

9.1.4. *Suppose $\delta_1 \cdots \delta_{n+1}=\delta(\lambda_1,\cdots,\lambda_k)$ and $\delta_1 \cdots \delta_n=0^n$, where $n>0$. Then if $\lambda_k=1$, then $k=2$, and if $\lambda_k>1$, then $k=1$.*

PROOF. It follows from 9.1.1 that if $\lambda_2>1$, then $0^{\lambda_1}1 \angle \delta(\lambda_1,\lambda_2)$. A direct computation shows that $\delta(\lambda_1,1,\lambda_3)=(0^{\lambda_1}1)^{\lambda_3}0^{\lambda_1}$. Also, if $1 \leqq p \leqq k$, then $\delta(\lambda_1,\cdots,\lambda_p) \angle \delta(\lambda_1,\cdots,\lambda_k)$, a consequence of the definitions of the functions F and δ. Hence a one is not present in $\delta_1 \cdots \delta_n$ in only two cases: 1) $k=1$; 2) $k=2$, $\lambda_k=1$. The proposition is proved.

9.2. Consider the equality $UV=F(A,B,\lambda_1,\cdots,\lambda_k)$. We say that U is the entire beginning of the word $F(A,B,\lambda_1,\cdots,\lambda_k)$ if there exists a word U_1 in the alphabet $\{\alpha,\beta\}$ such that $U_1 \angle F(\alpha,\beta,\lambda_1,\cdots,\lambda_k)$, $U=S_{AB}^{\alpha\beta}U_1|$.

9.2.1. *Suppose*

$$\mathscr{E}(A,B), \qquad UXUY=F(A,B,\lambda_0,\cdots,\lambda_k), \qquad [U^{\partial}>[F(A,B,\lambda_0,\cdots,\lambda_3)^{\partial}.$$

Then UX is the entire beginning of the word $F(A,B,\lambda_1,\cdots,\lambda_k)$.

PROOF. For brevity, write $F_0=A$ and $F_i=F(A,B,\lambda_0,\cdots,\lambda_{i-1})$, $i=1,\cdots,k+1$. Since $F_0 \angle F_1 \angle \cdots \angle F_{k+1}$ and $[U^{\partial}>[F_4^{\partial}$, there exists s, $4 \leqq s \leqq k$, such that $F_s \angle U \angle F_{s+1}$. Since $s>3$, by 5.12 the words F_s and F_{s-1} are simple. By 5.1, $F_{k+1}=F(F_s,F_{s-1},\lambda_s,\cdots,\lambda_k)$. Assume that $s<k$. Let $\delta_1 \cdots \delta_m$ denote the δ-procession of $(\lambda_{s+1},\cdots,\lambda_k)$. By 9.1.3 we have

$$UXUY=[F_s^{\lambda_s+\delta_i}F_{s-1}]_{i=1}^{m}\langle F_s\rangle_{k-s-1}.$$

Let $XY=A_1 \cdots A_j \cdots A_n$. We say that X cuts A_j if $A_j=PQ$, $X=A_1 \cdots A_{j-1}P$. There are two possible cases: 1) UX cuts F_{s-1}; 2) UX cuts F_s.

1) Since $[F_{s-1}^{\partial}<[U^{\partial}$, to the right of the cut factor F_{s-1} stands F_s. The cut gives $F_{s-1}=AB$ and $UY=BF_s\cdots$. Since $s>1$, we have $F_{s-1} \angle F_s \angle U$. Hence $A \angle F_s$, $BA \angle U$, $AB \angle U$ and $AB=BA=F_{s-1}$. Since F_{s-1} is a simple word, by 1.2 one of the words A or B is empty. It is obvious that in this case UX is the entire beginning of F_{k+1}.

2) Since $[F_s^{\partial}<[U^{\partial}$, to the right of the cut factor F_s stands F_s or F_{s-1}. If to the right stands F_s, then the cut gives $F_s=AB$ and $UY=BF_s\cdots=BA\cdots$. We have $AB \angle F_s \angle U$, $BA \angle U$, $AB=BA=F_s$ and, as in case 1), one of the words AB is empty, i.e. UX is the entire beginning. Suppose to the right of the cut factor F_s stands F_{s-1}. We have $F_s=AB$ and

$$UY = BF_{s-1}[F_s^{\lambda_i+\delta_i}F_{s-1}]_{i=n}^{m}\langle F_s\rangle_{s-k-1}, \qquad n \leqq m+1.$$

There are two possible subcases.

2.1) To the right of BF_{s-1} stands F_s. Since $s > 2$, we have $F_{s-2} \angle F_{s-1}$, $F_s = F_{s-1}^{\lambda_{s-1}}F_{s-2} \angle F_{s-1}^{\lambda_{s-1}}F_{s-1}F_{s-2} = F_{s-1}F_s$, i.e. $F_{s-1}F_s$ begins with F_s. Then $UY = BF_s \cdots = BA \cdots$. We again have $BA \angle U$, $AB \angle U$ and $AB = BA = F_s$, and UX is the entire beginning of F_{k+1}.

2.2) There are no factors to the right of BF_{s-1}. Then $UY = BF_{s-1}$. Consider in detail the cut $F_s = AB$. We have $F_{s-1}^{\lambda_{s-1}}F_{s-2} = AB$. Since $[U^\partial > [F_s^\partial$, this means that $[B^\partial > [F_{s-2}^\partial$. Hence A cuts F_{s-1}, i.e. $F_{s-1} = PQ$ and $B = Q(PQ)^cF_{s-2}$, $c \geqq 0$. Then $UY = Q(PQ)^cF_{s-2}F_{s-1}$. If $c > 0$, then $QP \angle U$. Also, $PQ = F_{s-1} \angle U$. Consequently, $QP = PQ = F_{s-1}$ and one of the words P or Q is empty. Obviously, UX is the entire beginning of F_{k+1}. Consider the case $c = 0$. We have $UY = QF_{s-2}F_{s-1}$. Since $s > 3$, arguing as in subcase 2.1), we obtain $F_{s-1} \angle F_{s-2}F_{s-1}$. Then $UY = QF_{s-1} \cdots = QP \cdots$. Again, $QP = PQ = F_{s-1}$, and UX is the entire beginning of F_{k+1}. The proposition is proved for the case $s < k$. If $s = k$, then $UXUY = F_s^{\lambda_s}F_{s-1}$ and the proof is analogous.

We write $U \sim V$ if one of the words U or V is a beginning of the other.

9.2.2. *Suppose* $F(A, B, \lambda_0, \cdots, \lambda_k) = URVD$, $[U^\partial, [V^\partial > [F(A, B, \lambda_0, \cdots, \lambda_3)^\partial$, $U \sim V$ *and* $\mathscr{E}(A, B)$. *Then* UR *is the entire beginning of the word* $F(A, B, \lambda_0, \cdots, \lambda_k)$.

Proof. Since $U \sim V$, either $U \angle V$ or $V \angle U$. If $U \angle V$, then $V = UZ$ and, by 9.2.1, UR is the entire beginning of F_{k+1}. If $V \angle U$, then $U = VZ$ and, by 9.2.1, VZR is the entire beginning of F_{k+1}. The proposition is proved.

Fix a two-letter alphabet $\{\alpha, \beta\}$. In 9.2.3 we use the following abbreviations. Suppose there is given a procession $(\lambda_0, \cdots, \lambda_{k-1})$. Put $F_{-1} = \beta$, $F_0 = \alpha$ and $F_i = F(\alpha, \beta, \lambda_0, \cdots, \lambda_{i-1})$, $i = 1, \cdots, k$. By definition of F we have $F_i = F_{i-1}^{\lambda_{i-1}}F_{i-2}$ for any $i = 1, \cdots, k$. It is also evident that if $0 \leqq p \leqq k$, then $F_p \angle F_q$.

The following proposition provides a representation of beginnings of the word $F(\alpha, \beta, \lambda_0, \cdots, \lambda_{k-1})$.

9.2.3. *Suppose* U *and* V *are nonempty words and* $F(\alpha, \beta, \lambda_0, \cdots, \lambda_{k-1}) = UV$. *Then there exist a word* Z *and sequences* $(t_1, \cdots, t_q)$ *and* $(c_1, \cdots, c_q)$ *such that* $q \geqq 1$, $0 \leqq t_q < \cdots < t_1 < k$, $1 \leqq c_i \leqq \lambda_{t_i}$ $(i = 1, \cdots, q)$, $U = F_{t_1}^{c_1} \cdots F_{t_q}^{c_q}$ *and* $V = F_{t_q}^{\lambda_{t_q} - c_q}F_{t_q-1}Z$, *where* $Z = \Lambda$, $Z = F_{t_q}$, $F_{t_q+1} \angle Z$ *or* $t_q = 0$.

Proof (by induction on k). Suppose $k = 1$. Then we have $\alpha^{\lambda_0}\beta = UV$. Since $U, V \neq \Lambda$, we have $U = \alpha^c$ and $V = \alpha^{\lambda_0 - c}\beta$, where $1 \leqq c \leqq \lambda_0$, and our assertion holds with $q = 1$, $t_q = 0$, $c_q = c$ and $Z = \Lambda$.

Suppose $m > 1$ and our assertion is true for all $k < m$. We will prove it for $k = m$. Suppose there is given a procession $(\lambda_0, \cdots, \lambda_{m-1})$ and suppose $F(\alpha, \beta, \lambda_0, \cdots, \lambda_{m-1}) = UV$. Since $F_m = F_{m-1}^{\lambda_{m-1}}F_{m-2}$, we have $UV - F_{m-1}^{\lambda_{m-1}}F_{m-2}$. The following three cases are possible.

Case 1. There exists c, $1 \leqq c \leqq \lambda_{m-1}$, such that $U = F_{m-1}^c$ and $V = F_{m-1}^{\lambda_{m-1} - c}F_{m-2}$. Then our assertion holds with $q = 1$, $t_q = m - 1$, $c_q = c$ and $Z = \Lambda$.

Case 2. There exists a nonempty word U_1 such that $U = F_{m-1}^{\lambda_{m-1}}U_1$ and $F_{m-2} = U_1V$. Since $U_1, V \neq \Lambda$, the induction assumption is applicable to the equality $F_{m-2} = U_1V$.

Applying the induction assumption to $F_{m-2} = U_1V$, we obtain

$$U_1 = F_{t_2}^{c_2} \cdots F_{t_q}^{c_q}, \qquad V = F_{t_q}^{\lambda t_q - c_q}F_{t_q-1}Z,$$

where $q \geqq 2$, $0 \leqq t_q < \cdots < t_2 < m-2$, $1 \leqq c_i \leqq \lambda_{t_i}$ $(i=2,\cdots,q)$, and where $Z=\Lambda$, $Z=F_{t_q}$, $F_{t_q+1} \angle Z$ or $t_q=0$. Then $U=F_{m-1}^{\lambda_{m-1}}F_{t_2}^{c_2}\cdots F_{t_q}^{c_q}$, and our assertion holds with $t_1=m-1$ and $c_1=\lambda_{m-1}$.

Case 3. There exist c, $1 \leqq c < \lambda_{m-1}$, and nonempty words U_1 and V_1 such that $U=F_{m-1}^cU_1$, $F_{m-1}=U_1V_1$ and $V=V_1F_{m-1}^{\lambda_{m-1}-c-1}F_{m-2}$. Applying the induction assumption to $F_{m-1}=U_1V_1$, we obtain

$$U_1=F_{t_2}^{c_2}\cdots F_{t_q}^{c_q}, \qquad V_1=F_{t_q}^{\lambda_{t_q}-c_q}F_{t_q-1}Z,$$

where $q \geqq 2$, $0 \leqq t_q < \cdots < t_2 < m-1$, $1 \leqq c_i \leqq \lambda_{t_i}$ $(i=2,\cdots,q)$, and where $Z_1=\Lambda$, $Z_1=F_{t_q}$, $F_{t_q+1} \angle Z_1$ or $t_q=0$. Then

$$U=F_{m-1}^cF_{t_2}^{c_2}\cdots F_{t_q}^{c_q}, \qquad V=F_{t_q}^{\lambda_{t_q}-c_q}F_{t_q-1}Z_1F_{m-1}^{\lambda_{m-1}-c-1}F_{m-2}.$$

Put $Z=Z_1F_{m-1}^{\lambda_{m-1}-c-1}F_{m-2}$. We will prove that if $t_q>0$, then $F_{t_q}=Z$ or $F_{t_q+1}\angle Z$. Suppose $t_q>0$. Consider the word Z_1. Since $t_q>0$, we have three possibilities: $Z_1=\Lambda$, $Z_1=F_{t_q}$ or $F_{t_q+1}\angle Z_1$. If $F_{t_q+1}\angle Z_1$, then certainly $F_{t_q+1}\angle Z$ and an alternative for Z holds. Suppose $Z_1=\Lambda$. Then $Z=F_{m-1}^{\lambda_{m-1}-c-1}F_{m-2}$. We have $t_q<m-1$, whence $F_{t_q+1}\angle F_{m-1}$. If now $\lambda_{m-1}-c-1>0$, then $F_{t_q+1}\angle Z$. Suppose $\lambda_{m-1}-c-1=0$. Then $Z=F_{m-2}$. Since $t_q \leqq m-2$, we have either $t_q=m-2$, $Z=F_{t_q}$ or $t_q<m-2$, and then $F_{t_q+1}\angle F_{m-2}=Z$ and an alternative for Z holds. There remains the case $Z_1=F_{t_q}$. Then $Z=F_{t_q}F_{m-1}^{\lambda_{m-1}-c-1}F_{m-2}$. If $\lambda_{m-1}-c-1>0$ or $\lambda_{m-1}-c-1=0$, $t_q<m-2$, then $F_{t_q+1}\angle F_{m-1}^{\lambda_{m-1}-c-1}F_{m-2}$, and then $F_{t_q}F_{t_q+1}\angle Z$. But $F_{t_q+1}=F_{t_q}^{\lambda_{t_q}}F_{t_q-1}$. Since $t_q>0$, we have $F_{t_q-1}\angle F_{t_q}$, and then $F_{t_q+1}\angle F_{t_q}^{\lambda_{t_q}+1}\angle F_{t_q}F_{t_q+1}\angle Z$. There remains the possibility that $\lambda_{m-1}-c-1=0$ and $t_q=m-2$. Then $c=\lambda_{m-1}-1$, $q=2$ and $U_1=F_{m-2}^{c_2}$. We have

$$\begin{aligned}F_m=UV&=F_{m-1}^{\lambda_{m-1}-1}F_{m-2}^{c_2}F_{m-2}^{\lambda_{m-2}-c_2}F_{m-3}F_{m-2}^2\\&=F_{m-1}^{\lambda_{m-1}-1}F_{m-2}^{\lambda_{m-2}}F_{m-3}F_{m-2}^2=F_{m-1}^{\lambda_{m-1}}F_{m-2}^2=F_mF_{m-2},\end{aligned}$$

which is impossible. The case $Z_1=F_{t_q}$, $t_q=m-2$, $\lambda_{m-1}-c-1=0$ is impossible. An alternative for Z holds. The proposition is proved.

Suppose there is given a procession $[A,B,\lambda_0,\cdots,\lambda_{k-1}]$. For brevity, put $F_{-1}=B$, $F_0=A$ and $F_i=F(A,B,\lambda_0,\cdots,\lambda_{i-1})$, $i=1,\cdots,k$.

9.2.4. *Suppose* $k>5$, $F(A,B,\lambda_0,\cdots,\lambda_{k-1})=URVD$, $U\sim V$, $\mathscr{L}(A,B)$ *and* $[D^\partial \leqq [A^\partial,[B^\partial$. *Then one of the following four assertions holds.*

1) $[U^\partial \leqq [F_p^\partial$, *where* $p \leqq \max\{5,[R^\partial\}$.

2) $[V^\partial \leqq [R_5^\partial$.

3) *There exist sequences* $(t_1,\cdots,t_q)$ *and* $(c_1,\cdots,c_q)$ *such that* $UR=F_{t_1}^{c_1}\cdots F_{t_q}^{c_q}$, $1\leqq c_i\leqq\lambda_{t_i}$ $(i=1,\cdots,q)$, $k>t_1>\cdots>t_q\geqq 0$, *where* $[V^\partial \leqq [F_{t_q+1}^\partial$.

4) $UR=F_p^c$, *where* $4\leqq p<k$ *and* $1\leqq c\leqq\lambda_p$.

Proof. By 9.2.2, it suffices to consider the case where UR is the entire beginning of the word F_k. From 9.2.3 and the definition of entire beginning, we obtain $UR=F_{t_1}^{c_1}\cdots F_{t_q}^{c_q}$, $VD=F_{t_q}^{\lambda_{t_q}-c_q}F_{t_q-1}Z$, where $q\geqq 1$, $k>t_1>\cdots>t_q\geqq 0$, $1\leqq c_i\leqq\lambda_{t_i}$ $(i=1,\cdots,q)$, and where $Z=\Lambda$, $Z=F_{t_q}$, $F_{t_q+1}\angle Z$ or $t_q=0$. We consider each of these four cases separately.

1) $Z=\Lambda$. Then $VD=F_{t_q}^{\lambda_{t_q}-c_q}F_{t_q-1}$. It is obvious that $[V^\partial\leqq[F_{t_q}^{\lambda_{t_q}}F_{t_q-1}^\partial=[F_{t_q-1}^\partial$, so that if $Z=\Lambda$ we have 3).

2) $Z=F_{t_q}$. Then $VD=F_{t_q}^{\lambda_{t_q}-c_q}F_{t_q-1}F_{t_q}$. Since $c_q\geqq 1$, we have $[V^\partial\leqq[F_{t_q}^{\lambda_{t_q}}F_{t_q-1}^\partial=[F_{t_q+1}^\partial$ and 3) holds.

3) $t_q=0$. Then $VD=A^{\lambda_0-c_q}BZ$ and $URA^{\lambda_0-c_q}BZ=F_k$. Since UR is the entire beginning of F_k, from the δ-representation of F (see 9.1.3) it follows that Z has the form $Z=[A^{\lambda_0+\delta_i}B]_{i=m}^{n-1}A^{\lambda_0}\langle A,B\rangle_k$, where $\delta_m\cdots\delta_n$ is an end of the δ-procession of $(\lambda_0,\cdots,\lambda_{k-1})$. If $[V^\partial\leqq[F_5^\partial$, then 2) holds. Suppose $[V^\partial>[F_5^\partial$. Then the equality $VD=A^{\lambda_0-c_q}BZ$ implies that $[Z^\partial\geqq[A^{\lambda_0}BA^\partial$. Hence from our representation of Z and from $VD=A^{\lambda_0-c_q}BZ$ we have $A^{\lambda_0-c_q}BA^{\lambda_0}\angle V$. Also, we have that $U\angle F_k$ and $F_1\not\angle F_k$. Hence either $U\angle F_1$ or $F_1\angle U$. Consider the case $F_1\angle U$. Then $A^{\lambda_0}B\angle U$, and from the relations $A^{\lambda_0-c_q}BA^{\lambda_0}\angle V$ and $U\sim V$ we obtain that $A^{\lambda_0}B\sim A^{\lambda_0-c_q}BA^{\lambda_0}$ and $A^{c_q}B=BA^{c_q}$, which by 1.1 contradicts the condition $\mathscr{L}(A,B)$. Thus the case $F_1\angle U$ is impossible. Consequently, $U\angle F_1$, and then 1) holds.

4) $F_{t_q+1}\angle Z$. Then $VD=F_{t_q}^{\lambda t_q-c_q}F_{t_q-1}F_{t_q+1}Z_1$ for some Z_1. We will prove that $F_{t_q}^{\lambda t_q-c_q}F_{t_q-1}F_{t_q}\angle V$. We may assume that $t_q>0$. If $t_q=1$, then the representation of VD has the form $VD=(A^{\lambda_0}B)^{\lambda_1-c_1}A(A^{\lambda_0}B)^{\lambda_1}AZ_1$. Assuming that $[V^\partial>[F_5^\partial$ (otherwise 2) holds), we obtain $(A^{\lambda_0}B)^{\lambda_1-c_1}AA^{\lambda_0}B\angle V$, i.e. $F_{t_q}^{\lambda t_q-c_q}F_{t_q-1}F_{t_q}\angle V$. If $t_q>1$, then $[F_{t_q-1}^\partial\geqq[F_1^\partial=[A^{\lambda_0}B^\partial\geqq[D^\partial$. Now from the equality $F_{t_q+1}=F_{t_q}^{\lambda t_q}F_{t_q-1}$ and the representation of VD we obtain $F_{t_q}^{\lambda t_q-c_q}F_{t_q-1}F_{t_q}\angle V$. Thus the relation $F_{t_q}^{\lambda t_q-c_q}F_{t_q-1}F_{t_q}\angle V$ is proved. Now consider the representation $UR=F_{t_1}^{c_1}\cdots F_{t_q}^{c_q}$. If $q=1$, then $UR=F_{t_1}^{c_1}$ and 4) holds. Suppose $q>1$. Since $c_1\geqq 1$, either $U\angle F_{t_1}$ or $F_{t_1}\angle U$.

Suppose $F_{t_1}\angle U$. Since $q>1$, we have $t_q<t_1$, and therefore $F_{t_q+1}\angle F_{t_1}\angle U$. We will prove by contradiction that the relation $F_{t_q+1}\angle U$ is impossible. Suppose $F_{t_q+1}\angle U$. Using another expression for F_{t_q+1}, we obtain $F_{t_q}^{\lambda t_q}F_{t_q-1}\angle U$. But we know $U\sim V$ and, by what has been proved, $F_{t_q}^{\lambda t_q-c_q}F_{t_q-1}F_{t_q}\angle V$. Hence $F_{t_q}^{\lambda t_q}F_{t_q-1}\sim F_{t_q}^{\lambda t_q-c_q}F_{t_q-1}F_{t_q}$, from which it follows that $F_{t_q}^{c_q}F_{t_q-1}\sim F_{t_q-1}F_{t_q}$. In view of 1.17, this last relation implies that $F_{t_q}F_{t_q-1}=F_{t_q-1}F_{t_q}$, which by 5.3 contradicts the condition $\mathscr{L}(A,B)$. Hence the relation $F_{t_q+1}\angle U$, and so also the case $F_{t_1}\angle U$, is impossible.

Suppose $U\angle F_{t_1}$. Then from the representation of UR we obtain $[F_{t_2}^\partial\leqq[R^\partial$, whence $t_2\leqq\max\{4,[R^\partial-1\}$. Since $t_2<t_1$, we have $F_{t_2+1}\angle F_{t_1}$, and since $U\angle F_{t_1}$, also $U\sim F_{t_2+1}$. If now $U\angle F_{t_2+1}$, then by the estimate for t_2 we have 1). If $F_{t_2+1}\angle U$, then $F_{t_q+1}\angle U$, since $t_q\leqq t_2$. But the case $F_{t_q+1}\angle U$, as was shown in the previous paragraph, is impossible. The proposition is proved.

We now consider the equation

$$(*)\qquad xAy\Phi(x,y,z)\to yBx\Psi(x,y,z),$$

where A and B are words in the alphabet $\mathbf{И}$, $A\neq B$, and $[A^\partial=[B^\partial$. Our results permit us to solve (*) when one of the words Φ and Ψ begins with z and the other with x (the general case is reduced to this one in 9.3). Then z can be expressed in terms of x and y and their beginnings. Indeed, since $[A^\partial=[B^\partial$, then any solution of (*) is a solution of the tail $\Phi=\Psi$, which in the situation under consideration has the form $\varphi(x,y)z\cdots=z\cdots$ (we assume that $x\angle\Phi$ and $z\angle\Psi$, and that z occurs in Φ and Ψ). By 1.16 we have $z=\varphi^t\varphi_1u$, where $t\geqq 0$, $\varphi_1\angle\varphi$, and u is a beginning of x or y (the case where z cuts a coefficient is trivial). The eliminability of the parameter t permits the restriction to a finite number of representations of z: $z=f_i(x,y)u$. Substituting them into (*), we arrive at the problem of determining the solutions (x,y,z) of (*) for which z is a beginning of x or y (u plays the role of z). We obtain that if $z\angle\xi$ (ξ is x or y), then $\xi=zRvD$, where $v\sim z$, and $[R^\partial$ and

$[D^\partial$ do not depend on x, y or z. Furthermore, in view of the comments following 4.9′, the interest lies in the solutions of (*) of sufficiently large rank. As is evident from 5.4, a solution $(\bar{x},\bar{y},\bar{z})$ of rank k of the equation (*) is obtained by means of the formulas $\bar{z}=z$, $\bar{x}=[F(xA,yB,\lambda_0,\cdots,\lambda_{k-1})*A]$, $\bar{y}=[F(xA,yB,\lambda_0,\cdots,\lambda_{k-2})*B]$ from the solutions (x,y,z) of the images of (*) of order k. Thus, the interesting case is where z is a beginning of the word $F(xA,yB,\lambda_0,\cdots,\lambda_{k-1})$ and k is sufficiently large.

The following Proposition 9.2.5 establishes that if (X,Y,Z) is a solution of (*) of sufficiently large rank and Z is a beginning of X or Y, then Z cannot cut X or Y arbitrarily deeply (U plays the role of Z). This permits us (9.2.6) to reduce the solution of the equation $xAyx\Phi(x,y,z)\to yBxz\Psi(x,y,z)$ to the solution of word equations in two unknowns.

9.2.5. *Suppose there is given a procession* $[XA,YB,\lambda_0,\cdots,\lambda_{k-1}]$, *where* $\mathscr{L}(XA,YB)$ *and* k *is even. For brevity, put* $F_i=F(XA,YB,\lambda_0,\cdots,\lambda_{i-1})$ $(i=-1,0,\cdots,k)$. *Suppose the words* U, V, R *and* D *satisfy the conditions* $U\sim V$, $U\neq\Lambda$, $V\neq\Lambda$, $[D^\partial\leqq[XA^\partial$, $[D^\partial\leqq[YB^\partial$ *and* $F_t=URVD$, *where* $k-1\leqq t\leqq k$, *and suppose also that*

$$VA\xi W=[\eta_iB_i]_{i=1}^m, \tag{1}$$

where $\xi,\eta_1,\cdots,\eta_m$ *are words from the list*[13] $[F_k*A]$, $[F_{k-1}*B]$, U, *and where among the words* $\eta_1,\cdots,\eta_m$ *is* $[F_k*A]$ *or* $[F_{k-1}*B]$. *Let* a $(a\geqq 0)$ *be the number of occurrences of* U *in the list* $\eta_1,\cdots,\eta_m$, *and suppose that* $\max\{[A_1^\partial,[B_i^\partial,[R^\partial,[D^\partial,5\}\leqq M$ *and* $k\geqq(a+2)M+9$. *Then there exist sequences* $(t_1,\cdots,t_q)$ *and* $(c_1,\cdots,c_q)$ *such that* $q>1$ *and* $UR=F_{t_1}^{c_1}\cdots F_{t_q}^{c_q}$, $1\leqq c_i\leqq\lambda_{t_i}$ $(i=1,\cdots,q)$, $k>t_1>\cdots>t_q>k-(a+1)M-9$.

PROOF. We have $t>5$, $F_t=URVD$, $U\sim V$, $\mathscr{L}(XA,YB)$ and $[D^\partial\leqq[XA^\partial,[YB^\partial$. In accordance with 9.2.4, we consider four cases. It is convenient to treat 1) and 4) together and 2) and 3) together.

Cases 1) *and* 4). We will first prove that if $[U^\partial\leqq[F_{p+1}^\partial$, then $p>k-(a+1)M-9$. Multiply (1) on the left by UR. We obtain for some W

$$[F_t*D]W=UR[\eta_iB_i]_{i=1}^m. \tag{2}$$

Among the words $\eta_1,\cdots,\eta_m$ is, by hypothesis, $[F_k*A]$ or $[F_{k-1}*B]$, so that the right-hand side of (2) has the form $UR[UB_i]_{i=1}^s\eta_{s+1}\cdots$, where $0\leqq s\leqq a$ and η_{s+1} is $[F_k*A]$ or $[F_{k-1}*B]$. It is obvious that each of the words $[F_t*D]$, $[F_k*A]$ and $[F_{k-1}*B]$ begins with the word F_{k-2}. Hence from (2) we have for some W_1

$$F_{k-2}W_1=UR[UB_i]_{i=1}^sF_{k-2}. \tag{3}$$

Let us estimate the length of the word $UR[UB_i]_{i=1}^s$. Since $s\leqq a$, since $[R^\partial,[B_i^\partial\leqq M$, and since $[U^\partial\leqq[F_{p+1}^\partial$, we have

$$\partial(UR[UB_i]_{i=1}^s)\leqq(a+1)([F_{p+1}^\partial+M).$$

Now 5.5.4 is applicable to (3). Indeed, the condition $\mathscr{L}(XA,YB)$, in view of 5.3, implies $\mathscr{L}(F_{p+1},F_p)$. Also, the estimate of the length of the word

---

[13] Since k is even, XA is an end of F_k and YB an end of F_{k-1}. The definition of the symbol $[*]$ precedes 5.4.

$UR[UB_i]_{i=1}^s$ can be written in the form $\partial(UR[UB_i]_{i=1}^s) \leqq (a+1)M[F_pF_{p+1}^\partial$, since $F_p \neq \Lambda$. If $k-3>p$, by 5.1 we have $F_{k-2}=F(F_{p+1},F_p,\lambda_{p+1},\cdots,\lambda_{k-3})$. If now $k-3-p \geqq (a+1)M+6$, then from (3), in view of 5.5.4, we obtain $UR[UB_i]_{i=1}^s=\Lambda$, which contradicts the condition $U\neq\Lambda$. Thus we have proved that if $[U^\partial \leqq [F_{p+1}^\partial$, then $k-3-p<(a+1)M+6$, i.e. $p>k-(a+1)M-9$. In particular, in case 4) of 9.2.4 we have $[U^\partial \leqq [F_{p+1}^\partial$. Therefore $p>k-(a+1)M-9$ and the assertion being proved holds with $q=1$, $t_q=p$ and $c_q=c$. Case 1) of 9.2.4 is impossible, since in this case $[U^\partial \leqq [F_p^\partial$, $p\leqq M$, but by hypothesis $k>(a+2)M+8$ and so the inequality $p>k-(a+1)M-9$ is impossible.

Cases 2) *and* 3). Here we will prove that if $[V^\partial \leqq [F_{p+1}^\partial$, then $p>k-M-9$. We argue by contradiction. Suppose $p\leqq k-M-9$ and $[V^\partial \leqq [F_{p+1}^\partial$. By hypothesis $[R^\partial,[D^\partial \leqq M$. Also, $p+1$ and M are small in comparison with $k-3$. In any case, $[RVD^\partial \leqq [F_{k-3}^\partial$. Since $t\geqq k-1$, we have $F_{k-2}^{\lambda_{k-3}}F_{k-3} \angle F_t$. Now $URVD=F_t$ implies that $F_{k-2}\angle U$. As noted above, F_{k-2} is a beginning of the words $[F_k * A]$ and $[F_{k-1}*B]$. Hence from (1) we have for some W

$$VA_1F_{k-2}=F_{k-2}W.$$

We know that $[VA_1^\partial \leqq [F_{p+1}^\partial+M \leqq M[F_pF_{p+1}^\partial$. Arguing as above following (3) and applying 5.5.4 (we know that $k-p-3\geqq M+6$), we obtain $VA_1=\Lambda$, which contradicts the condition $V\neq\Lambda$. Hence if $[V^\partial \leqq [F_{p+1}^\partial$, then $p>k-M-9$. In particular, in case 3) of 9.2.4 we have $t_q>k-M-9$, and case 2) of 9.2.4 is impossible. The proposition is proved.

The following proposition represents the goal of subsection 9.2. It allows us to reduce the solution of the equation $xAyx\Phi(x,y,z)\to yBxz\Psi(x,y,z)$, $A\neq B$, to the solution of word equations in two unknowns.

9.2.6. *Consider the equation in three unknowns*

$$xAy[\xi_iA_i]_{i=1}^n \to yBxzR[\eta_iB_i]_{i=1}^m, \tag{1}$$

where $n,m\geqq 2$, ξ_i *and* η_j *are unknowns,* $A,B,R,A_1,\cdots,A_n$, $B_1,\cdots,B_m$ *are coefficients,* $[A^\partial=[B^\partial$, $A\neq B$, $\xi_1=x$ *and* $\eta_1\neq y$. *Suppose*

$$M\geqq \max\{[A_i^\partial,[B_j^\partial,[A^\partial,[R^\partial,n,m,5\},$$

k *is even,* $k\geqq 2M^2$, *and suppose* $\mathfrak{A}_0,\cdots,\mathfrak{A}_k$ *is an infinite chain of* (1). *Then it is possible to construct a finite number* N $(N>0)$ *of triples* $\{\varphi_i(u,v),\ \psi_i(u,v),\ \chi_i(u,v)\}$ *of words in the alphabet* $\mathbf{H}\cup\{u,v\}$ *such that if* $\mathfrak{A}_k$ *has a solution, then* (1) *has a solution* (X,Y,Z) *such that* $X=\varphi_i(U,V)$, $Y=\psi_i(U,V)$, $Z=\chi_i(U,V)$, *where* $1\leqq i\leqq N$ *and* U *and* V *are words in the alphabet* $\mathbf{H}$.

PROOF. It is easy to see that any infinite chain of (1) is degenerate. Hence by 5.4 the equation $\mathfrak{A}_k$ has the form

$$xAy\,[\widetilde{\xi}_iA_i]_{i=1}^n \to yBxzR\,[\widetilde{\eta}_iB_i]_{i=1}^m, \tag{2}$$

where $\widetilde{x}$, $\widetilde{y}$, $\widetilde{z}$ are, respectively,

$$[F(xA,yB,\lambda_0,\cdots,\lambda_{k-1})*A],\qquad [F(xA,yB,\lambda_0,\cdots,\lambda_{k-2})*B],\qquad z.$$

Assume that (2) has a solution (X,Y,Z). For brevity, put

$$F_i=F(XA,YB,\lambda_0,\cdots,\lambda_{i-1}),\qquad i=1,\cdots,k.$$

Since (X, Y, Z) is a solution of (2) and $[A^\partial = [B^\partial$, in the alphabet $\mathbf{И}$ we have the equalities

$$XAY = YBX, \tag{3}$$

$$[\bar{\xi}_i A_i]_{i=1}^n = ZR[\bar{\eta}_i B_i]_{i=1}^m, \tag{4}$$

where $\bar{x}$, $\bar{y}$ and $\bar{z}$ are, respectively, $[F_k * A]$, $[F_{k-1} * B]$ and Z. We indicate in the word $\xi_1 \cdots \xi_n$ the first occurrence (starting from the left) of the unknown z (it is no loss of generality to assume that z occurs in $\xi_1 \cdots \xi_n$). Then (4) assumes the form

$$[\bar{\xi}_i A_i]_{i=1}^l Z A_{l+1} [\bar{\xi}_i A_i]_{i=l+2}^n = ZR[\bar{\eta}_i B_i]_{i=1}^m, \tag{5}$$

where $1 \leqq l < n$ and $\xi_1, \cdots, \xi_l$ are x or y.

Express Z from (5). By 1.16′ there exist integers t and r, where $t \geqq 0$ and $1 \leqq r \leqq l$, and a word $W \neq \Lambda$ such that

$$Z = ([\bar{\xi}_i A_i]_{i=1}^l)^t [\bar{\xi}_i A_i]_{i=1}^{r-1} W, \quad W \angle \bar{\xi}_r A_r. \tag{6}$$

Substitute the expression for Z given by (6) in place of Z in (5) and view the resulting equality as an exponential equation in the one unknown t. Denote it by $\mathfrak{A}(t)$. By hypothesis $\xi_1 = x$, so that $\partial([\bar{\xi}_i A_i]_{i=1}^l) \geqq [\bar{x}^\partial > [\bar{y}^\partial > 0$. Now applying Proposition 2.7 to the equation $\mathfrak{A}(t)$, we conclude that there exists a constant C, depending only on M, such that $\mathfrak{A}(t)$ is satisfied for some $t = t_0 \leqq C$. Put $Z_0 = Z(t_0)$ (see (6)). Consider the triple $(\bar{x}, \bar{y}, Z_0)$. Using (3) and the formulas for $\bar{x}$ and $\bar{y}$, it is easy to verify that the pair $(\bar{x}, \bar{y})$ satisfies the equation $xAy \to yBx$. Since $\mathfrak{A}(t_0)$ is true, the triple $(\bar{x}, \bar{y}, Z_0)$ satisfies the tail $[\xi_i A_i]_{i=1}^n = zR[\eta_i B_i]_{i=1}^m$ of (1). Hence $(\bar{x}, \bar{y}, Z_0)$ is a solution of (1). We now look at the word W. Substituting the expression for Z_0 in place of Z in (5) and cancelling the resulting equality by $([\xi_i A_i]_{i=1}^l)^{t_0} [\bar{\xi}_i A_i]_{i=1}^{r-1}$ on the left, we obtain

$$[\bar{\xi}_i A_i]_{i=r}^l [\bar{\xi}_i A_i]_{i=1}^{r-1} W \ldots = WR[\bar{\eta}_i B_i]_{i=1}^m, \tag{7}$$

where $\bar{z}$ is the right-hand side of (6) with $t = t_0$.

We know $r \leqq l$. Consider the words $\bar{\xi}_r$ and W. There are two possible cases.

1) $[\bar{\xi}_r^\partial \leqq [WR^\partial$. Since $W \angle \bar{\xi}_r A_r$ (see (6)), we have

$$|[\bar{\xi}_r^\partial - [W^\partial| \leqq \max\{[A_r^\partial, [R^\partial\} \leqq M,$$

so that there exist words V, S and T such that $\bar{\xi}_r = VS$, $W = VT$ and $[S^\partial, [T^\partial \leqq M$. If now ξ_r is x, then the solution $(\bar{x}, \bar{y}, Z_0)$ is obtained by means of the formulas

$$\bar{x} = vS, \; \bar{y} = u, \; Z_0 = ([\bar{\xi}_i A_i]_{i=1}^l)^{t_0} [\bar{\xi}_i' A_i]_{i=1}^{r-1} vT,$$

where $\bar{\xi}_1, \cdots, \bar{\xi}_l$ are vS or u. Analogous formulas are obtained if ξ_r is y.

2) $[\bar{\xi}_r^\partial > [WR^\partial$. Then it follows from (7) that there exists $V_1 \neq \Lambda$ such that

$$\bar{\xi}_r = WRV_1, \tag{8}$$

$$V_1 A_r [\bar{\xi}_i A_i]_{i=r+1}^l [\bar{\xi}_i A_i]_{i=1}^{r-1} W \ldots = [\bar{\eta}_i B_i]_{i=1}^m. \tag{9}$$

We will show that 9.2.5 is applicable to (9). Let us prove that $W \sim V_1$. This is obvious if $\bar{\eta}_1 = W$. If $\bar{\eta}_1 \neq W$, then $\bar{x} \angle \bar{\eta}_1$, since η_1 is x or z, and it is clear from (6) that if $Z \neq W$, then $\bar{\xi}_1 \angle Z$, and by hypothesis $\xi_1 = x$. Thus we assume $\bar{x} \angle \bar{\eta}_1$. By hypothesis ξ_r is x or y, so that $\bar{\xi}_r \angle \bar{x}$. By (8), $W \angle \bar{x}$. It is clear from (8) that $[V_1^\partial < [\bar{\xi}_r^\partial \leqq [\bar{x}^\partial$; hence from (9), taking into account that $\bar{x} \angle \bar{\eta}_1$, we have $V_1 \angle \bar{x}$. Thus $W \angle \bar{x}$ and $V_1 \angle \bar{x}$, whence $W \sim V_1$, as required. Since $A \neq B$ and $[A^\partial = [B^\partial$, it follows that $XAYB \neq YBXA$, so that $\mathscr{S}(XA, YB)$. Also, $k \geqq 2M^2 \geqq (a+2)M+9$, where a is the number of occurrences of z among $\eta_1, \cdots, \eta_m$. Since W plays the role of U in 9.2.5, the a just defined agrees with the a in 9.2.5. We may also assume that among $\eta_1, \cdots, \eta_m$ is x or y. Finally, since $\bar{\xi}_r$ is $[F_k * A]$ or $[F_{k-1} * B]$, we have $\bar{\xi}_r D = F_t = WRV_1D$, where D is A or B and t is k or $k-1$. By 9.2.5, there exist sequences $(t_1, \cdots, t_q)$ and $(c_1, \cdots, c_q)$ such that $WR = F_{t_1}^{c_1} \cdots F_{t_q}^{c_q}$, where $q \geqq 1$, $1 \leqq c_i \leqq \lambda_{t_i}$, $i = 1, \cdots, q$, and $k \geqq t_1 > \cdots > t_q > k-(a+1)M-9$. From the last chain of inequalities we have $q \leqq k - t_q < (a+1)M+9 \leqq 2M^2$. Since $k \geqq (a+2)M+9$, we have $t_q > M$, so certainly $[F_{t_q}^\partial > M$. Then it follows from the representation of WR that R is an end of F_{t_q}. The cases t_q even and t_q odd are handled analogously. Consider the case where t_q is even. Then, by 5.2, F_{t_q} ends with A, and F_{t_q-1} with B. By what has been proved, R is an end of F_{t_q}. The cases $[A^\partial \geqq [R^\partial$ and $[A^\partial \leqq [R^\partial$ are handled analogously. Consider the case $[A^\partial \leqq [R^\partial$. Since A and R are ends of F_{t_q}, we have $R = QA$ for some Q. Put $U = [F_{t_q} * R]$ and $V = [F_{t_q-1} * B]$, so that $F_{t_q} = UR$ and $F_{t_q-1} = VB$. Using 5.1, we obtain

$$\begin{aligned}\bar{x} &= [F_k * A] = [F(F_{t_q}, F_{t_q-1}, \lambda_{t_q}, \cdots, \lambda_{k-1}) * A]\\ &= G(F_{t_q}, F_{t_q-1}, \lambda_{t_q}, \cdots, \lambda_{k-1}) F_{t_q-1}[F_{t_q} * A]\\ &= G(UR, VB, \lambda_{t_q}, \cdots, \lambda_{k-1})\, VBUQ.\end{aligned}$$

Similarly, we obtain

$$\bar{y} = G(UR, VB, \lambda_{t_q}, \cdots, \lambda_{k-2})\, URV.$$

It remains to express Z_0. Since $F_{t_q} = UR$, from the representation of WR we have $W = [F_{t_i}^{c_i}]_{i=1}^{q-1}(UR)^{c_q-1}U$. By 5.1, $F_{t_i} = F(F_{t_q}, F_{t_q-1}, \lambda_{t_q}, \cdots, \lambda_{t_i-1}) = F(UR, VB, \lambda_{t_q}, \cdots, \lambda_{t_i-1})$. Thus (see (6)), the solution $(\bar{x}, \bar{y}, Z_0)$ is obtained (with $u = U$ and $v = V$) by means of the formulas

$$\begin{aligned}\bar{x} &= G(uR, vB, \lambda_{t_q}, \ldots, \lambda_{k-1})\, vBuQ = \varphi(u, v),\\ \bar{y} &= G(uR, vB, \lambda_{t_q}, \ldots, \lambda_{k-2})\, uRv = \psi(u, v),\\ w &= [F(uR, vB, \lambda_{t_q}, \ldots, \lambda_{t_i-1})^{c_i}]_{i=1}^{q-1} (uR)^{c_q-1} u,\\ Z_0 &= ([\bar{\xi}_i A_i]_{i=1}^{l})^{t_0} [\bar{\xi}_i A_i]_{i=1}^{r-1} w = \chi(u, v),\end{aligned}$$

where $\bar{\xi}_i$ is φ or ψ, depending on whether $\xi_i = x$ or $\xi_i = y$. Analogous formulas are obtained in the cases $[R^\partial \leqq [A^\partial$, t_q odd. It remains to observe that there are a finite number of triples of the form (φ, ψ, χ). This follows from the fact that $t_0 \leq C$, $q \leqq 2M^2$, $1 \leqq c_i \leqq \lambda_{t_i}$, and the numbers $\lambda_{t_q}, \cdots, \lambda_{k-1}$ are uniquely determined by the chain $\mathfrak{A}_0, \cdots, \mathfrak{A}_k$, being the parameters of the transformations in the chain $\mathfrak{A}_0, \cdots, \mathfrak{A}_k$. The proposition is proved.

9.2.6′. *There exists an algorithm recognizing the solvability of equations of the form* 9.2.6 (1).

PROOF. Suppose $\mathfrak{A}(x, y, z)$ is an equation of the form 9.2.6 (1). By 4.9′, for any $k > 0$ it is possible to construct a finite set Ω of one-stage exponential equations and word equations in two unknowns and a finite number of infinite images $\mathfrak{A}_1, \cdots, \mathfrak{A}_m$ of order k of the equation $\mathfrak{A}$ such that $\mathfrak{A}$ has solutions if and only if at least one equation of the set $\Omega \cup \bigcup_1^m A_i$ has. Fix an even number $k \geqq 2M^2$, where M is the number from 9.2.6. By 9.2.6, for each of the equations $\mathfrak{A}_i$ $(1 \leqq i \leqq m)$ it is possible to construct a finite set $\{\mathfrak{A}_i\}$ of word equations in two unknowns such that if $\mathfrak{A}_i$ has a solution, then at least one equation in the set $\{\mathfrak{A}_i\}$ has. Here each equation in the set $\mathfrak{M} = \bigcup_1^m \{\mathfrak{A}_i\}$ has the form $\mathfrak{A}(\varphi(u,v), \psi(u,v), \chi(u,v))$, where φ, ψ and χ are words in the alphabet $\mathbf{H} \cup \{u,v\}$. Using 7.7, we test whether the equations in $\mathfrak{M}$ have solutions. If at least one equation in $\mathfrak{M}$ has solutions, then $\mathfrak{A}$ has a solution (in accordance with the formulas $x = \varphi(u,v)$, $y = \psi(u,v)$ and $z = \chi(u,v)$). If no equation in $\mathfrak{M}$ has solutions, look at (with the aid of 2.3′ and 7.7) the equations in Ω. If Ω contains equations which have solutions, then $\mathfrak{A}$ has a solution; otherwise, $\mathfrak{A}$ has no solutions. The proposition is proved.

In conclusion, we emphasize that the possibility of applying the machinery connected with the function F (F-theory) and thereby proving 9.2.6′ is related to the possibility of expressing z in terms of x, y and "cuts" of F. The tail $x \cdots = z \cdots$ of equation 9.2.6 (1) was specially selected for this purpose.

9.3. In the main result of the previous subsection, Proposition 9.2.6, the reduction to equations in two unknowns was obtained, in particular, because we had expressions for z in terms of the function F and its "fractional" parts. In the general case, the unknown z in the tail $\Phi = \Psi$ may be shut off (on both sides) by prefixes $\varphi(x,y)$ and $\psi(x,y)$ containing x and y. It is possible to annul these prefixes in various ways, for example, by arguing in the spirit of Lemma 7.4, starting with the equation $\varphi(x,y)z \cdots = \psi(x,y)z \cdots$. Our method is based on combinatorial properties of the μ-images of an equation. A chain $\mathfrak{A}_0, \cdots, \mathfrak{A}_k$ of equations is called a chain of μ-images of $\mathfrak{A}_0$ if for any i, $1 \leqq i \leqq k$, the equation $\mathfrak{A}_i$ is (for some μ_i) a μ_i-image of $\mathfrak{A}_{i-1}$. Here $\mathfrak{A}_k$ is called a μ-image of order k of the equation $\mathfrak{A}_0$. It turns out that in μ-images of sufficiently high order the prefixes can be discarded simply by using 4.4.

We will now prove two lemmas of a combinatorial nature, which show how the prefix $\varphi(x,y) = \psi(x,y)$ is transformed in a chain of μ-images of the equation $xAy\varphi z \cdots \to yBx\psi z \cdots$.

Obviously, the projection of a μ-image of (*) on the alphabet $\{x,y\}$ agrees with the μ-image of the projection of (*) on $\{x,y\}$ (the μ-image of the projection $x \cdots \to y \cdots$ is obtained by means of the formulas $\bar{x} = (xy)^\mu x$ and $\bar{y} = xy$). In what follows we will be dealing with words in the alphabet $\{x,y\}$.

We define a function $h(\varphi,\psi)$ on pairs of words φ, ψ in the alphabet $\{x,y\}$. Suppose φ and ψ are words in $\{x,y\}$. Let $\{\varphi\}_l$ denote the beginning of length l of the word φ, and $(\varphi)_x$ the number of occurrences of x in φ. We will omit the braces and parentheses where possible. Put $l = \min\{[\varphi^\partial, [\psi^\partial\}$. By definition

$$h(\varphi, \psi) = \max_{1 \leqslant i \leqslant l} |(\{\varphi\}_i)_x - (\{\psi\}_i)_x|.$$

Note that if $h(\varphi,\psi) = 0$, then one of the words φ or ψ is a beginning of the other. If $h(\varphi,\psi) = 1$, then the equation $\varphi = \psi$ has the form $P_0[\xi_i Q_i \eta_i P_i]_{i=1}^n \xi P = P_0[\eta_i Q_i \xi_i P_i]_{i=1}^n \eta Q$, where P, Q, P_i and Q_i are words in the alphabet $\{x,y\}$, ξ_i and η_i are unknowns, and where $\xi_i \neq \eta_i$ and either $\xi = \eta = \Lambda$ or ξ and η are distinct unknowns; and one of the words P or Q is a beginning of the other.

Note, also, that if in the words φ and ψ we replace x by y, and y by x, then the h-function is unchanged. In other words,

$$h(\varphi(x,y),\psi(x,y)) = h(\varphi(y,x),\psi(y,x))$$

(this is the invariance of the function h with respect to the interchange of x and y). It is also obvious that $h(\varphi,\psi) = h(\psi,\varphi)$.

By the μ-image of the word $\varphi(x,y)$ we mean the word $\varphi(\bar{x},\bar{y})$, where $\bar{x} = (xy)^{\mu}x$ and $\bar{y} = xy$. The μ-image of φ is denoted by $\bar{\varphi}$ (the μ is "concealed" in the upper bar).

The following lemma shows that in the passage to μ-images the h-function decreases (if $h(\varphi,\psi) > 1$, then $h(\bar{\varphi},\bar{\psi}) < h(\varphi,\psi)$).

9.3.1. *For any $\mu \geqq 1$ we have the inequality*

$$h(\bar{\varphi},\bar{\psi}) \leqq \tfrac{1}{2}(h(\varphi,\psi)+1).$$

PROOF. Consider the words $\bar{\varphi}$ and $\bar{\psi}$. Put $d = \min\{[\bar{\varphi}^{\partial},[\bar{\psi}^{\partial}\}$. Choose an integer i from among $1,\cdots,d$ and compute $\delta_i = (\{\bar{\varphi}\}_i)_x - (\{\bar{\psi}\}_i)_x$. We will prove that $|\delta_i| \leqq \tfrac{1}{2}(h(\varphi,\psi)+1)$. Represent the words $\{\bar{\varphi}\}_i$ and $\{\bar{\psi}\}_i$ in the form

$$\{\bar{\varphi}\}_i = \varphi_j(\bar{x},\bar{y})u \qquad \{\bar{\psi}\}_i = \psi_\nu(\bar{x},\bar{y})v, \tag{1}$$

where in φ and ψ we have omitted the braces, $0 \leqq j \leqq [\varphi^{\partial}$, $0 \leqq \nu \leqq [\psi^{\partial}$, and u and v are proper beginnings of the words $\bar{x}$ or $\bar{y}$. Without loss of generality, we consider the case $j \leqq \nu$. Then ψ_j is defined. Put $m = (\varphi_j)_x - (\psi_j)_x$. If $m < 0$, then the word ψ_j contains $|m|$ more occurrences of x than does φ_j. Then the word $\psi_j(\bar{x},\bar{y})$ contains $|m|\,(2\mu-1)$ letters more than does $\varphi_j(\bar{x},\bar{y})$, since $[\bar{x}^{\partial} = 2\mu+1$ and $[\bar{y}^{\partial} = 2$. If we assume that $j < \nu$, then $\psi_\nu(\bar{x},\bar{y})$ is longer than $\varphi_j(\bar{x},\bar{y})$ by at least $2\mu+1$. On the other hand, $[u^{\partial} \leqq 2\mu$. Hence (see (1)) the case $m < 0$, $j < \nu$ is impossible. This allows us when considering the case $j \leqq \nu$ to restrict ourselves to the case $m \geqq 0$ (the case $j = \nu$ is symmetric with respect to the sign of m).

Hence we have $m \geqq 0$. Then the word φ_j contains m more occurrences of x than does ψ_j. Hence the word $\varphi_j(\bar{x},\bar{y})$ is longer than $\psi_j(\bar{x},\bar{y})$, namely

$$[\varphi_j(\bar{x},\bar{y})^{\partial} = [\psi_j(\bar{x},\bar{y})^{\partial} + m(2\mu-1). \tag{2}$$

Let α denote the end of the word ψ_ν which remains if we discard the beginning ψ_j (α is empty if $j = \nu$). Let $\bar{\alpha}$ denote the word $\alpha(\bar{x},\bar{y})$. Suppose that α contains a occurrences of x and b occurrences of y. Then $[\bar{\alpha}^{\partial} = a(2\mu+1)+2b$. We have (see (1) and Figure 5)

$$i = [\psi_j(\bar{x},\bar{y})^{\partial} + [\bar{\alpha}^{\partial} + [v^{\partial} = [\varphi_j(\bar{x},\bar{y})^{\partial} + [u^{\partial}. \tag{3}$$

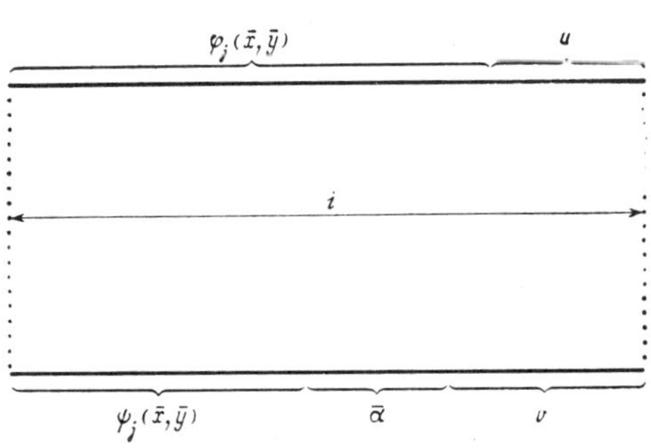

FIGURE 5

From (2), (3) and the expression for $[\bar{\alpha}^\partial$, we obtain

$$a(2\mu+1)+2b+[v^\partial=m(2\mu-1)+[u^\partial. \tag{4}$$

The word $\varphi_j(\bar{x},\bar{y})u$ contains $m\mu+u_x$ more occurrences of x than does the word $\psi_j(\bar{x},\bar{y})$. In the word $\bar{\alpha}v$ there are $a(\mu+1)+b+v_x$ occurrences of x. For the difference δ_i we obtain

$$\delta_i=m\mu+u_x-[a(\mu+1)+b+v_x].$$

On the other hand, from (4) we have

$$b=\tfrac{1}{2}[m(2\mu-1)+[u^\partial-[v^\partial-a(2\mu+1)].$$

The expression for δ_i assumes the form

$$\begin{aligned}\delta_i&=m\mu+u_x-a(\mu+1)-v_x-\left[m\mu-\frac{m}{2}+\frac{[u^\partial-[v^\partial}{2}-a\mu-\frac{a}{2}\right]\\&=\frac{1}{2}(m-a)+u_x-v_x-\frac{1}{2}([u^\partial-[v^\partial).\end{aligned} \tag{5}$$

Let us compute the "remainder term" $\epsilon=u_x-v_x-\tfrac{1}{2}([u^\partial-[v^\partial)$. We consider two cases, depending on the parity of the difference $[u^\partial-[v^\partial$.

1) $[u^\partial-[v^\partial$ is even. Then it is easy to see that $u_x-v_x=\tfrac{1}{2}([u^\partial-[v^\partial)$ (each of the words u and v has the form $(xy)^cx$, where $0\leqq c<\mu$, or the form $(xy)^c$ with different c for u and v). We obtain $\delta_i=\tfrac{1}{2}(m-a)$. It follows from (4) that $a\leqq m$, since $[u^\partial<2\mu+1$. Thus $0\leqq\delta_i=\tfrac{1}{2}(m-a)\leqq\tfrac{1}{2}m\leqq\tfrac{1}{2}(m+1)$.

2) $[u^\partial-[v^\partial$ is odd. Then one of the words u or v has the form $(xy)^cx$, and the other the form $(xy)^d$, where $0\leqq c<\mu$ and $0\leqq d\leqq\mu$. Considering various cases depending on relations among c, d, u and v, we see that $\epsilon=\pm\tfrac{1}{2}$. We also have $m>a$ (see (5)) (since $m\geqq a$ and δ_i is an integer). We obtain $\delta_i=\tfrac{1}{2}(m-a)\pm\tfrac{1}{2}$; hence $0\leqq\delta_i\leqq\tfrac{1}{2}(m-a+1)\leqq\tfrac{1}{2}(m+1)$. We have $0\leqq m\leqq h(\varphi,\psi)$, so that $|\delta_i|\leqq\tfrac{1}{2}(h(\varphi,\psi)+1)$. Since i was chosen arbitrarily, $1\leqq i\leqq d$, we obtain

$$h(\bar{\varphi},\bar{\psi})=\max_{1\leqslant i\leqslant d}|\delta_i|\leqslant\frac{1}{2}(h(\varphi,\psi)+1).$$

The lemma is proved.

A coefficient-free word equation in two unknowns is called canonical if it has the form

$$[\xi_i\eta_iA_i]_{i=1}^kP\to[\eta_i\xi_iA_i]_{i=1}^kQ,$$

where ξ_i and η_i are unknowns, $\xi_i\neq\eta_i$ $(i=1,\cdots,k)$, and at least one of the words P or Q is empty.

The following proposition establishes the structure of μ-images of sufficiently large order of equations in two unknowns.

9.3.2. *Suppose Φ and Ψ are words in the alphabet $\{x,y\}$. There exists $l(\Phi,\Psi)$ such that any μ-image of order l of the equation $xy\Phi\to yx\Psi$ is canonical.*

Proof. Suppose $\varphi\to\psi$ is a coefficient-free word equation in the two unknowns x and y. Denote it by $\mathfrak{A}$. Define the index $h(\mathfrak{A})$ of the equation $\mathfrak{A}$ by $h(\mathfrak{A})=h(\varphi,\psi)$. From the invariance property of the function h it follows that $h(\mathfrak{A}(x,y))=h(\mathfrak{A}(y,x))$, i.e. if x and y are interchanged, the index of the equation remains unchanged. The equation $xy\varphi(x,y)\to yx\psi(x,y)$ has two μ-images: $xy\bar{\varphi}\to yx\bar{\psi}$ and $yx\bar{\psi}\to xy\bar{\varphi}$. Interchanging x and y in the latter, we obtain an equation of the form $xy \cdots \to yx \cdots$ with the same index. We may therefore assume, without changing the indices of equations, that in a chain $\mathfrak{A}_0, \cdots, \mathfrak{A}_l$ of μ-images any equation $\mathfrak{A}_i$ has the form $xy\varphi(x,y) \to yx\psi(x,y)$. Then the μ-image $\mathfrak{A}_{i+1}$ of $\mathfrak{A}_i$ has the form $xy\bar{\varphi} \rightleftarrows yx\bar{\psi}$, where $\bar{\varphi}$ and $\bar{\psi}$ are the μ-images of φ and ψ. Hence, by 9.3.1, if $h(\mathfrak{A}_i) > 1$, then $h(\mathfrak{A}_{i+1}) < h(\mathfrak{A}_i)$. If now $q = h(\mathfrak{A}_0)$, then for any chain $\mathfrak{A}_0, \cdots, \mathfrak{A}_q$ of μ-images of $\mathfrak{A}_0$ we have $h(\mathfrak{A}_q) \leq 1$. Then $\mathfrak{A}_q$ has the form (possibly after interchanging x and y)

$$xyP_0\,[\xi_i Q_i \eta_i P_i]_{i=1}^n\, uPR \to yxP_0\,[\eta_i Q_i \xi_i P_i]_{i=1}^n vPS, \tag{1}$$

where $n \geq 0$; P_i, Q_i, P, R and S are words in the alphabet $\{x,y\}$; ξ_i and η_i are unknowns, $\xi_i \neq \eta_i$, and either $u = v = \Lambda$ or u and v are distinct unknowns; and at least one of the words R or S is empty.

Look at the μ-images of equation (1). They have the form

$$\alpha\beta\bar{P}_0\,[\bar{\xi}_i\bar{Q}_i\bar{\eta}_i\bar{P}_i]_{i=1}^n\,\bar{u}\bar{P}\bar{R} \to \beta\alpha\bar{P}_0\,[\bar{\eta}_i\bar{Q}_i\bar{\xi}_i\bar{P}_i]_{i=1}^n\bar{v}\bar{P}\bar{S}, \tag{2}$$

where $\bar{x} = (xy)^\mu x$, $\bar{y} = xy$, and α and β are x or y.

Consider the equation $\bar{\xi}_i\bar{Q}_i\bar{\eta}_i = \bar{\eta}_i\bar{Q}_i\bar{\xi}_i$. One of the letters ξ_i or η_i is x, the other is y. Let us say $\xi_i = x$ and $\eta_i = y$. Then the equation $\bar{\xi}_i\bar{Q}_i\bar{\eta}_i = \bar{\eta}_i\bar{Q}_i\bar{\xi}_i$ has the form

$$(xy)^\mu x[(xy)^{\mu+a_i}x]_{i=1}^m\,(xy)^{a+1} = xy\,[(xy)^{\mu+a_i}x]_{i=1}^m\,(xy)^{a+\mu}x, \tag{3}$$

where m is the number of occurrences of x in Q_i, and a_j, $a \geq 0$ and correspond to the occurrences of y in Q_i. Equation (3) has the form

$$A_0[\alpha_i\beta_i A_i]_{i=1}^k = A_0[\beta_i\alpha_i A_i]_{i=1}^k, \tag{4}$$

where $k > 0$; α_i and β_i are distinct unknowns, and $A_0, \cdots, A_n$ are words in the alphabet $\{x,y\}$. This becomes obvious if we write (3) as follows:

$$xy\,[(xy)^{\mu-1}x\,(xy)^{a_i+1}\,]_{i=1}^m\,(xy)^{\mu-1}x\,(xy)^{a+1}$$
$$= xy\,[(xy)^{\mu-1}x\,(yx)^{a_i+1}\,]_{i=1}^m\,(xy)^{\mu-1}x\,(yx)^{a+1}.$$

It is now clear that if in (1) we have $u = v = \Lambda$, then equation (2), i.e. $\mathfrak{A}_{q+1}$, is canonical.

Consider the equations $\mathfrak{A}_q$ for which $u \neq v$. In view of what has been said about the equations $\bar{\xi}_i\bar{Q}_i\bar{\eta}_i = \bar{\eta}_i\bar{Q}_i\bar{\xi}_i$, equation (2) has the form $\alpha\beta\Phi\bar{u}\bar{P}\bar{R} \to \beta\alpha\Psi\bar{v}\bar{P}\bar{S}$, where $\Phi = \Psi$ is an equation of the form (4). It is also clear that if an equation $\Phi = \Psi$ has the form (4), then $\bar{\Phi} = \bar{\Psi}$ has the form (4). The same is true if before putting the bar on Φ and Ψ we interchange (if $\alpha = y$) x and y in (2). Hence it suffices to prove the assertion for the equation $\alpha\beta\bar{u}\bar{P}\bar{R} \to \beta\alpha\bar{v}\bar{P}\bar{S}$. We will prove that μ-images of the equation $\alpha\beta\bar{u}\bar{P}\bar{R} \to \beta\alpha\bar{v}\bar{P}\bar{S}$ are canonical. The equation $\alpha\beta\bar{u}\bar{P}\bar{R} \to \beta\alpha\bar{v}\bar{P}\bar{S}$ has the form (we restrict ourselves to the case $u = x$, $v = y$)

$$\begin{aligned} &\alpha\beta xy\,[(xy)^{\mu-1}x\,(xy)^{a_i+1}\,]_{i=1}^m\,(xy)^{\mu-1}x\,(xy)^{a}\bar{R} \\ &\qquad \to \beta\alpha xy\,[(xy)^{\mu-1}x\,(yx)^{a_i+1}\,]_{i=1}^m\,(xy)^{a}\,\bar{S}, \end{aligned} \tag{5}$$

where m, a_i and a are defined as in (3), P plays the role of Q_i, and α and β are x or y. We consider the cases $R = \Lambda$ and $S = \Lambda$ separately.

Suppose $S = \Lambda$. Equation (5) has the form $\alpha\beta\Phi(xy)^{\mu-1}x(xy)^a\bar{R} \to \beta\alpha\Psi(xy)^a$, where $\Phi=\Psi$ is an equation of the form (4). If $\mu-1\geqq a$, the equation obtained is canonical. If $\mu-1<a$, it has the form $\alpha\beta\Phi(xy)^{\mu-1}x(xy)^{a-\mu}(xy)^{\mu}\overline{R}\to \beta\alpha\Psi(xy)^{m-1}x(yx)^{a-\mu}y$, i.e. the form $\alpha\beta\Phi_1(xy)^{\mu}\overline{R}\to\beta\alpha\Psi_1 y$, where the equation $\Phi_1=\Psi_1$ has the form (4). The μ-images of this last equation, as we easily compute (separately for the cases $\alpha=x$ and $\alpha=y$), are canonical.

Suppose $R=\Lambda$. Equation (5) has the form $\alpha\beta\Phi(xy)^{\mu-1}x(xy)^{a}\to\beta\alpha\Psi(xy)^{a}\overline{S}$. If x occurs in S, then $\overline{S}$ begins with the word $(xy)^{c+\mu}x$ ($c\geqq 0$ corresponds to the occurrences of y in S) and the equation is canonical. If x does not occur in S, then the equation has the form $\alpha\beta\Phi(xy)^{\mu-1}x(xy)^{a}\to\beta\alpha\Psi(xy)^{a+c}$, where $c\geqq 0$. If $c\geqq a$, this last equation is canonical. If $c<a$, it has the form $\alpha\beta\Phi_1(xy)^{\mu-c}\to\beta\alpha\Psi_1 y$, where the equation $\Phi_1=\Psi_1$ has the form (4). The μ-images of this last equation, as we easily compute, are canonical.

It follows that the μ-images of order $q+2$ of the equation $xy\Phi\to yx\Psi$ are canonical (we use the fact that μ-images of a canonical equation are canonical). The proposition is proved.

We now turn our attention to basic equations of characteristic zero. In other words, we consider the equation

$$(*)\qquad xAy\Phi(x,y,z)\to yBx\Psi(x,y,z),$$

where A and B are words in the alphabet $\mathbf{И}$, $[A^{\partial}=[B^{\partial}$, and Φ and Ψ are words in the alphabet $\mathbf{И}\cup\{x,y,z\}$. The equation (*) is said to be of type I if $A\neq B$, and of type II if $A=B$.

The equation (*) may be assumed to be indecomposable (the definition follows 4.2). Indeed, if (*) is a decomposable equation, then by 4.6 it is either equivalent to a disjunction of systems of the form $\mathfrak{A}(x,y)\,\&\,\mathfrak{B}(z)$, where $\mathfrak{A}$ is a word equation in two unknowns and $\mathfrak{B}$ in one, or equivalent to a disjunction of indecomposable equations of the form (*). In the first alternative an algorithm for solving (*) follows from 7.7. There remains the second alternative. Let us consider it.

We will first prove that μ-images of indecomposable equations are also indecomposable. Indeed, write (*) in the form

$$xAy\varphi(x,y)z\cdots\to yBx\psi(x,y)z\cdots,$$

where φ and ψ do not contain z, and put $k=\varphi_x-\psi_x$ and $l=\varphi_y-\psi_y$. By the indecomposability of the equation, one of the numbers k or l is different from zero. Putting, as usual, $\bar{x}=(xAyB)^{\mu}x$, $\bar{y}=xAy$ and $\overline{P}=\mathrm{S}^{xy}_{\bar{x}\bar{y}}P|$, we let $\bar{k}=\bar{\varphi}_x-\bar{\psi}_x$ and $\bar{l}=\bar{\varphi}_y-\bar{\psi}_y$. The indecomposability of μ-images means that at least one of the numbers $\bar{k}$ or $\bar{l}$ is different from zero. This is obvious if k and l have the same signs (we have $k^2+l^2>0$). Consider the case $kl<0$. Assuming, without loss of generality, that $k>0$ and $l<0$, and taking into account that $\bar{x}$ contains $\mu+1$ occurrences of x and μ occurrences of y, and that $\bar{y}$ contains just one x and one y, we obtain $\bar{k}=k(\mu+1)-l$ and $\bar{l}=k\mu-l$. The situation $\bar{k}=\bar{l}=0$ is impossible, which implies the indecomposability of μ-images of (*). Hence, μ-images of indecomposable equations are indecomposable.

The following proposition allows us to annul one of the prefixes φ or ψ in an indecomposable basic equation $xAy\varphi(x,y)z\cdots\to yBx\psi(x,y)z\cdots$ without changing the type of the equation.

9.3.3. *Suppose* $\mathfrak{A}_0$ *is an indecomposable basic equation of characteristic zero. Then it is possible to construct a tree of* $\mathfrak{A}_0$ *whose final vertices have the form*

$$(*) \qquad xAyx\Phi(x,y,z) \to yBxzR\eta\Psi(x,y,z),$$

where A, B *and* R *are coefficients,* $[A^\partial = [B^\partial$, *and* η *is an unknown different from* y. *Each final vertex is of the same type as* $\mathfrak{A}_0$.

PROOF. We construct a tree of μ-images of $\mathfrak{A}_0$ of height l. This is done as follows, starting with supporting trees. Construct a supporting tree of $\mathfrak{A}_0$ in accordance with 7.1. Its final vertices are, as noted in 7.7, μ-images $\mathfrak{B}_1, \cdots, \mathfrak{B}_k$ of $\mathfrak{A}_0$. They are also basic equations of characteristic zero of the same type as $\mathfrak{A}_0$. For each of the equations $\mathfrak{B}_1, \cdots, \mathfrak{B}_k$ we construct a supporting tree, and so on. Repeating this process l times, we obtain a tree Д of μ-images of $\mathfrak{A}_0$ of height l. For sufficiently large l this tree is easily extended to the desired one. Indeed, any path in Д joining the original equation $\mathfrak{A}_0$ to a final vertex is a chain $\mathfrak{A}_0, \cdots, \mathfrak{A}_l$ of μ-images of $\mathfrak{A}_0$. If an equation $\mathfrak{A}$ has the form $\varphi(x,y)z \cdots \to \psi(x,y)z \cdots$, let us agree to call the equation $\varphi \to \psi$ the prefix of $\mathfrak{A}$. Let $\mathfrak{A}_0^*, \cdots, \mathfrak{A}_l^*$ denote the projections of the prefixes of $\mathfrak{A}_0, \cdots, \mathfrak{A}_l$ on the alphabet $\{x,y\}$. The sequence $\mathfrak{A}_0^*, \cdots, \mathfrak{A}_l^*$ is a chain of μ-images of $\mathfrak{A}_0^*$. By 9.3.2, for sufficiently large l the equation $\mathfrak{A}_l^*$ has the form (to within designation of the unknowns) $xyPT_1 \to yxQT_2$, where $\nu(P) = \nu(Q)$, and at least one of the words T_1 or T_2 is empty. But the other is nonempty, a consequence of the indecomposability of $\mathfrak{A}_l$ (by hypothesis $\mathfrak{A}_0$ is indecomposable, and indecomposability is preserved in a chain of μ-images). Hence $\mathfrak{A}_l$ itself has the form $xAy\Phi(x,y)S_1(x,y)z \cdots \to yBx\Psi(x,y)S_2(x,y)z \cdots$, where $[A^\partial = [B^\partial$, $\nu(\Phi) = \nu(\Psi)$, and one of the words S_1 or S_2 contains the unknowns x and y, but the other does not (the projections of Φ, Ψ, S_1 and S_2 are P, Q, T_1 and T_2). By 4.5′, the equation $\mathfrak{A}_l$ is equivalent to a disjunction of equations of the form $xAyS_1'(x,y)z \cdots \to yBxS_2'(x,y)z \cdots$, where one of the words S_1' or S_2' contains the unknowns x and y, but the other does not. Hence, using, if necessary, the fact that in the equation $xAy\Phi \to yBx\Psi$ we can interchange Φ and Ψ and also multiply it on the right by an unknown, it is possible to construct a tree of $\mathfrak{A}_0$ whose final vertices have the form

$$(1) \qquad xAyC\xi\Phi(x,y,z) \to yBxDzE\eta\Psi(x,y,z),$$

where A, B, C, D and E are coefficients, ξ is x or y, and η is x, y or z. We complete the tree of $\mathfrak{A}_0$ in such a way that in (1) we have $\xi = x$ and $\eta \neq y$. Indeed, any transformation of (1) has the form $x = y \cdots$; hence any image of (1) has the form $yBxDzE\eta_1\Psi' \to xAyCy\Phi'$, where η_1 is y or z. The ends $DzE\eta_1\Psi'$ and $Cy\Phi'$ can be interchanged, the resulting equation being equivalent to the original. After this, the images of (1) have the form $yBxCy\Phi' \to xAyDzE\eta_1\Psi'$, i.e. the form (1) with $\xi = x$ and $\eta \neq y$, only x, y and A, B have been interchanged. Thus we may assume that $\xi = x$ and $\eta \neq y$ for the final vertices (1) of the tree of $\mathfrak{A}_0$. It now remains to achieve that in (1) we have $C = D = \Lambda$. This is done in the usual way, by applying the rule of removing the first coefficient to the equation $Cx\Phi = DzE\eta\Psi$. We restrict ourselves to the case $C \angle D$. Then $D = CQ$ and the equation $Cx \cdots = Dz \cdots$ assumes the form $x\cdots = Qz\cdots$. Make the change of variable $x = Qx_1$, and then $y = Qy_1$ (in the region where $x \angle Q$ or $y \angle Q$ the solution of (1) reduces to the solution of equations in two unknowns). After this replacement, (1) has the form $x_1AQy_1x_1\cdots \to y_1BQx_1\cdot zE_1\eta_1\cdots$, $\eta_1 \neq y_1$, i.e. to within designation of x, y, A and B the form (*). It is of the same type as the original equation $\mathfrak{A}_0$ (since $A = B$ if and only if $AQ = BQ$). The proposition is proved.

We can now complete the solution of basic equations of characteristic zero of type I.

9.3.4. *There exists an algorithm recognizing the solvability of basic equations of characteristic zero of type* I.

PROOF. By the comment following the proof of 9.3.2, it suffices to restrict ourselves to indecomposable equations. By 9.3.3 and 7.8, it suffices to consider equations of the form (*) of 9.3.3, where $A \neq B$ (type is preserved in the tree of 9.3.3). But this is the equation described in 9.2.6. Thus we need only apply 9.2.6′. The proposition is proved.

We now turn to the solution of equations of characteristic zero of type II.

9.3.5. *There exists an algorithm recognizing the solvability of basic equations of characteristic zero of type* II.

PROOF. By 9.3.3 and 7.8, it suffices to construct an algorithm recognizing the solvability of equations of the form

$$xAyT(x,y)z\Phi(x,y,z) \to yAxz\Psi(x,y,z), \tag{1}$$

where A is a word in the alphabet $\mathbf{И}$, T does not contain z and begins with x, and Φ and Ψ are words in the alphabet $\mathbf{И}\cup\{x,y,z\}$. The substitution $x = yAv$ reduces (1) to the equation

$$vAyT(yAv,y)z\Phi(yAv,y,z) = yAvz\Psi(yAv,y,z),$$

which in other notation has the form

$$xAyT(x,y)z\Phi(x,y,z) = yAxz\Psi(x,y,z) \tag{2}$$

(we have denoted v by x and introduced new T, Φ and Ψ; T does not contain z, but contains x and y). Equation (2) is equivalent to

$$T(x,y)z\Phi(x,y,z)xAy = z\Psi(x,y,z)yAx. \tag{3}$$

We can express z from (3). By 1.16, $z = T^tw$, where $w \angle T$, t is a natural unknown, and the symbol $\angle$ is a predicate. By 4.1, the parameter is eliminable in transformations of the equation $z\cdots \to T(x,y)z\cdots$. This means that if (3) has a solution (x,y,z), then it also has a solution $(x_0,y_0,z_0 = T(x_0,y_0)^tw_0)$ such that $t \leq C$, where C is a constant which does not depend on x, y or z. Hence to solve (3) it suffices to solve in the region $w \angle T$ a finite number of equations

$$TT^tw\Phi(x,y,T^tw)xAy = T^tw\Psi(x,y,T^tw)yAx,$$

where t is fixed, $t \leq C$. In other words, it suffices to construct an algorithm recognizing the existence of solutions (x,y,w) of the equation

$$Tw\Phi(x,y,T^tw) = w\Psi(x,y,T^tw) \tag{4}$$

which are such that $xAy = yAx$ and $w \angle T$. We express x and y from the equation $xAy = yAx$. By 1.23 we have

$$(5) \qquad x = (uv)^{\lambda}u, \qquad A = v(uv)^{\sigma}, \qquad y = (uv)^{\mu}u,$$

where λ, σ and μ are natural unknowns, and u and v are word unknowns. It remains for us to clarify whether (4) has solutions (x, y, w) where $w \angle T(x, y)$ and x and y are obtained by means of (5). The word $T(x, y)$ has the form $[\xi_i A_i]_{i=1}^{n}$, where $n \geqq 1$, $\xi_1, \cdots, \xi_n$ are x or y, and $A_1, \cdots, A_n$ are coefficients. We divide the domain of the variables u and v into two parts.

1. *The region* $\partial(uv) \leqq [A^{\partial}$. In this region, as is evident from (5), x and y are periodic words with period of bounded length: $x = S^{\lambda}S_1$, $y = S^{\mu}S_1$, $S_1 \angle S$ and $[S^{\partial} \leqq M$. In this region $T(x, y)$ has the form $[S^{\lambda_i}B_i]_{i=1}^{n}$, where the λ_i are λ or μ, and the B_i are words in the alphabet **И**.

Since we must have $w \angle T$, w is representable in one of the following forms:

1) w cuts a coefficient B_i:

$$w = [S^{\lambda_i}B_i]_{i=1}^{q-1}S^{\lambda_q}P, \qquad P \angle B_q, \qquad 1 \leqq q \leqq n;$$

2) w cuts a power of S:

$$w = [S^{\lambda_i}B_i]_{i=1}^{q-1}S^{\nu}P, \qquad \nu < \lambda_q, \qquad P \angle S, \qquad 1 \leqq q \leqq n,$$

where ν is a natural unknown.

In both cases equation (4) (t is fixed) is a one-stage exponential equation in the three unknowns λ, μ and ν. We need only apply 2.3′.

2. *The region* $\partial(uv) > [A^{\partial}$. In this region we must have $\sigma = 0$ and $v = A$. We obtain for x and y the expressions $x = (uA)^{\lambda}u$ and $y = (uA)^{\mu}u$. The word $T(x, y)$ has the form $[(uA)^{\lambda_i}uA_i]_{i=1}^{n}$, where the λ_i are λ or μ. From the condition $w \angle T$ we again obtain representations for w:

1) w cuts a coefficient A_i:

$$w = [(uA)^{\lambda_i}uA_i]_{i=1}^{q-1}(uA)^{\lambda_q}uP, \qquad P \angle A_q, \qquad 1 \leqq q \leqq n;$$

2) w cuts a power of uA, but does not cut u:

$$w = [(uA)^{\lambda_i}uA_i]_{i=1}^{q-1}(uA)^{\nu}uP, \qquad \nu < \lambda_q, \qquad P \angle A, \qquad 1 \leqq q \leqq n$$

(ν is a natural unknown);

3) w cuts occurrences of u:

$$w = [(uA)^{\lambda_i}uA_i]_{i=1}^{q-1}(uA)^{\nu}s, \qquad u = sv, \qquad \nu \leqq \lambda_q, \qquad 1 \leqq q \leqq n$$

(ν is a natural unknown, and s and v are word unknowns).

Substituting into (4) the expressions for x, y and w, we obtain in the first and second cases the equation

$$(6) \qquad f((uA)^{\lambda}u, (uA)^{\mu}u, (uA)^{\nu}u) = g((uA)^{\lambda}u, (uA)^{\mu}u, (uA)^{\nu}u),$$

where $f(x, y, z)$ and $g(x, y, z)$ are words in the alphabet **И** $\cup \{x, y, z\}$. In the third case we obtain the equation

$$(7) \qquad f((svA)^{\lambda}, (svA)^{\mu}, (svA)^{\nu}, s, v) = g((svA)^{\lambda}, (svA)^{\mu}, (svA)^{\nu}, s, v),$$

where $f(x, y, z, s, v)$ and $g(x, y, z, s, v)$ are words in the alphabet **И** $\cup \{x, y, z, s, v\}$. In view of 8.4, the solution of (6) and (7) reduces to the solution of word equations in two unknowns. We need only apply 7.7. The proposition is proved.

9.3.6. *There exists an algorithm recognizing the solvability of basic equations of characteristic zero.*

PROOF. This follows from 9.3.4 and 9.3.5.

§10. The main theorem

10.1. *There exists an algorithm recognizing the solvability of basic equations of characteristic* $\leqq 1$.

PROOF. This follows from 8.9 and 9.3.6.

By 7.8, 7.12 and 10.1, for a complete solution of basic equations it remains to solve equations of two kinds: 1) fundamental basic equations; 2) γ-equations. We will consider each of these.

Let us begin with fundamental basic equations. These (the definition precedes 7.12) are basic equations of the forms

$$xAzP \rightarrow zBxQ, \tag{1}$$

$$xAzP(xA)^n \rightarrow zAxQ(zA)^m \qquad (n+m>0), \tag{2}$$

where A, B, P and Q are words in the alphabet $\mathbf{H} \cup \{y\}$, and where $A \equiv B$ and $P \equiv Q$.

We consider separately the question of the existence of degenerate and nondegenerate solutions (see 8.7) of (1) and (2).

Let us begin with degenerate solutions. Arguing as in 8.7, we see that (1) has a degenerate solution if and only if at least one of the equations $xAzP \rightleftarrows zBxQ$ has a finite solution. Hence the determination of degenerate solutions of (1) reduces to solving word equations in two unknowns.

Now consider (2). We will prove that there exists a chain $\mathfrak{B}_0, \cdots, \mathfrak{B}_k$ of (2) possessing the following property: if Σ_0 is a degenerate solution of (2) and $(\mathfrak{A}_0, \Sigma_0), \cdots, (\mathfrak{A}_q, \Sigma_q)$ is a complete chain of the pair $(\mathfrak{A}_0, \Sigma_0)$, then $q < k$ and $\mathfrak{A}_i = \mathfrak{B}_i$ $(i = 1, \cdots, q)$. Denote equation (2) by $\mathfrak{A}_0$. Suppose $\mathfrak{A}_0$ has a degenerate solution Σ_0, and suppose $(\mathfrak{A}_0, \Sigma_0), \cdots, (\mathfrak{A}_q, \Sigma_q)$ is a complete chain of the pair $(\mathfrak{A}_0, \Sigma_0)$. We will prove that the chain $\mathfrak{A}_0, \cdots, \mathfrak{A}_q$ is uniquely determined by $\mathfrak{A}_0$. If $q = 0$, the assertion is trivial. Suppose $q > 0$. Consider the pair $(\mathfrak{A}_1, \Sigma_1)$. The equation $\mathfrak{A}_1$, being a degenerate image of $\mathfrak{A}_0$, is obtained from (2) under a transformation $x = (zA)^t v$, where $t > 0$ is the parameter of the transformation, v is a new unknown, and has the form $vAzP((zA)^t vA)^n \leftarrow zAvQ(zA)^m$. For $\mathfrak{A}_1$ to have solutions it is necessary that $m > nt$. Otherwise, the component v is empty and the equation has no solutions (because of the arrow). Then, cancelling the equation on the right by $(zA)^{nt}$ (after commuting zA and vA), we obtain the equation $vAzP(vA)^n \leftarrow zAvQ(zA)^{m-nt}$. This is also an equation of the form (2), but each of the exponents n and m has been decreased ($m - nt$ now plays the role of n, and n the role of m). For the new equation to have solutions it is necessary that $m - nt < n$, since, because of the arrow, we have $\partial(z) > \partial(v)$ for solutions (v, y, z) of $\mathfrak{A}_1$. In other words, the parameter t is equal to the quotient in the division of m by n and does not depend on Σ_0. Thus $\mathfrak{B}_1$ is uniquely determined by $\mathfrak{A}_0$ and has the form (2) with smaller m and n. Applying an analogous argument to the pair $(\mathfrak{A}_1, \Sigma_1)$, we obtain (by induction on q) that the chain $\mathfrak{A}_0, \cdots, \mathfrak{A}_q$ is uniquely determined by $\mathfrak{A}_0$. Let k denote the number of steps in the Euclidean algorithm for determining the greatest common divisor of m and n. It follows from our argument that if $q \geqq k$, then $\mathfrak{A}_k$ has the form (2), where one of the numbers m or n is zero and the other is positive. But in this case $\mathfrak{A}_k$ has no solutions, since one of the components x or z must be empty. Hence $q < k$. The existence of the chain $\mathfrak{B}_0, \cdots, \mathfrak{B}_k$ is proved. From the existence of this chain it follows that (2) has a degenerate solution if and only if at least one of the equations $\mathfrak{B}_0, \cdots, \mathfrak{B}_k$ has a finite solution. Thus, the determination of degenerate solutions of (2) reduces to the solution of word equations in two unknowns. Therefore, the determination of degenerate solutions of (1) and (2) reduces to the solution of word equations in two unknowns.

We now turn to the determination of nondegenerate solutions.

Let $\{\mathfrak{A}\}$ denote the set of degenerate images (of all orders) of $\mathfrak{A}$. It follows from the definitions (see the corresponding argument in 8.7) that $\mathfrak{A}$ has a nondegenerate solution if and only if at least one nondegenerate image of at least one equation in the set $\{\mathfrak{A}\}$ has a solution. For the equation (1) the set $\{\mathfrak{A}\}$ consists of the two equations $xAzP \rightleftarrows zBxQ$. For equation (2) we may assume that $\{\mathfrak{A}\}$ consists of a finite number of equations of the form (2) (the others have no solutions). Thus, the determination of nondegenerate solutions of (1) and (2) reduces to the determination of solutions of nondegenerate images of a finite number of equations of the forms (1) and (2). In view of the eliminability of the parameter, it suffices to consider a finite number of nondegenerate images (those which appear in a complete system of images). Thus, it remains to construct an algorithm recognizing the solvability of nondegenerate images of fundamental basic equations. By 6.26, any nondegenerate image of equations (1) and (2) possesses a tree Д whose final vertices are basic equations of characteristic $\leqq 2$. If the equations (1) and (2) themselves have characteristic two, by 6.27 we may assume that the final vertices of Д have characteristic $\leqq 1$. Thus, the determination of nondegenerate solutions of the equations (1) and (2) of characteristic > 2 reduces to the solution of basic equations of characteristic $\leqq 2$. Moreover, by 7.8 and 10.1, we have an algorithm recognizing the existence of nondegenerate solutions of fundamental basic equations of characteristic two. The determination of degenerate solutions of fundamental basic equations was reduced above to the solution of word equations in two unknowns. Therefore, we obtain

10.2. *There exists an algorithm recognizing the solvability of fundamental basic equations of characteristic two.*

10.3. *If there exists an algorithm recognizing the solvability of basic equations of characteristic* $\leqq 2$, *then there exists an algorithm recognizing the solvability of fundamental basic equations.*

To complete the solution of fundamental basic equations, it remains, therefore, to solve basic equations of characteristic two. For this purpose we consider γ-equations.

A γ-equation (the definition was given before 7.11) is a basic equation $xAz\Phi \to zBx\Psi$, each of whose solutions satisfies a condition $\tau(z, M)$ (M is some positive constant). In view of 10.1 the interest lies in γ-equations of characteristic $\geqq 2$. Then the words A and B contain y. Consider the μ-images of γ-equations. They have the form $xAz\overline{\Phi} \rightleftarrows zBx\overline{\Psi}$. The expressions relating a solution (x, y, z) of a μ-image to a solution $(\bar{x}, \bar{y}, \bar{z})$ of the γ-equation itself have the form $\bar{x} = (\bar{z}B)^{\mu}x$, $\bar{z} = xAz$. The condition $\tau(\bar{z}, M)$ implies that the solution (x, y, z) of the μ-image satisfies the condition $\tau(xAz, M)$. Since y occurs in A, we see that x, y and z are M-periodic words. Thus, μ-images of γ-equations have only M-periodic solutions. We can now prove

10.4. *There exists an algorithm recognizing the solvability of γ-equations of characteristic two.*

PROOF. It suffices to construct a supporting tree of a γ-equation of characteristic two in accordance with 7.2. The final vertices of this tree which are not μ-images are equations of characteristic $\leqq 1$. The vertices which are μ-images have only M-periodic solutions and reduce to one-stage exponential equations. It remains to apply 7.8 and 2.3′. The proposition is proved.

10.5. *There exists an algorithm recognizing the solvability of basic equations of characteristic $\leqq 2$.*

PROOF. This follows from 7.8, 7.12, 10.1, 10.2 and 10.4.

10.6. *There exists an algorithm recognizing the solvability of fundamental basic equations.*

PROOF. This follows from 10.3 and 10.5.

10.7. *There exists an algorithm recognizing the solvability of γ-equations.*

PROOF. By 10.5, it suffices to consider γ-equations of characteristic > 2. A supporting tree of such an equation has as final vertices basic equations of characteristic $\leqq 2$ and μ-images of the equation. But μ-images of γ-equations, as was noted above, reduce to one-stage exponential equations. It remains to apply 7.8, 10.5 and 2.3′.

10.8. THEOREM. *There exists an algorithm recognizing the solvability of basic equations.*

PROOF. This follows from 7.8, 7.12, 10.1, 10.6 and 10.7.

MAIN THEOREM. *There exists an algorithm recognizing the solvability of word equations in three unknowns.*

PROOF. This follows from 6.25, 2.3′, 7.7 and 10.8.

Chapter III

SYSTEMS OF EQUATIONS, EACH OF WHICH IS AN EQUATION IN TWO UNKNOWNS (W_2-SYSTEMS)

In the previous chapter we constructed an algorithm recognizing the solvability of word equations in three unknowns. The decision problem for arbitrary systems of word equations, as we observed in section 1 of the Introduction, reduces to the decision problem for systems consisting of equations in three unknowns (moreover, each equation contains at most three letters). The decision problem for such systems remains an open one. However, for systems consisting of equations in two unknowns it is possible to construct a decision algorithm. Systems consisting of equations in two unknowns will be called W_2-systems.

Consider an arbitrary W_2-system. By 1.28, we may assume that no two equations of the system contain the same pair of unknowns. Then the system has the form

$$\Phi_{ij}(x_i, x_j) = \Psi_{ij}(x_i, x_j), \tag{1}$$

where the indices i and j range independently over the values $1, \cdots, n$, and where $i < j$, but not all such pairs of indices are necessarily represented.

Our investigation of W_2-systems is carried out in two steps.

In the first step (this is the content of §§11 and 12) we consider a word equation in the two unknowns x and y. Here our aim is to obtain explicit formulas permitting us to find all solutions of the equation. The formulas have the form $x = \alpha$ and $y = \beta$. On the right-hand sides appear functions, defined in a nonempty region, whose values are words in the alphabet И (these functions are called predicate terms). In solving W_2-systems we seek the intersection of the ranges of these functions. Namely, if we solve each equation in the system (1) for x_i and x_j, we obtain a finite number of systems of the form

$$\mathop{\&}_{(i,j)} (x_i = \alpha_{ij},\ x_j = \beta_{ij}), \tag{2}$$

where α_{ij} and β_{ij} are predicate terms, and the conjunction extends over all pairs (i, j) corresponding to the equations in (1). The system (1) has solutions if and only if at least one of the systems (2) has. The solution of systems of the form (2) comprises the content of the second step. This is expounded in §18. Preliminary to §18 is the solution of equations whose left- and right-hand sides are predicate terms (T_2-equations). This material is covered in §§13–17.

Systems of the form (2) are called T-systems. In other notation, a T-system (2) has the form

$$\mathop{\&}_{i=1}^{n} (x_i = \alpha_{i1} = \alpha_{i2} = \ldots = \alpha_{iq_i}), \tag{3}$$

where $q_1, \cdots, q_n \geqq 1$ and the α_{ij} are predicate terms. One difficulty in solving T-systems is that some of the α_{ij} may depend on unknowns in the list $x_1, \cdots, x_n$. If a T-equation containing x_i is solved for x_i, we obtain, in general, functions different from those from which the original T-system was constructed. This leads to new forms of T-equations, then to new functions, and so on. Analysis shows that these difficulties can be anticipated by being prepared with a description of the general solution of equations in two unknowns. We obtain, in particular, the fact that the "dangerous" T-equations containing word unknowns have no solutions. For equations not containing word unknowns we show that the system $x = \alpha = \beta$ is equivalent, to within inessential details, to the equation $x = \alpha$. This allows us to successively eliminate from (3) the terms α_{ij} until $q_1 = \cdots = q_n = 1$.

The indicated description of the general solution, however, is successfully constructed only for W_2-systems consisting of equations whose left- and right-hand sides begin and end with different unknowns (normal equations). The reduction of arbitrary W_2-systems to W_2-systems consisting of normal equations is carried out in §19 (this is the so-called normalization of W_2-systems). An algorithm for solving W_2-systems is given at the end of §20 (the main theorem). The main theorem also contains a description of the structure of the general solution of W_2-systems.

Before turning to the construction of the description of the general solution of equations in two unknowns, we mention a simple example which characterizes one of the problems arising in this connection.

By an application of 1.16, functions describing the solutions have the form $\alpha^t\alpha_1$, where t is a natural unknown and α and α_1 are functions whose values are words in the alphabet И. The condition $\alpha_1 \angle \alpha$ is satisfied for any value of the unknowns occurring in α and α_1. The problem of the intersection of the ranges of such functions leads to the equation

$$\alpha^t\alpha_1 = \beta^s\beta_1, \tag{4}$$

where t and s are natural unknowns, and $\alpha_1 \angle \alpha$ and $\beta_1 \angle \beta$ for all values of the unknowns occurring in α, β, α_1 and β_1. With no additional restrictions on the unknowns occurring in α and β we have from (4) in the region $s, t > 1$ (see 1.7)

$$\alpha = z^\sigma, \qquad \beta = z^\tau, \qquad \sigma t - \tau s = \lambda, \qquad \beta_1 = z^\lambda \alpha_1 \tag{5}$$

(if $\lambda \geqq 0$), where σ, τ and λ are natural unknowns and z is a word unknown. We prepare a description of the solutions (a restriction $\mathscr{P}$ on the variables occurring in α and β) such that the values of the functions α and β in the region $\mathscr{P}$ are simple words. Then equation (4) in the region $\mathscr{P}\,\&\,(s, t > 1)$ is equivalent to the system $\alpha^2\alpha_1 = \beta^2\beta_1$, $t = s$ (see 1.7). The latter system (even in the region $\mathscr{P}$) is much simpler than the system (5).

The preparation of the description of the solutions is carried out in §11 (the analysis of the solutions of normal equations, properties of bounding predicates, the investigation of the simplicity of words depending on parameters, etc.). The description itself is given in §12.

§11. Analysis of the solutions of normal equations

To state our results we need the predicates τ and ρ. The predicate τ was defined after 4.2. We define ρ as follows. Suppose X is a word in the alphabet

И, and M and N are positive integers. By definition, the condition $\rho(X,M,N)$ holds if there exist t, n, C, R and a simple word S such that $X=(S^tC)^nS^tR$, where $n<N$, $[S^{\partial}\leqq M$; $[C^{\partial},[R^{\partial}\leqq 7M$; $\mathscr{S}(S,C)$, and $R\angle CS^{7M}$. It is obvious how, for given $M,N>0$ and a given word X, to determine whether or not $\tau(X,M)$ and/or $\rho(X,M,N)$.

Let us look at the properties of the predicate τ in detail. First of all, it follows from the definition of τ that $\tau(X,M)$ if and only if $AXV_1=XV_2$, where $0<[A^{\partial}\leqq M$. Indeed, if such an equality holds, then, by 1.16, $X=A^tA_1$, where $t\geqq 0$ and $A_1\angle A$, whence $\tau(X,M)$. On the other hand, if $X=A^tA_1$, where $t\geqq 0$ and $A_1\angle A$, then $A=A_1A_2$, and then $AX=XA_2A_1$. We have obtained the necessary equality ($V_1=\Lambda$). Note that after left cancellation the equality $AXV_1=XV_2$ has the form $AX=XB$, where $[A^{\partial}=[B^{\partial}$.

We list for reference the following obvious properties of τ.

1. Suppose that N and M are integers such that $N\geqq M>0$. Then $\tau(X,M)$ implies $\tau(X,N)$, and $\neg\tau(X,N)$ implies $\neg\tau(X,M)$.

2. Suppose $Y\angle X$. Then $\tau(X,M)$ implies $\tau(Y,M)$, and $\neg\tau(Y,M)$ implies $\neg\tau(X,M)$.

3. If $[X^{\partial}\leqq M$, then $\tau(X,M)$. Hence if $\neg\tau(X,M)$, then $[X^{\partial}>M$; in particular, if $\neg\tau(X,M)$, then $X\neq\Lambda$.

We will now prove two lemmas pertaining to τ.

Lemma 1. *Suppose $\neg\tau(U,M)$, $\neg\tau(V,M)$ and $[A^{\partial}\leqq M$, and suppose also that $UAV=VBU$ and $UPV=VAU$. Then $A=B$.*

Proof. The assertion is obvious if $[U^{\partial}=[V^{\partial}$. Furthermore, $[A^{\partial}=[B^{\partial}=[P^{\partial}$. The assertion is now obvious in the cases $[U^{\partial}\geqq[VA^{\partial}$ and $[V^{\partial}\geqq[UA^{\partial}$ (in the latter case $A=P$ and $UAV=VAU=VBU$). There remains the case $0<|[U^{\partial}-[V^{\partial}|=[A^{\partial}$. Here there are two possible subcases: 1) $[V^{\partial}<[U^{\partial}<[VA^{\partial}$; 2) $[U^{\partial}<[V^{\partial}<[UA^{\partial}$. Suppose $[V^{\partial}<[U^{\partial}<[VA^{\partial}$. Then $UAV=VBU$ implies that there exists R such that $U=RV$, $0<[R^{\partial}<[A^{\partial}$ and $RVAV=VBU$. Since $[A^{\partial}\leqq M$, the last two relations yield $\tau(V,M)$, which contradicts the hypothesis of the lemma. In the second subcase a contradiction is obtained analogously. The lemma is proved.

The following lemma clarifies the form of a solution of the equation $xAy=yBx$ when one component of the solution satisfies the condition τ.

Lemma 2. *Suppose the words A, B, U, V and an integer $M>0$ satisfy the conditions $[A^{\partial},[B^{\partial}\leqq M$; $[U^{\partial},[V^{\partial}>6M$; $\tau(U,M)$ and $\neg\tau(UAVB,M)$, and suppose also that $UAV=VBU$. Then there exist integers $\sigma\geqq 0$ and $\nu>0$, words C, E and H, and a simple word S such that $U=S^{\sigma}E$, $UA=S^{\sigma}C$, $V=(S^{\sigma}C)^{\nu-1}S^{\sigma}H$, $UAV=(S^{\sigma}C)^{\nu}S^{\sigma}H$, $HB\angle CS^{\sigma}$, $\mathscr{S}(S,C)$, $[S^{\sigma}\leqq M$ and*

$$\max\{[C^{\partial},[E^{\partial},[H^{\partial}\}\leqq 6M.$$

Proof. From the conditions $[U^{\partial}>6M$ and $\tau(U,M)$ we have for some P, Q and μ

$$U=(QP)^{\mu}Q,\qquad [QP^{\partial}\leqq M,\qquad [U^{\partial}>6M, \tag{1}$$

where the word QP is simple. By hypothesis, $UAV=VBU$. Substituting into this equality the expression for U from (1), we obtain

(2) $$(QP)^{\mu}QA\,V = VB(QP)^{\mu}Q.$$

Applying to the equality just obtained Proposition 1.16 (the roles of Y and X in 1.16 are played by $(QP)^{\mu}QA$ and V), we conclude that there exist $\lambda \geqq 0$ and t, $0 \leqq t \leqq \mu$, and a word R such that

(3) $$V = ((QP)^{\mu}QA)^{\lambda}(QP)^{t}R,$$

where $R \angle QA$ if $t = \mu$, and $R \angle QP$ if $t < \mu$. Since $[A^{\partial} \leqq M$ and $[QP^{\partial} \leqq M$, we have $[R^{\partial} \leqq 2M$. Substituting (3) into (2) and cancelling the resulting equality on the left by $((QP)^{\mu}QA)^{\lambda}(QP)^{t}$, we obtain

(4) $$(QP)^{\mu-t}QA\,(QP)^{t}R = RB(QP)^{\mu}Q.$$

Assume that $\urcorner\,\mathscr{S}(QP, QA)$. Then, in view of the simplicity of QP, the word QA is a power of QP, and from (4) we have for some $c \geqq 0$

(5) $$(QP)^{\mu+c}R = RB(QP)^{\mu}Q.$$

Let us estimate $[RBQP^{\partial}$. We know that $[R^{\partial} \leqq 2M$, $[B^{\partial} \leqq M$ and $[QP^{\partial} \leqq M$. From (1) we have $\mu[QP^{\partial} \leqq 5M$. Hence $[RBQP^{\partial} \leqq \mu[QP^{\partial}$. From (5) we obtain $RBQP \angle (QP)^{\mu}$, whence by 1.4 it follows that RB is a power of QP. Thus, QA and RB are powers of QP. Then from (1) and (3) it follows that $UA\,VB$ is a power of QP, which contradicts the assumption $\urcorner\,\tau(UA\,VB, M)$, since $[QP^{\partial} \leqq M$. From this contradiction we conclude that

(6) $$\mathscr{S}(QP, QA).$$

The rest of the argument is divided into three cases.

1) $t[QP^{\partial} \leqq 3M$. We have successively: $t < \mu$, $R \angle QP$, $[R^{\partial} \leqq M$ and $[QA\,(QP)^{t}R^{\partial} \leqq 6M < [(QP)^{\mu}Q^{\partial}$. From this last chain of inequalities and (4) we have $RB \angle (QP)^{\mu-t}$, which implies that $(QP)^{t}RB \angle (QP)^{\mu}$. Furthermore, we have $[(QP)^{t}R^{\partial} \leqq 4M$, and by hypothesis $[V^{\partial} > 6M$. Hence it follows from (3) that $\lambda > 0$ and

(7) $$V = ((QP)^{\mu}QA)^{\lambda-1}(QP)^{\mu}QA\,(QP)^{t}R.$$

Since $(QP)^{t}RB \angle (QP)^{\mu}$, we conclude from (1), (7) and (6) that our assertion holds with $\sigma = \mu$, $\nu = \lambda$, $S = QP$, $C = QA$, $E = Q$ and $H = QA\,(QP)^{t}R$.

2) $(\mu - t)\,[QP^{\partial} \leqq 3M$. Then $[(QP)^{\mu-t}QA^{\partial} \leqq 5M$. It follows from (6) that $\mathscr{S}(QP, (QP)^{\mu-t}QA)$. Since $[R^{\partial} \leqq 2M < [(QP)^{\mu}Q^{\partial}$, (4) implies that $RB \angle (QP)^{\mu-t}QA\,(QP)^{t}$. Furthermore, $(QP)^{\mu} = (QP)^{t}(QP)^{\mu-t}$. Taking all this into account, we conclude from (1) and (3) that our assertion holds with $\sigma = t$, $\nu = \lambda + 1$, $S = QP$, $C = (QP)^{\mu-t}QA$, $E = (QP)^{\mu-t}Q$ and $H = R$.

3) $t[QP^{\partial}$, $(\mu - t)\,[QP^{\partial} > 3M$. We have successively: $t < \mu$, $R \angle QP$, $[R^{\partial} \leqq M$ and $[RBQP^{\partial} < (\mu - t)\,[QP^{\partial}$. From the last inequality and (4) it follows by 1.4 that $RB = (QP)^{c}$ for some $c \geqq 0$. Cancelling (4) on the left by $(QP)^{\mu-t}$, we obtain

(8) $$QA\,(QP)^{t}R = (QP)^{t}(QP)^{c}Q.$$

Since $[QA\,QP^{\partial} \leqq 3M < t[QP^{\partial}$, it follows from (8), in view of 1.4, that QA is a power of QP, which contradicts (6). Hence case 3) is impossible. The lemma is proved.

A word equation in two unknowns is called normal if it has one of the following two forms:

$$\text{(I)} \qquad \xi\Phi_1 x\Psi = \eta\Phi_2 y\Psi,$$

$$\text{(II)} \qquad \xi\Phi_1\alpha P\Psi = \eta\Phi_2\alpha\Psi, \quad P \neq \Lambda,$$

where x, y, ξ, η and α are unknowns, $\xi \neq \eta$; Φ_1, Φ_2 and Ψ are words in the alphabet $\text{И} \cup \{x,y\}$, and P is a word in the alphabet И.

The following proposition is a fundamental one in the study of the structure of solutions of normal equations.

11.1. *Consider the word equation in two unknowns*

$$(*) \qquad xP_1[yP_i]_{i=2}^{l} xP_{l+1}\xi\Phi = ([yQ_i]_{i=1}^{k})^a xQ_{k+1}\eta\Psi,$$

where a, k and l are positive integers, ξ and η are unknowns, P_i and Q_j are coefficients, and Φ and Ψ are words in the alphabet $\text{И} \cup \{x,y]$. Suppose that $(*)$ is a normal equation, that the lengths of its coefficients do not exceed an integer $M > 0$, and that the pair (X, Y) with $[X^\partial \geqq [Y^\partial > 0$ is a solution of this equation. Then for any integers $N \geqq 15M$ and $L > 1$ one of the following five assertions holds:

a) $\tau(Y,N) \vee \rho(Y,M,L)$.

b) *There exist integers $\lambda \geqq 0$ and r, $0 \leqq r < k$, such that*

$$X = ([YQ_i]_{i=1}^{k})^\lambda [YQ_i]_{i=1}^{r} Y.$$

c) *There exist integers σ, ν, λ, ϵ, r and words S, C, R such that S is simple, $\mathscr{C}(S,C)$, $R \angle CS^\sigma$, $[S^\sigma C^\partial > N$, $\nu \geqq L$, $\epsilon = \pm 1$, $[S^\partial \leqq M$; $[C^\partial, [R^\partial \leqq 7M$; $\lambda \geqq 0$, $0 \leqq r < k$, and*

$$x = ([YQ_i]_{i=1}^{k})^\lambda [YQ_i]_{i=1}^{r} (S^\sigma C)^{\nu+\epsilon} S^\sigma R, \qquad Y = (S^\sigma C)^\nu S^\sigma R.$$

In this connection, if $\epsilon = 1$, then there exist an integer d and a word B such that $B \angle C$, $S^d B \perp\!\!\!\perp R$ and $RQ_{r+1} \perp\!\!\!\perp CS^d$.[14)]

d) *There exist integers m and n, $m > n > 0$, and a word Z such that ZP_1 is simple, $X = (ZP_1)^m Z$ and $Y = (ZP_1)^n A$.*

e) *The pair (X, Y) satisfies the equation $xP_1 y = yQ_1 x$. In this case we also have that $P_1 \neq Q_1$, $Q_1 = \cdots = Q_k$, and the system consisting of $(*)$ and the equation $xP_1 y \to yQ_1 x$ is equivalent to the equation $xP_1 y \to yQ_1 x$.*

PROOF. Suppose the hypotheses of the theorem are satisfied. The condition on the lengths of the coefficients yields

$$(0) \qquad [P_1^\partial, \cdots, [P_{l+1}^\partial, [Q_1^\partial, \cdots, [Q_{k+1}^\partial \leqq M.$$

Since (X, Y) is a solution of $(*)$, for certain V_1 and V_2 we have $XV_1 = ([YQ_i]_{i=1}^{k})^a XV_2$. By 1.16″ (where the role of Y is played by $[YQ_i]_{i=1}^{k}$), this equality implies that there exist an integer $\lambda \geqq 0$ and words W_1 and W_2 such that

$$(1) \qquad X = ([YQ_i]_{i=1}^{k})^\lambda W_1,$$

$$(2) \qquad [YQ_i]_{i=1}^{k} = W_1 W_2, \qquad W_1 \neq \Lambda.$$

14) The definition of the symbol $\perp\!\!\!\perp$ precedes 1.21.

Let N and L be integers such that

(3) $$N \geqq 15M, \quad L > 1.$$

We will prove that one of the assertions a) – e) holds. If $\tau(Y,N) \vee \rho(Y,M,L)$, then we have a). We will assume that

(4) $$\neg\, \tau(Y,N),$$

(5) $$\neg\, \rho(Y,M,L).$$

Since (*) is a normal equation, it is an equation of one of the forms (I) or (II). We will prove that it is of the form (I). Assume that (*) has the form (II). Consider the case where α is y. Then $YP = AY$ for some A. We have $0 < [A^{\partial} \leqq N$, since $P \neq \Lambda$, $M \leqq N$ and, in view of the condition on the coefficients, $[P^{\partial} \leqq M$. We obtain $\tau(Y,N)$. Similarly, if α is x, then $\tau(X,N)$, whence we again have $\tau(Y,N)$, since $Y \angle X$. But $\tau(Y,N)$ contradicts (4). From this contradiction we conclude that our equation is not of the form (II). Consequently, it has the form (I). But then

(6) $$Z_1 X = Z_2 Y$$

for some Z_1 and Z_2. Let us consider (2). There are two possible cases.

Case 1. There exist an integer r, $0 \leqq r < k$, and a word P such that

(7) $$W_1 = [YQ_i]_{i=1}^{r} YP, \qquad P \angle Q_{r+1}.$$

From (1) and (7) we have

(8) $$X = ([YQ_i]_{i=1}^{k})^{\lambda} [YQ_i]_{i=1}^{r} YP.$$

It follows from (6) and (8) that $YP = AY$ for some A. Since $P \angle Q_{r+1}$, we have $[P^{\partial} \leqq [Q_{r+1}^{\partial} \leqq M \leqq N$. Thus, $[P^{\partial} \leqq N$, and then from (4) and the equality $YP = AY$ it follows that $P = \Lambda$. From (8) we conclude that b) holds.

Case 2. There exist an integer r, $0 \leqq r < k$, and words Z and W such that

(9) $$Y = ZW, \qquad Z \neq \Lambda, \qquad W \neq \Lambda,$$

(10) $$W_1 = [YQ_i]_{i=1}^{r} Z.$$

From (1) and (10) we have

(11) $$X = ([YQ_i]_{i=1}^{k})^{\lambda} [YQ_i]_{i=1}^{r} Z.$$

Substituting the right-hand sides of (9) and (11) into (*) in place of x and y and cancelling the resulting equality on the left by $([YQ_i]_{i=1}^{k})^{\lambda} [YQ_i]_{i=1}^{r} Z$, we have for some A_1 and A_2

(12) $$P_1 ZWA_1 = WQ_{r+1} ZA_2.$$

We will prove that

(13) $$[ZP_1^{\partial}, \ [W^{\partial} > N.$$

It follows from (6), (9) and (11) that $CZ = ZW$ for some C; hence $CZW = ZWW$ and $CY = YW$. From this and the conditions $W \neq \Lambda$ and $\neg\, \tau(Y,N)$ we conclude that $[W^{\partial} > N$. Multiplying (12) on the left by Z and taking into account (9), we have $ZP_1 YA_1 = YQ_{r+1} ZA_2$. It follows from this and the conditions $Z \neq \Lambda$ and $\neg\, \tau(Y,N)$ that $[ZP_1^{\partial} > N$. The inequalities (13) are proved.

We will now prove that there exist words U, V, C_1 and C_2 such that

$$Z = UC_1, \qquad W = P_1VC_2, \tag{14}$$

$$UC_1P_1V = VC_2Q_{r+1}U, \tag{15}$$

$$[C_1P_1^\partial, \quad [C_2Q_{r+1}^\partial \leqq M, \tag{16}$$

where either $C_1 = \Lambda$ or $C_2 = \Lambda$. It follows from (13) and (3) that $[W^\partial \geqq 2M$. Then (12) implies that there exists V_1 such that $W = P_1V_1$, $[V_1^\partial \geqq M$ and

$$ZP_1V_1A_1 = V_1Q_{r+1}ZA_2. \tag{17}$$

Assume that $[P_1^\partial \geqq [Q_{r+1}^\partial$. Then it follows from (17) that there exists D such that $ZP_1V_1 = V_1Q_{r+1}ZD$. Since $[D^\partial \leqq [P_1^\partial \leqq M$ and $[V_2^\partial \geqq M$, the last equality implies that there exists V such that $V_1 = VD$ and $ZP_1 = VDQ_{r+1}Z$. Putting $C_1 = \Lambda$, $C_2 = D$, $U = Z$ and taking into account that $W = P_1V_1$ and $[DQ_{r+1}^\partial = [P_1^\partial \leqq M$, we obtain (14), (15) and (16).

Assume that $[Q_{r+1}^\partial \geqq [P_1^\partial$. Then it follows from (17) that there exists D such that $ZP_1V_1D = V_1Q_{r+1}Z$. We have $[D^\partial \leqq [Q_{r+1}^\partial \leqq M \leqq N - M \leqq [Z^\partial$. Hence the last equality implies that there exists U such that $Z = UD$ and $UDP_1V_1 = V_1Q_{r+1}U$. Putting $C_1 = D$, $C_2 = \Lambda$ and $V = W_1$, and taking into account that $W = P_1V_1$ and $[DP_1^\partial = [Q_{r+1}^\partial \leqq M$, we obtain (14), (15) and (16). Thus the relations (14), (15) and (16) are proved. Now from (9), (14) and (11) we have

$$Y = UC_1P_1VC_2, \tag{18}$$

$$X = ([YQ_i]_{i=1}^k)^\lambda [YQ_i]_{i=1}^r UC_1. \tag{19}$$

From (13), (3), (14) and (16) we conclude that

$$[U^\partial, \quad [V^\partial > 6M. \tag{20}$$

The rest of the argument is divided into three subcases.

Subcase 2.1. $\tau(U, M)$. We will prove that if $\epsilon = 1$, then c) holds. From (4) and (18), using properties 1 and 2 of τ, we obtain $\urcorner\,\tau(UC_1P_1VC_2Q_{r+1}, M)$. From (15), (16) and (20), in accordance with Lemma 2 (the roles of A and B in Lemma 2 are played by C_1P_1 and C_2Q_{r+1}), we obtain that there exist $\sigma > 0$, $\nu > 0$, C, E, H and a simple word S such that

$$U = S^\sigma E, \qquad UC_1P_1 = S^\sigma C, \tag{21}$$

$$UC_1P_1V = (S^\sigma C)^\nu S^\sigma H, \qquad HC_2Q_{r+1} \angle CS^\sigma, \tag{22}$$

$$[S^\partial \leqslant M; \quad [C^\partial,\ [E^\partial,\ [H^\partial \leqslant 6M; \quad \mathscr{E}(S, C). \tag{23}$$

Put $B = EC_1$ and $R = HC_2$. Then from (18), (22) and (19) we have

$$Y = (S^\sigma C)^\nu S^\sigma R, \tag{24}$$

$$X = ([YQ_i]_{i=1}^k)^\lambda [YQ_i]_{i=1}^r S^\sigma B. \tag{25}$$

It follows from (21) that $EC_1 \angle C$, so that $B \angle C$. From (23) and (16) we conclude that $[R^\partial \leqq 7M$. Since $HC_2 \angle CS^\sigma$ by (22), we have $R \angle CS^\sigma$. From (5), (24) and (23) we have $\nu \geqq L$. From (13) and (14) we have $[UC_1P_1^\partial > N$, so that $[S^\sigma C^\partial > N$ (21). From this, (3) and (23) we see that $\sigma[S^\partial \geqq 9M$ and $(\sigma - 1)[S^\partial \geqq 8M$. We now consider two cases.

1) $r = 0$. Then from (9), (11) and the condition $[X^\partial \geqq [Y^\partial$ it follows that $\lambda > 0$. Since k, $\lambda > 0$, $r = 0$ and $\nu \geqq L > 1$, from (6), (24) and (25) we have

for some V_1 and V_2

$$V_1 S^\sigma C S^\sigma R Q_k S^\sigma B = V_2 S^\sigma C S^\sigma C S^\sigma R. \tag{26}$$

Let us estimate $|[V_1^\partial - [V_2^\partial|$. Using (26) and the inequalities $(\sigma - 1)[S^\partial \geqq 8M$, $[Q_k^\partial \leqq M$ and $[B^\partial \leqq [C^\partial \leqq 6M$, we have

$$|[V_1^\partial - [V_2^\partial| = |[RQ_k S^\sigma B^\partial - [CS^\sigma R^\partial| \leqq 8M.$$

Moreover, $\mathscr{E}(S, C)$. Hence from (26) and 1.15 it follows that $RQ_k S^\sigma B = CS^\sigma R$. But then from (25) and (24) we have (since k, $\lambda > 0$ and $r = 0$)

$$X = ([YQ_i]_{i=1}^k)^{\lambda-1} [YQ_i]_{i=1}^{k-1} (S^\sigma C)^{\nu+1} S^\sigma R. \tag{27}$$

We know $(\sigma - 1)[S^\partial \geqq 8M \geqq |[RQ_k^\partial - [C^\partial|$. Hence from the equality $RQ_k S^\sigma B = CS^\sigma R$ and 1.21 it follows that there exists an integer d such that $RQ_k \perp\!\!\perp CS^d$ and $S^d B \perp\!\!\perp R$. By what has been proved, $\nu \geqq L$, $[S^\sigma C^\partial > N$, $R \angle CS^\sigma$ and $[R^\partial \leqq 7M$. Now from (27), (24) and (23) we conclude that c) holds. The λ, r and ϵ appearing in c) are $\lambda - 1$, $k - 1$ and 1, respectively.

2) $r > 0$. By an analogous argument, we see that c) holds in this case also. The λ, r and ϵ appearing in c) are λ, $r - 1$ and 1, respectively.

Subcase 2.2. $\rceil\, \tau(U, M)$ and $\tau(V, M)$. We will prove that in this case c) holds with $\epsilon = -1$. We will first show that if $\rceil\, \tau(U, M)$, then $C_1 = C_2 = \Lambda$. Indeed, from (6), (19), (18) and (15) it follows that $A_1 U C_1 = A_2 U C_2$ for some A_1 and A_2. We have $[C_1^\partial, [C_2^\partial \leqq M$ by (16). If $\rceil\, \tau(U, M)$, then, obviously, $C_1 = C_2$. But since either $C_1 = \Lambda$ or $C_2 = \Lambda$, then $C_1 = C_2 = \Lambda$, as required. Thus, $C_1 = C_2 = \Lambda$. From this and (4), (18), (15) and properties 1 and 2 of τ we conclude that $\rceil\, \tau(VQ_{r+1}UP_1, M)$. Since $\tau(V, M)$, from (15) and (20), in accordance with Lemma 2 (the roles of U, V, A and B in Lemma 2 are played by V, U, Q_{r+1} and P_1), we see that there exist σ, ν, C, E, H and a simple word S such that $VQ_{r+1} = S^\sigma C$, $U = (S^\sigma C)^{\nu-1} S^\sigma H$, $UP_1 V = (S^\sigma C)^\nu S^\sigma H$, $H \angle CS^\sigma$, $\mathscr{E}(S, C)$, $[S^\partial \leqq M$, and $[C^\partial, [E^\partial, [H^\partial \leqq 6M$. Putting $R = H$ and taking into account that $C_1 = C_2 = \Lambda$, we conclude from (18), (19) and (5) that c) holds with $\epsilon = -1$.

Subcase 2.3. Suppose now that

$$\rceil\tau(U, M) \qquad \rceil\tau(V, M). \tag{28}$$

It was shown in Subcase 2.2 that if $\rceil\, \tau(U, M)$, then $C_1 = C_2 = \Lambda$. But then from (15), (18) and (19) we have

$$Y = UP_1V = VQ_{r+1}U, \tag{29}$$

$$X = ([YQ_i]_{i=1}^k)^\lambda [YQ_i]_{i=1}^r U. \tag{30}$$

We will prove that if $P_1 = Q_{r+1}$, then d) holds, and if $P_1 \neq Q_{r+1}$, then e) holds. We will first show that either $Q_1 = \cdots = Q_k$ or $Q_1 = \cdots = Q_{r+1} = P_1$. Substituting the expressions for X and Y in (30) and (29) in place of x and y in the equation (*) and cancelling the resulting equality on the left by $([YQ_i]_{i=1}^k)^\lambda [YQ^i]_{i=1}^r UP_1$, we obtain

$$\begin{aligned} &[YP_i]_{i=2}^l ([YQ_i]_{i=1}^k)^\lambda [YQ_i]_{i=1}^r UP_{l+1}\overline{\xi}\overline{\Phi} \\ &= VQ_{r+1} [UP_1VQ_i]_{i=r+2}^k [UP_1VQ_i]_{i=1}^r ([YQ_i]_{i=r+1}^k [YQ_i]_{i=1}^r)^{a-1} UQ_{k+1}\overline{\eta}\overline{\Psi}, \end{aligned} \tag{31}$$

where the bar denotes the result of substituting X and Y in place of x and y.

Consider first the case $\lambda > 0$. Then the left-hand side of (31) begins with the word $[YP_i]_{i=2}^{l}[YQ_i]_{i=1}^{k}U$, which by (29) is equal to the word

$$[VQ_{r+1}UP_i]_{i=2}^{l}[VQ_{r+1}UQ_i]_{i=1}^{k}U.$$

Cancelling (31) on the left by $VQ_{r+1}U$, we obtain for some T_1 and T_2

$$(32)\qquad [P_iVQ_{r+1}U]_{i=2}^{l}[Q_iVQ_{r+1}U]_{i=1}^{k-1}Q_kUT_1 = [P_1VQ_iU]_{i=r+2}^{k}[P_1VQ_iU]_{i=1}^{r}T_2.$$

The left-hand side of (32) begins with a word of the form $[A_iVQ_{r+1}U]_{i=1}^{t}$, $t \geqq k-1$, $[A_i^\partial \leqq M$. Using (28) and (0), we obtain $A_i = P_1$ $(i = 1, \cdots, k-1)$ and $Q_i = Q_{r+1}$ $(i = r+2, \cdots, k;\ i = 1, \cdots, r)$. The second group of equalities yields $Q_1 = \cdots = Q_k$. Hence if $\lambda > 0$, then $Q_1 = \cdots = Q_k$.

Now consider the case $\lambda = 0$. Then $r > 0$, since $[X^\partial \geqq [Y^\partial$. It follows from (29) and (30) that $V \angle \overline{\eta\Psi}$, so that the right-hand side of (31) has the form $VQ_{r+1}[UP_1VQ_i]_{i=r+2}^{k}[UP_1VQ_i]_{i=1}^{r}UAVT$, where A is P_1 or Q_{k+1}, and T is a word in the alphabet И. The left-hand side of (31) begins with the word $[YP_i]_{i=2}^{l}[YQ_i]_{i=1}^{r}UP_{l+1}VQ_{r+1}U$ (in view of (29) and (30), $VQ_{r+1}U \angle \overline{\xi\Phi}$), which by (29) is equal to the word $[VQ_{r+1}UP_i]_{i=2}^{l}[VQ_{r+i}UQ_i]_{i=1}^{r}UP_{l+1}VQ_{r+1}U$. Cancelling (31) on the left by $VQ_{r+1}U$, we have for some T_2

$$(33)\qquad \begin{aligned}[P_iVQ_{r+1}U]_{i=2}^{l}[Q_iVQ_{r+1}U]_{i=1}^{r-1}Q_rUP_{l+1}VQ_{r+1}UT_2 \\ = [P_1VQ_iU]_{i=r+2}^{k}[P_1VQ_iU]_{i=1}^{r}AVT.\end{aligned}$$

If $l-1+r-1 \geqq k-1$, then, arguing as in the case $\lambda > 0$, we obtain $Q_i = Q_{r+1}$ $(i = r+2, \cdots, k;\ i = 1, \cdots, r)$, whence $Q_1 = \cdots = Q_k$.

Suppose $l-1+r-1 < k-1$. Here we see that $Q_i = P_1$ $(i = 1, \cdots, r-1)$ and that (33) can be cancelled on the left by $[P_iVQ_{r+1}U]_{i=2}^{l}[Q_iVQ_{r+1}U]_{i=1}^{r-1}$. After cancellation we obtain $Q_rUB_1 = P_1VB_2$ for some B_1 and B_2. Using (28), the estimates $[P_1^\partial \leqq M$ and $[Q_r^\partial \leqq M$, and the fact that one of the words U or V is a beginning of the other, we obtain $Q_r = P_1$. We have proved that $Q_1 = \cdots = Q_r = P_1$; in particular, $Q_1 = P_1$, since $r > 0$. We will now prove that $P_1 = Q_{r+1}$. The right-hand side of (33) begins with a product of $k-(r+1)$ factors of the form P_1VQ_iU, after which comes the word P_1VQ_1U (since $r > 0$), and the left-hand side begins with a product of $l-1+r-1$ factors of the form $A_iVQ_{r+1}U$, where $[A_i^\partial \leqq M$. Hence if $l-1+r-1 > k-(r+1)$, then $Q_1 = Q_{r+1}$. Since we have proved that $Q_1 = P_1$, this means that $P_1 = Q_{r+1}$. Suppose $l-1+r-1 \leqq k-(r+1)$. Then $k \geqq l+2r-1$. The equality (33) can be cancelled on the left by $[P_iVQ_{r+1}U]_{i=2}^{l}[Q_iVQ_{r+1}U]_{i=1}^{r-1}$. After cancellation we obtain

$$(34)\qquad Q_rUP_{l+1}VQ_{r+1}UT_2 = [P_1VQ_iU]_{i=2r+l}^{k}[P_1VQ_iU]_{i=1}^{r}AVT.$$

Since $k \geqq l+2r-1$, there are two possible cases.

1) $k = l+2r-1$. Then $[P_1VQ_iU]_{i=2r+l}^{k}$ is empty and the right-hand side of (34) begins with the word P_1VQ_1UDV, where $D = P_1$ if $r > 1$, and $D = A$ if $r = 1$ (so that $[D^\partial \leqq M)$. Therefore, $Q_rUP_{l+1}VQ_{r+1}UT_2 = P_1VQ_1UDVT_3$ for some T_3. Since we have proved that $Q_r = P_1$, on cancelling the last equality on the left by Q_r we obtain

$$(35)\qquad UP_{l+1}VQ_{r+1}UT_2 = VQ_1UDVT_3.$$

It follows from (29) that $[P_1^\partial = [Q_{r+1}^\partial$. We have proved that $P_1 = Q_1$. Hence $[Q_1^\partial = [Q_{r+1}^\partial$. Since $[P_{l+1}^\partial \leqq M$ and $[D^\partial \leqq M$, we have

$$|[UP_{l+1}VQ_{r+1}^{\partial} - [VQ_1UD^{\partial}| \leq M.$$

Using (28), the last inequality, and the fact that one of the words U or V is a beginning of the other, we see from (35) that $UP_{l+1}VQ_{r+1} = VQ_1UD$. One of the words U or V is an end of the other (see (29)). Moreover, $[D^{\partial}, [Q_{r+1}^{\partial} \leq M$. Hence from (28) and the equality $UP_{l+1}VQ_{r+1} = VQ_1UD$ it follows that $Q_{r+1} = D$, so that $UP_{l+1}V = VQ_1U$. Since $Q_1 = P_1$, we have $UP_{l+1}V = VP_1U$. Now from (29) and (28), in accordance with Lemma 1, we obtain $P_1 = Q_{r+1}$.

2) $k > l + 2r - 1$. Then the right-hand side of (34) begins with the word $P_1VQ_{2r+l}UP_1V$, and after cancelling (34) on the left by Q_r we have $UP_{l+1}VQ_{r+1}UT_2 = VQ_{2r+l}UP_1VT_3$ for some T_3. From the last equality, in the same way as from (35), we see that $UP_{l+1}VQ_{r+1} = VQ_{2r+l}UP_1$, from which it follows that $P_1 = Q_{r+1}$ (since $[P_1^{\partial} = [Q_{r+1}^{\partial}$ by (29)). We have proved that $P_1 = Q_{r+1}$.

Hence, we have proved that either $Q_1 = \cdots = Q_{r+1} = P_1$ or $Q_1 = \cdots = Q_k$, and if $\lambda > 0$, then $Q_1 = \cdots = Q_k$. Putting $\mu = k\lambda + r$, we have from (29) and (30)

$$(36) \qquad Y = UP_1V = VQ_1U,$$

$$(37) \qquad X = (YO_1)^{\mu}U, \quad \mu > 0.$$

We consider two subcases.

Subcase 2.3.1. $P_1 = Q_1$. Then from (36) and (37) we have

$$(38) \qquad YP_1 = UP_1VP_1 = VP_1UP_1,$$

$$(39) \qquad XP_1 = (UP_1VP_1)^{\mu}UP_1.$$

It follows from (28) that $U, V \neq \Lambda$. From (38) and 1.2, there exist integers $a > 0$, $b > 0$ and a simple word W such that

$$(40) \qquad UP_1 = W^a, \qquad VP_1 = W^b.$$

From (28), (40) and the definition of τ we conclude that $[W^{\partial} > M$, so that $[W^{\partial} > [P_1^{\partial}$. But then (40) implies that $W = ZP_1$ for some Z. Now from (38), (39) and (40) we have $YP_1 = (ZP_1)^{n+1}$ and $XP_1 = (ZP_1)^{m+1}$, where $n = a + b - 1$ and $m = (a+b)\mu + a - 1$. Since $a, b, \mu > 0$, we have $m > n > 0$, so that d) holds.

Subcase 2.3.2. $P_1 \neq Q_1$. Then we must have $Q_1 = \cdots = Q_k$. From (36) and (37) we have $XP_1Y = YQ_1X$. Hence from (36), (37), (28) and 7.6 it follows (the role of the tail $\Phi = \Psi$ is played by the equation (*)) that the system consisting of the equation $xP_1y \to yQ_1x$ and (*) is equivalent to the equation $xP_1y \to yQ_1x$. Hence e) holds. The proposition is proved.

We apply our results to equations of the form $\Phi(x) = \Psi(y)$. In particular, in the case of the equation $\Phi(x) = y^n$, $n > 1$, we can find a region $\omega(x)$ in which the values of the function $\Phi(x)$ are simple words.

We define a predicate ω_1. Let M be a positive integer. The predicate ω_1 is defined by the equivalence

$$\omega_1(x, M) \Leftrightarrow \neg\tau(x, (6M)^3) \,\&\, \neg\rho(x, M, M).$$

From the definitions of the predicates τ and ω_1 it follows that if $\omega_1(X, M)$ and $X \angle Z^n$, then $[Z^{\partial} > (6M)^3$. In particular, if $\omega_1(X, M)$, then $[X^{\partial} > (6M)^3 > 0$.

11.2. *Suppose the integers $n>0$, $m>0$, $M>1$ and the words $X, Y, P_1, \cdots, P_{n-1}, Q_1, \cdots, Q_{m-1}$ satisfy the conditions $[X^\partial \geqq [Y^\partial$, $\omega_1(Y,M)$, and also*

$$[XP_i]_{i=1}^{n-1}X = [YQ_i]_{i=1}^{m-1}Y, \tag{1}$$

$$\max\{n, m, [P_1^\partial, \dots, [P_{n-1}^\partial, [Q_1^\partial, \dots, [Q_{m-1}^\partial\} \leqslant M. \tag{2}$$

Then one of the following two assertions holds:

1) *There exists an integer r, $0 \leqq r < m$, such that $X = [YQ_i]_{i=1}^r Y$.*

2) *There exist integers a and b, $a > b \geqq 0$, and a word V such that $a < M$, $X = (VP_1)^a B$ and $Y = (VP_1)^b V$.*

PROOF. If $n=1$, then for $r=m-1$ we have 1). Assume that $n>1$. It follows from (1) that

$$n[X^\partial - m[Y^\partial = [Q_1 \dots Q_{m-1}^\partial - [P_1 \dots P_{n-1}^\partial. \tag{3}$$

Consider the equation

$$[xP_i]_{i=1}^{n-1}xxx = [yQ_i]_{i=1}^{m-1}yxx. \tag{4}$$

Equation (4) is a normal equation of the form 11.1 (*) (the a, k, l, ξ, η, P_1 and Q_k appearing in 11.1 (*) are 1, m, 1, x, x, P_1 and Λ, respectively). The coefficient lengths of (4) do not exceed M (see (2)). The condition $\omega_1(Y,M)$ implies that $[Y^\partial > 0$, so that $[X^\partial \geqq [Y^\partial > 0$. Finally, it is evident from (1) that the pair (X,Y) is a solution of (4). Hence, by 11.1, for any integers N and L such that $N \geqq 15M$ and $L>1$ one of 11.1 a)–e) holds. Choose $N = (6M)^3$, $L=M$ and consider each of these cases.

Case 11.1 a). This case is impossible, since by assumption $\omega_1(Y,M)$ and, consequently, $\urcorner\,\tau(Y,(6M)^3)$ and $\urcorner\,\rho(Y,M,M)$.

Before turning to the analysis of cases b) and c), we note the following. Since $n>1$ and $[X^\partial > 0$, from (1) we obtain $[X^\partial < [[UQ_i]_{i=1}^{m-1}Y^\partial$. But then, obviously, the λ appearing in 11.1 b), c) is equal to zero.

Case 11.1 b). Since $\lambda = 0$, we see that in this case 1) holds.

Case 11.1 c). Since $\lambda=0$, $N=(6M)^3$ and $L=M$, we have

$$X = [YQ_i]_{i=1}^r (S^\sigma C)^{\nu+\epsilon} S^\sigma R, \qquad Y = (S^\sigma C)^\nu S^\rho R,$$

where $0 \leqq r < m$, $[S^\sigma C^\partial > (6M)^3$, $\nu \geqq M$, $\epsilon = \pm 1$, $[S^\partial \leqq M$ and $[R^\partial, [C^\partial \leqq 7M$. We will prove that this is impossible. We obtain the contradiction in the following way. Let us compute the lengths of the words X and Y. From the expressions for X and Y we have

$$[Y^\partial = \nu[S^\sigma C^\partial + [S^\sigma R^\partial \geqslant M[S^\sigma C^\partial + 2M^2 + M^3 + 1, \tag{5}$$

$$[X^\partial = (r+1)[Y^\partial + [Q_1 \dots Q_r^\partial + \varepsilon[S^\sigma C^\partial. \tag{6}$$

From (3) and (6) we obtain

$$[n(r+1)-m][Y^\partial = [Q_1 \cdots Q_{m-1}^\partial - [P_1 \cdots P_{n-1}^\partial - n[Q_1 \cdots Q_r^\partial - n\epsilon[S^\sigma C^\sigma. \tag{7}$$

Denote the right-hand side of (7) by d. Let us estimate $|d|$. From the estimates (2) and the inequality $r<m$ it follows that the sum of the absolute values of the first three summands of d do not exceed $2M^2 + M^3$, i.e. the sum is less than $[S^\sigma C^\partial$. For the absolute value of the fourth summand we have $[S^\sigma C^\partial \leqq |n\epsilon|\,[S^\sigma C^\partial \leqq M[S^\sigma C^\partial$. Hence $d \neq 0$ and $|d| < M[S^\sigma C^\partial + 2M^2 + M^3 + 1$. Taken together with (5), this gives $d \neq 0$ and $|d| < [Y^\partial$, which is impossible, since the square brackets on the left-hand side of (7) enclose an integer. From this contradiction we conclude that c) is impossible.

Case 11.1 e). This case is also impossible. Indeed, in this case $P_1 \neq Q_1$, the equation $xP_1y \to yQ_1x$ has a solution, and any of its solutions satisfies (4). As is easily seen from 5.9, for fixed Y the component X of a solution (X, Y) of the equation $xP_1y \to yQ_1x$ can have arbitrarily large length. The contradiction comes from the fact that for equation (4) the length of the component X for fixed Y is bounded (see (3)).

Case 11.1 d). In this case there exist integers λ and μ, $\lambda > \mu > 0$, and a word Z such that

$$X = (ZP_1)^\lambda Z, \qquad Y = (ZP_1)^\mu Z. \tag{8}$$

From (8) we have $n[XP_1^\partial = n(\lambda+1)\,[ZP_1^\partial$, $m[YP_1^\partial = m(\mu+1)\,[ZP_1^\partial$. From this and (3) we obtain

$$[n(\lambda+1) - m(\mu+1)][ZP_1^\partial = [Q_1 \cdots Q_{m-1}^\partial - [P_1 \cdots P_{n-1}^\partial + (n-m)\,[P_1^\partial. \tag{9}$$

Since by hypothesis $\omega_1(Y, M)$, we have $\neg\,\tau(Y, (6M)^3)$. From this and (8) it follows that $[ZP_1^\partial > (6M)^3$. Using (2), we see that the absolute value of the right-hand side of (9) is less than $[ZP_1^\partial$, from which, as in c), we deduce that it is equal to zero. Then

$$n(\lambda+1) = m(\mu+1). \tag{10}$$

It is now easy to prove that 2) holds. Let (a, b) denote the greatest common divisor of the numbers a and b. Suppose that $(n, m) = d$, so that there exist integers $n_1, m_1 > 0$ such that $n = dn_1$, $m = dm_1$ and $(n_1, m_1) = 1$. From (10) we have $n_1(\lambda+1) = m_1(\mu+1)$. The last equality implies that there exists an integer $t > 0$ such that $\lambda + 1 = m_1 t$ and $\mu + 1 = n_1 t$. Since $\lambda > \mu > 0$, we have $m_1 > n_1 > 0$. Taking (2) into account, we obtain $1 \leqq n_1 < m_1 \leqq M$. From (8) we have

$$XP_1 = (ZP_1)^{\lambda+1} = (ZP_1)^{m_1 t} = ((ZP_1)^t)^m, \tag{11}$$

$$YP_1 = (ZP_1)^{\mu+1} = (ZP_1)^{n_1 t} = ((ZP_1)^t)^n. \tag{12}$$

Putting $a = m_1 - 1$, $b = n_1 - 1$ and $V = (ZP_1)^{t-1}Z$, we obtain from (11) and (12) that $X = (VP_1)^a V$ and $Y = (VP_1)^b V$, where $0 \leqq b < a < M$, so that 2) holds. The proposition is proved.

The following lemma follows from 1.16 and the definition of τ.

LEMMA 3. *Suppose $\Phi(x)$ and $\Psi(x)$ are words in the alphabet И $\cup \{x\}$, the lengths of the coefficients of the equation $\Phi = \Psi$ do not exceed an integer $M > 0$, and for some X such that $\neg\,\tau(X, M)$ we have $\Phi(X) = \Psi(X)$. Then in the alphabet* И $\cup \{x\}$ *we have the (graphical) equality $\Phi(x) = \Psi(x)$.*

The relation ∇. Suppose there are given words $Q_1, \cdots, Q_m$, $m > 1$, and suppose the letter x does not occur in the word $Q_1 \cdots Q_m$. We write $\nabla(Q_1, \cdots, Q_m)$ if the word $[xQ_i]_{i=1}^m$ is simple.

11.3. *Suppose the integers $m > 1$ and $M > 1$ and the words $Q_1, \cdots, Q_m$ and Y satisfy the conditions $\max\{m, [Q_1^\partial, \cdots, [Q_m^\partial\} \leq M$, $\nabla(Q_1, \cdots, Q_m)$, $\omega_1(Y, M)$. Then the word $[YQ_i]_{i=1}^m$ is simple.*

PROOF (BY CONTRADICTION). Assume that our assumptions are satisfied and the word $[YQ_i]_{i=1}^m$ is not simple. Then there exist an integer $n > 1$ and a simple word Z such that

$$[YQ_i]_{i=1}^m = Z^n \qquad (m, n > 1). \tag{1}$$

It follows from (1) that $Y \angle Z^n$. From this and the condition $\omega_1(Y, M)$ we obtain $[Z^\partial > (6M)^3$. In particular, $[Z^\partial > \max\{[Q_1^\partial, \cdots, [Q_m^\partial\}$. Consequently, Q_m is an end of Z. We have for some X that $Z = XQ_m$ and

$$[YQ_i]_{i=1}^m = (XQ_m)^n, \tag{2}$$

$$[YQ_i]_{i=1}^{m-1} Y = (XQ_m)^{n-1} X. \tag{3}$$

Three cases can arise: $[Y^\partial > [XQ_m^\partial$, $[X^\partial < [Y^\partial \leqq [XQ_m^\partial$ or $[X^\partial \geqq [Y^\partial$. We will prove that each of these three cases is impossible.

Suppose $[Y^\partial > [XQ_m^\partial$. Then $[Y^\partial > [Z^\partial > \max\{[Q_1^\partial, \cdots, [Q_m^\partial\}$, and from (1) and 1.27 we obtain $Q_1 = \cdots = Q_m$. This contradicts the condition $\nabla(Q_1, \cdots, Q_m)$, since $m > 1$.

Suppose $[X^\partial < [Y^\partial \leqq [XQ_m^\partial$. Then (3) implies that there exists P such that $Y = XP$ and $0 < [P^\partial \leqq [Q_m^\partial \leqq M$. Comparing the ends of (3), we see that there exists a word A such that $XP = AX$. We have $XP^2 = AXP$ and $YP = AY$. The latter equality contradicts the condition $\omega_1(Y, M)$, since $0 < [P^\partial \leqq (6M)^3$.

There remains the case $[X^\partial \geqq [Y^\partial$. It follows from $\omega_1(Y, M)$ that $[Y^\partial > (6M)^3$. Then $[X^\partial \geqq [Y^\partial > [Q_1 \cdots Q_m^\partial$. It is now evident from (3) that $n \leqq m$, so that $\max\{n, m\} \leqq M$. The condition $\omega_1(Y, M)$ also implies that $[Q_1^\partial, \cdots, [Q_m^\partial \leqq M$. Moreover, $[X^\partial \geqq [Y^\partial$. Hence from (3) and 11.2 it follows that one of the following two assertions holds: 1) There exists r, $0 \leqq r < m$, such that $X = [YQ_i]_{i=1}^r Y$. 2) There exist a and b, $a > b \geqq 0$, and V such that $X = (VQ_m)^a V$, $Y = (VQ_m)^b V$ and $b < M$. In case 1) we have (see (2)) $[YQ_i]_{i=1}^m = ([YQ_i]_{i=1}^r YQ_m)^n$. Since $\omega_1(Y, M)$, then $\neg\,\tau(Y, M)$. By hypothesis, $[Q_1^\partial, \cdots, [Q_m^\partial \leqq M$. Then, according to Lemma 3, in the alphabet $\text{И} \cup \{y\}$ we have the equality $[yQ_i]_{i=1}^m = ([yQ_i]_{i=1}^m yQ_m)^n$, which contradicts $\nabla(Q_1, \cdots, Q_m)$, since $n > 1$. Thus 1) is impossible. Consider case 2). From (2) we have $[(VQ_m)^b VQ_i]_{i=1}^m = ((VQ_m)^a VQ_m)^n = (VQ_m)^{(a+1)n}$. We know $Y = (VQ_m)^b V$, $b < M$. Moreover, $\omega_1(Y, M)$ holds. From this and the definitions of ω_1, τ and ρ it is not difficult to see that $\neg\,\tau(V, M)$. Then, by Lemma 3, in the alphabet $\text{И} \cup \{v\}$ we have the equality $[(vQ_m)^b vQ_i]_{i=1}^m = (vQ_m)^{(a+1)n}$. Hence $Q_1 = \cdots = Q_m$, which contradicts $\nabla(Q_1, \cdots, Q_m)$, since $m > 1$. Thus 2) is also impossible. The proposition is proved.

We define a predicate τ_1. We will write $\tau_1(X, M)$ if $X = S^t R$, where S is a simple word and $[R^\partial, [S^\partial \leqq M$. Obviously, $\tau(X, M) \supset \tau_1(X, M)$.

11.4. *Suppose a simple word S, integers $k > 1$, $M > 1$, σ and ν, and words $R, C, Q_1, \cdots, Q_k$ satisfy the conditions $[S^\partial \leqq 30M$; $[R^\partial, [C^\partial \leqq 90M$; $\mathscr{E}(S, C)$; $R \angle CS^{[R^\partial}$; $\nu \geqq 4M^2$; $[S^\sigma C^\partial > (6M)^3$; $k, [Q_1^\partial, \cdots, [Q_k^\partial \leqq M$; and $\nabla(Q_1, \cdots, Q_k)$. Then the word $[(S^\sigma C)^\nu S^\sigma R Q_i]_{i=1}^k$ is simple.*

PROOF. This follows from 11.3 and the following lemma (with $N = M$).

LEMMA 4. *Suppose σ, M, S, R, C are the same as in 11.4, N is an integer, $\nu > \max\{N, 3\}$, and $X = (S^\sigma C)^\nu S^\sigma R$. Then $\neg\,\tau_1(X, (6M)^3) \,\&\, \neg\,\rho(X, M, N)$.*

PROOF OF THE LEMMA (BY CONTRADICTION). Assume that

$$\neg(\neg\tau_1(X,(6M)^3)\ \&\ \neg\rho(X,M,N)).$$

Then one of the following two assertions holds:

1) $\tau_1(X,(6M)^3)$;

2) $\neg\tau_1(X,(6M)^3)\ \&\ \rho(X,M,N)$.

We will prove that both of these cases are impossible. First note that the conditions $R\angle CS^{[R^\partial}$, $[C^\partial, [R^\partial \leqq 90M$ and $[S^\sigma C^\partial > (6M)^3$ imply $\sigma > 0$ and $R\angle CS^\sigma$. Then it is easy to see that there exist an integer n, $\nu\leqq n\leqq \nu+1$, and a word A such that

$$(1)\qquad X=(S^\sigma C)^n A, \qquad \Lambda\neq A\angle S^\sigma C.$$

Moreover, by 1.10, the word $S^\sigma C$ is simple.

Case 1. Suppose $\tau_1(X,(6M)^3)$. Then $X=T^tB$, where $[T^\partial, [B^\partial\leqq(6M)^3$ and T is a simple word. From (1) we have

$$(2)\qquad (S^\sigma C)^n A = T^t B.$$

From (2) and the estimates for n, $[T^\partial$, $[B^\partial$, $[S^\sigma C^\partial$ and $[A^\partial$ we deduce that $n[S^\sigma C^\partial, t[T^\partial \geqq [S^\sigma CT^\partial$. By 1.5, $S^\sigma C=T$, which is impossible, since $[T^\partial\leqq(6M)^3 < [S^\sigma C^\partial$. Thus Case 1 is impossible.

Case 2. $\neg\tau_1(X,(6M)^3)\ \&\ \rho(X,M,N)$. We will prove that there exist m, t, B, D and a simple word T such that

$$(3)\qquad X=(T^tD)^mB, \quad \Lambda\neq B\angle T^tD, \quad m\leqq N,$$

$$(4)\qquad [T^\partial\leqq M, \quad [D^\partial\leqq 7M, \quad \mathscr{S}(T,D); \quad [T^tD^\partial > (6M)^3.$$

It follows from the definition of ρ that $X=(T^tD)^\mu T^tQ$, where $\mu<N$, $[T^\partial\leqq M$, $[D^\partial\leqq 7M$, $\mathscr{S}(T,D)$ and $Q\angle DT^t$. There are two possible cases: $[Q^\partial\leqq[D^\partial$ or $[Q^\partial>[D^\partial$. If $[Q^\partial\leqq[D^\partial$, then $Q\angle DT^t$ implies that $Q\angle D$, so that $T^tQ\angle T^tD$. Since $\neg\tau(X,(6M)^3)$, we have $[T^tD^\partial>(6M)^3$; in particular, $T^t\neq\Lambda$. Putting $m=\mu$ and $B=T^tQ$, we obtain (3) and (4). If $[Q^\partial>[D^\partial$, then $Q=DB$, where $\Lambda\neq B\angle T^t$, and then $X=(T^tD)^{\mu+1}B$, and (3) and (4) hold with $m=\mu+1$. Thus, (3) and (4) are proved. From (1) and (3) we have

$$(5)\qquad (S^\sigma C)^n A = (T^tD)^m B.$$

Let us consider (5). Note that by 1.10 the word T^tD is simple. If $m>1$, then from (5) and 1.7 it follows that $n=m$, which is impossible, since $n\geqq \nu>N$ and $m\leqq N$. Consequently, $m\leqq 1$. Since $m\leqq 1$ and $B\angle T^tD$, from (5) we have $(S^\sigma C)^n\angle(T^tD)^2$. Since $n\geqq 4$, we have $(S^\sigma C)^2\angle T^tD$. Obviously, $[D^\partial<(\sigma-1)[S^\partial$. Hence $S^\sigma CS\angle T^t$, so certainly $S^\sigma\angle T^t$. From the latter relation we obtain $S^\sigma=T^kT_1$, where $T_1\angle T$. We have $\sigma[S^\partial, k[T^\partial\geqq[ST^\partial$. Hence, by 1.5, $S=T$. We have $S^\sigma CS\angle S^t$ and $CS\angle S^{t-\sigma}$, which by 1.4 contradicts the condition $\mathscr{S}(S,C)$. Thus Case 2 is also impossible. The lemma is proved.

We now define a predicate ω by means of the equivalence

$$\omega(x,M)\Leftrightarrow\neg\tau_1(x,(6M)^3)\ \&\ \neg\rho(x,M,4M^2).$$

From the definitions of τ, τ_1, ω and ω_1 we have the implication

$$\omega(x,M)\supset[\omega_1(x,M)\ \&\ \neg\tau(x,M)\ \&\ (\partial(x)>(6M)^3)].$$

11.5. *For any integer $M > 0$ and any word A in the alphabet* И *such that $[A^{\partial} \leq M$ there exists a word X such that $\omega(X, M)$, $\neg \tau(X, M)$, and XA is a simple word.*

PROOF. Put $X = (a^{\sigma} b)^{\nu} a^{\sigma}$, where $\sigma = (6M)^3$, $\nu = 4M^2 + 1$, and a and b are distinct letters of the alphabet И with $b \neq A$. We have that $\mathscr{L}(a, b)$, a is a simple word, and $[a^{\sigma} b^{\partial} > (6M)^3$. Then, by Lemma 4, $\neg \tau_1(X, (6M)^3) \;\&\; \neg \rho(X, M, 4M^2)$, i.e. $\omega(X, M)$. Certainly we have $\neg \tau(X, M)$. The simplicity of the word XA follows from 1.13. The proposition is proved.

We now consider the solutions (X, Y) of equation 11.1 (*) for which $\tau(Y, N) \vee \rho(Y, M, L)$. The component Y of such solutions has the form $[S^t P_i]_{i=1}^n$, where $1 \leq n \leq L$, $t \geq 0$, S is simple, and the lengths of the words S and $P_1, \cdots, P_n$ are bounded by constants depending on N and M. Substituting this expression for Y in 11.1 (*), we obtain $X \cdots = [S^t A_i]_{i=1}^m X$, from which it is possible to express X. By 1.16, $X = ([S^t A_i]_{i=1}^m)^{\lambda} V$, where $V \angle [S^t A_i]_{i=1}^m$. It is easy to show that V cannot cut the word S^t deeply. This yields a "good" representation for X and Y. The other problem is to clarify for which t the word $[S^t A_i]_{i=1}^m$ is simple and to find the simple period $\varphi(t)$ of $[S^t A_i]_{i=1}^m$. (If $P = S^a$, where S is a simple word, then S is the simple period of P.) The interest lies in values of t which are sufficiently large in comparison with $[A_1 \cdots A_n S^{\partial}$.

The symbol $\oplus$. Suppose k and M are positive integers and $S, P_1, \cdots, P_k$ are words in the alphabet И. We write $\oplus(S, P_1, \cdots, P_k, M)$ if the word $[S^t P_i]_{i=1}^k$ is simple for any t such that $t[S^{\partial} > M$.

It turns out that if $[S^t P_i]_{i=1}^k$ is simple for some sufficiently large t, then it is simple for any sufficiently large t. Let us prove this. We require

LEMMA 5. *Suppose S is a simple word, $m, n \geq 1$, $k = \min\{n, m\}$, and suppose also that*

$$S^{\mu} A_1 [S^{\mu+\nu} A_i]_{i=2}^n S V_1 = B_1 [S^{\mu+\nu} B_i]_{i=2}^m S V_2, \tag{1}$$

$$(\mu - 1)[S^{\partial}, (\nu - 1)[S^{\partial} > \max\{[A_1 \cdots A_k^{\partial}, [B_1 \cdots B_k^{\partial}\}. \tag{2}$$

Then: a) *if $m > n$, then $A_1, \cdots, A_n$ are powers of S*; b) *if $m \leq n$, then $B_1, \cdots, B_m$ are powers of S.*

PROOF. We proceed by induction on $k = \min\{n, m\}$. By hypothesis, $k \geq 1$. From (1) and (2) it follows that $B_1 S \angle S^{\mu}$ and, by 1.4, the word B_1 is a power of S. If $m = 1$, the assertion is proved [statement a) is trivially true, since the inequality $n < 1$ is impossible]. Suppose $m > 1$. Then, cancelling (1) on the left by S and taking into account that B_1 is a power of S, we obtain

$$A_1 [S^{\mu+\nu} A_i]_{i=2}^n S V_1 = S^{\nu} B_1 B_2 [S^{\mu+\nu} B_i]_{i=3}^m S V_2. \tag{3}$$

From (3) and (2) it follows that $A_1 S \angle S^{\nu}$ and, by 1.4, the word A is a power of S. If $n = 1$, the assertion is proved. The basis for the induction is also proved. Now suppose $n > 1$. Cancelling (3) on the left by S^{ν}, we obtain

$$S^{\mu} A_1 A_2 [S^{\mu+\nu} A_i]_{i=3}^n S V_1 = B_1 B_2 [S^{\mu+\nu} B_i]_{i=3}^m S V_2. \tag{4}$$

The induction assumption is applicable to (4). Applying it and using the fact that A_1 and B_1 are powers of S, we see that the lemma is true.

The following proposition allows us to compute the simple periods of words of the form $[S^tP_i]_{i=1}^k$, where t is sufficiently large in comparison to $[P_1\cdots P_kS^\partial$.

11.6. *Suppose there is given a parametric word* $\varphi(\sigma)=[S^\sigma P_i]_{i=1}^k$, *where* $k\geqq 1$, *S is a simple word, and $P_1,\cdots,P_k$ are not all powers of S. Call σ admissible if* $\sigma[S^\partial > N=9k[P_1\cdots P_kSS^\partial$. *There exist integers* $a\geqq 0$, n, $1\leqq n\leqq k$, *and* d, $1\leqq d\leqq k$, *and words* $A_1,\cdots,A_n$ *such that for any admissible σ we have* $\varphi(\sigma)=([S^{\sigma-a}A_i]_{i=1}^n)^d$; *in this connection we have* $\oplus(S,A_1,\cdots,A_n,N)$ *and*

$$\max\{[A_1^\partial,\cdots,[A_n^\partial,2a[S^\partial\}\leqq 4[S^2P_1\cdots P_k^\partial.$$

If $\varphi(\sigma)$ is simple for some admissible σ, then $\varphi(\sigma)$ is simple for any admissible σ.

PROOF. Suppose σ is an admissible number and the word $\varphi(\sigma)$ is not simple. Then there exist an integer $d>1$ and a simple word Y such that

$$[S^\sigma P_i]_{i=1}^k=Y^d\qquad(k>0,\ d>1,\ \sigma>1).\tag{1}$$

We will prove that $[Y^\partial>(\sigma-1)[S^\partial$. It follows from (1) that $[S^\partial\leqq(d-1)[Y^\partial$. If we now assume that $[S^\partial\geqq[Y^\partial$, then from (1) and 1.5 it follows that $Y=S$, so that $[S^\sigma P_i]_{i=1}^k=S^d$. From this last equality and 1.4 we deduce that $P_1,\cdots,P_k$ are powers of S, which contradicts the hypothesis. From this contradiction we conclude that $[Y^\partial>[S^\partial$. If we now assume that $[Y^\partial\leqq(\sigma-1)[S^\partial$, then by 1.5 we again obtain $Y=S$, which contradicts $[Y^\partial>[S^\partial$. Hence we have proved that $[Y^\partial>(\sigma-1)[S^\partial$. Since σ is admissible, (1) implies that $k>1$. For brevity, put $N=[P_1\cdots P_kS^\partial$. Then the admissibility condition for σ can be written in the form

$$\sigma[S^\partial>9k(N+[S^\partial).\tag{2}$$

Choose an integer $a>0$ such that $(a-1)[S^\partial\leqq N$ and $a[S^\partial>N$ (such an a obviously exists). Then $a[S^\partial\leqq N+[S^\partial$, and from (2) we have

$$(\sigma-2a)[S^\partial>(9k-2)(N+[S^\partial),\tag{3}$$

so that, in particular, $\sigma>2a$. Now put

$$R_1=S^{2a}P_1S^a,\qquad R_i=S^aP_iS^a\quad(i=2,\cdots,k-1),\qquad R_k=S^aP_k.\tag{4}$$

Obviously,

$$\forall\lambda\,[(\lambda\geqslant 2a)\supset(\varphi(\lambda)=[S^{\lambda-2a}R_i]_{i=1}^k)].\tag{5}$$

Let us express Y from (1). We will prove that there exist a number q, $1\leqq q\leqq k$, and a word Q such that

$$Y=[S^{\sigma-2a}R_i]_{i=1}^{q-1}S^{\sigma-2a}Q,\qquad Q\angle R_q.\tag{6}$$

We have proved that $[Y^\partial>(\sigma-1)[S^\partial$. If $[Y^\partial\leqq[S^\sigma P_1S^{a\partial}$, then (6) obviously follows from (1) and (4); here $q=1$.

Consider the case $[Y^\partial>[S^\sigma P_1S^{a\partial}$. Then it follows from (1) that there exist integers r, $0<r<k$, and τ, $0\leqq\tau\leqq\sigma$, and a word A such that

$$Y=[S^\sigma P_i]_{i=1}^rS^\tau A,\tag{7}$$

where if $\tau<\sigma$, then $A\angle S$, and if $\tau=\sigma$, then $A\angle P_{r+1}$. We will now prove that either $\tau<a$ or $\sigma-\tau<a$. The proof is by contradiction.

Assume that $\tau, \sigma-\tau \geqq a$. Then $(\tau-1)[S^\partial, (\sigma-\tau-1)[S^\partial \geqq (a-1)[S^\partial > N-[S^\partial \geqq [P_1\cdots P_k^\partial$. In particular, $\sigma-\tau-1>0$ and $\tau<\sigma$. Since $\tau<\sigma$, we have $A \angle S$, so that $S=AB$ for some $B\neq\Lambda$. Cancelling (1) on the left by Y and using (7), we obtain

$$BS^{\sigma-\tau-1}P_{r+1}[S^\sigma P_i]_{i=r+2}^k = ([S^\sigma P_i]_{i=1}^r S^\tau A)^{d-1} \qquad (r>0, d>1). \tag{8}$$

Since $\sigma-\tau-1>0$ and $[B^\partial \leqq [S^\partial$, (8) implies that $BS \angle S^\sigma$; hence by 1.4 the word B is a power of S. Since $S=AB$ and $B\neq\Lambda$, we have $B=S$ and $A=\Lambda$. Now from (8) we have

$$S^{\sigma-\tau}P_{r+1}[S^\sigma P_i]_{i=r+2}^k S = ([S^\sigma P_i]_{i=1}^r S^\tau)^{d-1}S \qquad (r>0, d>1). \tag{9}$$

We know that $(\tau-1)[S^\partial > [P_1\cdots P_k^\partial$. Then it follows from (9) that $k-r-1\geqq r$. Now (9) is an equality of the form (1) of Lemma 5 (the roles of B_1, μ, ν, n and m in Lemma 5 are played by Λ, $\sigma-\tau$, τ, $k-r$ and $r+1$). Since $k-r\geqq r+1$ and $(\tau-1)[S^\partial, (\sigma-\tau-1)[S^\partial > [P_1\cdots P_k^\partial$, by Lemma 5b) the words $P_1,\cdots,P_r$ are powers of S. Then (7) contradicts the simplicity of Y (since $A=\Lambda$). We have obtained the required contradiction. Hence either $\tau<a$ or $\sigma-\tau<a$. Let us consider each case separately.

Suppose $\tau<a$. Since $\sigma>a$ (see (3)), we have $\tau<\sigma$, $A\angle S$ and $S^\tau A\angle S^a$. Now (6) follows from (7) and (4) (here $q=r>1$ and $Q=S^aP_rS^\tau A$).

Suppose $\sigma-\tau<a$. Then $(\tau-a)-(\sigma-2a)=\tau-\sigma+a>0$. We have $A\angle S^{\sigma-\tau}P_{r+1}$. Put $c=(\tau-a)-(\sigma-2a)$. Then $c>0$ and $S^cA\angle S^{c+\sigma-\tau}P_{r+1}=S^aP_{r+1}\angle R_{r+1}$. From (7) and (4) we have (since $\sigma>2a$ and $\tau>a$) that $Y=[S^{\sigma-2a}R_i]_{i=1}^r S^{\tau-a}A=[S^{\sigma-2a}R_i]_{i=1}^r S^{\sigma-2a}S^cA$, and (6) holds with $q=r+1$ and $Q=S^cA$. The representation (6) is proved.

We will prove that the integers d, q, a and the words $R_1,\cdots,R_{q-1},Q$ are the desired ones. We have $a[S^\partial \leqq N+[S^\partial$ and $[Q^\partial \leqq [R_q^\partial$. From this and (4) we obtain

$$\max\{[R_1^\partial,\cdots,[R_k^\partial,[Q^\partial\} \leqq 4[S^2P_1\cdots P_k^\partial. \tag{10}$$

Put $X(\sigma)=[S^{\sigma-2a}R_i]_{i=1}^{q-1}S^{\sigma-2a}Q$. From (1), (5) and (6) we have

$$[S^{\sigma-2a}R_i]_{i=1}^k = (X(\sigma))^d. \tag{11}$$

It follows from (3) and (10) that $(\sigma-2a)[S^\partial > k\max\{[R_1^\partial,\cdots,[R_k^\partial,[Q^\partial\}+[S^\partial$. By 1.18, the equality (11) holds for any admissible σ. It follows from (5) that $\varphi(\sigma)=(X(\sigma))^d$ for any admissible σ. We have proved that if $\varphi(\sigma)$ is not simple for some admissible σ, then it is not simple for any admissible σ. The second part of our assertion is obtained from this by arguing by contradiction. Let us complete the proof of the first part. Let σ_0 denote the smallest admissible σ. If the word $\varphi(\sigma_0)$ is simple, then by the second part we have

$$\oplus\,(S,P_1,\cdots,P_k,N)$$

and the first part holds with $a=0$, $d=1$, $n=k$ and $A_i=P_i$. If the word $\varphi(\sigma_0)$ is not simple, then by what has been proved (with $\sigma=\sigma_0$) there exist $c>1$ and a word $Z(\sigma)=[S^{\sigma-2a}A_i]_{i=1}^n$, the same as $X(\sigma)$, but with different q and Q, such that $Z(\sigma_0)$ is a simple word and $\varphi(\sigma)=(Z(\sigma))^c$ for any admissible σ. It remains to prove that $\oplus\,(S,A_1,\cdots,A_n,N)$, i.e. that $Z(\sigma)$ is simple for any admissible σ. Suppose $\tilde\sigma$ is admissible. Then $\varphi(\tilde\sigma)$ is not simple. By what has been proved (with $\sigma=\tilde\sigma$) there exist $d>1$ and a word $X(\sigma)$ such that

$X(\tilde{\sigma})$ is a simple word and $\varphi(\sigma) = (X(\sigma))^d$ for any admissible σ. We have $(X(\sigma))^d = (Z(\sigma))^c$ for any admissible σ. Since the words $Z(\sigma_0)$ and $X(\tilde{\sigma})$ are simple, by 1.1, $X(\sigma_0)$ is a power of $Z(\sigma_0)$, and $Z(\tilde{\sigma})$ is a power of $X(\tilde{\sigma})$. The first fact yields $c \geqq d$, the second $d \geqq c$, from which we conclude that $c = d$, $Z(\tilde{\sigma}) = X(\tilde{\sigma})$ and $Z(\tilde{\sigma})$ is a simple word, as required. The proposition is proved.

REMARK. From the second part of 11.6 we obviously obtain an algorithm recognizing, for given $S, P_1, \cdots, P_k$ and M, whether or not $\oplus\,(S, P_1, \cdots, P_k, M)$.

We now turn to an analysis of case 11.1 a). We are interested in sufficiently long solutions.

11.7. *Suppose that all of the conditions of* 11.1 *are satisfied, and also (see* 11.1) *that* $\max\{k, l\} \leqq M$, $[Y^\partial > (5M)^9$, $[X^\partial > [Y^\partial + (8M)^4$, *and* $\tau_1(Y, (6M)^3) \vee \rho(Y, M, 4M^2)$. *Then one of the following two assertions holds:*

a) *There exist integers* λ *and* μ, *words* A *and* B, *and a simple word* S *such that* $X = S^\lambda A$, $Y = S^\mu B$, $[S^\partial \leqq (6M)^3$ *and* $[A^\partial, [B^\partial \leqq (8M)^4$.

b) *There exist integers* n, m, r, λ, t, *words* $A_1, \cdots, A_n$, $B_1, \cdots, B_m$, *and a simple word* S *such that* $X = ([S^tA_i]_{i=1}^n)^\lambda [S^tA_i]_{i=1}^r S^tA$, $Y = [S^tB_i]_{i=1}^m$, $0 < m \leqq 4M^2$, $0 \leqq r < n \leqq 4M^3$, $[S^\partial \leqq (6M)^3$; $[A_1^\partial, \cdots, [A_n^\partial, [B_1^\partial, \cdots, [B_m^\partial, [A^\partial \leqq (5M)^7$, $\oplus\,(S, A_1, \cdots, A_n, (5M)^7)$, *and* $A \angle A_{r+1}S^{[A^\partial}$.

PROOF. From the condition $\tau_1(Y, (6M)^3) \vee \rho(Y, M, 4M^2)$ we obtain the representation for Y

$$Y = [S^\sigma R_i]_{i=1}^{m-1} S^\sigma R, \tag{1}$$

where either $m = 1$, $[S^\partial, [R^\partial \leqq (6M)^3$ or $m \leqq 4M^2$, $[S^\partial \leqq M$, $[R_1^\partial, \cdots, [R_{m-1}^\partial$, $[R^\partial \leqq 7M$, and in both cases S is a simple word.

Since (X, Y) is a solution of 11.1 (*), we have

$$XV_1 = ([[S^\sigma R_i]_{i=1}^{m-1} S^\sigma RQ_i]_{i=1}^k)^a XV_2 \tag{2}$$

for some V_1 and V_2. We consider two cases.

Case 1. The words $R_1, \cdots, R_{m-1}$, $RQ_1, \cdots, RQ_k$ are powers of S. Then (1) implies that there exists $\mu > 0$ such that $Y = S^\mu R$. It follows from (2) that $XV_1 = S^cXV_2$ for some $c > 0$. From $XV_1 = S^cXV_2$ and 1.16, there exist $\lambda \geqq 0$ and a word A such that $X = S^\lambda A$ and $A \angle S$. Thus a) holds for the pair (X, Y).

Case 2. Not all words in the list $R_1, \cdots, R_{m-1}$, $RQ_1, \cdots, RQ_k$ are powers of S. Put

$$N(m) = \begin{cases} 3(6M)^3 + M, & \text{if} \quad m = 1, \\ 10M, & \text{if} \quad m > 1. \end{cases}$$

We have

$$\max\{[R_1^\partial, \ldots, [R_{m-1}^\partial, [RQ_1^\partial, \ldots, [RQ_k^\partial\} + 2\,[S^\partial \leqslant N(m).$$

Furthermore, $mN(m) < 650M^3$, $kmN(m) \leqq 650M^4$,

$$\sigma\Big[S^\partial \doteq \frac{1}{m}([Y^\partial - [R_1 \ldots R_{m-1}R^\partial) > \frac{(4M)^9 - mN(m)}{4M^2} > 36\cdot 650M^7$$
$$> 9kmkmN(m) \geqslant 9km\,[[R_1 \ldots R_{m-1}RQ_i]_{i=1}^{k\ \partial} + 2\,[S^\partial.$$

Moreover, $(5M)^7 > 36 \cdot 650M^7$. Hence by 11.6 there exist integers d $(1 \leqq d$

$\leqq km$), n $(1 \leqq n \leqq km)$, a, $\mu \geqq 0$ and words $D_1, \cdots, D_n$ such that $\sigma = \mu + a$, $\oplus(S, D_1, \cdots, D_n, (5M)^7)$ and

$$(3)\qquad \begin{gathered} [[S^\sigma R_i]_{i=1}^{m-1} S^\sigma R Q_i]_{i=1}^{k} = ([S^\mu D_i]_{i=1}^{n})^d, \\ \max\{[D_1^\partial, \dots, [D_n^\partial, a\,[S^\partial\} \leqslant 4kmN(m) \leqslant (8M)^4 - N(m). \end{gathered}$$

From (2) and (3) we have $XV_1 = ([S^\mu D_i]_{i=1}^n)^{ad} X V_2$, from which, using 1.16, we conclude that there exist $\lambda \geqq 0$, q $(0 \leqq q < n)$, τ $(0 \leqq \tau \leqq \mu)$ and D such that

$$(4)\qquad X = ([S^\mu D_i]_{i=1}^n)^\lambda [S^\mu D_i]_{i=1}^q S^\tau D,$$

where $D \angle S$ if $\tau < \mu$, and $D \angle D_{q+1}$ if $\tau = \mu$. Hence $[D^\partial \leqq (8M)^4$. Let us express Y in terms of μ. We have $\sigma = \mu + a$. Putting $C_i = S^a R_i$ $(i = 1, \cdots, m-1)$ and $C_m = S^a R$, we have from (1) and the estimates for $a[S^\partial, [R_i^\partial$ and $[R^\partial$

$$(5)\qquad Y = [S^\mu C_i]_{i=1}^m,\ [C_1^\partial, \cdots, [C_m^\partial \leqq (8M)^4.$$

Substituting the expressions for X and Y in (4) and (5) into the left-hand side of 11.1 (*) and the expressions for X and Y in (4) and (1) into the right-hand side and using (3), we see that the resulting equality can be cancelled on the left by $([S^\mu D_i]_{i=1}^n)^\lambda [S^\mu D_i]_{i=1}^q S^\tau$. After cancellation we obtain

$$(6)\qquad \begin{aligned} DP_1\,[[S^\mu C_i]_{i=1}^m P_i]_{i=2}^l ([S^\mu D_i]_{i=1}^n)^\lambda\, [S^\mu D_i]_{i=1}^q S^\tau Z_1 \\ = S^{\mu-\tau} D_{q+1}\,[S^\mu D_i]_{i=q+2}^n\,[S^\mu D_i]_{i=1}^q S^\tau Z_2. \end{aligned}$$

for some Z_1 and Z_2. It is not difficult to calculate that

$$(7)\qquad \begin{aligned} (l-1)\,[C_1 \cdots C_m^\partial + [DP_1 \cdots P_l^\partial + [D_1 \cdots D_n^\partial \\ \leqq (5M)^7 - 2(8M)^4 - (6M)^3 = L(M). \end{aligned}$$

It remains to eliminate τ in the representation (4). We consider three subcases.

Subcase 2.1. $(\tau - 1)\,[S^\partial$, $(\mu - \tau - 1)\,[S^\partial > L(M)$. We will prove that in this subcase a) holds. Apply Lemma 5 to (6). The roles of B_1, μ, ν, n and m in Lemma 5 are played by DP_1, $\mu - \tau$, τ, n and $(l-1)m + n\lambda + q + 1$, and the applicability of the lemma is guaranteed by the estimates (7). If $(l-1)m + n\lambda + q + 1 > n$, then by Lemma 5 a) the words $D_{q+1}, \cdots, D_n$, $D_1, \cdots, D_q$ are powers of S, which contradicts the condition $\oplus(S, D_1, \cdots, D_n, (5M)^7)$. Consequently, $(l-1)m + n\lambda + q + 1 \leqq n$. Then $\lambda = 0$, whence by Lemma 5 b) the words DP_1, C_i, P_j and $D_1, \cdots, D_q$ are powers of S. Since $\lambda = 0$, from (4) we have $X = S^t D$ for some t. Since $Y \angle X$, we have $Y = S^\nu B$ for some ν and B such that $[B^\partial \leqq \max\{[D^\partial, [S^\partial\}$. We have proved that $[D^\partial \leqq (8M)^4$. Hence a) holds for the pair (X, Y).

Subcase 2.2. $(\mu - \tau - 1)\,[S^\partial \leqq L(M)$. Then $(\mu - \tau)\,[S^\partial \leqq (5M)^7 - (8M)^4$. Put $A_i = S^{\mu-\tau} D_i$ $(i = 1, \cdots, n)$, $A = D$, $B_i = S^{\mu-\tau} C_i$ $(i = 1, \cdots, m)$, and $r = q$. The condition $\oplus(S, D_1, \cdots, D_n, (5M)^7)$ implies that $\oplus(S, A_1, \cdots, A_n, (5M)^7)$. Since $D \angle S^{\mu-\tau} D_{r+1}$, we have $A \angle A_{r+1}$. Now from (4), (5) and our estimates we conclude that b) holds for the pair (X, Y) with $t = \tau$.

Subcase 2.3. $(\tau - 1)\,[S^\partial \leqq L(M)$. Then $\tau[S^\partial \leqq (5M)^7 - 2 \cdot (8M)^4$. From (5) and the estimate for $[Y^\partial$ we obtain $(\mu - \tau)\,[S^\partial > (8M)^4$. Since $D \angle S^{\mu-\tau} D_{q+1}$ and $[D^\partial \leqq (8M)^4$, we have $D \angle S^{\mu-\tau}$, whence $S^\tau D \angle S^\mu$. There are two possible subcases.

2.3.1). $q>0$. Then, putting $A_i=D_i$ $(i=1,\cdots,n)$, $A=D_qS^\tau D$, $r=q-1$, $t=\mu$ and $B_i=C_i$ $(i=1,\cdots,m)$, we see that b) holds.

2.3.2). $q=0$. Then from (4), (5) and the inequalities $\tau\leqq\mu$, $[D^\partial\leqq(8M)^4$ and $[X^\partial>[Y^\partial+(8M)^4$ it follows that $\lambda>0$ and

$$X=([S^\mu D_i]_{i=1}^n)^{\lambda-1}[S^\mu D_i]_{i=1}^{n-1}S^\mu D_i S^\tau D.$$

Putting $A_i=D_i$ $(i=1,\cdots,n)$, $A=D_nS^\tau D$, $r=n-1$, $t=\mu$ and $B_i=C_i$ $(i=1,\cdots,m)$, and using the fact that $S^\tau D\angle S^\mu$, we see that b) holds. The proposition is proved.

In order to complete the preparation of the description of the solutions of normal equations, it remains to consider a predicate related to the function G. This predicate distinguishes in the domain of G that part in which the solution of equations containing G has the necessary simple form (cf. 17.1.5). In solving equations without additional restrictions on the variables processions appear on many occasions and the arguments become cumbersome. The predicate $\mathscr{B}$ divides the domain of G into two parts: $\mathscr{B}$ and $\urcorner\mathscr{B}$. In the region $\urcorner\mathscr{B}$ it is convenient to solve equations containing the function G. In the region $\mathscr{B}$ the values of G are described by means of parametric words of type 11.1 c).

The predicate $\mathscr{B}$. We define a relation $\mathscr{B}$ between processions and positive integers. Suppose M is a positive integer and $[A,B,\lambda_0,\cdots,\lambda_{k-1}]$ a procession. By definition, we write $\mathscr{B}([A,B,\lambda_0,\cdots,\lambda_{k-1}],M)$ in two cases:

a) There exist integers $a>0$ and $p\geqq0$ and a simple word S such that $F(A,B,\lambda_0,\cdots,\lambda_{p-1})=S^a$, $[S^\partial\leqq30M$, where $(\lambda_p,\cdots,\lambda_{k-1})$ is a procession of one of the following four forms: (t,μ), $(t,1,\mu)$, $(t,\mu,1)$, $(t,1,\mu,1)$, where t and μ are positive integers.

b) There exist integers $a>0$ and $p\geqq0$ and a simple word S such that $F(A,\dot{B},\lambda_0,\cdots,\lambda_{p-1})=S^a$, $[S^\partial\leqq(6M)^3$, where $(\lambda_p,\cdots,\lambda_{k-1})$ is a procession of one of the following three forms: Λ, (t), $(t,1)$, where t is a positive integer.

It is obvious, for a given procession $[A,B,\lambda_0,\cdots,\lambda_{k-1}]$ and integer $M>0$, how to determine whether or not $\mathscr{B}([A,B,\lambda_0,\cdots,\lambda_{k-1}],M)$ holds.

The following properties of $\mathscr{B}$ follow from the definitions of $\mathscr{B}$ and F. We use the usual abbreviations: $F_i=F(A,B,\lambda_0,\cdots,\lambda_{i-1})$, $i=0,1,\cdots,k$, if there is given a procession $[A,B,\lambda_0,\cdots,\lambda_{k-1}]$.

1. Suppose $[A^\partial,[B^\partial\leqq M$ and $\urcorner\mathscr{B}([A,B,\lambda_0,\cdots,\lambda_{k-1}],M)$. Then $[F_{k-2}^\partial>30M$, $k>2$; if one of the numbers λ_{k-1} or λ_{k-2} is equal to one, then $[F_{k-3}^\partial>30M$, $k>3$; if $\lambda_{k-1}=\lambda_{k-3}=1$, then $[F_{k-4}^\partial>30M$, $k>4$.

PROOF. The inequalities involving the $[F_i^\partial$ follow from the definition of $\mathscr{B}$. The inequalities for k follow from those for the $[F_i^\partial$ and the conditions $[A^\partial,[B^\partial\leqq M$.

2. Suppose $[A^\partial,[B^\partial\leqq M$, $\mathscr{L}(A,B)$ and $k>1$. Then

$$\urcorner\mathscr{B}([A,B,\lambda_0,\cdots,\lambda_{k-1}],M)\supset\urcorner\mathscr{B}([A,B,\lambda_0,\cdots,\lambda_{k-1},t],M),$$

where t is a positive integer.

This is proved by contradiction. We assume that $\mathscr{B}[(A,B,\lambda_0,\cdots,\lambda_{k-1},t],M)$ and, using the definition, show that $\mathscr{B}([A,B,\lambda_0,\cdots,\lambda_{k-1}],M)$. The conditions $\mathscr{L}(A,B)$ and $[A^\partial,[B^\partial\leqq M$ are used when in b) the procession $(\lambda_p,\cdots,\lambda_{k-1},t)$ is empty. Then $F(A,B,\lambda_0,\cdots,\lambda_{k-1},t)=S^a$, where $a>0$, $[S^\partial\leqq(6M)^3$ and S is a simple word. Here we argue as follows. Since $[A^\partial,[B^\partial\leqq M$

and $\mathscr{L}(A,B)$, by 5.12′ we have $[F(A,B,\lambda_0,\cdots,\lambda_{k-1},t)^\partial \leqq (6M)^3$. Hence $[F(A,B,\lambda_0,\cdots,\lambda_{k-1})^\partial \leqq (6M)^3$, and then $F(A,B,\lambda_0,\cdots,\lambda_{k-1}) = T^b$, where $b > 0$, T is a simple word and $[T^\partial \leqq (6M)^3$, i.e. $\mathscr{B}([A,B,\lambda_0,\cdots,\lambda_{k-1}],M)$.

3. Suppose $[A^\partial, [B^\partial \leqq M$, $\mathscr{L}(A,B)$ and $k \geqq 1$. Then

$$\neg\, \mathscr{B}([A,B,\lambda_0,\cdots,\lambda_{k-1},1],M)$$

if and only if

$$\neg\, \mathscr{B}([A,B,\lambda_0,\cdots,\lambda_{k-2},\lambda_{k-1}+1],M).$$

It suffices to prove that $\mathscr{B}([A,B,\lambda_0,\cdots,\lambda_{k-1},1],M)$ if and only if $\mathscr{B}([A,B,\lambda_0,\cdots,\lambda_{k-2},\lambda_{k-1}+1],M)$. In the proof we consider 14 assumptions: seven in connection with the condition $\mathscr{B}([A,B,\lambda_0,\cdots,\lambda_{k-1},1],M)$ and seven in connection with the condition $\mathscr{B}([A,B,\lambda_0,\cdots,\lambda_{k-1}+1],M)$. Only two of these 14 use the conditions $\mathscr{L}(A,B)$ and $[A^\partial, [B^\partial \leqq M$. They correspond to the case where in b) the procession $(\lambda_p,\cdots,\lambda_{k-1})$ is empty. This case is handled as in the proof of property 2, and depends on the equality

$$[F(A,B,\lambda_0,\cdots,\lambda_{k-1},1)^\partial = [F(A,B,\lambda_0,\cdots,\lambda_{k-2},\lambda_{k-1}+1)^\partial$$

(see 5.11).

4. Suppose $k > 1$, $t > 1$, $[A^\partial, [B^\partial \leqq M$ and $\mathscr{L}(A,B)$. Then

$$\neg\, \mathscr{B}([A,B,\lambda_0,\cdots,\lambda_{k-1},t],M)$$

if and only if

$$\neg\, \mathscr{B}([A,B,\lambda_0,\cdots,\lambda_{k-1},2],M).$$

The argument is the same as in the proof of properties 2 and 3. An exception to this statement is the assumption that $F(A,B,\lambda_0,\cdots,\lambda_{k-1},2) = S^a$, where $a > 0$, S is simple and $[S^\partial \leqq (6M)^3$. Here we obtain

$$[F(A,B,\lambda_0,\cdots,\lambda_{k-1})^\partial \leqq [F(A,B,\lambda_0,\cdots,\lambda_{k-1},2)^\partial \leqq (6M)^3,$$

and then $F(A,B,\lambda_0,\cdots,\lambda_{k-1}) = T^b$, where $b > 0$, T is simple, $[T^\partial \leqq (6M)^3$, and we have $\mathscr{B}([A,B,\lambda_0,\cdots,\lambda_{k-1},t],M)$.

We will now prove that in the region $\mathscr{B}([A,B,\lambda_0,\cdots,\lambda_{k-2}],M)$ the values of the pair $[G(A,B,\lambda_0,\cdots,\lambda_{k-1}), G(A,B,\lambda_0,\cdots,\lambda_{k-2})]$ are described by parametric words of type 11.1 c).

11.8. *Suppose* $k > 1$,

$$X = G(QP,QR,\lambda_0,\cdots,\lambda_{k-1})Q, \qquad Y = G(QP,QR,\lambda_0,\cdots,\lambda_{k-2})Q,$$

$[PQR^\partial \leqq M$, $\mathscr{L}(QP,QR)$, $\mathscr{B}([QP,QR,\lambda_0,\cdots,\lambda_{k-2}],M)$ *and* $\neg\, \tau(Y,(6M)^3)$. *Then there exist* σ, ν, λ, ϵ, C, Q_1, R_1 *and a simple word* S *such that* $\mathscr{L}(S,C)$, $R_1 \angle CS^2$, $R_1 \angle Q_1$, $Q_1 \neq C$, $[S^\partial \leqq 30M$, $[C^\partial, [Q_1^\partial, [R_1^\partial \leqq 90M$, $\epsilon = \pm 1$, *and*

$$(*) \qquad X = ((S^\sigma C)^\nu S^\sigma Q_1)^\lambda (S^\sigma C)^{\nu+\epsilon} S^\sigma R_1, \qquad Y = (S^\sigma C)^\nu S^\sigma R_1.$$

If $\epsilon = 1$, *then there exist* d *and* B *such that* $B \angle C$, $Q_1 \perp\!\!\!\perp CS^d$ *and* $S^d B \perp\!\!\!\perp R_1$.

PROOF. We have $\mathscr{B}([\cdots],M)$. For brevity, put

$$F_i = F(QP,QR,\lambda_0,\cdots,\lambda_{i-1}), \qquad i = 1,\cdots,k.$$

In accordance with the definition of $\mathscr{B}$ we consider two cases.

Case a). There exist integers $a > 0$, $p \geqq 0$ and a simple word S such that

$$(1) \qquad F_p = S^a, \qquad [S^\partial \leqq 30M,$$

where $(\lambda_p, \cdots, \lambda_{k-2})$ is a procession of one of the following four forms: (t, μ), $(t, 1, \mu)$, $(t, \mu, 1)$ or $(t, 1, \mu, 1)$. Correspondingly, we consider four subcases. We use the estimate

$$(2) \qquad [F_p^\partial, [F_{p-1}^\partial \leqq 30M,$$

which follows from the conditions $[PQR^\partial \leqq M$, $\mathscr{L}(QP, QR)$ and (1) in view of 5.12′. We also note that by 5.13 there exists a word A such that

$$(3) \qquad X = G(F_p, F_{p-1}, \lambda_p, \ldots, \lambda_{k-1})A,$$
$$(4) \qquad Y = G(F_p, F_{p-1}, \lambda_p, \ldots, \lambda_{k-2})A,$$
$$(5) \qquad A \angle F_{p-1}F_p, \quad A \angle F_pF_{p-1}, \quad A \angle F_p^2.$$

Subcase a$_1$). $(\lambda_p, \cdots, \lambda_{k-2}) = (t, \mu)$. Put

$$(6) \qquad \begin{gathered} C = F_pF_{p-1}, \quad Q_1 = F_pF_{p-1}F_p, \quad \sigma = a(t-1), \quad \nu = \mu - 1, \\ \lambda = \lambda_{k-1} - 1, \quad R_1 = F_pA, \quad B = A, \quad d = a. \end{gathered}$$

Calculating the value of the function G and using (3), (4), (1) and (6), we obtain

$$\begin{aligned} X &= G(F_p, F_{p-1}, t, \mu, \lambda_{k-1})A = ((F_p^t F_{p-1})^\mu F_p)^{\lambda_{k-1}} F_p^{t-1} A \\ &= ((F_p^tF_{p-1})^{\mu-1} F_p^t F_{p-1}F_p)^{\lambda_{k-1}-1} (F_p^t F_{p-1})^\mu F_p^t A = ((S^\sigma C)^\nu S^\sigma Q_1)^\lambda (S^\sigma C)^{\nu+1} S^\sigma R_1; \\ Y &= G(F_p, F_{p-1}, t, \mu) = (F_p^t F_{p-1})^{\mu-1} F_p^t A = (S^\sigma C)^\nu S^\sigma R_1. \end{aligned}$$

From (6), (5), (1) and (2) we obtain $B \angle C$, $Q_1 = CS^a$, $S^aB = R_1$, $R_1 \angle Q_1$, $Q_1 \neq C$, $[S^\partial \leqq 30M$ and $[Q_1^\partial, [R_1^\partial, [C^\partial \leqq 90M$. If $t = 1$, then $\sigma = 0$, $Y = C^\nu R_1$ and $R_1 \angle C^2$, which contradicts the condition $\neg\, \tau(Y, (6M)^3)$. Hence $t > 1$. Then $\sigma \geqq a$ and $R_1 \angle CS^a \angle CS^\sigma$. Finally, using 5.3 and $\mathscr{L}(QP, QR)$, we obtain $\mathscr{L}(F_p, F_{p-1})$, which implies $\mathscr{L}(S, C)$. Thus the assertion being proved holds with $\epsilon = 1$.

Subcase a$_2$). $(\lambda_p, \cdots, \lambda_{k-2}) = (t, 1, \mu)$. Put

$$(7) \qquad \begin{gathered} C = F_pF_{p-1}F_p, \quad Q_1 = F_pF_{p-1}, \quad \sigma = a(t-1), \quad \nu = \mu, \\ \lambda = \lambda_{k-1} - 1, \quad R_1 = A, \quad B = S^aA, \quad d = -a. \end{gathered}$$

Calculating the value of G, we have (see (3), (4), (1) and (7))

$$\begin{aligned} X &= G(F_p, F_{p-1}, t, 1, \mu, \lambda_{k-1})A = ((F_p^t F_{p-1} F_p)^\mu F_p^t F_{p-1})^{\lambda_{k-1}} F_p^t A \\ &= ((F_p^{t-1} F_pF_{p-1} F_p)^\mu F_p^{t-1} F_pF_{p-1})^{\lambda_{k-1}-1} (F_p^{t-1} F_pF_{p-1}F_p)^{\mu+1} F_p^{t-1} A \\ &= ((S^\sigma C)^\nu S^\sigma Q_1)^\lambda (S^\sigma C)^{\nu+1} S^\sigma R_1, \\ Y &= G(F_p, F_{p-1}, t, 1, \mu)A = (F_p^t F_{p-1} F_p)^\mu F_p^{t-1} A = (S^\sigma C)^\nu S^\sigma R_1. \end{aligned}$$

We also have $B \angle C$, $C = Q_1S^a$ and $B = S^aR_1$, so that $CS^d \perp\!\!\!\perp Q_1$ and $S^dB \perp\!\!\!\perp R_1$. Furthermore, $R_1 \angle Q_1 \angle C \angle CS^\sigma$, $C \neq Q_1$ and $[C^\partial, [Q_1^\partial, [R_1^\partial \leqq 90M$. Arguing as in Subcase a$_1$), we see that $\mathscr{L}(S, C)$ and our assertion holds with $\epsilon = 1$.

Subcase a$_3$). $(\lambda_p, \cdots, \lambda_{k-2}) = (t, \mu, 1)$. Put

$$(8) \qquad C = F_{p-1}, \quad Q_1 = F_pF_{p-1}, \quad \sigma = at, \quad \nu = \mu, \quad \lambda = \lambda_{k-1}, \quad R_1 = A.$$

We have (see (3), (4), (8) and (1))

$$X = G(F_p, F_{p-1}, t, \mu, 1, \lambda_{k-1}) A = ((F_p^t F_{p-1})^\mu F_p F_p^t F_{p-1})^{\lambda_{k-1}} (F_p^t F_{p-1})^{\mu-1} F_p^t A$$
$$= ((S^\sigma C)^\nu S^\sigma Q_1)^\lambda (S^\sigma C)^{\nu-1} S^\sigma R_1,$$

$$Y = G(F_p, F_{p-1}, t, \mu, 1) A = (F_p^t F_{p-1})^\mu F_p F_p^{t-1} A = (S^\sigma C)^\nu S^\sigma R_1.$$

Furthermore, $C \neq Q_1$, $R_1 \angle Q_1$, $R_1 \angle CS^a \angle CS^\sigma$ and $\mathscr{L}(S, C)$, and our assertion holds with $\epsilon = -1$.

Subcase a$_4$). $(\lambda_p, \cdots, \lambda_{k-2}) = (t, 1, \mu, 1)$. Put

(9) $$C = F_p F_{p-1} F_p, \quad Q_1 = F_{p-1} F_p, \quad \sigma = a(t-1), \quad \nu = \mu + 1, \quad \lambda = \lambda_{k-1}, \quad R_1 = A.$$

We have (see (3), (4), (9) and (1))

$$X = G(F_p, F_{p-1}, t, 1, \mu, 1, \lambda_{k-1}) A = ((F_p^t F_{p-1} F_p)^\mu F_p^t F_{p-1} F_p^t F_{p-1} F_p)^{\lambda_{k-1}}$$
$$\times (F_p^t F_{p-1} F_p)^\mu F_p^{t-1} A = ((F_p^t F_{p-1} F_p)^{\mu+1} F_p^{t-1} F_{p-1} F_p)^{\lambda_{k-1}} (F_p^t F_{p-1} F_p)^\mu F_p^{t-1} A$$
$$= ((S^\sigma C)^\nu S^\sigma Q_1)^\lambda (S^\sigma C)^{\nu-1} S^\sigma R_1,$$

$$Y = G(F_p, F_{p-1}, t, 1, \mu, 1) A = (F_p^t F_{p-1} F_p)^{\mu+1} F_p^{t-1} A = (S^\sigma C^\nu) S^\sigma R_1.$$

Furthermore, $Q_1 \neq C$, $R_1 \angle Q_1 \angle CS^\sigma$ and $\mathscr{L}(S, C)$, and our assertion holds with $\epsilon = -1$.

Case b). There exist integers $a > 0$, $p \geqq 0$ and a simple word S such that

(10) $$F_p = S^a, \qquad [S^\partial \leqq (6M)^3,$$

where $(\lambda_p, \cdots, \lambda_{k-2})$ is a procession of one of the forms Λ, (t) or $(t, 1)$.

We will prove that this case contradicts the condition $\neg\, \tau(Y, (6M)^3)$. Indeed, if $(\lambda_p, \cdots, \lambda_{k-2}) = \Lambda$, then $p = k-1$,

$$Y = G(QP, QR, \lambda_0, \cdots, \lambda_{p-1}) Q \angle F_p,$$

and it follows from (10) that $\tau(Y, (6M)^3)$. If $(\lambda_p, \cdots, \lambda_{k-2}) = (t)$, then (see (4))

$$Y = G(F_p, F_{p-1}, t) A = F_p^{t-1} A \angle F_p^{t+1}$$

(see (5)), and (10) implies that $\tau(Y, (6M)^3$. If $(\lambda_p, \cdots, \lambda_{k-2}) = (t, 1)$, then

$$Y = G(F_p, F_{p-1}, t, 1) A = F_p^t A \angle F_p^{t+2},$$

and (10) implies that $\tau(Y, (6M)^3)$. Thus b) is impossible. The proposition is proved.

The relation $\mathscr{A}$. Suppose h and g are processions. By definition, $\mathscr{A}(h, g)$ if and only if one of the processions h or g is obtained from the other by dropping the last component.

11.9. *For any words A and B such that $\mathscr{L}(A, B)$ and $[A^\partial, [B^\partial \leqq M$ there exist processions h and g such that $\neg\, \mathscr{B}([A, B, h], M)$, $\neg\, \mathscr{B}([A, B, g], M)$, $\mathscr{A}(h, g)$.*

PROOF. We first note that if $k > (6M)^3 + 4$, then

$$\neg\, \mathscr{B}([A, B, \lambda_0, \cdots, \lambda_k], M).$$

Indeed, if $k > (6M)^3 + 4$, $0 \leqq i \leqq 4$, then $F(A, B, \lambda_0, \cdots, \lambda_{k-1})^\partial > (6M)^3$ and, by 5.12, $F(A, B, \lambda_0, \cdots, \lambda_{k-i})$ cannot be a power of any S such that $[S^\partial \leqq (6M)^3$. But this means that $\neg\, \mathscr{B}([A, B, \lambda_0, \cdots, \lambda_k], M)$. Our assertion is now obvious.

§12. Representation of the general solution of normal equations in two unknowns by predicate terms

We define "terms" and "predicate terms", by means of which we construct the general solution of normal word equations in two unknowns.

Terms. We begin with unknowns. We define three kinds of unknowns.

1) Word unknowns, whose values are words in the alphabet **И**.

2) Positive unknowns, whose values are positive integers.

3) Processional unknowns: even and odd. Sequences of positive integers will be called processions. The values of an even (odd) processional unknown are nonempty processions containing an even (odd) number of components.

Word unknowns will be denoted by the letters x, y, z, u, v, positive unknowns by $\lambda, \mu, \nu, \sigma, \tau, t$, and processional unknowns by h, g, f, with or without indices.

We now define terms. The inductive definition of terms has the following form.

1) If T is a word in the alphabet **И**, then T is a term.

2) If T is a word unknown, then T is a term.

3) If A and B are nonempty words in the alphabet **И** and h is a processional unknown, then $G(A,B,h)$ and $F(A,B,h)$ are terms.

4) If T is a term and λ is a positive unknown, then $(T)^\lambda$ is a term.

5) If T_1 and T_2 are terms, then T_1T_2 is a term.

6) There are no terms other than those defined in 1)–5).

Any term can be viewed as a function of the unknowns occurring in that term. The values of this function are words in the alphabet **И**. We write $T(\xi_1, \cdots, \xi_m)$ if the list $\xi_1, \cdots, \xi_m$ of unknowns contains all of the unknowns occurring in the term T. We denote by $T(\overline{\xi}_1, \cdots, \overline{\xi}_m)$ the word in the alphabet **И** which is obtained as a result of substituting into T the values $\overline{\xi}_1, \cdots, \overline{\xi}_m$ in place of the unknowns $\xi_1, \cdots, \xi_m$.

Suppose $T_1(\xi_1, \cdots, \xi_m)$ and $T_2(\xi_1, \cdots, \xi_m)$ are terms. The equality $T_1 = T_2$ is called an equation. By a solution of this equation we mean any set $(\overline{\xi}_1, \cdots, \overline{\xi}_m)$ of values of the unknowns $\xi_1, \cdots, \xi_m$ such that in the alphabet **И** we have the equality $T_1(\overline{\xi}_1, \cdots, \overline{\xi}_m) = T_2(\overline{\xi}_1, \cdots, \overline{\xi}_m)$.

Predicate terms. By predicate terms we mean pairs $[T; \mathscr{P}]$, where T is a term and $\mathscr{P}$ a predicate. The term T is called the left-hand side of $[T; \mathscr{P}]$, the predicate $\mathscr{P}$ the right-hand side. A predicate term $[T; \mathscr{P}]$ is called admissible if $\mathscr{P}$ is a nonempty decidable predicate (a predicate $\mathscr{P}(\xi_1, \cdots, \xi_m)$ is called nonempty if it is satisfied by at least one set of values of the unknowns $\xi_1, \cdots, \xi_m$).

Suppose

$$\alpha = [T_1(\xi_1, \cdots, \xi_m);\ \mathscr{P}(\xi_1, \cdots, \xi_m)]$$

and

$$\beta = [T_2(\xi_1, \cdots, \xi_m);\ \mathscr{R}(\xi_1, \cdots, \xi_m)]$$

are predicate terms. The equality $\alpha = \beta$ is called an equation. By a solution of this equation we mean any set $(\overline{\xi}_1, \cdots, \overline{\xi}_m)$ of values of the unknowns $\xi_1, \cdots, \xi_m$ such that $\mathscr{P}(\overline{\xi}_1, \cdots, \overline{\xi}_m) \,\&\, \mathscr{R}(\overline{\xi}_1, \cdots, \overline{\xi}_m)$ and in the alphabet **И** we have the equality $T_1(\overline{\xi}_1, \cdots, \overline{\xi}_m) = T_2(\overline{\xi}_1, \cdots, \overline{\xi}_m)$. It is obvious that the equation $\alpha = \beta$ is equivalent to the system

$$T_1 = T_2, \qquad \mathscr{P}(\xi_1, \cdots, \xi_m) \,\&\, \mathscr{R}(\xi_1, \cdots, \xi_m).$$

This is called the expanded form of the equation $\alpha = \beta$.

We now define special predicate terms, by means of which we construct the general solution of a normal word equation in two unknowns.

Special predicate terms of characteristic M

Suppose M is an integer greater than two. We define predicate V_1-, $\widetilde{V}_1$-, Θ_1-, Θ_2-, Ψ_1-, Ψ_2-, and G-terms of characteristic M.

1. A V_1-term of characteristic M (a V_1^M-term) is a pair

$$[([xA_i]_{i=1}^k)^\lambda [xA_i]_{i=1}^r x;\quad \omega(x, M) \mathbin{\&} \lambda \geqq (4M)^6],$$

where x is a word unknown, λ is a positive unknown, k and r are integers, $A_1, \cdots, A_k$ are words in the alphabet И, and $0 \leqq r < k$, $1 < k \leqq M$, $\nabla(A_1, \cdots, A_k)$ and $[A_1^\partial, \cdots, [A_k^\partial \leqq M$.

2. A $\widetilde{V}_1$-term of characteristic M (a $\widetilde{V}_1^M$-term) is a pair

$$[(xA)^\lambda x;\quad \neg\, \tau(x, M) \mathbin{\&} \mathscr{S}(xA)],$$

where x is a word unknown, λ is a positive unknown, A is a word in the alphabet И, $[A^\partial \leqq M$, and $\mathscr{S}(x)$ is the predicate "x is a simple word". The word A is called the base of the $\widetilde{V}_1^M$-term.

3. A Θ_1-term of characteristic M (a Θ_1^M-term) is a pair

$$[([(S^\sigma C)^\nu S^\sigma RQ_i]_{i=1}^k)^\lambda [(S^\sigma C)^\nu S^\sigma RQ_i]_{i=1}^r (S^\sigma C)^{\nu+\varepsilon} S^\sigma R;$$
$$\lambda \geqq M,\qquad \nu \geqslant 4M^2,\qquad \partial(S^\sigma C) > (6M)^3],$$

where σ, ν, λ are distinct positive unknowns, ϵ, r, k are integers, S, R, C, $Q_1, \cdots, Q_k$ are words in the alphabet И, S is simple, $0 \leqq r < k \leqq M$, $\epsilon = \pm 1$, $\mathscr{S}(S, C)$, $\nabla(Q_1, \cdots, Q_k)$, $\neg\,(RQ_1 = \cdots = RQ_k = C)$, $R \angle CS^{[R^\partial}$, $[S^\partial \leqq 30M$ and $[C^\partial, [RQ_1^\partial, \cdots, [RQ_k^\partial \leqq 90M$. The following conditions also hold: α) if $\epsilon = 1$, then there exist an integer d and a word B such that $B \angle C$, $RQ_{r+1} \amalg CS^d$ and $S^d B \amalg R$; β) if $k > 1$, then $[Q_1^\partial, \cdots, [Q_k^\partial \leqq M$. The word $(S^\sigma C)^\nu S^\sigma$ is called the base of the Θ_1^M-term.

4. A Θ_2-term of characteristic M (a Θ_2^M-term) is a pair

$$[[(S^\sigma C)^\nu S^\sigma A_i]_{i=1}^{m-1} (S^\sigma C)^{\nu+\varepsilon} S^\sigma A_m;\qquad \nu \geqslant 4M^2,\qquad \partial(S^\sigma C) > (6M)^3],$$

where σ and ν are distinct positive unknowns, $\epsilon = \pm 1$, $1 < m \leqq M^2$, S, C, $A_1, \cdots, A_m$ are words in the alphabet И, S is simple, $\mathscr{S}(S, C)$, $\neg\,(A_1 = \cdots = A_{m-1} = C)$, $[S^\partial \leqq 30M$ and $[C^\partial, [A_1^\partial, \cdots, [A_m^\partial \leqq 90M$. The word $(S^\sigma C)^\nu S^\sigma$ is called the base of the Θ_2^M-term.

5. Ψ_1-terms of characteristic M (Ψ_1^M-terms). Here we distinguish $\tilde{\Psi}_1^M$-terms and $\overline{\Psi}_1^M$-terms. A $\tilde{\Psi}_1^M$-term is a pair

$$[([S^\sigma A_i]_{i=1}^n)^\lambda [S^\sigma A_i]_{i=1}^r S^\sigma A;\quad \lambda \geqq 9M^6,\quad \sigma[S^\partial > (6M)^{13}],$$

where σ and λ are distinct positive unknowns, $0 \leqq r < n \leqq 4M^3$, S, A and A_i are words in the alphabet И, S is simple, $A \angle A_{r+1} S^{[A^\partial}$, $\oplus(S, A_1, \cdots, A_n, (6M)^7)$, $[S^\partial \leqq (6M)^3$, and $[A_1^\partial, \cdots, [A_n^\partial, [A^\partial \leqq (6M)^7$. The word S^σ is called the base of the $\tilde{\Psi}_1^M$-term.

A $\overline{\Psi}_1^M$-term is a pair $[(S^\sigma C)^\nu S^\sigma R;\ \nu \geqq 4M^2,\ \partial(S^\sigma C) > (6M)^3]$, where σ and ν are distinct positive unknowns, S, C and R are words in the alphabet И, S is simple, $\mathscr{S}(S, C)$, $R \angle CS^{[R^\partial}$, $[S^\partial \leqq 30M$, and $[R^\partial, [C^\partial \leqq 90M$. The word $(S^\sigma C)^\nu S^\sigma$ is called the base of the $\overline{\Psi}_1^M$-term.

Any Ψ_1^M-term is a $\tilde{\Psi}_1^M$- or $\bar{\Psi}_1^M$-term.

6. A Ψ_2-term of characteristic M (a Ψ_2^M-term) is a pair

$$[[S^\sigma A_i]_{i=1}^m; \quad \sigma[S^\partial > (6M)^{13}],$$

where σ is a positive unknown, $0 < m \leqq 4M^2$, $S, A_1, \cdots, A_m$ are words in the alphabet **И**, S is simple, $[S^\partial \leqq (6M)^3$ and $[A_i^\partial \leqq (6M)^7$. The word S^σ is called the base of the Ψ_2^M-term.

7. A G-term of characteristic M (a G^M-term) is a pair

$$[G(QP, QR, h)Q; \quad \neg \mathscr{B}([QP, QR, h], M)],$$

where h is a processional unknown, P, Q and R are words in the alphabet **И** such that $\mathscr{S}(QP, QR)$, $Q \neq \Lambda$, and $[PQR^\partial \leqq M$. The triple $[P, Q, R]$ is called the base of the G^M-term.

REMARK 1. Special predicate terms of characteristic M are admissible for any M. This is obvious for Θ_1^M-, Θ_2^M-, Ψ_1^M- and Ψ_2^M-terms; for V_1^M- and $\widetilde{V}_1^M$-terms this follows from 11.5; for G^M-terms, from 11.10.

The general solution of any normal equation in two unknowns consists of a finite number of formulas of the form $x = \alpha$ and $y = \beta$, where α and β are special predicate terms. The variables occurring in α and β are related in some way. It is convenient to consider this relation separately and not introduce it into the right-hand sides of the predicate terms α and β. The solutions will be represented by triples $[\alpha, \beta, \pi]$, where α and β are predicate terms and π is a predicate.

Suppose there are given predicate terms $\alpha = [T_1(\xi_1, \cdots, \xi_m); \ \mathscr{P}(\xi , \cdots, \xi_m)]$, $\beta = [T_2(\xi_1, \cdots, \xi_m); \ \mathscr{Q}(\xi_1, \cdots, \xi_m)]$ and a predicate $\pi(\xi_1, \cdots, \xi_m)$. Consider the triple $[\alpha, \beta, \pi]$. This triple is called admissible if the predicates $\mathscr{P}$, $\mathscr{Q}$ and π are decidable and the predicate $\mathscr{P} \,\&\, \mathscr{Q} \,\&\, \pi$ is nonempty. A set $(\bar{\xi}_1, \cdots, \bar{\xi}_m)$ of values of the unknowns is called admissible for the triple if it satisfies the predicate $\mathscr{P} \,\&\, \mathscr{Q} \,\&\, \pi$. Suppose $\mathfrak{A}$ is a word equation in the two unknowns x, y. We say that $[\alpha, \beta, \pi]$ is a solution of $\mathfrak{A}$ if whenever a set $(\bar{\xi}_1, \cdots, \bar{\xi}_m)$ is admissible for the triple, then the pair

$$(X = T_1(\bar{\xi}_1, \cdots, \bar{\xi}_m), \ Y = T_2(\bar{\xi}_1, \cdots, \bar{\xi}_m))$$

is a solution of $\mathfrak{A}$. Suppose X and Y are words in the alphabet **И**. We say that the pair (X, Y) is representable by the triple $[\alpha, \beta, \pi]$ if there exists an admissible set $(\bar{\xi}_1, \cdots, \bar{\xi}_m)$ for the triple such that $X = T_1(\bar{\xi}_1, \cdots, \bar{\xi}_m)$ and $Y = T_2(\bar{\xi}_1, \cdots, \bar{\xi}_m)$.

We now define the triples which will enable us to construct the general solution of a normal word equation in two unknowns.

Suppose γ is a symbol in the list V_1^M, $\widetilde{V}_1^M$, Θ_1^M, Θ_2^M, Ψ_1^M. We define γ-triples.

1. A V_1^M-triple is a triple $[\alpha(y, \lambda), \beta, 0 = 0]$, where α is a V_1^M-term, and β the predicate term $[y; 0 = 0]$.

2. A $\widetilde{V}_1^M$-triple is a triple $[\alpha(y, \lambda), \beta, \omega(y, M) \,\&\, \lambda \geqq (4M)^6]$, where α is a $\widetilde{V}_1^M$-term, and β the predicate term $[y; 0 = 0]$.

3. A Θ_1^M-term is a triple $[\alpha, \beta, 0 = 0]$, where α is a Θ_1^M-term, β a $\bar{\Psi}_1^M$-term, and α and β have the same base.

4. A Θ_2^M-triple is a triple $[\alpha, \beta, 0 = 0]$, where α is a Θ_2^M-term, β a $\bar{\Psi}_1^M$-term, and α and β have the same base.

5. A Ψ_1^M-triple is a triple $[\alpha,\beta,0=0]$, where α is a $\tilde{\Psi}_1^M$-term, β a Ψ_2^M-term, and α and β have the same base.

6. If $[\alpha,\beta,\pi]$ is a γ-triple, then $[\beta,\alpha,\pi]$ is also a γ-triple. This completes the definition of γ-triples.

We also define four other kinds of triples. Suppose $\mathfrak{A}$ is a word equation in two unknowns. We define $G_{\mathfrak{A}}$-, $V_{\mathfrak{A}}$-, $\Psi'_{\mathfrak{A},M}$- and $\Psi''_{\mathfrak{A},M}$-triples. We assume that $\mathfrak{A}$ has the form 11.1 (*). Let $\Phi(x,y)$ and $\Psi(x,y)$ denote the left- and right-hand sides of $\mathfrak{A}$.

7. $\Psi'_{\mathfrak{A},M}$-triples. Suppose S is a simple word, $[S^\partial \leqq (6M)^3$, and $[A^\partial, [B^\partial \leqq (8M)^4$. Consider the exponential equation $\Phi(S^\lambda A, S^\mu B) = \Psi(S^\lambda A, S^\mu B)$. By 2.3, this equation is equivalent in the region $\lambda[S^\partial, \mu[S^\partial > (4M)^{10}$ to some LDR $\mathscr{D}'_{\mathfrak{A},S,A,B}(\lambda,\mu)$. The $\Psi'_{\mathfrak{A},M}$-triples are all triples of the form $[\alpha,\beta,\pi]$, where $\alpha=[S^\lambda A;\ \lambda[S^\partial > (4M)^{10}]$, $\beta=[S^\mu B;\ \mu[S^\partial > (4M)^{10}]$, $\pi = \mathscr{D}_{\mathfrak{A},S,A,B}(\lambda,\mu)$, for which the LDR $\mathscr{D}_{\mathfrak{A},S,A,B}$ has a solution.

8. $\Psi''_{\mathfrak{A},M}$-triples. Suppose S is a simple word, $\mathscr{S}(S,C)$, $[S^\partial \leqq 30M$, $[C^\partial$, $[R^\partial \leqq 90M$ and $R \angle CS^{[R^\partial}$. Consider the exponential equation

$$\Phi\left((S^\sigma C)^\lambda S^\sigma R,\ (S^\sigma C)^\mu S^\sigma R\right) = \Psi\left((S^\sigma C)^\lambda S^\sigma R,\ (S^\sigma C)^\mu S^\sigma R\right).$$

By 2.10, this equation is equivalent in the region λ, $\mu \geqq 4M^2$, $\partial(S^\sigma C) > (6M)^3$ to some LDR $\mathscr{D}_{\mathfrak{A},S,C,R}(\sigma,\lambda,\mu)$. The $\Psi''_{\mathfrak{A},M}$-triples are all triples of the form $[\alpha(\sigma,\lambda),\alpha(\sigma,\mu),\mathscr{D}_{\mathfrak{A},S,C,R}(\sigma,\lambda,\mu)]$, where $\alpha(\sigma,\lambda)=[(S^\sigma C)^\lambda S^\sigma R;\ \lambda \geqq 4M^2$, $\partial(S^\sigma C) > (6M)^3]$, for which the LDR $\mathscr{D}_{\mathfrak{A},S,C,R}$ has a solution.

9. $V_{\mathfrak{A}}$-triples. Consider in the alphabet $\text{И} \cup \{v\}$ the exponential equation $\Phi((vP_1)^\lambda v, (vP_1)^\mu v) = \Psi((vP_1)^\lambda v, (vP_1)^\mu v)$, where λ and μ are positive unknowns, and P_1 is the P_1 of 11.1 (*). By 2.3, this equation is equivalent to some LDR $\mathscr{D}_{\mathfrak{A}}(\lambda,\mu)$. If the LDR $\mathscr{D}_{\mathfrak{A}}$ has no solutions, then there are no $V_{\mathfrak{A}}$-triples. If $\mathscr{D}_{\mathfrak{A}}$ has a solution, then a $V_{\mathfrak{A}}$-triple is a triple $[\alpha(v,\lambda),\alpha(v,\mu),\ \mathscr{D}_{\mathfrak{A}}(\lambda,\mu)]$, where $\alpha(v,\lambda)$ is a $\widetilde{V}_1^M$-term with base P_1.

10. $G_{\mathfrak{A}}$-triples. Suppose $P_1 \neq Q_1$. Suppose the system $\mathfrak{B}$ consisting of $\mathfrak{A}$ and the equation $xP_1y \to yQ_1x$ is equivalent to the equation $xP_1y \to yQ_1x$ (this can be tested by means of 7.5), and suppose the equation $xP_1y \to yQ_1x$ has solutions (this can be tested by means of 5.10). Then the $G_{\mathfrak{A}}$-triples are all triples of the form $[\alpha(h),\alpha(g),\mathscr{A}(g,h)]$, where h and g are processional unknowns, $\mathscr{A}$ is the predicate defined before 11.9, and $\alpha(h)$ is a G^M-term with base $[P,Q,R]$. The words P, Q, R, P_1, Q_1 and the unknown h are connected by the relations: 1) $P_1=PQR$, $Q_1=RQP$, h is an even unknown, or 2) $P_1=RQP$, $Q_1=PQR$, h is an odd unknown.

The ten kinds of triples defined above are called special triples of characteristic M.

Note that any special triple is admissible. This is obvious for Θ_1^M-, Θ_2^M-, Ψ_1^M-, Ψ_2^M-, $V_{\mathfrak{A}}$-, $\Psi'_{\mathfrak{A},M}$- and $\Psi''_{\mathfrak{A},M}$-triples. For V_1^M- and $\widetilde{V}_1^M$-triples this follows from 11.5, and for $G_{\mathfrak{A}}$-triples, from 11.10. We note also that any special triple of characteristic M has the form $[\alpha,\beta,\mu]$, where each of the terms α and β is either a special predicate term of characteristic M or a predicate term of the form $[y;0=0]$, where y is a word unknown.

In solving W_2-systems we are not interested in all solutions of the equations of the system. The solutions of interest are described by a predicate $\epsilon(x,y,M)$. Suppose $M>2$ is an integer, and X and Y are words in the alphabet И.

By definition, $\epsilon(X, Y, M)$ holds if $[S^\partial \leq (6M)^{15}$ or $[Y^\partial \leq (6M)^{15}$ or there exist integers n and m and words $V, A_0, \cdots, A_n$, $B_0, \cdots, B_m$ such that $X = A_0[VA_i]_{i=1}^n$, $Y = B_0[VB_i]_{i=1}^m$, $n, m \leq (2M)^8$, and $[A_i^\partial, [B_j^\partial \leq (6M)^{10}$. A solution (X, Y) of a word equation in two unknowns is called an ϵ_M-solution ($\overline{\epsilon}_M$-solution) if $\epsilon(X, Y, M)$ $(\neg\epsilon(X, Y, M))$. Note that the determination of the ϵ_M-solutions of equations in two unknowns reduces to the solution of equations in one unknown.

Lastly, we say that an equation has characteristic M if $M \geq \max\{k, l, N, 3\}$, where N is the maximum coefficient length of the equation, and k and l are the number of occurrences of the unknowns on the left- and right-hand sides of the equation after the equation has been reduced to the form 11.1 (*) (by multiplying on the right by suitable unknowns).

We now turn to the proof of the theorem which, by means of special triples of characteristic M, describes the structure of the $\overline{\epsilon}_M$-solutions of normal equations of characteristic M.

12.1. **Theorem**. *Suppose $\mathfrak{A}$ is a normal equation in two unknowns of characteristic M. Then it is possible to construct a list Γ of special triples of characteristic M such that*

a) *if Γ is empty, then $\mathfrak{A}$ has no ϵ_M-solutions;*

b) *any triple in Γ is a solution of $\mathfrak{A}$;*

c) *any ϵ_M-solution of $\mathfrak{A}$ is representable by at least one triple in Γ.*

Proof. Multiplying, if necessary, $\mathfrak{A}$ on the right by suitable unknowns, we write it in the form

(1) $$xP_1[yP_i]_{i=2}^l xP_{l+1}\xi\Phi_1 = ([yQ_i]_{i=1}^k)^a xQ_{k+1}\eta\Psi_1,$$

where a, k and l are positive integers, ξ and η are unknowns, P_i and Q_j are coefficients, and $\nabla(Q_1, \cdots, Q_k)$. Equation (1), like $\mathfrak{A}$, is normal of characteristic M.

We construct a list Γ' of special triples as follows: 1) in the list Γ' we put all V_1^M-, $\widetilde{V}_1^M$-, Θ_1^M-, Θ_2^M-, and Ψ_1^M-triples (obviously, there are a finite number of these); 2) in the list Γ' we put all $V_{\mathfrak{A}}$-, $G_{\mathfrak{A}}$-, $\Psi'_{\mathfrak{A},M}$-, and $\Psi''_{\mathfrak{A},M}$-triples (it is obvious how to find all such triples; there are a finite number of these); 3) no other triples appear in Γ'.

We will now prove the following two assertions.

I. *Any ϵ_M-solution of $\mathfrak{A}$ is representable by at least one triple in Γ'.*

II. *If a triple in Γ' satisfies $\mathfrak{A}$ for some admissible values of the unknowns, then it satisfies $\mathfrak{A}$ for any admissible values of the unknowns.*

Proof of I. Suppose (X, Y) is an $\overline{\epsilon}_M$-solution of $\mathfrak{A}$. In view of the symmetry of the list Γ' we can restrict ourselves to the case $[X^\partial \geq [Y^\partial$. The condition $\neg\epsilon(X, Y, M)$ implies that $[Y^\partial > 0$ and hence, by 11.1, for any $N \geq 15M$ and $L > 1$ one of the five assertions 11.1 a)–e) holds. Choose $N = (6M)^3$ and $L = 4M^2$. We consider each of the five cases separately.

Case a). $\tau(Y, (6M)^3) \vee \rho(Y, M, 4M^2)$. Look at the more general situation $\tau_1(Y, (6M)^3) \vee \rho(Y, M, 4M^2)$. Here 11.7 is applicable. Indeed, $\neg\epsilon(X, Y, M)$ implies $[X^\partial > [Y^\partial + (8M)^4$, $[Y^\partial > (6M)^{15}$. Since the characteristic of (1) is M, we have $k, l \leq M$. Hence by 11.7 we have 11.7 a) or 11.7 b). If 11.7 a), then the pair (X, Y) is representable by a $\Psi'_{\mathfrak{A},M}$-triple; the estimates for

$\lambda[S^\partial$ and $\mu[S^\partial$ follow from the estimates for $[X^\partial$ and $[Y^\partial$. If 11.7 b), then $\neg\epsilon(X,Y,M)$ implies $\lambda \geqq 9M^5$ and the pair (X,Y) is representable by a Ψ_1^M-triple; the necessary estimate for $\sigma[S^\partial$ follows from the estimates for $[Y^\partial$, m and $[B_i^\partial$.

In analyzing the remaining cases we may assume that

$$(2) \qquad \neg\tau_1(Y,(6M)^3)\ \&\ \neg\rho(Y,M,4M^2).$$

Case c). If $RQ_1 = \cdots = RQ_k = C$, then $X = (S^\sigma C)^\mu S^\sigma R$, where $\mu \geqq \nu$, and the pair (X,Y) is representable by a $\Psi''_{\mathfrak{A},M}$-triple. Suppose $\neg(RQ_1 = \cdots = RQ_k = C)$. If $\lambda \geqq M$, then (X,Y) is representable by a Θ_1^M-triple. Suppose $\lambda < M$. Then there exist an integer $m \geqq 1$ and words $A_1,\cdots,A_{m-1}$ such that $X = [(S^\sigma C)^\nu S^\sigma A_i]_{i=1}^{m-1}(S^\sigma C)^{\nu+\iota}S^\sigma R$, $m \leqq M^2$, $[A_1^\partial,\cdots,[A_{m-1}^\partial \leqq 90M$. If either $m=1$ or $m>1$, $A_1 = \cdots = A_{m-1} = C$, then $X = (S^\sigma C)^\mu S^\sigma R$, where $\mu \geqq \nu$, and (X,Y) is representable by a $\Psi''_{\mathfrak{A},M}$-triple. If $m>1$, $\neg(A_1 = \cdots = A_{m-1} = C)$, then (X,Y) is representable by a Θ_2^M-triple.

Case d). Assume that $\tau(Z,M)$. Then $Z = S^\sigma S_1$, where S is a simple word, $S_1 \angle S$ and $[S^\partial \leqq M$. Then $X = (S^\sigma S_1 P_1)^m S^\sigma S_1$, $Y = (S^\sigma S_1 P_1)^n S^\sigma S_1$. But $[S^\partial \leqq M$ and $\neg\tau(Y,(6M)^3)$. Hence S_1P_1 is not a power of S, so that $\mathscr{L}(S,S_1P_1)$. Also, it follows from (2) that $[S^\sigma S_1 P_1^\partial > (6M)^3$, $n \geqq 4M^2$. Since $[X^\partial \geqq [Y^\partial$, we have $m \geqq n$. The pair (X,Y) is representable by a $\Psi''_{\mathfrak{A},M}$-triple. There remains the case $\neg\tau(Z,M)$. If here $n=0$, then $\neg\epsilon(X,Y,M)$ implies $m > (4M)^6$, and then, as is easily seen using (2), the pair (X,Y) is representable by a $\widetilde{V}_1^M$-triple. If $n>0$, then (X,Y) is representable by a $V_{\mathfrak{A}}$-triple.

Case b). Here we consider two subcases.

1) $k=1$. Then $X = (YQ_1)^\lambda Y$. Since $YQ_1 \neq \Lambda$, we have $YQ_1 = V^t$, where $t>0$ and V is simple. Since $Y \angle V^t$ and $\neg\tau(Y,(6M)^3)$, it follows that $[V^\partial > (6M)^3 > [Q_1^\partial$. Hence there exists Z such that $V = ZQ_1$. We have $Y = (ZQ)^{t-1}Z$ and $X = (ZQ_1)^{\lambda t+t-1}Z$. The rest of the argument is like Case d).

2) $k>1$. From $\neg\epsilon(X,Y,M)$ it follows that $\lambda \geqq (4M)^6$, and then, as is easily seen using (2), the pair (X,Y) is representable by a V_1^M-triple.

Case e). In this case (X,Y) satisfies the equation $xP_1y \to yQ_1x$, where $P_1 \neq Q_1$ and $[P_1^\partial = [Q_1^\partial$. By 5.9 b) there exist words P, Q, R and a procession $(\lambda_0,\cdots,\lambda_{k-1})$ such that $k>1$, $QP_1 = \langle QP,QR\rangle_k$, $QQ_1 = \langle QP,QR\rangle_{k-1}$, $Q \neq \Lambda$, $X = G(QP,QR,\lambda_0,\cdots,\lambda_{k-1})Q$, $Y = G(QP,QR,\lambda_0,\cdots,\lambda_{k-2})Q$.

Since $P_1 \neq Q_1$, we have $QP_1 \neq QQ_1$ and $QPQR \neq QRQP$, so that $\mathscr{L}(QP,QR)$. Also, we have $[PQR^\partial = [P_1^\partial \leqq M$. If $\neg\mathscr{B}([QP,QR,\lambda_0,\cdots,\lambda_{k-2}],M)$, then by property 2 of the predicate $\mathscr{B}$ we have $\neg\mathscr{B}([QP,QR,\lambda_0,\cdots,\lambda_{k-1}],M)$ and the pair (X,Y) is representable by a $G_{\mathfrak{A}}$-triple. If

$$\mathscr{B}([QP,QR,\lambda_0,\cdots,\lambda_{k-2}],M),$$

then by 11.8 we have for (X,Y) the representation (*) of 11.8. From $R_1 \angle CS^\sigma$ and $\neg\tau(Y,(6M)^3)$ it follows that in this representation $[S^\sigma C^\partial > (6M)^3$. Let us look at the possibilities for ν and λ. If $\nu \geqq 4M^2$ and $\lambda \geqq M$, then (X,Y) is representable by a Θ_1^M-triple; if $\nu \geqq 4M^2$ and $0<\lambda<M$, then by a Θ_2^M-triple; and if $\nu \geqq 4M^2$ and $\lambda = 0$, then by a $\Psi''_{\mathfrak{A},M}$-triple. There remains the case $\nu < 4M^2$. Then $\neg\epsilon(X,Y,M)$ implies $\lambda \geqq 9M^5$, and from the estimates for $[Y^\partial$, ν, $[S^\partial$, $[C^\partial$ and $[R_1^\partial$ it follows that $\sigma[S^\partial > (6M)^{13}$, and (X,Y) is representable by a Ψ_1^M-triple. Thus I is proved.

PROOF OF II. Choose some triple in Γ'. There are nine possible cases.

1) A $G_{\mathfrak{A}}$-triple. Then the assertion follows from 5.9 and the construction.

2) A $\Psi'_{\mathfrak{A},M}$-triple. Then the assertion is true by construction.

3) A $\Psi''_{\mathfrak{A},M}$-triple. The assertion is true by construction.

4) A $V_{\mathfrak{A}}$-triple. The assertion follows from the fact that if a $V_{\mathfrak{A}}$-triple $[\alpha(z,\lambda),\alpha(z,\mu),\mathscr{D}]$ satisfies the equation $\Phi=\Psi$ for some λ, μ and Z, then (see the condition $\neg\tau(z,M)$ in α and Lemma 3 of §11) in the alphabet $И\cup\{z\}$ we have the graphical equality

$$\Phi((zP_1)^\lambda z,(zP_1)^\mu z)=\Psi((zP_1)^\lambda z,(zP_1)^\mu z).$$

5) A V_1^M-triple. If for some λ and Y the triple $[\alpha(y,\lambda),y,0=0]$, where α is a V_1^M-term, satisfies the equation $\Phi=\Psi$, then from the condition $\omega(Y,M)$ and Lemma 3 of §11 we have in the alphabet $И\cup\{y\}$ the graphical equality $(A_1,\cdots,A_k$ are coefficients)

(3) $$\Phi(([yA_i]_{i=1}^k)^\lambda[yA_i]_{i=1}^r\, y,\, y)=\Psi\,(([yA_i]_{i=1}^k)^\lambda\,[yA_i]_{i=1}^r,\, y).$$

If we view (3) as an exponential equation in the alphabet $И\cup\{y\}$ in the one unknown λ, then for the height h and characteristic H of this equation we have $h\leqq M$ and $H\leqq 2M$. Now the assertion follows from 2.7.

6) A $\widetilde{V}_1^M$-triple. This is handled in the same way as a V_1^M-triple.

7) A Θ_1^M-triple. The assertion is a consequence of the following lemma.

Lemma. *We define parametric words A, B, T_1 and T_2. Put*

$$T_2=(S^\sigma C)^{\mu+1}S^\sigma R,\ A=[T_2Q_i]_{i=1}^k,\ B=(T_2Q_i]_{i=1}^r(S^\sigma C)^{\mu+\varepsilon}S^\sigma R,\ T_1=A^\lambda B,$$

where σ, ν, S, C, R, Q_i, k and r are the same as in the definition of a Θ_1^M-term, $\mu=\nu-1$, and ϵ is 0 or 2. Consider the exponential equation

(1) $$\Phi(T_1,T_2)=\Psi(T_1,T_2).$$

If (1) *has a solution (σ,μ,λ) such that $\lambda\geqq M$, $\mu\geqq 4$ and $\partial(S^\sigma C)>(6M)^3$, then* (1) *is satisfied for any σ, μ and λ such that $\lambda\geqq M$, $\mu\geqq 4$ and $\partial(S^\sigma C)>(6M)^3$.*

Proof. Denote equation (1) by $\mathfrak{A}(\sigma,\mu,\lambda)$. Fix $\bar\sigma$ and $\bar\mu$ such that $\partial(S^\sigma C)>(6M)^3$, $\bar\mu\geqq 4$, and consider the exponential equation in one unknown $\mathfrak{A}(\bar\sigma,\bar\mu,\lambda)$. Since the equation $\Phi=\Psi$ has characteristic M, for the characteristic G of $\mathfrak{A}(\bar\sigma,\bar\mu,\lambda)$ we obtain $G\leqq M-1$. Hence, by 2.7, if $\mathfrak{A}(\bar\sigma,\bar\mu,\lambda)$ is satisfied for some $\lambda\geqq M$, then it is satisfied for any $\lambda\geqq 0$. Because of the arbitrariness of $\bar\sigma$ and $\bar\mu$, we conclude that the lemma will be proved if we prove that the equation in two unknowns $\mathfrak{A}(\sigma,\mu,M)$ is satisfied for any σ and μ such that $\partial(S^\sigma C)>(6M)^3$ and $\mu\geqq 4$. Let us prove this. Fix $\bar\sigma$ such that $[S^{\bar\sigma}C^\partial>(6M)^3$, and consider the equation in one unknown $\mathfrak{A}(\bar\sigma,\mu,M)$. For the characteristic G of this equation, by an easy calculation, we have $G\leqq 3$. Hence, by 2.7, if $\mathfrak{A}(\bar\sigma,\mu,M)$ is satisfied for some $\mu\geqq 4$, then it is satisfied for any $\mu\geqq 0$. Because of the arbitrariness of $\bar\sigma$, the assertion concerning $\mathfrak{A}(\sigma,\mu,M)$ will be proved if we prove that the equation $\mathfrak{A}(\sigma,4,M)$ is satisfied for any σ such that $\partial(S^\sigma C)>(6M)^3$. Let us prove this. We estimate the characteristic G of the equation $\mathfrak{A}(\sigma,4,M)$. By an easy calculation,

$$G\leqq((6M)^3-90M)/[S^\partial-1,$$

although this estimate is excessive. If we now assume that the triple $(\tilde\sigma,\tilde\mu,\tilde\lambda)$ implied by the hypothesis of the lemma satisfies $\mathfrak{A}(\sigma,\mu,\lambda)$, then, by our argument, the triple $(\tilde{\sigma}, 4, M)$ also satisfies $\mathfrak{A}(\sigma, \mu, \lambda)$, so that $\tilde{\sigma}$ satisfies $\mathfrak{A}(\sigma, 4, M)$. Now from $\tilde{\sigma} > ((6M)^3 - [C^\partial)/[S^\partial \geq G+1$ and 2.7 it follows that $\mathfrak{A}(\sigma, 4, M)$ is satisfied for any σ such that $[S^\sigma C^\partial > (6M)^3$, as required. The lemma is proved.

8) A Θ_2^M-triple. The assertion is proved as in case 7).

9) A Ψ_1^M-triple. The assertion is a consequence of the following lemma.

LEMMA. *We define parametric words T, B, T_2 and T_1. Put $T = [S^\sigma A_i]_{i=1}^n$, $B = [S^\sigma A_i]_{i=1}^r S^\sigma A$, $T_2 = [S^\sigma B_i]_{i=1}^m$, $T_1 = T^\lambda B$, where σ, r, n, S, A and A_i are the same as in the definition of a $\tilde{\Psi}_1^M$-term, $1 \leq m \leq 4M^2$ and $[B_1^\partial, \cdots, [B_m^\partial \leq (6M)^7$. Consider the equation $\Phi(T_1, T_2) = \Psi(T_1, T_2)$. If this equation has a solution (σ, λ) such that $\lambda \geq 9M^5$ and $\sigma[S^\partial \geq (6M)^{13}$, then it is satisfied for any such σ and λ.*

The proof is analogous to that of the lemma in case 7). Thus II is proved.

The desired list Γ is obtained if we discard from Γ' those Ψ_1^M-, Θ_1^M-, Θ_2^M-, V_1^M- and $\widetilde{V}_1^M$-triples which do not satisfy $\mathfrak{A}$ on the first occurring admissible set. This can be found because of the admissibility of the triples indicated. The theorem is proved.

The description of the general solution contained in Theorem 12.1 may seem cumbersome. This description can be formulated in a simpler way. The proof of Theorem 12.1 shows how this can be done. For example, we can remove the "bounding" predicates (the right-hand sides of the predicate terms), leaving only the predicates relating x and y (the right components of the triples); however, the bounding predicates are needed for §§13–18. Furthermore, we can show that the general solution can be written in terms of only one special triple (or two in the case of the equation $xAy = yAx$). However in principle the description of the general solution contained in Theorem 12.1 cannot be improved, in the sense that none of the seven special predicate terms can be discarded in the description of the general solution. Indeed, consider the equations

$$(1)\qquad xQ_{r+1}[yQ_i]_{i=r+2}^k [yQ_i]_{i=1}^r y = [yQ_i]_{i=1}^k x,$$

$$(2)\qquad \Phi S y = \Psi y S,$$

$$(3)\qquad xAy = yAx,$$

$$(4)\qquad x[CS^{a_i}y]_{i=r+1}^k [CS^{a_i}y]_{i=1}^r = [yS^{a_i}C]_{i=1}^k x$$

$$(5)\qquad x[S^{a_i}Cy]_{i=r+1}^k [S^{a_i}Cy]_{i=1}^r = [yCS^{a_i}]_{i=1}^k x$$

$$(6)\qquad x\,PQ\,Ry = yR\,QPx, \qquad (PQ\,R \neq PQP),$$

where $k > 1$, $0 \leq r < k$, $Q_1, \cdots, Q_k$, S, A, C, P, Q, R are words in the alphabet $\mathcal{U}$, $a_1, \cdots, a_k > 0$, and $\Phi = \Psi$ is equation (1). It can be verified directly that (for brevity, identically true predicates in the representation of predicate terms have been omitted; σ, λ, μ and ν denote natural variables, y and v word variables, and h, g and f processional variables):

1) a solution of (1) is the triple

$$x = ([yQ_i]_{i=1}^k)^\lambda\ [yQ_i]_{i=1}^r y, \qquad y = y,\ 0 = 0;$$

2) a solution of (2) is the triple

$$x = ([S^\sigma Q_i]_{i=1}^k)^\lambda\ [S^\sigma Q_i]_{i=1}^r\ S^\sigma, \quad y = S^\sigma,\ 0 = 0;$$

3) solutions of (3) are the triples

$$[x=(vA)^{\lambda}v,\quad y=(vA)^{\mu}v,\quad 0=0],\quad [x=A^{\lambda},\quad y=A^{\mu},\quad 0=0];$$

4) a solution of (4) is the triple

$$x=([(S^{\sigma}C)^{\nu}S^{\sigma}Q_i]_{i=1}^{k})^{\lambda}[(S^{\sigma}C)^{\nu}S^{\sigma}Q_i]_{i=1}^{r}(S^{\sigma}C)^{\nu-1}S^{\sigma},\qquad y=(S^{\sigma}C)^{\nu}S^{\sigma},\ \nu>0,$$

where $Q_i=S^{a_i}C$ $(i=1,\cdots,k)$;

5) a solution of (5) is the triple

$$x=([(S^{\sigma}C)^{\nu}S^{\sigma}Q_i]_{i=1}^{k})^{\lambda}[(S^{\sigma}C)^{\nu}S^{\sigma}Q_i]_{i=1}^{r}(S^{\sigma}C)^{\nu+1}S^{\sigma},\qquad y=(S^{\sigma}C)^{\nu}S^{\sigma},\qquad 0=0,$$

where $Q_i=CS^{a_i}$ $(i=1,\cdots,k)$;

6) by 5.9 solutions of (6) are the triples

$$[x=G(QP,QR,h)Q,\quad y=G(QP,QR,g)Q,\qquad \mathcal{A}(h,g)],$$
$$[x=G(QR,QP,f)Q,\quad y=G(QP,QR,g)\,Q,\qquad \mathcal{A}\,(f,g)],$$

where h is an even processional unknown and f an odd processional unknown.

In the description of the general solution of (1) there will appear V_1^M-terms; in that of (2), $\tilde{\Psi}_1^M$- and Ψ_2^M-terms; in that of (3), $\widetilde{V}_1^M$-terms; in that of (6), G^M-terms; in those of (4) and (5), Θ_1^M-, Θ_2^M- and $\bar{\Psi}_1^M$-terms. It can be proved that the general solution of each of the equations (1) – (6) cannot be described by means of the other terms, i.e. the general solution of (1) cannot be described by means of $\widetilde{V}_1^M$-, Ψ_1^M-, Ψ_2^M- Θ_1^M-, Θ_2^M- and G^M-terms. Analogous statements hold for the other equations.

We will now give another formulation of Theorem 12.1 which is more convenient from the point of view of the concluding §20. For this we need the notion of equivalence in a region.

DEFINITION. Suppose $(\xi_1,\cdots,\xi_n)$, $(\eta_1,\cdots,\eta_m)$ and $(\zeta_1,\cdots,\zeta_k)$ are lists of unknowns (possible empty), where none of the letters $\xi_1,\cdots,\xi_n$ occur in the list $(\eta_1,\cdots,\eta_m,\zeta_1,\cdots,\zeta_k)$, and suppose

$$\mathscr{P}_1(\xi_1,\cdots,\xi_n,\eta_1,\cdots,\eta_m),\qquad \mathscr{P}_2(\xi_1,\cdots,\xi_n,\zeta_1,\cdots,\zeta_k),\qquad \mathscr{R}(\xi_1,\cdots,\xi_n)$$

are decidable predicates. We say that $\mathscr{P}_1$ is equivalent in the region $\mathscr{R}(\xi_1,\cdots,\xi_n)$ to $\mathscr{P}_2$ if we have the equivalence

$$\exists\,\eta_1\ldots\eta_m\,(\mathscr{P}_1\,\&\,\mathscr{R})\Leftrightarrow\exists\,\zeta_1\ldots\zeta_k\,(\mathscr{P}_2\,\&\,\mathscr{R}).$$

We say that $\mathscr{P}_1$ is equivalent to $\mathscr{P}_2$ with respect to the unknowns $\xi_1,\cdots,\xi_n$ if we have the equivalence

$$\exists\,\eta_1\ldots\eta_m\,\mathscr{P}_1\Leftrightarrow\exists\,\zeta_1\ldots\zeta_k\,\mathscr{P}_2.$$

Note that if predicates are equivalent with respect to the unknowns $\xi_1,\cdots,\xi_n$, then they are equivalent in any region $\mathscr{R}(\xi_1,\cdots,\xi_n)$.

Suppose $\omega=[\alpha,\beta,\pi]$ is a special triple of characteristic M. Let π^* be the predicate defined as follows: if ω is not a $\widetilde{V}_1^M$-triple, then π^* is π, and if ω is a $\widetilde{V}_1^M$-triple, then π^* is the predicate $0=0$.

12.2. THEOREM. *Suppose $\mathfrak{A}(x,y)$ is a normal word equation in the two unknowns x, y of characteristic M. Then one of the following two assertions holds:*

1) *$\mathfrak{A}$ has no $\bar{\epsilon}_M$-solutions;*

2) *$\mathfrak{A}$ is equivalent in the region $\neg\epsilon(x,y,M)$ to the disjunction*

$$\bigvee_{i=1}^{N} (x=\alpha_i,\ y=\beta_i,\ \pi_i),$$

where $N \geqq 1$, *and the* $[\alpha_i, \beta_i, \pi_i]$ *are special triples of characteristic* M.

In this connection: 1) *we have the implications*

$$(x=\alpha_i^*,\ y=\beta_i^*,\ \pi_i^*) \supset \mathfrak{A}(x,y) \qquad (i=1,\cdots,N),$$

where α_i^* *and* β_i^* *are the left-hand sides of the predicate terms* α_i *and* β_i; 2) *if* $[\alpha_i, \beta_i, \pi_i]$ *is a* V_1^M*-triple, then a word unknown occurring in* α_i *is one of the letters* x *or* y; 3) *if one of the terms* α_i *or* β_i *is a* $\tilde{\Psi}_1^M$*-term, then the other is a* Ψ_2^M*-term, and* α_i *and* β_i *have the same base.*

PROOF. The theorem is a reformulation of Theorem 12.1 using the terminology of equivalence in a region. The supplementary statements 1) – 3) follow from the proof of 12.1.

Theorem 12.2, in the following way, reduces W_2-systems to T-systems (see the beginning of this chapter). Suppose there is given a W_2-system

$$(*) \qquad \& \ \mathfrak{A}_{ij}(x_i, x_j),$$

where the $\mathfrak{A}_{ij}$ are normal equations of characteristic M, and the conjunction & extends over some set of pairs (i,j) such that $1 \leqq i < j \leqq n$. In the region where $\epsilon(x_i, x_j, M)$ holds for at least one pair (i,j) the unknowns x_i and x_j are expressed by means of a finite number of formulas of the form $x_i = \alpha(v)$ and $x_j = \beta(v)$, where α and β are words in the alphabet $H \cup \{v\}$. This allows us to lower the number of unknowns in the system (*). In the region $\Omega = \&_{i,j} \neg \epsilon(x_i, x_j, M)$ the solution of each equation $\mathfrak{A}_{ij}$ is, by 12.2, representable in the form

$$\bigvee_{k=1}^{N_{ij}} [x_i = \alpha_{ijk},\ x_j = \beta_{ijk},\ \pi_{ijk}],$$

where $N_{ij} \geqq 1$, and the $[\alpha_{ijk}, \beta_{ijk}, \pi_{ijk}]$ are special triples of characteristic M. Using the well-known permutability of the symbols & and $\vee$ (see [**5**]), we obtain that the system (*) is equivalent in Ω to a disjunction of a finite number of T-systems.

To solve T-systems it is necessary to solve equations of the form $\alpha = \beta$, where α and β are special predicate terms of characteristic M. We consider these equations in §§13 – 17. For convenience, the special predicate terms are divided into three groups: V^M-, Σ^M- and G^M-terms. The V^M-terms are the V_1^M- and $\tilde{V}_1^M$-terms; the Σ^M-terms are the Θ_1^M-, Θ_2^M-, Ψ_1^M- and Ψ_2^M-terms. If α and β are symbols in the list Σ^M, V^M, G^M, then we say that an equation has type $\alpha\beta$ if one of its sides is an α-term, the other a β-term. In §13 we consider equations of type $\Sigma^M\Sigma^M$, in §14 equations of type $\Sigma^M V^M$, in §15 equations of type $V^M V^M$, in §16 equations of types $\Sigma^M G^M$ and $V^M G^M$, and, finally, in §17 equations of type $G^M G^M$. In §18 we consider T-systems.

"Terms" do not appear in explicit form in what is to follow. In accordance with this and a number of abbreviations we will say simply "term" in place of "predicate term" if no ambiguity is possible.

Comparing the methods of solving equations of the various types, we note that in solving equations of type $\Sigma^M\Sigma^M$ a fundamental role is played by the exponential equations of §2. In solving equations of type $V^M V^M$ we rely on the results of §11 pertaining to the equation $\varphi(x) = f(y)$. In solving equations containing G^M-terms essential roles are played both by exponential equations and the more subtle properties of the functions F and G, in particular, the δ-representation of these functions of §9.

§13. Equations of type ΣΣ

In this section we consider equations of type ΣΣ. We will prove that any equation of type ΣΣ is equivalent to some LDR.

Before proving this, we will go into some of the details characterizing the system of special predicate terms. These details will help to avoid repetition in the arguments to follow and will suggest, to some extent, the method of solving the equations.

Consider predicate terms α and β with the same right-hand side

$$\alpha = [T_1(\xi_1, \ldots, \xi_n);\ \mathscr{P}(\xi_1, \ldots, \xi_n)];\quad \beta = [T_2(\xi_1, \ldots, \xi_n);\ \mathscr{P}].$$

The pair (α, β) is called normal if for any set $(\bar{\xi}_1, \cdots, \bar{\xi}_n)$ of values of the unknowns which satisfies $\mathscr{P}$ the word $T_1(\bar{\xi}_1, \cdots, \bar{\xi}_n)$ is simple and $\Lambda \neq T_2(\bar{\xi}_1, \cdots, \bar{\xi}_n) \angle T_1(\bar{\xi}_1, \cdots, \bar{\xi}_n)$. If λ is a positive unknown and a and N are natural numbers, we denote by $\alpha^{\lambda+a}\beta$ the predicate term $[T_1^{\lambda+a}T_2;\ \mathscr{P}]$, by $\alpha^\lambda\beta \,\&\, \lambda \geqq N$ the predicate term $[T_1^\lambda T_2;\ \mathscr{P} \,\&\, (\lambda \geqq N)]$, and by $\alpha^a\beta$ the predicate term $[T_1^a T_2;\ \mathscr{P}]$. We write $\alpha \gg N$ if $[T_1(\bar{\xi}_1, \cdots, \bar{\xi}_n)^\partial > N$ for any set $(\bar{\xi}_1, \cdots, \bar{\xi}_n)$ satisfying $\mathscr{P}$. The predicate terms α and β are called equivalent if for any set $(\bar{\xi}_1, \cdots, \bar{\xi}_n)$ which satisfies $\mathscr{P}$ we have in the alphabet И the equality $T_1(\bar{\xi}_1, \cdots, \bar{\xi}_n) = T_2(\bar{\xi}_1, \cdots, \bar{\xi}_n)$.

Now consider the special predicate terms. For the time being we put the G^M-terms aside. Then we have terms with lower index 1, the $\widetilde{V}_1^M$-, V_1^M-, Θ_1^M- and Ψ_1^M-terms, and terms with lower index 2, the Θ_2^M- and Ψ_2^M-terms. It turns out (and we aim at this in selecting the right-hand side $\mathscr{P}$) that the terms with index 1 have the form (or are equivalent to terms of the form) $\alpha^{\lambda+a}\beta \,\&\, \lambda \geqq N$, where λ is a positive unknown, a and N are natural numbers with $N > 1$, and (α, β) is a normal pair of predicate terms. The following proposition is used in solving equations whose left- and right-hand sides are terms of such a form.

13.1. *Consider the system*

(1) $$\alpha^\lambda\alpha_1 = \beta^\mu\beta_1, \qquad \lambda \geqq N_1, \qquad \mu \geqq N_2,$$

where λ and μ are positive unknowns, $N_1, N_2 > 1$, and (α, α_1) and (β, β_1) are normal pairs of predicate terms. The system (1) *is equivalent to the system*

(2) $$\alpha^2\alpha_1 = \beta^2\beta_1, \qquad \lambda = \mu, \qquad \lambda \geqq N_1, \qquad \mu \geqq N_2.$$

PROOF. We write (1), (2) in expanded form. We obtain the systems

(1′) $$T^\lambda T_1 = S^\mu S_1, \qquad \mathscr{P} \,\&\, \mathscr{Q}, \qquad \lambda \geqq N_1, \qquad \mu \geqq N_2,$$

(2′) $$T^2 T_1 = S^2 S_1, \qquad \mathscr{P} \,\&\, \mathscr{Q}, \qquad \lambda \geqq N_1, \qquad \mu \geqq N_2, \qquad \lambda = \mu,$$

where $T^\lambda T_1$ and $S^\mu S_1$ are the left-hand sides and $\mathscr{P}$ and $\mathscr{Q}$ the right-hand sides of the predicate terms $\alpha^\lambda\alpha_1$ and $\beta^\mu\beta_1$. Suppose the unknowns occurring in $\alpha^\lambda\alpha_1$ and $\beta^\mu\beta_1$ comprise the list $(\xi_1, \cdots, \xi_n)$. If the set $(\bar{\xi}_1, \cdots, \bar{\xi}_n)$ satisfies the predicates $\mathscr{P}$ and $\mathscr{Q}$, then by the normality of the pairs (α, α_1) and

(β,β_1) the words $\overline{T}$ and $\overline{S}$ are simple, $\Lambda \neq \overline{T}_1 \angle \overline{T}$, and $\Lambda \neq \overline{S}_1 \angle \overline{S}$ (a bar over a term denotes the value of the term on the set $(\bar{\xi}_1,\cdots,\bar{\xi}_n)$). If now $(\bar{\xi}_1,\cdots,\bar{\xi}_n)$ is a solution of the system (1′), then $\overline{T}^{\bar{\lambda}}\overline{T}_1 = \overline{S}^{\bar{\mu}}\overline{S}_1$, whence, by 1.7, $\bar{\lambda}=\bar{\mu}$, $\overline{T}=\overline{S}$, $\overline{T}_1=\overline{S}_1$ and the set $(\bar{\xi}_1,\cdots,\bar{\xi}_n)$ satisfies (2′). If, conversely, $(\bar{\xi}_1,\cdots,\bar{\xi}_n)$ satisfies (2′), then $\bar{\lambda}=\bar{\mu}$ and $\overline{T}^2\overline{T}_1=\overline{S}^2\overline{S}_1$. From this last equality, by 1.7, we have $\overline{S}=\overline{T}$ and $\overline{S}_1=\overline{T}_1$, and since $\bar{\lambda}=\bar{\mu}$, also $\overline{T}^{\bar{\lambda}}\overline{T}_1=\overline{S}^{\bar{\mu}}\overline{S}_1$ and $(\bar{\xi}_1,\cdots,\bar{\xi}_n)$ satisfies (1′). The proposition is proved.

In solving equations of type ΣΣ it is convenient to introduce $\overline{\Psi}_2^M$- and $\overline{\Theta}_2^M$-terms.

A $\overline{\Theta}_2^M$-term is defined just like a Θ_2^M-term, only without the restriction $m \leq M^2$ (the restriction $m>1$ is retained). A $\overline{\Psi}_2^M$-term is defined just like a Ψ_2^M-term, only without the restrictions on m and the lengths of the coefficients A_i. The restriction $[S^\partial \leq (6M)^3$ is retained. The restriction on σ has the form $\sigma[S^\partial \geq N$, where N is a natural number. We allow the case where the left-hand side of the $\overline{\Psi}_2^M$-term is a word in the alphabet И.

Obviously, any Θ_2^M-term is a $\overline{\Theta}_2^M$-term, and any Ψ_2^M-term is a $\overline{\Psi}_2^M$-term.

13.2. *Any Θ_1^M-term is equivalent to a predicate term of the form $\alpha^{\lambda+a}\beta$, $\lambda \geq M$, where λ is a positive unknown, a is a natural number, and (α,β) is a normal pair of predicate terms such that* 1) *$\alpha \gg (6M)^3$, and* 2) *for any $c>0$ the predicate term $\alpha^c\beta$ is equivalent to some $\overline{\Theta}_2^M$-term or, if $c=2$, to a Θ_2^M-term.*

PROOF. Suppose there is given a Θ_1^M-term of §12. Denote it by γ. The relations below which are part of the definition of a Θ_1^M-term are used without specific references. Since $\epsilon=\pm 1$, there are two possible cases.

Case 1. $\epsilon=-1$. We define predicate terms α and β:

$$\alpha = [[(S^\sigma C)^\nu S^\sigma RQ_i]_{i=1}^k; \qquad \nu \geqslant 4M^2, \quad \partial(S^\sigma C) > (6M)^3]. \tag{1}$$

$$\beta = [[(S^\sigma C)^\nu S^\sigma RQ_i]_{i=1}^r (S^\sigma C)^{\nu+\epsilon} S^\sigma R; \qquad \nu \geqslant 4M^2, \quad \partial(S^\sigma C) > (6M)^3]. \tag{2}$$

Then the predicate term γ has the form $\alpha^{\lambda+a}\beta$, $\lambda \geq M$, with $a=0$. For any $c>0$ the predicate term $\alpha^c\beta$, as is easily seen, is a $\overline{\Theta}_2^M$-term (here $m = ck+r+1$). If $c=2$, then $m \leq 3M \leq M^2$, so that $\alpha^2\beta$ is a Θ_2^M-term. Thus 2) is satisfied. It remains to prove 1) and the normality of (α,β).

Let $\alpha'(\sigma,\nu)$ and $\beta'(\sigma,\nu)$ denote the left-hand sides of the predicate terms α and β. Suppose σ and ν are positive integers such that $\nu \geq 4M^2$ and $[S^\sigma C^\partial > (6M)^3$. Since $k>0$ and $[S^\sigma C^\partial > (6M)^3$, we have $[\alpha'(\sigma,\nu)^\partial > (6M)^3$, so that 1) holds. It is obvious that $\beta'(\sigma,\nu) \neq \Lambda$. Since $[R^\partial \leq 90M$, $R \angle CS^{[R^\partial}$ and $[S^\sigma C^\partial < 90M$, we have $R \angle CS^\sigma$. Hence $(S^\sigma C)^{\nu+\epsilon}S^\sigma R \angle (S^\sigma C)^\nu S^\sigma$, since $\epsilon=-1$. Now, since $r<k$, we conclude from (1) and (2) that $\beta'(\sigma,\nu) \angle \alpha'(\sigma,\nu)$. Thus $\Lambda \neq \beta'(\sigma,\nu) \angle \alpha'(\sigma,\nu)$. It remains to prove that $\alpha'(\sigma,\nu)$ is a simple word. If $k>1$, then $[Q_1^\partial,\cdots,[Q_k^\partial \leq M$ and the simplicity of $\alpha'(\sigma,\nu)$ follows from 11.4. If $k=1$, then $\alpha'(\sigma,\nu)$ is simple by 1.13 (since $RQ_1 \neq C$). This proves the normality of (α,β). Our assertion is proved in the case $\epsilon=-1$.

Case 2. $\epsilon=1$. Then there exist an integer d and word B such that $B \angle C$, $RQ_{r+1} \amalg CS^d$ and $S^dB \amalg R$. Suppose σ and ν are the same as in Case 1. We have $RQ_{r+1}S^\sigma B \amalg CS^{\sigma+d}B \amalg CS^\sigma R$. Since $RQ_{r+1}S^\sigma B$ and $CS^\sigma R$ are words in the alphabet И, we have $RQ_{r+1}S^\sigma B = CS^\sigma R$. Consequently,

$$[(S^\sigma C)^\nu S^\sigma RQ_i]_{i=1}^r (S^\sigma C)^{\nu+\epsilon} S^\sigma R = [(S^\sigma C)^\nu S^\sigma RQ_i]_{i=1}^{r+1} S^\sigma B. \tag{3}$$

Since $r < k$, there are two possible subcases.

2.1) $r < k-1$. Then, defining predicate terms α and β as in (1) and (2) and arguing as in Case 1, we see that our assertion is true; only the proof of the relation $\beta'(\sigma,\nu) \angle \alpha'(\sigma,\nu)$ changes. Now it follows from (3), since $B \angle C$.

2.2) $r = k-1$. We define predicate terms α, β_1 and β. The term α is defined as in (1), β_1 like β in (2); β is defined as follows:

$$\beta = [S^\sigma B;\ \nu \geqq 4M^2;\ \partial(S^\sigma C) > (6M)^3]. \tag{4}$$

Using the relation $B \angle C$, we see that (α,β) is normal. Also, it follows from (1)–(4) that the term β_1 is equivalent to the term $\alpha\beta$. The Θ_1^M-term γ has the form $\alpha^\lambda\beta_1 \& \lambda \geqq N$ and is equivalent to the term $\alpha^{\lambda+1}\beta \& \lambda \geqq M$. Furthermore, for any $c > 0$ the term $\alpha^c\beta$ is equivalent to the term $\alpha^{c-1}\beta_1$, which is a $\overline{\Theta}_2^M$-term, and for $c = 2$ a Θ_2^M-term. The pair (α,β) is the desired one. The proposition is proved.

13.3. *Any Ψ_1^M-term is a predicate term of the form $\alpha^{\lambda+a}\beta \& \lambda \geqq N$, where λ is a positive unknown, $a \geqq 0$, $N \geqq 4M^2$, and (α,β) is a normal pair of predicate terms such that* 1) $\alpha \gg (6M)^3$, *and* 2) *for any $c > 0$ the term $\alpha^c\beta$ is a $\overline{\Psi}_2^M$-term.*

PROOF. Suppose there is given a Ψ_1^M-term γ. It is either a $\tilde{\Psi}_1^M$-term or a $\overline{\Psi}_1^M$-term. In both cases γ has the form

$$[([S^\sigma A_i]_{i=1}^n)^\lambda [S^\sigma A_i]_{i=1}^r S^\sigma A;\ \lambda \geqq N,\ \sigma[S^\partial \geqq N_1],$$

where $0 \leqq r < n$, $1 \leqq n \leqq 4M^2$, $N \geqq 4M^2$, S is a simple word, $A \angle A_{r+1}S^{[A^\partial}$ and $\oplus\,(S, A_1, \cdots, A_n, N)$. If γ is a $\tilde{\Psi}_1^M$-term, then $n = 1$, $r = 0$, $R = A$, $C = A_1$ and $N_1 = (6M)^3$. In this case the condition $\oplus$ follows from $\mathscr{C}(S,C)$ by 1.10. For a $\overline{\Psi}_1^M$-term, $N_1 = (6M)^{13}$. Put

$$\alpha = [[S^\sigma A_i]_{i=1}^n;\ \sigma[S^\partial \geqq N_1].$$

We have $\alpha \gg (6M)^3$. The simplicity of α follows from the condition $\oplus$. We consider two cases.

1. $A \angle A_{r+1}$. Then, putting

$$\beta = [[S^\sigma A_i]_{i=1}^r S^\sigma A;\ \sigma[S^\partial \geqq N_1], \tag{1}$$

we see that our assertion is true.

2. $[A^\partial > [A_{r+1}^\partial$. Then $A = A_{r+1}B$, where $\Lambda \neq B \angle S^{[A^\partial}$. Since $[B^\partial \leqq [A^\partial \leqq \sigma[S^\partial$, we have $B \angle S^\sigma$ (for admissible σ). If $r < n-1$, then, defining β by (1), we see that our assertion is true. If $r = n-1$, then put $\beta = [B$ and $\sigma[S^\partial \geqq N_1]$. Then γ has the form $(\alpha^{\lambda+1}\beta)$, $\lambda \geqq N$, and our assertion holds with $a = 1$. The proposition is proved.

We now turn to equations of type $\Sigma\Sigma$. Here we have equations of type $\alpha_i\beta_j$, where α and β are symbols from the list Ψ^M, Θ^M, and the indices i and j assume the values 1 and 2. Propositions 13.1–13.3 allow us to reduce equations of type $\alpha_1\beta_1$ to equations of type $\overline{\alpha}_2\overline{\beta}_2$. The solution of equations of type $\overline{\alpha}_2\overline{\beta}_2$ is described in 13.6 and 13.7. There remains the type $\alpha_1\beta_2$. Equations of this type reduce to equations of type $\overline{\alpha}_2\beta_2$ by means of the following two propositions.

13.4. *Suppose γ is a Ψ_2^M-term, λ is a positive unknown, and (α,β) is a normal pair of predicate terms, where $\alpha \gg (6M)^3$. Then the system $(\alpha^\lambda\beta=\gamma)\ \&\ \lambda \geqq 9M^2$ has no solutions.*

PROOF. Expressing the Ψ_2^M-term γ in more detail, we obtain a system

$$(1)\qquad [S^tP_i]_{i=1}^n=\alpha^\lambda\beta,\quad \lambda\geqq 9M^2,\quad t[S^\partial>(6M)^{13},$$

where t is a positive unknown, n is an integer, $S, P_1,\cdots,P_n$ are words in the alphabet И, S is a simple word, $[S^\partial\leqq(6M)^3$, $[P_1^\partial,\cdots,[P_n^\partial\leqq(6M)^7$ and $0<n\leqq 4M^2$. Assume that (1) has a solution. This means, in view of the normality of (α,β) and the condition $\alpha\gg(6M)^3$, that there exist integers λ and t, a simple word A and a word B such that

$$(2)\qquad [S^tP_i]_{i=1}^n=A^\lambda B,\quad \lambda\geqq 9M^2,\quad t[S^\partial>(6M)^{13},$$

where $[A^\partial>(6M)^3$. Since $[S^\partial\leqq(6M)^3$, we have $[S^\partial$. If $[A^\partial<(t-1)[S^\partial$, then from (2) and 1.5 it follows that $A=S$, which contradicts $[S^\partial<[A^\partial$. This contradiction implies that $(t-1)[S^\partial\leqq[A^\partial$. Then $t[S^\partial\leqq[AS^\partial<2[A^\partial$. We have $\lambda[A^\partial\leqq nt[S^\partial+[P_1\cdots P_n^\partial\leqq 4M^2t[S^\partial+4M^2(6M)^7<(4M^2+1)t[S^\partial\leqq(8M^2+2)[A^\partial<\lambda[A^\partial$, which is impossible. The proposition is proved.

13.5. *Suppose β is a Θ_2^M-term, λ is a positive unknown, and (α,α_1) is a normal pair of predicate terms. Then the system $(\alpha^\lambda\alpha_1=\beta)\ \&\ \lambda>M^2$ has no solutions.*

PROOF. Expressing the Θ_2^M-term β in more detail, we obtain a system $\mathfrak{A}$

$$(1)\qquad [(S^\sigma C)^\nu S^\sigma A_i]_{i=1}^{m-1}(S^\sigma C)^{\nu+\varepsilon}S^\sigma A_m=\alpha^\lambda\alpha_1,$$

$$(2)\qquad \nu\geqslant 4M^2,\qquad \partial(S^\sigma C)>(6M)^3,\qquad \lambda>M^2,$$

where σ and ν are unknowns, $\epsilon=\pm1$, $1<m\leqq M^2$, S, C and A_i are words in the alphabet И, S is a simple word, $\mathscr{L}(S,C)$, $[S^\partial\leqq 30M$, $[C^\partial\leqq 90M$, $[A_1^\partial,\cdots,[A_m^\partial\leqq 90M$, and $\urcorner(A_1=\cdots=A_{m-1}=C)$. We prove the assertion by contradiction.

Suppose the system $\mathfrak{A}$ has a solution. Then, considering the normality of (α,α_1) and avoiding repetition of notation, we have in the alphabet И the equality (1), where σ, ν and λ are numbers, α and α_1 are words in И, α is simple, $\alpha_1\angle\alpha$ and the inequalities (2) hold. Comparing the lengths of the left- and right-hand sides of (1) and using the estimates for m, λ, σ, ν, $[S^\partial$, $[C^\partial$ and $[A_i^\partial$, we obtain

$$(3)\qquad (\sigma-1)[S^\partial>\max\{[C^\partial,[A_1^\partial,\ldots,[A_m^\partial\},$$

$$(4)\qquad [S^\sigma C^\partial\leqslant(\lambda-1)[\alpha^\partial,$$

$$(5)\qquad [\alpha^\partial\leqslant(\nu-1)[S^\sigma C^\partial.$$

From the condition $\mathscr{L}(S,C)$ and the inequality (3), it follows by 1.10 that $S^\sigma C$ is a simple word. From (1), (4), (5) and 1.6 we have $\alpha=S^\sigma C$. The equality (1) assumes the form

$$(6)\qquad [(S^\sigma C)^\nu S^\sigma A_i]_{i=1}^{m-1}(S^\sigma C)^{\nu+\varepsilon}S^\sigma A_m=(S^\sigma C)^\lambda\alpha_1,$$

where $\alpha_1\angle S^\sigma C$. We will now prove by induction on m that $A_1=\cdots=A_{m-1}=C$.

Since $\epsilon=\pm1$, $\nu>3$ and $\alpha_1\angle S^\sigma C$, (6) implies that $\lambda>\nu+2$. Cancelling

(6) on the left by $(S^\sigma C)^\nu S^\sigma$, we obtain

(7) $$A_1\,[(S^\sigma C)^\nu S^\sigma A_i]_{i=2}^{m-1}\,(S^\sigma C)^{\nu+\varepsilon}\,S^\sigma A_m = C\,(S^\sigma C)^{\lambda-\nu-1}\,\alpha_1.$$

Since ν, $\nu+\epsilon$, $\lambda-\nu-1>1$, the left-hand side of (7) begins with the word $A_1S^\sigma CS^\sigma$, and the right-hand side with $CS^\sigma CS^\sigma$. Hence from (7) and 1.15 it follows that $A_1 = C$. If $m=2$, the assertion is proved. Suppose $m>2$. Since $A_1=C$, on cancelling (7) on the left by A_1, we obtain

(8) $$[(S^\sigma C)^\nu S^\sigma A_i]_{i=2}^{m-1}\,(S^\sigma C)^{\nu+\varepsilon}\,S^\sigma A_m = (S^\sigma C)^{\lambda-\nu-1}\,\alpha_1.$$

Applying the induction assumption to (8), we obtain $A_2=\cdots=A_{m-1}=C=A_1$, which contradicts the relation $\neg(A_1=\cdots=A_{m-1}=C)$. The proposition is proved.

13.6. *Suppose $n>1$, $m>1$, $M>1$, and ϵ and δ are integers. Consider the system*

(1) $$\begin{gathered}[(S^\sigma C)^\nu S^\sigma A_i]_{i=1}^{n-1}\,(S^\sigma C)^{\nu+\varepsilon}\,S^\sigma A_n = [(T^\tau D)^\mu T^\tau B_i]_{i=1}^{m-1}\,(T^\tau D)^{\mu+\delta}T^\tau B_m,\\ \nu,\ \mu \geqslant 4M^2;\qquad \partial\,(S^\sigma C),\qquad \partial\,(T^\tau D) > (6M)^3,\end{gathered}$$

where σ, ν, τ and μ are positive unknowns, $\epsilon=\pm1$, $\delta=\pm1$, S, C, T, D, A_i and B_j are words in the alphabet $\mathfrak{A}$, $\mathscr{L}(S,C)$, $\mathscr{L}(T,D)$, S and T are simple words, $[S^\partial, [T^\partial \leqq 30M$, and $[C^\partial$, $[D^\partial$, $[A_1^\partial, \cdots, [A_n^\partial$, $[B_1^\partial, \cdots, [B_m^\partial \leqq 90M$. The system (1) is equivalent to some LDR.

Proof. There are two possible cases: $[D^\partial \geqq [C^\partial$ or $[C^\partial \geqq [D^\partial$. Without loss of generality, we consider the case $[C^\partial \geqq [D^\partial$.

Suppose (σ,ν,τ,μ) is a solution of the system (1). Then, since $n,m>1$ and $\nu,\mu\geqq 3$, we have for some V_1 and V_2

(2) $$(S^\sigma C)^3 V_1 = (T^\tau D)^3 V_2.$$

It is obvious that $\sigma[S^\partial, \tau[T^\partial \geqq 2[SCTD^\partial$. By hypothesis, $\mathscr{L}(S,C)$, $\mathscr{L}(T,D)$, and S and T are simple. Applying 1.14 to (2), we obtain

(3) $$S=T,$$

(4) $$S^\sigma C = T^\tau D.$$

Since by assumption $[C^\partial \geqq [D^\partial$, it follows from (3) and (4) that there exists $a\geqq 0$ such that $\tau=\sigma+a$ and

(5) $$C = S^a D.$$

Thus, if (1) has a solution, (3) and (5) hold. On the other hand, it is obvious that if (3) and (5) hold, (1) is equivalent to the system

(6) $$\begin{gathered}[(S^\sigma C)^\nu S^\sigma A_i]_{i=1}^{n-1}\,(S^\sigma C)^{\nu+\varepsilon}S^\sigma A_n = [(S^\sigma C)^\mu S^\sigma C^a B_i]_{i=1}^{m-1}(S^\sigma C)^{\mu+\delta}S^\sigma S^a B_m,\\ \tau=\sigma+a,\qquad \nu,\mu \geqslant 4M^2,\qquad \partial\,(S^\sigma C) > (6M)^3.\end{gathered}$$

Therefore, either the conjunction of the conditions (3) and (5) does not hold, and then the system (1) has no solutions, or (3) and (5) hold, and then (1) is equivalent to (6). In the first case $\nu=\nu+1$ is the desired LDR.

Let us consider the second case. The upper equation of the system (6) is the equation described in 2.8 (here $\lambda_1=\cdots=\lambda_{n-1}=\nu$, $\lambda_n=\nu+\epsilon$, $\mu_1=\cdots$

$=\mu_{m-1}=\mu$ and $\mu_m=\mu+\delta$). It follows from 2.8 that the system (6), and hence also (1), is equivalent to some LDR. The proposition is proved.

13.7. *Suppose $n>0$, $m>1$ and $M>1$ are integers. Consider the system $\mathfrak{A}$:*

$$[T^tP_i]_{i=1}^n=[(S^\sigma C)^\nu S^\sigma Q_i]_{i=1}^{m-1}(S^\sigma C)^{\nu+\varepsilon}S^\sigma Q_m, \tag{1}$$

$$\partial(S^\sigma C)>(6M)^3,\qquad \nu\geqslant 4M^2, \tag{2}$$

where t, σ and ν are positive unknowns, $\epsilon=\pm1$, S, C, T, P_i and Q_j are words in the alphabet И, *S and T are simple, $\mathscr{B}(S,C)$, and*

$$[T^\partial\leqq(6M)^3,\quad [S^\partial\leqq 30M,\quad [C^\partial\leqq 90M. \tag{3}$$

The system $\mathfrak{A}$ is equivalent to some LDR.

PROOF. For small t the assertion is obvious. We choose $N=3[T^\partial+30M$ and prove the assertion under the additional assumption that $t[T^\partial>N$. Suppose the triple (σ,ν,t) is a solution of $\mathfrak{A}$. Then we have in the alphabet И the equality (1), where the numbers σ and ν satisfy (2), and $t[T^\partial>N$. We will prove that $S=T$. There are three possible cases.

1) $[T^\partial\leqq[S^\partial$. From (2), (3) and $t[T^\partial>N$ we have $(t-1)[T^\partial>[S^\partial$, $t,\sigma>1$. From (1) and 1.5 we obtain $S=T$.

2) $(\sigma-1)[S^\partial\geqq[T^\partial>[S^\partial$. By 1.5 we obtain analogously $S=T$.

3) $(\sigma-1)[S^\partial<[T^\partial$. Using (2), (3), we have $[S^\sigma C^\partial=(\sigma-1)[S^\partial+[SC^\partial<2(\sigma-1)[S^\partial<2[T^\partial\leqq(t-1)[T^\partial$. It follows from (2) and (3) that $[S^\sigma C^\partial>[T^\partial$. Hence $[T^\partial<[S^\sigma C^\partial<(t-1)[T^\partial$. From $\mathscr{B}(S,C)$, $[S^\sigma C^\partial>2[SC^2$ and 1.10 we see that $S^\sigma C$ is simple. Now applying 1.5 to (1), we obtain $S^\sigma C=T$, which contradicts $[T^\partial<[S^\sigma C^\partial$. Thus case 3) is impossible. We have proved that $S=T$.

If now $t[T^\partial>[S^\sigma CS^\partial$, then from (1) we have $S^\sigma CS\angle T^t=S^t$, which by 1.4 contradicts $\mathscr{B}(S,C)$. Consequently, $t[S^\partial<[S^\sigma CS^\partial$, so that $(t-1)[T^\partial<[S^\sigma C^\partial$. Then it follows from (1) that $\nu\leqq n+n[T^\partial+[P_1\cdots P_n^\partial$. We have proved that there exists a constant l ($l=n+n[T^\partial+[P_1\cdots P_n^\partial$) such that for any solution (t,σ,ν) of $\mathfrak{A}$ we must have $\nu\leqq l$. The assertion being proved follows in an obvious way from 2.3 if we recognize that for fixed ν the equation (1) is a one-stage exponential equation. The proposition is proved.

We will now prove that any equation of type $\Sigma^M\Sigma^M$ is equivalent to some LDR.

13.8. *Any equation of type $\Sigma^M\Sigma^M$ is equivalent to some LDR.*

PROOF. The Σ^M-terms are the Ψ_1^M-, Ψ_2^M-, Θ_1^M- and Θ_2^M-terms. An equation of type $\Sigma^M\Sigma^M$ has one of the following ten types: $\Psi_1^M\Psi_1^M$, $\Psi_1^M\Psi_2^M$, $\Psi_2^M\Psi_2^M$, $\Theta_1^M\Theta_1^M$, $\Theta_1^M\Theta_2^M$, $\Theta_2^M\Theta_2^M$, $\Theta_1^M\Psi_1^M$, $\Theta_2^M\Psi_2^M$, $\Theta_1^M\Psi_2^M$, $\Theta_2^M\Psi_1^M$, or, more concisely, the type $\alpha_i\beta_j$, where α and β are symbols in the list Ψ^M, Θ^M, and the indices i and j assume the values 1 and 2. We consider three cases.

Case 1. $i=j=2$. We will prove the more general assertion that any equation of type $\overline{\alpha}_2\overline{\beta}_2$ is equivalent to a LDR (any Ψ_2^M-term (Θ_2^M-term) is a $\overline{\Psi}_2^M$-term ($\overline{\Theta}_2^M$-term)). An equation of type $\overline{\Psi}_2^M\overline{\Psi}_2^M$ in expanded form is a system consisting of a one-stage exponential equation in the two unknowns σ, t and two inequalities of the form $\sigma\geqq N_1$, $t\geqq N_2$. For equations of type $\overline{\Psi}_2^M\overline{\Psi}_2^M$ the assertion follows from 2.3. An equation of type $\overline{\Theta}_2^M\overline{\Theta}_2^M$ in expanded form is the system described in 13.6, and an equation of type $\overline{\Psi}_2^M\overline{\Theta}_2^M$ the system in 13.7.

Case 2. $i=j=1$. Then by 13.2 and 13.3 the term α_1 is equivalent to a predicate term of the form $\gamma^{\lambda+a}\gamma_1$, $\lambda \geqq N_1$, and the term β_1 to a predicate term of the form $\pi^{\mu+b}\pi_1$, $\mu \geqq N_2$, where λ and μ are positive unknowns; a, b, N_1 and N_2 are natural numbers, $N_1 > 1$ and $N_2 > 1$; (γ, γ_1) and (π, π_1) are normal pairs of predicate terms, and for any $c > 1$ the term $\gamma^c\gamma_1$ is equivalent to some $\bar{\alpha}_2$-term, and $\pi^c\pi_1$ to some $\bar{\beta}_2$-term. The equation $\alpha_1 = \beta_1$ has the form $\gamma^{\lambda+a}\gamma_1 = \pi^{\mu+b}\pi_1$, $\lambda \geqq N_1$, $\mu \geqq N_2$, and, by 13.1, is equivalent to the system $(\gamma^2\gamma_1 = \pi^2\pi_1) \,\&\, \mathscr{D}$, where $\mathscr{D}$ is a LDR. The equation $\gamma^2\gamma_1 = \pi^2\pi_1$ is equivalent to an equation of type $\bar{\alpha}_2\bar{\beta}_2$ and, by what was proved in Case 1, is equivalent to some LDR. The assertion is proved in the case $i=j=1$.

Case 3. One of the indices i, j is 1, the other 2. It suffices to consider the case $i=1$, $j=2$. Then the term α_1 is equivalent to a predicate term of the form $\gamma^{\lambda+a}\gamma_1$, $\lambda \geqq N_1$, where γ, γ_1, a, λ and N_1 are the same as in Case 2. The equation $\alpha_1 = \beta_2$ is equivalent to the system

$$\gamma^{\lambda+a}\gamma_1 = \beta_2, \qquad \lambda \geqq N_1. \tag{1}$$

In view of 13.4 and 13.5, the equation $\gamma^{\lambda+a}\gamma_1 = \beta_2$ has no solutions in the region $\lambda \geqq 9M^2$. Hence the system (1) is equivalent to a disjunction of a finite number of equations of the form $\gamma^c\gamma_1 = \beta_2$. The equation $\gamma^c\gamma_1 = \beta_2$ is equivalent to an equation of type $\bar{\alpha}_2\beta_2$, which, by what was proved in Case 1, is equivalent to some LDR. The proposition is proved.

§14. Equations of type $V\Sigma$

In this section we solve equations of type $V\Sigma$. In solving a T-system the same word unknown can occur in different equations of the T-system. If an equation of type $V\Sigma$ is solved with respect to a word unknown, on substituting the resulting expression into the other V-terms containing this unknown we obtain, in general, terms not occurring among the seven kinds of special predicate terms of characteristic M (see §12), which, in turn, leads to the consideration of new types of equations. This "danger", however, can be circumvented. Namely, in all "dangerous" cases (in particular, the case $V\Theta$) we can prove that equations of type $V\Sigma$ have no solutions. A major role in this regard is played by the restrictions given by the right-hand sides of the predicate terms.

Just as in solving equations containing Θ- and Ψ-terms it was convenient to introduce $\bar{\Theta}_2^M$- and $\bar{\Psi}_2^M$-terms, in solving equations containing V-terms it is convenient to introduce V_2^M-terms.

A V_2^M-term is a predicate term of the form $[[yA_i]_{i=1}^{m-1};\ \omega(y, M)]$, where y is a word unknown, m is an integer such that $1 < m \leq 3M$, the A_i are words in the alphabet $И$, and $[A_i^? \leq M$ $(i = 1, \cdots, m-1)$.

14.1. *Any* V_1^M-*term is a predicate term of the form* $\alpha^\lambda\beta \,\&\, \lambda \geqq (4M)^6$, *where* λ *is a positive unknown, and* (α, β) *is a normal pair of predicate terms such that* $\alpha \gg (6M)^3$. *Here* $\alpha^2\beta$ *is a* V_2^M-*term*.

PROOF. Suppose there is given a V_1^M-term (§12). Denote it by γ. We define predicate terms α and β. Put

$$\alpha = [[xA_i]_{i=1}^{k};\ \omega(x, M)], \qquad \beta = [[xA_i]_{i=1}^{r}x;\ \omega(x, M)].$$

Then the predicate term γ has the form $\alpha^\lambda\beta \,\&\, \lambda \geqq (4M)^6$. The predicate term $\alpha^2\beta$ is the term $[([xA_i]_{i=1}^k)^2[xA_i]_{i=1}^r x;\ \omega(x,M)]$, i.e. a V_2^M-term with $m = 2k+r+1$. We will prove that $\alpha \gg (6M)^3$ and the pair (α,β) is normal. Suppose X is a word in the alphabet И such that $\omega(X,M)$ holds. The condition $\omega(X,M)$ implies $\omega_1(X,M)$. Since $1 < k \leqq M$, $[A_1^\partial, \cdots, [A_k^\partial \leqq M$ and $\nabla(A_1,\cdots,A_k)$, by 11.3 the word $[XA_i]_{i=1}^k$ is simple. Since $r < k$,

$$[XA_i]_{i=1}^r X \angle [XA_i]_{i=1}^k.$$

Since $\omega_1(X,M)$ holds, $[X^\partial > (6M)^3 > 0$. Hence $[XA_i]_{i=1}^r X \neq \Lambda$ and $\partial([XA_i]_{i=1}^k) > (6M)^3$. We have shown that (α,β) is normal and $\alpha \gg (6M)^3$. The proposition is proved.

Thue V_1^M-terms fall into our scheme, using 13.1, 13.4 and 13.5.

14.2. *Suppose $\alpha(x,\mu)$ is a $\widetilde{V}_1^M$-term with base A, and (β,β_1) is a normal pair of predicate terms. Suppose $N > 1$ and one of the conditions $N > M+1$ or $\beta_1 \gg [A^\partial$ holds. Then the system*

$$\alpha(x,\mu) = \beta^\lambda\beta_1, \qquad \lambda > N,$$

is equivalent to the system

$$\beta = xA, \quad \beta_1 = x, \quad \neg\tau(x,M), \quad \lambda = \mu > N.$$

The proof is analogous to that of 13.1, only instead of 1.7 we use 1.26 and the implication $\neg\tau(x,M) \supset (\partial(x) > M \geqq [A^\partial)$.

Equations of type $V\Sigma$ can be of two types: $V\Psi$ and $V\Theta$. Let us begin with equations of type $V\Psi$. Here we will prove three assertions. The first of these pertains to equations of type $V^M\Psi_2^M$ and $V^M\overline{\Psi}_1^M$, and states that under certain additional assumptions such equations have no solutions.

14.3. *Consider the system*

$$\text{(1)} \qquad \alpha(x,\lambda) = \beta, \qquad \omega(x,M), \qquad \lambda \geqq 9M^2,$$

where α is a V^M-term, x is a word unknown, λ is a positive unknown, and β is a Ψ_2^M-term or a $\overline{\Psi}_1^M$-term. The system (1) has no solutions.

PROOF. We consider four cases.

Case 1. α is a V_1^M-term and β is a Ψ_2^M-term. The assertion follows from 14.1 and 13.4.

Case 2. α is a $\widetilde{V}_1^M$-term and β is a Ψ_2^M-term. Then α has the form

$$[(xA)^\lambda x;\ \mathscr{G}(xA) \,\&\, \neg\tau(x,M)].$$

We define predicate terms α_1 and α_2. Put

$$\alpha_1 = [xA;\quad \mathscr{E}(xA) \,\&\, \neg\tau(x,M) \,\&\, \omega(x,M)];$$

$$\alpha_2 = [x;\quad \mathscr{E}(xA) \,\&\, \neg\tau(x,M) \,\&\, \omega(x,M)].$$

Then $\alpha_1 \gg (6M)^3$ (see the condition ω) and the pair (α_1,α_2) is normal. The system (1) can be written in the form $(\alpha_1^\lambda\alpha_2 = \beta) \,\&\, \lambda \geqq 9M^2$. Now our assertion follows from 13.4.

Case 3. α is a $\widetilde{V}_1^M$-term and β is a $\overline{\Psi}_1^M$-term. The system (1) expanded has the form

$$(xA)^\lambda x = (S^\sigma C)^\mu S^\sigma R,$$
$$(2)\qquad \mathscr{E}(xA),\quad \neg\tau(x,M),\quad \omega(x,M),\quad \lambda \geqslant 9M^2,\quad \mu \geqslant 4M^2,$$
$$\partial(S^\sigma C) > (6M)^3,$$

where σ and μ are unknowns, A, S, C and R are words in the alphabet $И$, $[A^\partial \leqq M$, S is simple, $\mathscr{L}(S,C)$, $R \angle CS^{[R^\partial}$, $[S^\partial \leqq 30M$, and $[C^\partial, [R^\partial \leqq 90M$. Assuming that the system (2) has a solution (X,σ,μ,λ), we obtain in the alphabet $И$ the equality $(XA)^\lambda X = (S^\sigma C)^\mu S^\sigma R$, where $\lambda,\mu > 1$, XA is simple, $\omega(X,M)$, $X \neq \Lambda$ and $[S^\sigma C^\partial > (6M)^3$. By 1.10, the word $S^\sigma C$ is simple. We have $R \angle CS^{[R^\partial}$; hence $R \angle CS^\sigma$, since $[R^\partial < \sigma[S^\partial$. If now $R \angle C$, then from $(XA)^\lambda X = (S^\sigma C)^\mu S^\sigma R$ and 1.7 it follows that $XA = S^\sigma C$, whence $X \angle S^\sigma C$. If $[R^\partial > [C^\partial$, then we have $(XA)^\lambda X = (S^\sigma C)^{\mu+1}B$, where $\Lambda \neq B \angle S^\sigma$, and, again by 1.7, $XA = S^\sigma C$. We have proved that $X \angle S^\sigma C$, which contradicts the condition $\omega(X,M)$.

Case 4. α is a V_1^M-term and β is a $\bar{\Psi}_1^M$-term. Here the argument is almost the same as in the previous case. Assuming that the equation $\alpha = \beta$ has a solution $(X,\lambda,\cdots)$, we obtain in the alphabet $И$ the equality

$$(3)\qquad ([XA_i]_{i=1}^k)^\lambda [XA_i]_{i=1}^r X = (S^\sigma C)^\mu S^\sigma R,$$

where $1 < k \leqq M$, $0 \leqq r < k$; X, λ, σ, μ, S, C and R are the same as in Case 3, and $A_1,\cdots,A_k$ are words in the alphabet $И$ such that $[A_1^\partial,\cdots,[A_k^\partial \leqq M$ and $\nabla(A_1,\cdots,A_k)$. By 11.3, the word $[XA_i]_{i=1}^k$ is simple. Considering the cases $R \angle C$ and $[R^\partial > [C^\partial$ and applying Proposition 1.7 to (3), we obtain $[XA_i]_{i=1}^k = S^\sigma C$; hence $X \angle S^\sigma C$, which contradicts the condition $\omega(X,M)$. The proposition is proved.

14.4. *Consider the system*

$$(1)\qquad \alpha(x,\lambda) = \beta,\quad \omega(x,M),\quad z = \gamma(\sigma),$$

where x and z are word unknowns, σ and λ are positive unknowns, α is a V^M-term, β is a $\tilde{\Psi}_1^M$-term, and γ is a Ψ_2^M-term with the same base as β. If $(X,Z,\cdots)$ is a solution of (1), *then $\epsilon(X,Z,M)$ (the definition of the predicate ϵ precedes Theorem* 12.1).

PROOF. We consider two cases.

Case 1. α is a V_1^M-term. Then by 14.1 and 13.3 the equation $\alpha = \beta$ has the form $\alpha_1^\lambda\alpha_2 = \beta_1^{\mu+a}\beta_2$, $\lambda \geqq (4M)^6$, $\mu \geqq N$, where (α_1,α_2) and (β_1,β_2) are normal pairs of predicate terms, $a \geqq 0$, $N > 1$, and where $\alpha_1^2\alpha_2$ is a V_2^M-term, and the term $\beta_1^2\beta_2$ satisfies all conditions for a Ψ_2^M-term, only instead of $m \leqq 4M^2$ we now have $m \leqq 12M^3$. By 13.1, any solution of the equation $\alpha = \beta$ satisfies the equation $\alpha_1^2\alpha_2 = \beta_1^2\beta_2$. Thus any solution of (1) satisfies the system $(z = \gamma)\,\&\,(\alpha_1^2\alpha_2 = \beta_1^2\beta_2)$. The latter expanded has the form

$$(2)\quad [xA_i]_{i=1}^{n-1}x = [S^\sigma P_i]_{i=1}^k,\qquad z = [S^\sigma Q_i]_{i=1}^m,\qquad \omega(x,M),\qquad \sigma[S^\partial > (6M)^{13},$$

where $1 < n \leqq 3M$, $1 \leqq k \leqq 12M^3$, $1 \leqq m \leqq 4M^2$, S is simple, $[S^\partial \leqq (6M)^3$, $[A_1^\partial,\cdots,[A_{n-1}^\partial \leqq M$, and $[P_i^\partial, [Q_j^\partial \leqq (6M)^7$. If we assume that the system (2) has a solution (X,Z,σ), then in the alphabet $И$ we have the equalities

$$(3)\qquad [XA_i]_{i=1}^{n-1}X = [S^\sigma P_i]_{i=1}^k,$$
$$(4)\qquad Z = [S^\sigma Q_i]_{i=1}^m,$$

where $\omega(X,M)$, and the inequalities

(5) $$\sigma[S^\partial > (6M)^{13} > 9k[P_1 \cdots P_kSS^\partial.$$

Let us consider (3). Since $\omega(X,M)$, we have $\urcorner\tau(X,(6M)^3)$. Since $[S^\partial \leqq (6M)^3$, X is not a beginning of the word S^σ. Hence (3) implies that $[X^\partial > \sigma[S^\partial$; in particular, $[X^\partial > (\sigma-1)[S^\partial$. Taking into account (5) and arguing from (3) as we did in 11.6 from (1) (from 11.6 (1) to 11.6 (10)), but making those changes necessitated by the fact that the left-hand side of (3) has the form $[XA_i]_{i=1}^{n-1}X$, and not X^d as in 11.6 (1), we see that there exist integers $a \geqq 0$ and l, $1 \leqq l \leqq k$, and words $B_1, \cdots, B_l$ such that (cf. (6) and (10) of 11.6)

(6) $$X = [S^{\sigma-a}B_i]_{i=1}^{l},$$

(7) $$[B_1^\partial, \ldots, [B_l^\partial,\ a[S^\partial \leqslant 4[P_1\ldots P_kSS^\partial \leqslant (6M)^{10} - (6M)^7.$$

Putting $C_i = S^aQ_i$, we have from (4) and our estimates for $[Q_i^\partial$ and $a[S^\partial$

(8) $$Z = [S^{\sigma-a}C_i]_{i=1}^m, \qquad [C_1^\partial, \cdots, [C_m^\partial \leqq (6M)^{10}.$$

We have $m \leqq 4M^2$ and $l \leqq k \leqq 12M^3$. Now from (6), (7) and (8) we conclude that $\epsilon(X,Z,M)$ holds.

Case 2. α is a $\widetilde{V}_1^M$-term. Arguing here as in Case 1 (only instead of 13.1 we now use 14.2), we obtain (3) and (4) with $n=1$, from which $\epsilon(X,Z,M)$ follows. The proposition is proved.

14.5. *Consider the equation*

(1) $$\alpha(x,\lambda) = \beta(\sigma,\mu),$$

where α is a $\widetilde{V}_1^M$-term, β is a Ψ^M-term, x is a word unknown, and σ, λ and μ are positive unknowns. Equation (1) is equivalent with respect to the unknowns x, σ, λ and μ[15] *to the system $(x=\gamma)\,\&\,\mathscr{D}$, where $\mathscr{D}$ is a LDR and γ is a $\overline{\Psi}_2^M$-term.*

PROOF. We require the following lemma.

LEMMA. *Suppose $n \geqq 1$ and $M \geqq 1$ are integers, $S, A_1, \cdots, A_n$ are words in the alphabet $\mathfrak{A}$, S is simple, and $X(t) = [S^tA_i]_{i=1}^n$. Then one of the following two assertions holds*:

1) $\tau(X(t),M)$ *for any* $t \geqq 0$;

2) $\urcorner\tau(X(t),M)$ *for any* $t > 1 + M/\partial(S)$.

PROOF OF THE LEMMA. There are three possible subcases.

a) $A_n \angle S^{\partial(A_n)}$, and either $n=1$ or $A_1, \cdots, A_{n-1}$ are powers of S. Then $X(t) = S^{\lambda(t)}S_1$, where $S_1 \angle S$ and $\lambda(t) \geqq t$. If $[S^\partial \leqq M$, then 1) holds. Suppose $[S^\partial > M$. We will prove by contradiction that in this case 2) holds. Assume that $t>1$ and $\tau(X(t),M)$. From $\tau(X(t),M)$ we have $X(t) = T^mT_1$, where $T_1 \angle T$, $[T^\partial \leqq M$ and T is simple. Since $X(t) = S^{\lambda(t)}S_1$, we have $S^{\lambda(t)}S_1 = T^mT_1$. We have $\lambda(t) \geqq t > 1$ and $[S^\partial > [T^\partial$. Hence $m>1$. Then by 1.7 we have $S=T$, which is impossible, since $[S^\partial > [T^\partial$. We have obtained the required contradiction. This completes the consideration of a).

[15] The definition of this relative equivalence precedes Theorem 12.2.

b) $\neg(A_n \angle S^{\partial(A_n)})$, and either $n=1$ or $A_1, \cdots, A_{n-1}$ are powers of S. Then $X(t) = S^{\lambda(t)}A_n$, where $\lambda(t) \geqq t$. We will prove by contradiction that in this case 2) holds. Assuming $\tau(X(t), M)$, where $t > 1 + M/\partial(S)$, and arguing as in a), we arrive at the equality

$$(*) \qquad S^{\lambda(t)}A_nQ = T^{m+1},$$

where $[T^\partial \leqq M$, T is simple and Q is an end of T. From (*) and the estimates for $\lambda(t)$ and $[T^\partial$ we have $m \geqq 1$, $[S^\partial \leqq m[T^\partial$ and $[T^\partial \leqq (\lambda(t)-1)[S^\partial$. By 1.6, $S = T$. Then A_nQ is a power of S, which contradicts the assumption b). This contradiction implies that 2) holds.

c) $n > 1$ and at least one of the words $A_1, \cdots, A_{n-1}$ is not a power of S. Then $X(t) = S^{\lambda(t)}A_qB$, where $1 \leqq q \leqq n-1$, $\lambda(t) \geqq t$ and $\mathscr{L}(A_q, S)$. Assuming $\tau(X(t), M)$ and $t > 1 + M/\partial(S)$, and arguing as in b), we arrive at the equality

$$S^{\lambda(t)}A_qBQ = T^{m+1}.$$

Arguing as in b), we obtain $m \geqq 1$, $[S^\partial \leqq m[T^\partial$, $[T^\partial \leqq (\lambda(t)-1)[S^\partial$ and $S = T$. Then $A_qS \angle S^{m+1-\lambda(t)}$, which, by 1.4, contradicts $\mathscr{L}(A_q, S)$. Thus in case c) we have 2). The lemma is proved.

We will now prove 14.5. There are two possible cases.

Case 1. β is a Ψ_1^M-term. From 13.3 and 14.2 we conclude that the equation $\alpha = \beta$ is equivalent to the system

$$(2) \qquad xA = \beta_1, \quad x = \beta_2, \quad \lambda = \mu + a, \quad \mu \geqq N, \quad \neg\tau(x, M),$$

where $N \geqq 4M^2$ and β_1 and β_2 are the predicate terms α and β described in 13.3. The system (2) is equivalent to the system

$$(3) \qquad x = \beta_2, \quad \beta_2 A = \beta_1, \quad \lambda = \mu + a, \quad \mu \geqq N, \quad \neg\tau(x, M).$$

The equation $\beta_1 = \beta_2 A$ in expanded form is a system consisting of a one-stage exponential equation in the one unknown σ and the inequality $\sigma \geqq N_1$. By 2.3, the equation $\beta_1 = \beta_2 A$ is equivalent to some LDR $\mathscr{D}_1(\sigma)$. Hence (3) is equivalent to the system

$$(4) \qquad x = \beta_2, \quad \neg\tau(x, M), \quad \mathscr{D}_1(\sigma), \quad \lambda = \mu + a, \quad \mu \geqq N.$$

The equation $x = \beta_2$ expanded has either the form (see 13.3) $x = [S^\sigma A_i]_{i=1}^n$, $\sigma[S^\partial \geqq N'$, where $n > 0$, $S, A_1, \cdots, A_n$ are words in the alphabet $\mathbf{И}$, S is simple and $N' \geqq [S^\partial + 2M$, or the form $x = P$, where P is a word in $\mathbf{И}$. Applying the lemma, we see that either the system $x = \beta_2$, $\neg\tau(x, M)$ has no solutions or we have the implication $(x = \beta_2) \supset \neg\tau(x, M)$. In both cases the system (4) is equivalent to the system $(x = \beta_2)\,\&\,\mathscr{D}$, where $\mathscr{D}$ is a suitable LDR. It remains to observe that β_2 is a $\overline{\Psi}_2^M$-term.

Case 2. β is a Ψ_2^M-term. Then equation (1) expanded has the form

$$(5) \qquad (xA)^\lambda x = [S^\sigma P_i]_{i=1}^k, \quad \sigma[S^\partial > (6M)^{13}, \quad \mathscr{L}(xA), \quad \neg\tau(x, M),$$

where $0 \leqq k \leqq 4M^2$, $A, S, P_1, \cdots, P_k$ are words in the alphabet $\mathbf{И}$, S is simple, $[A^\partial \leqq M$, $[S^\partial \leqq (6M)^3$ and $[P_1^\partial, \cdots, [P_k^\partial \leqq (6M)^7$. We have the estimate

$$(6) \qquad (6M)^{13} > 9k[P_1 \cdots P_kSSA^\partial + 2[AS^\partial + 2M.$$

The system (5) is equivalent to the system

$$(7) \qquad (xA)^{\lambda+1} = [S^\sigma P_i]_{i=1}^k A, \quad \sigma[S^\partial > (6M)^{13}, \quad \mathscr{L}(xA), \quad \neg\tau(x, M).$$

There are two possible subcases.

2.1) The words $P_1, \cdots, P_{k-1}$ and P_kA are powers of S. Then (7) is equivalent to the system

$$(8) \qquad (xA)^{\lambda+1} = S^{k\sigma+c}, \quad \mathscr{L}(xA), \quad \neg\,\tau(x,M), \quad \sigma[S^\sigma > (6M)^{13},$$

where $c \geqq 0$. By 1.3, the system $(xA)^{\lambda+1} = S^{k\sigma+c}$, $\mathscr{L}(xA)$ is equivalent to the system $xA = S$, $\lambda+1 = k\sigma + c$. It is now clear that either the system (8) has no solutions or it is equivalent to the system $(x = B)\,\&\,\mathscr{D}$, where the word B satisfies the relations $BA = S$, $\neg\,\tau(B,M)$, and $\mathscr{D}$ is a LDR. Put $\gamma = [B; 0 = 0]$. Then γ is a $\overline{\Psi}_2^M$-term, and (1) is equivalent to the system $(x = \gamma)\,\&\,\mathscr{D}$.

2.2) Not all of the words $P_1, \cdots, P_{k-1}, P_kA$ are powers of S. Then by 11.6 there exist $d \geqq 1$, $a \geqq 0$, and n, $1 \leqq n \leqq k$, and words $A_1, \cdots, A_n$ such that (see (6))

$$(9) \qquad \forall\sigma t\,[((\sigma - t = a)\ \&\sigma\,[S^\partial > (6M)^{13}) \supset ([S^\sigma P_i]_{i=1}^k A = ([S^t A_i]_{i=1}^n)^d)],$$

where $\max\{[A_1^\partial, \cdots, [A_n^\partial, a[S^\partial\} \leqq 4[P_1 \cdots P_kSA^\partial$ and we have the implication

$$(10) \qquad (t[S^\partial > (6M)^{13}) \supset \mathscr{L}([S^tA_i]_{i=1}^n).$$

Using 1.3, (9) and (10), we see that (7) is equivalent with respect to the unknowns x, σ and λ to the system

$$(11) \qquad \begin{gathered} xA = [S^tA_i]_{i=1}^n, \\ \lambda + 1 = d, \qquad \sigma = a + t, \qquad t\,[S^\partial > (6M)^{13} - a\,[S^\partial, \qquad \neg\,\tau\,(x, M). \end{gathered}$$

It is not difficult to express x from (11). Choose $c > 0$ such that $(c-1)\,[S^\partial \leqq [A^\partial$ and $c[S^\partial > [A^\partial$. Such a c clearly exists. Note that $c[S^\partial \leqq [AS^\partial$, so that $(a+c)\,[S^\partial < (6M)^{13}$. Put $B_i = S^cA_i$ $(i = 1, \cdots, n)$. Then $[B_n^\partial \geqq c[S^\partial > [A^\partial$. If A is not an end of B_n, the system (11) has no solutions and the assertion is trivially true. As $\mathscr{D}$ we may take the LDR $0 = 1$, and as γ any $\overline{\Psi}_2^M$-term. Suppose A is an end of B_n. Then $B_n = BA$ for some B. The first equation of (11) implies that $x = [S^\nu B_i]_{i=1}^{n-1} S^\nu B$, where $\nu = t - c$. Put

$$\gamma = ([S^\nu B_i]_{i=1}^{n-1}S^\nu B;\ [S^{\nu\partial} > (6M)^{13} - (a+c)\,[S^\partial).$$

The term γ is a $\overline{\Psi}_2^M$-term. Applying the lemma, we obtain that either the system $(x = \gamma)\,\&\,\neg\,\tau(x,M)$ has no solutions or we have the implication $(x = \gamma) \supset \neg\,\tau(x,M)$. In both cases (11) is equivalent to a system $(x = \gamma)\,\&\,\mathscr{D}$, where $\mathscr{D}$ is a suitable LDR. The proposition is proved.

We now consider equations of type $V^M\Theta^M$. We will prove that such equations have no solutions. The following proposition is fundamental here.

14.6. *Consider the equation*

$$(*) \qquad [xA_i]_{i=1}^{n-1} x = \gamma,$$

where γ is a Θ_2^M-term, x is a word unknown, $A_1, \cdots, A_{n-1}$ are words in the alphabet И, *$[A_1^\partial, \cdots, [A_{n-1}^\partial \leqq M$ and $1 < n \leqq M^2$. The equation* (*) *has no solutions.*

PROOF (BY CONTRADICTION). Write (*) in expanded form. We obtain the system

$$(1) \qquad [xA_i]_{i=1}^{n-1}x = [(S^\sigma C)^\nu S^\sigma Q_i]_{i=1}^{m-1}(S^\sigma C)^{\nu+\varepsilon} S^\sigma Q_m, \quad \nu \geqq 4M^2, \quad \partial(S^\sigma C) > (6M)^3,$$

where σ and ν are positive unknowns, $1 < m \leqq M^2$, $\epsilon = \pm 1$, S, C, $Q_1, \cdots, Q_m$ are words in the alphabet $\mathit{И}$, S is simple, $\mathscr{S}(S, C)$, $[S^\partial \leqq 30M$; $[C^\partial, [Q_1^\partial, \cdots, [Q_m^\partial \leqq 90M$, and $\urcorner(Q_1 = \cdots = Q_{m-1} = C)$.

The proof is carried out in accordance with the following scheme. We assume that the system (1) has a solution (X, σ, ν). Then in $\mathit{И}$ we have the equality

$$[XA_i]_{i=1}^{n-1}X = [(S^\sigma C)^\nu S^\sigma Q_i]_{i=1}^{m-1}(S^\sigma C)^{\nu+\varepsilon}S^\sigma Q_m, \tag{2}$$

where $[S^\sigma C^\partial > (6M)^3$ and $\nu \geqq 4M^2$. We express X from (2). It turns out that $X = (S^\sigma C)^\mu S^\sigma Q_m$ for some $\mu \geqq 3$. Substituting this expression in place of X in (2), we obtain a two-stage exponential equation which is a special case of the equation considered in 2.8. The triple (σ, ν, μ) satisfies this equation. We prove that this is possible only if $Q_1 = \cdots = Q_{m-1} = C$. This contradicts the condition $\urcorner(Q_1 = \cdots = Q_{m-1} = C)$. We use the following lemma.

Lemma. *If for our σ, ν, m, S, C, $Q_1, \cdots, Q_m, X$ we have the equality*

$$XA = [(S^\sigma C)^r S^\sigma Q_i]_{i=1}^r (S^\sigma C)^d B, \tag{3}$$

where $[A^\partial, [B^\partial \leqq 91M$, $0 \leqq r < m$ and $0 \leqq d \leqq \nu$, then $4 \leqq d \leqq \nu - 3$.

Proof of the Lemma. We compute the length of the word $[XA_i]_{i=1}^{n-1}X$. Computing $[X^\partial$ from (3), we obtain

$$\begin{aligned}[[XA_i]_{i=1}^{n-1}X^\partial &= n[X^\partial + [A_1 \ldots A_{n-1}^\partial = n[r(\nu+1)+d][S^\sigma C^\partial \\ &\quad - nr[C^\partial + n([Q_1 \ldots Q_r^\partial + [B^\partial - [A^\partial) + [A_1 \ldots A_{n-1}^\partial.\end{aligned} \tag{4}$$

On the other hand, we have trom (2)

$$[[XA_i]_{i=1}^{n-1}X^\partial = m(\nu+1)[S^\sigma C^\partial - m[C^\partial + \varepsilon[S^\sigma C^\partial + [Q_1 \ldots Q_m^\partial. \tag{5}$$

Equating the right-hand sides of (4) and (5) and transposing to the left the summands containing the factor $[S^\sigma C^\partial$, we obtain

$$\begin{aligned}[(\nu+1)(nr-m)+nd-\varepsilon][S^\sigma C^\partial &= Q_1 \ldots Q_m + (nr-m)[C^\partial \\ &\quad - [A_1 \ldots A_{n-1}^\partial - n([Q_1 \ldots Q_r B^\partial - [A^\partial].\end{aligned} \tag{6}$$

Comparing (2) and (3) and using $\nu \geqq 4M^2$, we have $nr \leqq m$. Estimating the right-hand side of (6), we see that it is less in absolute value than $[S^\sigma C^\partial$. The square brackets in the left-hand side of (6) enclose an integer. Hence this integer is zero, i.e. $(\nu+1)(nr-m)+nd-\epsilon = 0$, from which it follows that

$$d = \frac{(\nu+1)(m-nr)+\varepsilon}{n}. \tag{7}$$

Since $n > 1$, $\epsilon = \pm 1$, $0 \leqq d \leqq \nu$, $\nu \geqq 4n$ and d, m, n, r are integers, (7) implies that $1 \leqq m - nr \leqq n - 1$. Hence $d \geqq (\nu + 1 + \epsilon)/n \geqq 4 + (1+\epsilon)/n \geqq 4$, so that $d \geqq 4$.

Put $a = \nu - d$. Then $d = \nu - a$ and we have from (7)

$$na = (\nu+1)(nr-m) - \epsilon + n\nu. \tag{8}$$

We know that $m - nr \leqq n - 1$, i.e. $nr - m \geqq 1 - n$. Since $\nu \geqq 4n$, we have $(\nu+1)(nr-m) + n\nu - \epsilon \geqq (\nu+1)(1-n) + n\nu - \epsilon = \nu + 1 - n - \epsilon \geqq 3n$ and it follows from (8) that $a \geqq 3$. Hence $d = \nu - a$, $a \geqq 3$. Therefore, $d \leqq \nu - 3$. We have proved above that $d \geqq 4$. The lemma is proved.

We continue the proof of 14.6.

Consider (2). We express X from (2). Since $n > 1$, XA_1 is a beginning of the right-hand side of (2) and, as is easily seen,

$$[XA_1^\partial < \partial([(S^\sigma C)^\nu S^\sigma Q_i]_{i=1}^{m-1}(S^\sigma C)^2).$$

Hence there exist integers r, $0 \leqq r < m$, and d, $0 \leqq d \leqq \nu$, and words U and V such that $V \neq \Lambda$ and

$$XA_1 = [(S^\sigma C)^\nu S^\sigma Q_i]_{i=1}^{r}(S^\sigma C)^d U, \tag{9}$$

where $UV = S^\sigma C$ if $d < \nu$, and $UV = S^\sigma Q_{r+1}$, $r < m-1$, if $d = \nu$. We will prove that $U = \Lambda$. Since $0 \leqq d \leqq \nu$, there are three possible cases.

Case 1. $d = \nu$. Then $UV = S^\sigma Q_{r+1}$, $r < m-1$. Cancelling (2) on the left by XA_1 and using (9), we conclude that the left-hand side of the resulting equality begins with X, and hence with $(S^\sigma C)^3$ (since $(S^\sigma C)^3 \angle (S^\sigma C)^\nu \angle X$), and the right-hand side with $VS^\sigma C$ (since $r < m-1$ and $\nu, \nu+\epsilon > 0$). Also, from the equality $UV = S^\sigma Q_{r+1}$ and consideration of the quantities $\sigma[S^\partial$ and $[Q_{r+1}^\partial$ we obtain $[V^\partial < [S^\sigma Q_{r+1}^\partial < 2[S^\sigma C^\partial$. It follows that $VS^\sigma C \angle (S^\sigma C)^3$. By 1.10, the word $S^\sigma C$ is simple. Then from $VS^\sigma C \angle (S^\sigma C)^3$ and 1.4 we obtain that V is a power of $S^\sigma C$. Since $V \neq \Lambda$ and $[V^\partial < 2[S^\sigma C^\partial$, we have $V = S^\sigma C$. Thus $S^\sigma Q_{r+1} = UV = US^\sigma C$, from which it follows that $[U^\partial \leqq [Q_{r+1}^\partial \leqq 90M$. Applying the lemma to (9) (the roles of A and B in the lemma are played by A_1 and U), we conclude that $d \leqq \nu - 3$ despite the assumption $d = \nu$. Thus Case 1 is impossible.

Case 2. $d = \nu - 1$. Then $UV = S^\sigma C$ and, clearly, $r < m-1$. Cancelling (2) on the left by XA_1, we see that the left-hand side of the resulting equality begins with $S^\sigma C$ and the right-hand side with $VS^\sigma Q_{r+1}S^\sigma C$, so that for some Z_1 and Z_2 we have

$$S^\sigma CZ_1 = VS^\sigma Q_{r+1}Z_2 = UVZ_1. \tag{10}$$

Assume that $[U^\partial > \sigma[S^\partial$. Since $UV = S^\sigma C$ and $[C^\partial \leqq 90M$, we have $[V^\partial < 90M$. Since $[A_1^\partial \leqq M$, we get $[A_1V^\partial < 91M$. Multiplying (9) on the right by V and using $UV = S^\sigma C$ and $d = \nu - 1$, we obtain $XA_1V = [(S^\sigma C)^\nu S^\sigma Q_i]_{i=1}^{r}(S^\sigma C)^\nu$. Applying the lemma to this equality (the roles of A and B in the lemma are now played by A_1V and A), we obtain $\nu \leqq \nu - 3$, which is impossible. From this contradiction we conclude that $[U^\partial \leqq \sigma[S^\partial$. Then $U \angle S^\sigma$, $VU \angle VS^\sigma$ and we obtain from (10) that $VU = UV = S^\sigma C$. Since $S^\sigma C$ is simple and $V \neq \Lambda$, by 1.2′ we have $U = \Lambda$, as required.

Case 3. $0 \leqq d < \nu - 1$. Then $UV = S^\sigma C$, so that $U \angle S^\sigma C$. Cancelling (2) on the left by XA_1, we see that the left-hand side of the resulting equality begins with X and the right-hand side with $VS^\sigma C$. But, as is easily seen, $[X^\partial > [S^\sigma C^\partial$, whence $S^\sigma C \angle X$. We obtain $UV = S^\sigma C = VU$. From the last two equalities, the relation $V \neq \Lambda$ and 1.2′ it follows that $U = \Lambda$. Thus we have proved that $U = \Lambda$. Now from (9) we have

$$XA_1 = [(S^\sigma C)^\nu S^\sigma Q_i]_{i=1}^{r}(S^\sigma C)^d. \tag{11}$$

Applying the lemma to (11), we conclude that $4 \leqq d \leqq \nu - 3$. In particular, $[X^\partial > 3[S^\sigma C^\partial$. Comparing (11) and (2), we obtain that XA_1 ends with $(S^\sigma C)^2$ and with $S^\sigma CS^\sigma Q_mA_1$, so that $Z_1S^\sigma CS^\sigma C = Z_2S^\sigma CS^\sigma Q_mA_1$ for some Z_1 and Z_2. From the last equality and 1.15 (the applicability of 1.15 is obvious) it follows that $C = Q_mA_1$. From (11) we obtain

$$X = [(S^\sigma C)^\nu S^\sigma Q_i]_{i=1}^r (S^\sigma C)^{d-1} S^\sigma Q_m. \tag{12}$$

We will prove that $Q_1 = \cdots = Q_r = C$.

Comparing the lengths of the left-and right-hand sides of (2) and using (12) and the fact that $d \geqq 4$, we obtain $r < m-1$. Cancelling (2) on the left by XA_1, we have for some W (cf. (11))

$$[(S^\sigma C)^\nu S^\sigma Q_i]_{i=1}^r (S^\sigma C)^{d-1} W = (S^\sigma C)^{\nu-d} S^\sigma Q_{r+1} [(S^\sigma C)^\nu S^\sigma Q_i]_{i=r+2}^{m-1} (S^\sigma C)^{\nu+\varepsilon} S^\sigma Q_m. \tag{13}$$

We will prove that if $r > 0$, then $Q_1 = \cdots = Q_r = C$. Cancelling (13) on the left by $(S^\sigma C)^{\nu-d} S^\sigma$, we obtain

$$C (S^\sigma C)^{d-1} S^\sigma Q_1 [(S^\sigma C)^\nu S^\sigma Q_i]_{i=2}^r (S^\sigma C)^{d-1} W = Q_{r+1} [(S^\sigma C)^\nu Q_i]_{i=r+2}^{m-1} (S^\sigma C)^{\nu+\varepsilon} S^\sigma Q_m. \tag{14}$$

Since $d \geqq 4$ and $\nu, \nu+\epsilon > 1$, it follows from (14) and 1.15 that $C = Q_{r+1}$. Then (14) can be cancelled on the left by $(CS^\sigma)^d$, and after cancellation we obtain $Q_1(S^\sigma C)^2 V_1 = C(S^\sigma C)^2 V_2$ for some V_1 and V_2 (we know $d \leqq \nu - 3$ and $\nu + \epsilon \geqq d+2$). From this last equality and 1.15 we have $Q_1 = C$. If $r = 1$, our assertion is proved. If $r > 1$, it is easy to calculate that $r < m-2$. Cancelling (14) on the left by $C(S^\sigma C)^{d-1} S^\sigma Q_1$ and using $Q_1 = Q_{r+1} = C$, we obtain

$$[(S^\sigma C)^\nu S^\sigma Q_i]_{i=2}^r (S^\sigma C)^{d-1} W = (S^\sigma C)^{\nu-d} S^\sigma Q_{r+2} [(S^\sigma C)^\nu S^\sigma Q_i]_{i=r+3}^{m-1} (S^\sigma C)^{\nu+\varepsilon} S^\sigma Q_m. \tag{15}$$

From (15), in view of the induction assumption, $Q_2 = \cdots = Q_r = C$. Hence we have proved that $Q_1 = \cdots = Q_r = C$. Putting $\mu = r(\nu+1) + d - 1$, we obtain from (12)

$$X = (S^\sigma C)^\mu S^\sigma Q_m, \tag{16}$$

where $\mu \geqq 3$. This is the desired representation for X.

The equality (2) assumes the form

$$[(S^\sigma C)^\mu S^\sigma Q_m A_i]_{i=1}^{n-1} (S^\sigma C)^\mu S^\sigma Q_m = [(S^\sigma C)^\nu S^\sigma Q_i]_{i=1}^{m-1} (S^\sigma C)^{\nu+\varepsilon} S^\sigma Q_m. \tag{17}$$

It follows from our argument that if we view (17) as an exponential equation ("two-stage") in the four unknowns σ, μ, ν and $\nu_1 = \nu + \epsilon$, then it has a solution $(\sigma, \mu, \nu, \nu_1)$ such that

$$[S^\sigma C^\partial \geqslant \max\{[Q_1^\partial, \ldots, [Q_m^\partial, [Q_m A_1^\partial, \ldots, [Q_m A_{n-1}^\partial\} + 2[SC^\partial, \qquad \mu, \nu, \nu_1 \geqslant 3.$$

By 2.8 (in the present application of 2.8, for the λ_i and μ_j we have $\lambda_1 = \cdots = \lambda_n = \mu$, $\mu_1 = \cdots = \mu_{m-1} = \nu$ and $\mu_m = \nu + \epsilon$), there exist sequences $(n_1, \cdots, n_k)$, $(m_1, \cdots, m_k)$ and $(\epsilon_1, \cdots, \epsilon_k)$ of integers such that

1) $k \geqq 1$, $1 < n_1 < \cdots < n_k = n$, $1 \leqq m_1 < \cdots < m_k = m$, $|\epsilon_1|, \cdots, |\epsilon_k| \leqq 1$, $\epsilon_k = 0$;

2) $Q_i = C$ for any i different from $m_1, \cdots, m_k$;

3) the triple (σ, ν, μ) is a solution of the system

$$n_i\mu - m_i\nu = m_i - n_i + \varepsilon_i \quad (i = 1, \ldots, k-1),$$
$$n_k\mu - m_k\nu - \varepsilon = m_k - n_k + \varepsilon_k.$$

We will prove that $k = 1$. Assume $k > 1$. Then we have

$$n_1(\mu+1)-m_1(\nu+1)=\epsilon_1, \qquad n(\mu+1)-m(\nu+1)=\epsilon,$$

since $n_k=n$, $m_k=m$ and $\epsilon_k=0$, so that the system of linear equations

$$n_1x-m_1y=\epsilon_1, \qquad nx-my=\epsilon, \tag{18}$$

has the solution $(x=\mu+1,\ y=\nu+1)$. If the determinant Δ of the system (18) is different from zero, then by Cramer's rule

$$|\nu+1|=\frac{|n_1\varepsilon-\varepsilon_1 n|}{|\Delta|}\leqslant 2n,$$

since $|\epsilon|, |\epsilon_1|\leqq 1$, $n_1<n$, and Δ is an integer. But we have $\nu\geqq 4M^2>2n$. We have obtained a contradiction, from which it follows that $\Delta=0$. Since (18) has a solution, we have $\begin{vmatrix} -m_1 & \epsilon_1 \\ -m & \epsilon \end{vmatrix}=0$, whence $m_1\epsilon=\epsilon_1 m$, which is impossible, since $0<m_1<m$, $\epsilon=\pm 1$, $|\epsilon_1|\leqq 1$, and m_1, m, ϵ_1 and ϵ are integers. This contradiction shows that $k=1$. Hence the sequence $(m_1,\cdots,m_k)$ consists of one element $m_k=m$. From 2) we obtain $Q_1=\cdots=Q_{m-1}=C$. Contradiction. The proposition is proved.

14.7. *Suppose α is a V^M-term and β is a Θ^M-term. Then the equation $\alpha=\beta$ has no solutions.*

PROOF. The proof is divided into four cases.

1) α is a V_1^M-term and β is a Θ_2^M-term. Then our assertion follows from 13.5 and 14.1.

2) α is a V_1^M-term and β is a Θ_1^M-term. Then, using 14.1 and 13.2, we obtain that the equation $\alpha=\beta$ is equivalent to the system

$$\alpha_1^\lambda\alpha_2=\beta_1^{\mu+a}\beta_2, \qquad \lambda\geqq(4M)^6, \qquad \mu\geqq M, \tag{1}$$

where $a\geqq 0$, (α_1,α_2) and (β_1,β_2) are normal pairs of predicate terms, and where $\beta_1^2\beta_2$ is equivalent to a Θ_2^M-term γ, and $\alpha_1^2\alpha_2$ is a V_2^M-term. By 13.1, any solution of the system (1) satisfies the equation $\alpha_1^2\alpha_2=\beta_1^2\beta_2$ or, in view of the equivalence of the terms γ and $\beta_1^2\beta_2$, the equation $\alpha_1^2\alpha_2=\gamma$. The latter expanded has the form $[xA_i]_{i=1}^{n-1}x=\gamma\,\&\,\omega(x,M)$, where $1<n\leqq 3M\leqq M^2$ and $[A_1^\partial,\cdots,[A_{n-1}^\partial\leqq M$. Since γ is a Θ_2^M-term, by 14.6 the equation $\alpha_1^2\alpha_2=\gamma$ has no solutions. But then the equation $\alpha=\beta$ also has no solutions.

3) α is a $\widetilde{V}_1^M$-term and β is a Θ_2^M-term. Expressing the term α in more detail, we obtain a system $(xA)^\lambda x=\beta$, $\mathscr{E}(xA)$, $\urcorner\tau(x,M)$, where $[A^\partial\leqq M$. This system is equivalent to the disjunction

$$[(xA)^\lambda x=\beta, \quad \lambda\geqslant M^2, \quad \mathscr{E}(x,A), \quad \urcorner\tau(x,M)]\vee\bigvee_{n=2}^{M^2}[(xA)^{n-1}x=\beta, \tag{2}$$
$$\mathscr{E}(xA), \quad \urcorner\tau(x,M)].$$

The disjunction (2) has no solutions. Indeed, the right-hand summand has no solutions, since by 14.6 the equation $(xA)^{n-1}x=\beta$ has no solutions if $1<n\leqq M^2$. Consider the left summand. We introduce predicate terms α_1 and α_2. Put $\alpha_1=[xA;\mathscr{E}(xA)\,\&\urcorner\tau(x,M)]$, $\alpha_2=[x;\mathscr{E}(xA)\,\&\urcorner\tau(x,M)]$. The pair (α_1,α_2) is normal. The left summand of (2) can be written in the form

$$\alpha_1^\lambda\alpha_2=\beta, \qquad \lambda\geqq M^2. \tag{3}$$

By 13.5, the system (3) has no solutions in the region $\lambda>M^2$. But also when $\lambda=M^2$ the equation $\alpha_1^\lambda\alpha_2=\beta$ has no solutions. The proof is exactly the same as that of 13.5, except that the inequality $[\alpha_1^\partial \leqq (\nu-1)[S^\sigma C^\partial$ (see 13.5 (5)) now follows from the fact that $\lambda = M^2$ and α_2 differs from α_1 in length by the constant $[A^\partial \leqq M$. Thus (2) has no solutions. Then the equation $\alpha=\beta$ also has no solutions.

4) α is a $\widetilde{V}_1^M$-term and β is a Θ_1^M-term. By 13.2, the equation $\alpha=\beta$ is equivalent to the system $\alpha = \beta_1^{\mu+a}\beta_2$, $\mu \geqq M$, where $a \geqq 0$ and (β_1,β_2) is a normal pair of predicate terms. Here the term $\beta_1^2\beta_2$ is equivalent to a Θ_2^M-term γ and satisfies the condition $\beta_2 \gg M$. By 14.2, any solution of the system $\alpha = \beta_1^{\mu+a}\beta_2$, $\mu \geqq M$, satisfies the equation $(xA)^2x = \beta_1^2\beta_2$ or, in view of the equivalence of the terms $\beta_1^2\beta_2$ and γ, the equation $(xA)^2x=\gamma$. The last equation, by 14.6, has no solutions. Then the equation $\alpha=\beta$ also has no solutions. The proposition is proved.

§15. Equations of type VV

We now turn to equations of type V^MV^M. We also consider systems made up of such equations.

We will first prove that equations of type V^MV^M have no solutions such that $\neg\epsilon(x,y,M)$ (x and y are the word unknowns occurring in the left- and right-hand sides of the equation).

15.1. *Consider the equation $\alpha(x,\lambda)=\beta(y,\mu)$, where α and β are V_1^M-terms, x and y are word unknowns, and λ and μ are positive unknowns. If (X,Y,λ,μ) is a solution of the equation $\alpha=\beta$, then $\epsilon(X,Y,M)$.*

PROOF. We write the equation $\alpha=\beta$ in expanded form. We obtain a system

(1) $$([xA_i]_{i=1}^n)^\lambda\,[xA_i]_{i=1}^r x = ([yB_i]_{i=1}^m)^\mu\,[yB_i]_{i=1}^s y,$$

(2) $$\omega(x,M), \qquad \omega(y,M), \qquad \lambda,\mu \gg (4M)^6,$$

where

$$1<n\leqq M, \quad 1<m\leqq M, \quad 0\leqq r<n, \quad 0\leqq s<m,$$
$$\nabla(A_1,\cdots,A_n), \quad \nabla(B_1,\cdots,B_m), \quad [A_i^\partial, [B_j^\partial \leqq M.$$

If (X,Y,λ,μ) is a solution of $\alpha=\beta$, then in the alphabet $\boldsymbol{\mathit{И}}$ we have the equality $([XA_i]_{i=1}^n)^\lambda[XA_i]_{i=1}^rX = ([YB_i]_{i=1}^m)^\mu[YB_i]_{i=1}^sY$, where $\lambda,\mu>1$, $\omega(X,M)$ and $\omega(Y,M)$. By 11.3, the words $[XA_i]_{i=1}^n$ and $[YB_i]_{i=1}^m$ are simple. Using 1.7, we obtain $[XA_i]_{i=1}^rX=[YB_i]_{i=1}^sY$. From this last equality, using 11.2, $\omega(X,M)$ and $\omega(Y,M)$, we conclude that $\epsilon(X,Y,M)$. The proposition is proved.

15.2. *Consider the equation*

(1) $$\alpha(u,\lambda)=\beta(v,\mu),$$

where α and β are $\widetilde{V}_1^M$-terms, λ and μ are positive unknowns, and u and v are word unknowns. If the terms α and β have different bases, then (1) *has no solutions; if the bases of α and β are the same and equal to A, then* (1) *is equivalent to the system*

(2) $$u=v, \qquad \lambda=\mu, \qquad \neg\tau(u,M), \qquad \mathscr{B}(uA).$$

PROOF. Let A and B be the bases of α and β. Then equation (1) expanded has the form

(3) $$(uA)^\lambda u=(vB)^\mu v, \quad \mathscr{B}(uA), \quad \mathscr{B}(vB), \quad \neg\tau(u,M), \quad \neg\tau(v,M),$$

where $[A^\partial, [B^\partial \leqq M$. Assume that the quadruple (U, V, λ, μ) is a solution of (3). Then $\neg\tau(U,M)$, $\neg\tau(V,M)$, UA and VB are simple words, and $(UA)^\lambda U = (VB)^\mu V$. If one of the numbers λ or μ is greater than one, then by 1.26 we have $U = V$, $A = B$ and $\lambda = \mu$. If $\lambda = \mu = 1$, then $UAU = VBV$, which, as is easily seen, contradicts the condition $\neg\tau(U,M) \,\&\, \neg\tau(V,M)$. Hence if (U, V, λ, μ) is a solution of (3), then $A = B$, $U = V$ and $\lambda = \mu$. Now our assertion is obvious.

15.2′. *If* (U, V, λ, μ) *is a solution of the equation* 15.2 (1), *then* $\epsilon(U, V, M)$.

This follows from 15.2 and the implication $(u = v) \supset \epsilon(u, v, M)$.

In §12 we defined the predicate $\epsilon(x, y, M)$. We now define a predicate $\epsilon(x_1, \cdots, x_n, M)$ by means of the equivalence

$$\varepsilon(x_1, \ldots, x_n, M) \Leftrightarrow \bigvee_{i \neq j} \varepsilon(x_i, x_j, M),$$

where the disjunction extends over all possible pairs of indices i, j, $i \neq j$.

Let us agree on the following notation. If some symbol π (it makes no difference what kind of symbol—it can even be a numeral in parentheses) is used to denote the system

$$\mathop{\&}_{i=1}^{n} (x_i = \alpha_{i1} = \ldots = \alpha_{iq_i}) \,\&\, \mathscr{D},$$

where $n \geqq 1$, $q_1, \cdots, q_n \geqq 1$, $x_1, \cdots, x_n$ are unknowns, the α_{ij} are predicate terms, and $\mathscr{D}$ is a LDR, then the symbol π^* will denote the system

$$\mathop{\&}_{i=1}^{n} (x_i = \overset{*}{\alpha}_{i1} = \ldots = \overset{*}{\alpha}_{iq_i}) \,\&\, \mathscr{D},$$

where α_{ij}^* is the left-hand side of α_{ij}.

15.3. *Suppose* $n > 1$, $1 \leqq k \leqq n$; $v, x_1, \cdots, x_n$ *are distinct word unknowns, and* x *is a letter in the list* $x_1, \cdots, x_n$. *Consider the system*

$$\begin{aligned} x_1 &= \alpha(x, \mu) = \beta(v, \mu_1), \\ x_i &= \beta(v, \mu_i) \qquad (i = 2, \ldots, k), \end{aligned} \tag{1}$$

where $\mu, \mu_1, \cdots, \mu_k$ *are positive unknowns,* α *is a* V_1^M*-term, and* β *is a* $\widetilde{V}_1^M$*-term. Then one of the following two assertions holds*:

1) *The system* (1) *has no solutions.*

2) *The system* (1) *is equivalent in the region* $\neg\epsilon(x_1, \cdots, x_n, M)$ *to the system*

$$\mathop{\&}_{i=1}^{k} (x_i = \alpha(x, \mu_i)) \,\&\, (\mu = \mu_1). \tag{2}$$

In this connection we have the implication $(2)^* \supset \exists v (1)^*$.

PROOF. The expanded form of (1) is the system

$$\begin{aligned} & x_1 = ([xA_i]_{i=1}^m)^\mu [xA_i]_{i=1}^r \, x = (vA)^{\mu_1} v, \\ & x_i = (vA)^{\mu_i} v \qquad (i = 2, \ldots, k), \\ & \omega(x, M), \qquad \mu \geqslant (4M)^6, \qquad \mathscr{E}(vA), \qquad \neg\tau(v, M), \end{aligned} \tag{1′}$$

where $0 \leqq r < m$, $m > 1$, and $A_1, \cdots, A_m, A$ are words in the alphabet И; we also have

$$\max\{m, [A_1^\partial, \ldots, [A_m^\partial, [A^\partial\} \leqslant M, \qquad \nabla(A_1, \ldots, A_m). \tag{3}$$

The system (2) expanded has the form

$$(2') \quad \mathop{\&}_{i=1}^{k} (x_i = ([xA_i]_{i=1}^{m})^{\mu_i} [xA_i]_{i=1}^{r} x) \,\&\, \omega(x, M) \,\&\, \mathop{\&}_{i=1}^{k} (\mu_i \geqslant (4M)^6) \quad \& \, \mu = \mu_1.$$

Suppose $(X_1, \cdots, X_n, V, \mu, \mu_1, \cdots, \mu_k)$ is a solution of (1′) such that $\neg\epsilon(X_1, \cdots, X_n, M)$. Then

$$(4) \qquad ([XA_i]_{i=1}^{m})^{\mu} [XA_i]_{i=1}^{r} X = (VA)^{\mu_1} V, \quad \mu \geqslant (4M)^6, \quad \mu_1 > 0,$$

$$(5) \qquad X_i = (VA)^{\mu_i} V \quad (i = 1, \ldots, k),$$

$$(6) \qquad \omega(X, M),$$

$$(6') \qquad \neg\tau(V, M),$$

and VA is a simple word. Arguing as in 15.1, we obtain that $[XA_i]_{i=1}^m$ is a simple word. From (6′) we have $[V^{\partial} > M \geqq [A^{\partial}$. Hence from (4) and 1.26 (the roles of T, V, λ and X in 1.26 are now played by $[XA_i]_{i=1}^m$, $[XA_i]_{i=1}^r X$, μ and V) we obtain

$$(7) \qquad [XA_i]_{i=1}^m = VA, \qquad [XA_i]_{i=1}^r X = V,$$

$$(8) \qquad \mu = \mu_1.$$

From (5) and (7) we have

$$(9) \qquad X_i = ([XA_i]_{i=1}^m)^{\mu_i} [XA_i]_{i=1}^r X \qquad (i = 1, \cdots, k).$$

Also, we know that the quantities m, $[X^{\partial}$ and $\mu_1, \cdots, \mu_k$ are positive. Hence from (9) we conclude that x is different from $x_1, \cdots, x_k$. Then $\neg\epsilon(X_1, \cdots, X_n, M)$ implies that $\neg\epsilon(X_i, X, M)$ for $i = 1, \cdots, k$. Recalling the definition of the predicate ϵ (preceding Theorem 12.1), we conclude from (3) and (9) that $\mu_1, \cdots, \mu_k \geqq (4M)^6$. From (9), (6) and (8) it follows that $(X_1, \cdots, X_n, \mu, \mu_1, \cdots, \mu_k)$ is a solution of the system (2′). We have proved that if $(X_1, \cdots, X_n, V, \mu, \mu_1, \cdots, \mu_k)$ is a solution of (1′) such that $\neg\epsilon(X_1, \cdots, X_n, M)$, then $(X_1, \cdots, X_n, \mu, \mu_1, \cdots, \mu_k)$ is a solution of (2′). From (6) it follows that $[X^{\partial} > M \geqq [A^{\partial}$. Hence in (7) we must have $r = m - 1$ and $A = A_m$. We have proved that if we do not have $r = m - 1$ and $A = A_m$, then (1′) has no solutions.

Suppose $r = m - 1$, $A = A_m$, and suppose (2′) has a solution $(X_1, \cdots, X_n, \mu, \mu_1, \cdots, \mu_k)$. Then $\mu_1, \cdots, \mu_k \geqq (4M)^6$ and we have (9), (8) and (6). Put $V = [XA_i]_{i=1}^{m-1} X$. We will prove that $(X_1, \cdots, X_n, V, \mu, \mu_1, \cdots, \mu_k)$ is a solution of (1′). Since $r = m - 1$ and $A_m = A$, we have $VA = [XA_i]_{i=1}^m$. Hence (9) and (8) imply (5) and (4). Since $VA = [XA_i]_{i=1}^m$ and the word $[XA_i]_{i=1}^m$ is simple, VA is simple. Finally, since $X \angle V$, (6) implies (6′). Hence $(X_1, \cdots, X_n, V, \mu, \mu_1, \cdots, \mu_k)$ is a solution of (2′). Similarly, if $r = m - 1$, $A_m = A$ and $(X_1, \cdots, X_n, \mu, \mu_1, \cdots, \mu_k)$ is a solution of (2)*, then there exists V such that $(X_1, \cdots, X_n, V, \mu, \mu_1, \cdots, \mu_k)$ is a solution of (1)* ((1)* is obtained from (1′) by discarding the bottom row, and (2)* from (2′) by discarding the factor $\omega(x, M)$).

We summarize our argument. We have proved: a) if we do not have $A_m = A$ and $r = m - 1$, then (1) has no solutions; b) if $A_m = A$ and $r = m - 1$, then for any solution $(X_1, \cdots, X_n, \mu, \mu_1, \cdots, \mu_k)$ of the system (2′) (the system (2)*) there exists V such that $(X_1, \cdots, X_n, V, \mu, \mu_1, \cdots, \mu_k)$ is a solution of the system (1′) (the system (1)*); c) if $(X_1, \cdots, X_n, V, \mu, \mu_1, \cdots, \mu_k)$ is a solution of (1′) such that $\urcorner\epsilon(X_1,\cdots,X_n,M)$, then $(X_1,\cdots,X_n,\mu,\mu_1,\cdots,\mu_k)$ is a solution of (2′). The assertion being proved follows from a), b) and c). The proposition is proved.

15.3′. *If (X, V,μ,μ_1) is a solution of the equation $\alpha(x,\mu)=\beta(v,\mu_1)$ of* 15.3, *then $\epsilon(X, V, M)$.*

Indeed, if (X, V,μ,μ_1) is a solution of $\alpha=\beta$, then we have 15.3 (3) and (7), from which $\epsilon(X, V, M)$ follows.

15.4. *Consider the system*

$$(1)\qquad \begin{aligned} x_1&=\alpha(u,\mu)=\beta(v,\mu_1)\\ x_i&=\beta(v,\mu_i)\qquad (i=2,\cdots,k),\end{aligned}$$

where α and β are $\widetilde{V}_1^M$-terms, x_i, u and v are word unknowns, μ and $\mu_1,\cdots,\mu_k$ are positive unknowns, and v does not occur in the list $x_1,\cdots,x_k$. If the bases of α and β are different, then (1) *has no solutions. If the bases of α and β are the same, then* (1) *is equivalent in the region $\urcorner\epsilon(x_1,\cdots,x_k,M)$ to the system*

$$(2)\qquad \mathop{\&}_{i=1}^{k}[x_i=\alpha(u,\mu_i)]\;\&\;\mu=\mu_1.$$

In this connection we have the implication $(2)^*\supset\exists v(1)^*$.

Proof. If the bases of α and β are different, then by 15.2 the equation $\alpha=\beta$ has no solutions. Hence the system (1) also has no solutions. Suppose the bases of α and β are the same and equal to A. By 15.2, the equation $\alpha=\beta$ is equivalent to the system 15.2 (2). Hence (1) is equivalent to the system

$$(3)\qquad x_i=\alpha(u,\mu_i)\quad(i=1,\cdots,k),\quad u=v,\quad \mu=\mu_1,\quad \mathscr{E}(uA),\quad \urcorner\tau(u,M).$$

We have the implication $(x=\alpha(u,\mu))\supset[\mathscr{E}(uA)\,\&\urcorner\tau(u,M)]$. Hence (3) is equivalent to the system $\&_{i=1}^{k}(x_i=\alpha(u,\mu_i))\,\&\,u=v\,\&\,\mu=\mu_1$. Since v does not occur in the list $x_1,\cdots,x_k$, this last system is equivalent with respect to the unknowns $x_1,\cdots,x_k$, so certainly in the region $\urcorner\epsilon(x_1,\cdots,x_k,M)$, to the system (2). But then (1) is equivalent in the region $\urcorner\epsilon(x_1,\cdots,x_k,M)$ to (2). The implication $(2)^*\supset\exists v(1)^*$ is obvious. The proposition is proved.

15.5. *Consider the system*

$$(1)\qquad x=\alpha(u,\lambda)=\beta(u,\mu),$$

where α and β are V^M-terms, x and u are word unknowns, and λ and μ are positive unknowns. One of the following two assertions holds:

1) *The system* (1) *has no solutions.*
2) (1) *is equivalent to the system* $(x=\alpha)\,\&\,(\lambda=\mu)$.

Proof. There are three possible cases.

Case 1. α and β are $\widetilde{V}_1^M$-terms. Then the assertion follows in an obvious way from the argument in 15.4.

Case 2. α and β are V_1^M-terms. Then (1) expanded has the form (cf. 15.1 (1) and (2))

$$(2)\quad x=([uA_i]_{i=1}^{n})^{\lambda}[uA_i]_{i=1}^{r}u=([uB_i]_{i=1}^{m})^{\mu}[uB_i]_{i=1}^{s}u,\quad \omega(u,M),\quad \lambda,\mu\geqq(4M)^6,$$

where r, n, s, m, A_i and B_i are the same as in 15.1. Arguing as in 15.1 and relying on Lemma 3 preceding 11.3 and the implication $\omega(u,M)\supset\neg\tau(u,M)$, we see that one of the following two assertions holds: 1) the system (2) has no solutions; 2) $r=s$, $n=m$ and $A_i=B_i$ $(i=1,\cdots,n)$. In the first case we have 1); in the second, 2).

Case 3. One of the terms α or β is a V_1^M-term, and the other is a $\widetilde{V}_1^M$-term. Without loss of generality, we may assume that α is a V_1^M-term and β is a $\widetilde{V}_1^M$-term. Then (1) has no solutions. It is easy to see this if we observe that (1) is a system of the form 15.3 (1) (with $k=1$ and $v=x=u$). Assuming that (U,λ,μ) is a solution of (1) and arguing as in 15.3, we arrive at the equality $UA=[UA_i]_{i=1}^m$, where $m>1$ and $[U^\partial>M\geqq[A^\partial$, which is impossible. The proposition is proved.

§16. Equations of types ΣG and VG

In solving equations of type ΣG there arises, first of all, the question of which beginnings of the word $F(A,B,\lambda_0,\cdots,\lambda_{k-1})$ can be a (sufficiently high) power of a fixed simple word. More precisely, we are speaking of the equality $S^tW=F(A,B,\lambda_0,\cdots,\lambda_{k-1})$, where S is simple and $t[S^\partial$ is large in comparison with $[A^\partial$ and $[B^\partial$. We will prove that in this case there exists $p<k$ such that $F(A,B,\lambda_0,\cdots,\lambda_{p-1})$ is a power of S. This result is the key to the solution of equations of type ΣG. Combining this result with the δ-representation of the function F of §9, we will reduce equations of type ΣG to the exponential equations of §2.

The material is presented in the following order. We first study the equality $S^tW=F(A,B,\lambda^0,\cdots,\lambda_{k-1})$ (Propositions 16.1–16.3). We then solve equations whose left-hand side is a parametric word and whose right-hand side is an F-term (16.4 and 16.5). From these last two results we derive the solution of equations of type ΣG (16.6 and 16.7). Finally, using 16.5 and 11.1, we prove that equations of type VG have no solutions (16.10–16.12).

We now turn to the equality $S^tW=F(A,B,\lambda_0,\cdots,\lambda_{k-1})$. We will prove a preliminary lemma.

LEMMA. *Suppose S is a simple word, $[U^\partial\geq[S^\partial$, $V\neq\Lambda$, and $S^m=UVU^2A$. Then U is a power of S.*

PROOF. It follows from our assumptions that there exist k, $0<k<m$, and words A and B such that $U=S^kA$, $S=AB$ and $B\neq\Lambda$. Substituting these expressions for U and S into $S^m=UVU^2Z$ and cancelling the resulting equality on the left by U, we have

$$B(AB)^{m-k-1}=V(AB)^kA(AB)^kAZ,\qquad k>0, \tag{1}$$

from which it follows that $VAB\angle B(AB)^{m-k-1}$. By 1.4″, there exists $c\geqq 0$ such that $V=B(AB)^c$. Cancelling (1) on the left by $V(AB)^kA$, we obtain $B(AB)^{m-2k-c-2}=(AB)^kAZ$, which easily implies that either $A=\Lambda$ or $BA=AB$. In the latter case, by 1.2′, we also have $A=\Lambda$. Thus, $A=\Lambda$. Then $U=S^k$, as required.

16.1. *Suppose S is a simple word, $k>1$, and*

$$S^tW=F(A,B,\lambda_0,\cdots,\lambda_{k-1}), \tag{*}$$

$$\max\{[A^\partial,[B^\partial\}\leqq M, \tag{1}$$

(2) $$t[S^\partial > 9[S^\partial + \max\{6M, 6[S^\partial\}.$$

Then there exist $a > 0$ and p, $0 \leqq p \leqq k-1$, such that $F(A, B, \lambda_0, \cdots, \lambda_{p-1}) = S^a$.

PROOF. For brevity, put

$$F_{-1} = B, \quad F_i = F(A, B, \lambda_0, \cdots, \lambda_{i-1}) \quad (i = 0, 1, \cdots, k).$$

Since $k > 1$, we have $F_k = F_{k-1}^{\lambda_{k-1}} F_{k-2}$, and (*) assumes the form

(3) $$S^t W = F_{k-1}^{\lambda_{k-1}} F_{k-2}.$$

Since $k > 1$, we also have $[F_{k-2}^\partial < [F_{k-1}^\partial$, and hence

(4) $$[F_{k-1}^{\lambda_{k-1}} F_{k-2}^\partial < (\lambda_{k-1} + 1)\, [F_{k-1}^\partial.$$

Since $k > 1$, $[A^\partial < [F_{k-1}^\partial$. The following three cases are possible: 1) $[F_{k-1}^\partial < [S^\partial$; 2) $[A^\partial < [S^\partial \leqq [F_{k-1}^\partial$; 3) $[S^\partial \leqq [A^\partial$.

Case 1. $[F_{k-1}^\partial < [S^\partial$. Let us estimate $[S^\partial$ from above. It follows from (3) and (4) that $t[S^\partial < (\lambda_{k-1} + 1)\, [F_{k-1}^\partial$. Since $t > 3$ and $[F_{k-1}^\partial < [S^\partial$, we have $[S^\partial < (t-2)\, [S^\partial > (\lambda_{k-1} + 1)\, [F_{k-1}^\partial - 2[S^\partial < (\lambda_{k-1} - 1)\, [F_{k-1}^\partial$. Thus, $[F_{k-1}^\partial < [S^\partial < (\lambda_{k-1} - 1)\, [F_{k-1}^\partial$ and $t \geqq 2$. Then from (3) and 1.5 and the simplicity of S it follows that F_{k-1} is a power of S. The definition of the function F implies that all of the words F_i $(i = -1, 0, \cdots, k)$ are nonempty. Hence F_{k-1} is a positive power of S and our assertion holds with $p = k - 1$.

Case 2. Now suppose $[A^\partial < [S^\partial \leqq [F_{k-1}^\partial$. In this case it follows from $[A^\partial < [F_0^\partial < [F_1^\partial < \cdots < [F_{k-1}^\partial$ that there exists q, $1 \leqq q \leqq k-1$, such that

(5) $$[F_0^\partial < [F_1^\partial < \cdots < [F_{q-1}^\partial < [S^\partial \leqq [F_q^\partial.$$

Subcase 2.1. Assume that

(6) $$(\lambda_{q-1} - 1)\, [F_{q-1}^\partial \geqq [S^\partial.$$

Since $1 \leqq q \leqq k-1$, we have $F_{q-1}^{\lambda_{q-1}} F_{q-2} = F_q \angle F_{k-1} \angle F_{k-1}^{\lambda_{k-1}}$, so that $F_{q-1}^{\lambda_{q-1}} \angle F_{k-1}^{\lambda_{k-1}}$. Then from (3) we have for some V

(7) $$S^t W = F_{q-1}^{\lambda_{q-1}} V.$$

We know that $[F_{q-1}^\partial < [S^\partial \leqq (\lambda_{q-1} - 1)\, [F_{q-1}^\partial$ (see (5) and (6)) and $t \geqq 2$. Hence from (7) and 1.5 it follows that F_{q-1} is a power of S and our assertion holds with $p = q - 1$.

Subcase 2.2. Now assume that

(8) $$(\lambda_{q-1} - 1)\, [F_{q-1}^\partial < [S^\partial.$$

We will prove that

(9) $$[F_q^\partial \leqq 2[S^\partial + \max\{M, [S^\partial\}.$$

Indeed, since $q \geqq 1$, we have $F_q = F_{q-1}^{\lambda_{q-1}} F_{q-2}$. By (1), $[F_{q-2}^\partial \leqq \max\{M, [F_{q-1}^\partial\}$. By (5), $[F_{q-1}^\partial < [S^\partial$. Hence, in view of (8), we have

$$[F_q^\partial = (\lambda_{q-1} - 1)\, [F_{q-1}^\partial + [F_{q-1}^\partial + [F_{q-2}^\partial < [S^\partial + [S^\partial + \max\{M, [S^\partial\},$$

from which (9) follows.

We now compare Subcase 2.2 and Case 3. As a consequence of (5) and (9), Subcase 2.2 is characterized by the fact that there exists q, $1 \leqq q \leqq k-1$, such that $[F_{q-1}^\partial < [S^\partial$, $[F_q^\partial \geqq [S^\partial$ and $[F_q^\partial \leqq 2[S^\partial + \max\{M, [S^\partial\}$. Case 3 is that

$[S^\partial \leqq [A^\partial$. Taking into account $A = F_0$, $B = F_{-1}$ and (1), we conclude that Case 3 is characterized by the fact that for $q = 0$ we have $[F^\partial_{q-1} \leqq M$, $[F^\partial_q \geqq [S^\partial$ and $[F^\partial_q \leqq M$. We now consider a new Case 4, which consists of the assumption that there exists q, $0 \leqq q \leqq k-1$, such that

$$[F^\partial_{q-1} \leqq \max\{M, [S^\partial\}, \tag{10}$$

$$[F^\partial_q \geqq [S^\partial, \tag{11}$$

$$[F^\partial_q \leqq 2[S^\partial + \max\{M, [S^\partial\}. \tag{12}$$

Once we prove our assertion for Case 4, we will have proved it both for Case 3 and for Subcase 2.2.

So suppose we have (10) – (12), where $0 \leqq q \leqq k-1$.

Subcase 4.1. Suppose $\lambda_q > 1$. Since $0 \leqq q \leqq k-1$, it is easy to see that $F_q^{\lambda_q} \angle F_{k-1}^{\lambda_{k-1}}$. Then from (3) we have for some V

$$S^t W = F_q^{\lambda_q} V. \tag{13}$$

We have $\lambda_q > 1$ and $[F^\partial_q \geqq [S^\partial$. It follows from (2) and (12) that $[F^\partial_q < (t-1)[S^\partial$. From (13), 1.5 and the simplicity of S we obtain that F_q is a power of S and our assertion holds with $p = q$.

Subcase 4.2. Now suppose

$$\lambda_q = 1. \tag{14}$$

Let us estimate $[F^\partial_{q+1}$. From (14), (12), (10) and (2) we have

$$[F^\partial_{q+1} = [F_q^{\lambda_q} F^\partial_{q-1} = [F_q F^\partial_{q-1} \leqslant 2[S^\partial + 2\max\{M, [S^\partial\} < (t-1)[S^\partial. \tag{15}$$

From this and (*) we conclude that $q+1 < k$, whence $q < k-1$.

Subcase 4.2.1. Assume that $\lambda_{q+1} > 1$. We have $q \geqq 0$ and $q < k-1$. Then it is easy to see that $F_{q+1}^{\lambda_{q+1}} \angle F_{k-1}^{\lambda_{k-1}}$. But then from (3) we have for some V

$$S^t W = F_{q+1}^{\lambda_{q+1}} V. \tag{16}$$

We know $\lambda_{q+1} > 1$. From (11) we have $[S^\partial \leqq [F^\partial_q < [F^\partial_{q+1}$. From (16), (15) and 1.5 we conclude that F_{q+1} is a power of S and our assertion holds with $p = q+1$.

Subcase 4.2.2. Now suppose that

$$\lambda_{q+1} = 1. \tag{17}$$

We have shown above that $q < k-1$, i.e. $q \leqq k-2$.

If $q = k-2$, then from (4), (17), (15) and (2) we have

$$\begin{aligned}[F_{k-1}^{\lambda_{k-1}} F^\partial_{k-2} &< (\lambda_{k-1}+1)[F^\partial_{k-1} \\ &= (\lambda_{q+1}+1)[F^\partial_{q+1} \leqslant 4[S^\partial + 4\max\{M, [S^\partial\} < (t-1)[S^\partial,\end{aligned}$$

which contradicts (3). Consequently, $q < k-2$.

If $q = k-3$, $\lambda_{q+2} = 1$, then from (4), (17), (15), (12) and (2) we have

$$\begin{aligned}[F_{k-1}^{\lambda_{k-1}} F^\partial_{k-2} &< (\lambda_{k-1}+1)[F^\partial_{k-1} = (\lambda_{q+2}+1)[F^\partial_{q+2} = 2[F^\partial_{q+2} = 2[F_{q+1}^{\lambda_{q+1}} F^\partial_{q+2} \\ &= 2[F_{q+1} F^\partial_q \leqq 8[S^\partial + 6\max\{M, [S^\partial\} < t[S^\partial,\end{aligned}$$

which contradicts (3). Hence either $q < k-3$ or $q = k-3$, $\lambda_{q+2} > 1$.

But if $q < k-3$, then $F_{q+2}F_{q+1} \angle F_{q+3} \angle F_{k-1} \angle F_{k-1}^{\lambda_{k-1}}$. If $q = k-3$ and $\lambda_{q+2} > 1$, then $F_{q+2}F_{q+1} \angle F^2_{q+2} \angle F_{q+2}^{\lambda_{q+2}} = F_{k-1}^{\lambda_{k-1}}$. Thus $F_{q+2}F_{q+1} \angle F_{k-1}^{\lambda_{k-1}}$, and hence

$F_{q+1}^{\lambda_{q+1}} F_q F_{q+1} \angle F_{k-1}^{\lambda_{k-1}}$. Now from (3), (14) and (17) we have for some V

$$S^t W = F_{q+1} F_q F_{q+1} V = F_q F_{q-1} F_q F_q F_{q-1} V. \tag{18}$$

From (2), (10) and (12) it follows that $[F_q F_{q-1} F_q F_q^\partial < t[S^\partial$. Hence from (18) there exists Z such that $S^t = F_q F_{q-1} F_q^2 Z$. Since $[F_q^\partial \geqq [S^\partial$, it follows from this last equality, in accordance with the lemma, that F_q is a power of S and our assertion holds with $p = q$. The proposition is proved.

If $[S^\partial$ is large in comparison with $[A^\partial$ and $[B^\partial$ and the natural requirement $\mathscr{L}(A,B)$ is satisfied, then we obtain the following extension of 16.1.

16.2. *Suppose S is a simple word, $[A^\partial, [B^\partial \leqq M$, $[S^\partial > 6M$, $\mathscr{L}(A,B)$, $t > 14$, $k > 1$, and*

$$S^t W = F(A, B, \lambda_0, \cdots, \lambda_{k-1}). \tag{*}$$

Then there exists p, $0 \leqq p \leqq k-1$, such that $F(A, B, \lambda_0, \cdots, \lambda_{p-1}) = S$.

Proof. It is easy to see that 16.1 is applicable to (*); hence there exists p, $0 \leqq p \leqq k-1$, such that $F(A, B, \lambda_0, \cdots, \lambda_{p-1})$ is a positive power of S. But $[S^\partial > 6M$ by hypothesis. Therefore $[F(A, B, \lambda_0, \cdots, \lambda_{p-1})^\partial \geqq [S^\partial > 6M$. Since by hypothesis $\mathscr{L}(A,B)$ and $[A^\partial, [B^\partial \leqq M$, by 5.12 the word

$$F(A, B, \lambda_0, \cdots, \lambda_{p-1})$$

is simple. Hence $F(A, B, \lambda_0, \cdots, \lambda_{p-1}) = S$, as required.

The following result is used in solving equations of types VG and ΘG. We use the usual abbreviations $F_i = F(A, B, \lambda_0, \cdots, \lambda_{i-1})$, $i = -1, 0, \cdots, k$.

16.3. *Suppose S is a simple word, $\mathscr{L}(A,B)$, $\mathscr{L}(S,C)$, $k > 1$, and*

$$(S^\sigma C)^{15} W = F(A, B, \lambda_0, \cdots, \lambda_{k-1}), \tag{*}$$

$$[A^\partial, [B^\partial \leqq M; \quad [S^\partial \leqq 30M, \quad [C^\partial \leqq 90M, \tag{1}$$

$$[S^\sigma C^\partial > (6M)^3. \tag{2}$$

Then there exist integers a, p and q, $0 < a < \sigma$, $0 \leqq p < q \leqq k-1$, such that $F_p = S^a$, $[F_p^\partial \leqq 30M$ and $F_q = S^\sigma C$. Here either $q = p+1$ or $q = p+2$, $\lambda_{p+1} = 1$.

Proof. The left-hand side of (*) begins with the word S^σ. Hence (*) is an equality of the form 16.1 (*) (the role of t in 16.1 (*) is now played by σ). It follows from (1) and (2) that $\sigma[S^\partial > 8[S^\partial + \max\{6M, 6[S^\partial\}$. Hence by 16.1 there exist integers $a > 0$ and p, $0 \leqq p \leqq k-1$, such that

$$F_p = S^a. \tag{3}$$

From $\mathscr{L}(S,C)$, (1), (2) and 1.10 it follows that $S^\sigma C$ is a simple word. Furthermore, $[S^\sigma C^\partial > 6M$. Applying Proposition 16.2 to (*) (the roles of S and t in 16.2 are played by $S^\sigma C$ and 15), we conclude that there exists q, $0 \leqq q \leqq k-1$, such that

$$F_q = S^\sigma C. \tag{4}$$

From (3) and 5.12′ we have $[F_p^\partial \leqq \max\{6M, [S^\partial\} \leqq 30M$. Now from (1) – (4) it follows that $a < \sigma$ and $[F_p^\partial < [F_q^\partial$. Since $p, q \geqq 0$, we see that $[F_p^\partial < [F_q^\partial$ implies $p < q$. Thus we have proved that there exist integers a, p and q, $0 < a < \sigma$, $0 \leqq p < q \leqq k-1$, such that (3) and (4) hold and $[F_p^\partial \leqq 30M$. It remains to prove that either $q = p+1$ or $q = p+2$, $\lambda_{p-1} = 1$. Let us prove this. Applying 5.1, we obtain

$$F_q = F(F_{p+1}, F_p, \lambda_{p+1}, \cdots, \lambda_{q-1})$$
$$= F(F_p^{\lambda_p} F_{p-1}, F_p, \lambda_{p+1}, \cdots, \lambda_{q-1}) = F(S^{a\lambda_p} F_{p-1}, S^a, \lambda_{p+1}, \cdots, \lambda_{q-1}).$$

Thus,

$$F_q = F(S^{a\lambda_p} F_{p-1}, S^a, \lambda_{p+1}, \cdots, \lambda_{q-1}). \tag{5}$$

Now assume that $q > p+1$ and $\lambda_{p+1} > 1$. In this case the right-hand side of (5) begins with the word $(S^{a\lambda_p}F_{p-1})^2S^a$. Then it follows from (4) and (5) that for some Z

$$S^\sigma C = (S^{a\lambda_p}F_{p-1})^2S^aZ.. \tag{6}$$

If $\sigma[S^\partial < [S^{a\lambda_p}F_{p-1}S^\partial$, then from (6) we have $[C^\partial \geqq [S^{a\lambda_p}F_{p-1}^\partial > (\sigma-1)[S^\partial$, which contradicts (1) and (2). Consequently, $\sigma[S^\partial \geqq [S^{a\lambda_p}F_{p-1}S^\partial$. Then (6) implies that $F_{p-1}S \angle S^{\sigma - a\lambda_p}$. From this last relation and 1.4 it follows that F_{p-1} is a power of S, which, together with (3) and the condition $\mathscr{L}(A,B)$, contradicts 5.3. From this contradiction we conclude that the case $q > p+1$, $\lambda_{p+1} > 1$ is impossible. Since $q > p$, two possibilities remain: either $q = p+1$ or $q > p+1$, $\lambda_{p+1} = 1$. If $q > p+2$, $\lambda_{p+1} = 1$, then the right-hand side of (5) begins with the word $S^{a\lambda_p}F_{p-1}S^aS^{a\lambda_p}F_{p-1}$. Then from (4), (5) we have for some Z that $S^\sigma C = S^{a\lambda_p}F_{p-1}S^aS^{a\lambda_p}F_{p-1}Z$. From this equality we derive a contradiction in the same way as from (6). Thus, the case $q > p+2$, $\lambda_{p+1} = 1$ is impossible. Hence either $q = p+1$ or $q = p+2$, $\lambda_{p+1} = 1$, as required. The proposition is proved.

We can now turn to the solution of equations of type ΣG. To formulate the results we need the notion of a processional-Diophantine relation (PDR).

Recall that a procession is defined to be an ordered set of positive integers. If $h_1 = (\lambda_0, \cdots, \lambda_{k-1})$ and $h_2 = (\mu_0, \cdots, \mu_{s-1})$ are processions, then we denote by (h_1, h_2) the procession $(\lambda_0, \cdots, \lambda_{k-1}, \mu_0, \cdots, \mu_{s-1})$. In a similar way we define the procession $(h_0, \cdots, h_k)$, where $k \geqq 0$ and $h_0, \cdots, h_k$ are processions, in particular, positive integers.

Consider the conjunction

$$h = (h_0, \lambda_1, h_1, \cdots, \lambda_k, h_k) \;\&\; \mathscr{D}(\sigma_1, \cdots, \sigma_n, \lambda_1, \cdots, \lambda_k), \tag{*}$$

where h is a processional unknown, $h_0, h_1, \cdots, h_k$ are fixed processions, $\lambda_1, \cdots, \lambda_k$, $\sigma_1, \cdots, \sigma_n$ are positive unknowns, and $\mathscr{D}$ is a LDR. By a solution of the conjunction (*) we mean any set $(\bar{h}, \bar{\lambda}_1, \cdots, \bar{\lambda}_k, \bar{\sigma}_1, \cdots, \bar{\sigma}_n)$ of values of the unknowns $h, \lambda_1, \cdots, \lambda_k$, $\sigma_1, \cdots, \sigma_n$ such that $\mathscr{D}(\bar{\sigma}_1, \cdots \bar{\sigma}_n, \bar{\lambda}_1, \cdots, \bar{\lambda}_k)$ holds and $\bar{h}$ is equal to the procession $(h_0, \bar{\lambda}_1, h_1, \cdots, \bar{\lambda}_k, h_k)$.

We now define a PDR. By definition, any PDR has the form

$$\bigvee_{i=1}^{N} [\exists \lambda_1 \ldots \lambda_k \,(h = (h_0^i, \lambda_1, h_1^i, \ldots, \lambda_k, h_k^i) \;\&\; \mathscr{D}_i(\sigma_1, \ldots, \sigma_n, \lambda_1, \ldots, \lambda_k))],$$

where $k, n, N \geqq 0$, the h_j^i are fixed processions, h is a processional unknown, $\lambda_1, \cdots, \lambda_k$, $\sigma_1, \cdots, \sigma_n$ are positive unknowns, and each $\mathscr{D}_i$ is a LDR. We say that this is a PDR among the unknowns $h, \sigma_1, \cdots, \sigma_n$. We write $\mathscr{K}(h; \sigma_1, \cdots, \sigma_n)$ if $\mathscr{K}$ is a PDR among the unknowns h, $\sigma_1, \cdots, \sigma_n$.

Our main result concerning equations of type $\Sigma^M G^M$ is that any such equation is equivalent to some PDR among the unknowns occurring in the left- and right-hand sides of the equation.

We first consider two equations in which the left-hand side is a parametric word and the right-hand side has the form $F(A,B,h)$. We then pass from F to G.

16.4. *Consider the system in two unknowns*

$$(*)\quad [S^tP_i]_{i=1}^n = F(A,B,h),\quad t > N+6M+1,\quad \neg\mathscr{B}([A,B,h],M),$$

where $n \geqq 1$, t is a positive unknown, h is a processional unknown, $\mathscr{L}(A,B)$, $[A^\partial, [B^\partial \leqq M$, S is a simple word, $[S^\partial \leqq (6M)^3$, and

$$N > \max\{14[S^\partial + n[S^\partial + [P_1\cdots P_n^\partial, (6M)^3\}.$$

The system (*) *is equivalent to some PDR.*

PROOF. A pair of processions $(h_1; h_2)$ is called admissible for the system (*) if the following conditions are satisfied: 1) $F(A,B,h_1)$ is a power of S; 2) the length of the procession $\delta(h_2)$ (see 9.1) does not exceed n; 3) $h_2 \neq \Lambda$, $h_2 \neq (1)$, and if $[F(A,B,h_1)^\partial \leqq 30M$, then h_2 is not a procession of one of the forms (μ), $(1,\mu)$, $(\mu,1)$ or $(1,\mu,1)$.

The number of pairs which are admissible for a given system (*) is finite. Indeed, if $F(A,B,h_1)$ is a power of S, then from $\mathscr{L}(A,B)$, $[A^\partial, [B^\partial \leqq M$ and 5.12′ it follows that $[F(A,B,h_1)^\partial \leqq \max\{6M, [S^\partial\}$, from which it follows that the number of processions h_1 is finite. The finiteness of the number of processions h_2 follows from 9.1.2 b) and the estimate $[\delta(h_2)^\partial \leqq n$. It is also obvious how to construct a list of all pairs which are admissible for the system (*).

Suppose there is given a system (*), and let $(h_i; g_i)$ $(i = 1,\cdots,l;\ l \geqq 0)$ be a list of all pairs which are admissible for (*). We will prove that (*) is equivalent to a disjunction

$$(1)\qquad \bigvee_{i=1}^{l} \exists\lambda\,[h = (h_i,\lambda,g_i)\,\&\,\mathscr{D}_i(t,\lambda)],$$

where each of $\mathscr{D}_1,\cdots,\mathscr{D}_l$ is a LDR.

Consider the solutions of (*). Assume that the pair $(t, h = (\lambda_0,\cdots,\lambda_{k-1}))$ is a solution of (*). Then

$$(2)\qquad [S^tP_i]_{i=1}^n = F(A,B,\lambda_0,\cdots,\lambda_{k-1}),\qquad n \geqq 1,$$

$$(3)\qquad \neg\mathscr{B}([A,B,\lambda_0,\cdots,\lambda_{k-1}],M),$$

where $t > N+6M+1$. Equality (2) has the form 16.1 (*). It follows from (3) and the inequalities $[A^\partial, [B^\partial \leqq M$ that $k > 1$. From the estimates for t and N we obtain

$$(4)\qquad t[S^\partial > (t-1)\,[S^\partial > N+6M > n[S^\partial + [P_1\cdots P_n^\partial + 14[S^\partial + 6M;$$

in particular, $t[S^\partial > 6M + 14[S^\partial$. Hence by 16.1 there exist p, $0 \leqq p \leqq k-1$, and $a > 0$ such that

$$(5)\qquad F(A,B,\lambda_0,\cdots,\lambda_{p-1}) = S^a.$$

We will prove that the pair of processions $((\lambda_0,\cdots,\lambda_{p-1}),(\lambda_{p+1},\cdots,\lambda_{k-1}))$ is admissible for (*).

For brevity, put $F_i = F(A,B,\lambda_0,\cdots,\lambda_{i-1})$ $(i = -1,0,\cdots,k)$. Since $0 \leqq p \leqq k-1$, by 5.1

(6) $F(A, B, \lambda_0, \ldots, \lambda_{k-1}) = F(F_p, F_{p-1}, \lambda_p, \ldots, \lambda_{k-1}) = F(S^a, F_{p-1}, \lambda_p, \ldots, \lambda_{k-1})$.

By hypothesis, $[S^\partial \leqq (6M)^3$. Then from (3) and (5) it follows that the procession $(\lambda_{p+1}, \cdots, \lambda_{k-1})$ is nonempty and is different from the procession "(1)". By 9.1.2, there exists a δ-procession $\delta_1 \cdots \delta_{m+1}$ such that

$$(7) \qquad \delta_1 \cdots \delta_{m+1} = \delta(\lambda_{p+1}, \cdots, \lambda_{k-1}), \qquad m > 0, \quad \delta_1 = 0.$$

Then by 9.1.3 we have (for any λ_p)

$$(8) \qquad F(S^a, F_{p-1}, \lambda_p, \ldots, \lambda_{k-1}) = [S^{a(\lambda_p+\delta_i)} F_{p-1}]_{i=1}^m S^{a\lambda_p} \langle S^a, \quad F_{p-1}\rangle_{k-p}.$$

From (2), (6) and (8) we obtain

$$(9) \qquad [S^t P_i]_{i=1}^n = [S^{a(\lambda_p+\delta_i)} F_{p-1}]_{i=1}^m S^{a\lambda_p} \langle S^a, \quad F_{p-1}\rangle_{k-p}.$$

We will prove that $m < n$. For this purpose we estimate $t[S^\partial$. Assume that $t[S^\partial \geqq [S^{a\lambda_p} F_{p-1} S^\partial$. Then, since $\delta_1 = 0$, $n, m > 0$ and $a\lambda_p > 0$, we have from (9) that $S^{a\lambda_p} F_{p-1} S \angle S^t$ and $F_{p-1} S \angle S^{a\lambda_p - t}$. From the latter relation and 1.4 it follows that F_{p-1} is a power of S, which, together with (5) and the condition $\mathscr{B}(A, B)$, contradicts 5.3. From this contradiction we conclude that $[S^\partial < [S^{a\lambda_p} F_{p-1} S^\partial$, and hence

$$(10) \qquad (t-1)[S^\partial < [S^{a\lambda_p} F_{p-1}^\partial.$$

Using 5.1, we have $[\langle F_{p-1}, S^a\rangle_{k-p}^\partial = [F_{p-1} S^{a\partial} > [F_{p-1}^\partial$. Hence the length of the right-hand side of (9) is greater than $(m+1)[S^{a\lambda_p} F_{p-1}^\partial$. Then the left-hand side of (9) is also longer than $(m+1)[S^{a\lambda_p} F_{p-1}^\partial$. Therefore,

$$[[S^t P_i]_{i=1}^n{}^\partial$$
$$= n(t-1)[S^\partial + n[S^\partial + [P_1 \cdots P_n^\partial > (m+1)[S^{a\lambda_p} F_{p-1}^\partial > (m+1)(t-1)[S^\partial$$

(see (10)). Since $(t-1)[S^\partial > n[S^\partial + [P_1 \cdots P_n^\partial$ (see (4)), it follows from our chain of inequalities that $n > m$.

Thus the inequality $n > m$ is proved. It remains to show that if

$$[F(A, B, \lambda_0, \cdots, \lambda_{p-1})^\partial \leqq 30M,$$

then $(\lambda_{p+1}, \cdots, \lambda_{k-1})$ is not a procession of one of the forms (μ), $(1, \mu)$, $(\mu, 1)$ or $(1, \mu, 1)$. But this follows from (3). We have proved that the pair $((\lambda_0, \cdots, \lambda_{p-1}), (\lambda_{p+1}, \cdots, \lambda_{k-1}))$ is admissible for the system (*).

Let us estimate $a\lambda_p[S^\partial$. We first estimate $[F_p^\partial$. From $\mathscr{B}(A, B)$, $[A^\partial, [B^\partial \leqq M$, (5) and 5.12′ it follows that $[F_p^\partial \leqq \max\{6M, [S^\partial\} \leqq (6M)^3$. It is also obvious that $[F_{p-1}^\partial \leqq \max\{M, [F_p^\partial\}$. Then from (10) we have $a\lambda_p[S^\partial > (t-1)[S^\partial - [F_{p-1}^\partial > N[S^\partial > (6M)^3$. We know $[F_p^\partial \leqq (6M)^3$. Hence $a[S^\partial \leqq (6M)^3$, $\lambda_p > 1$.

We summarize our argument. We have proved that if the procession $(\lambda_0, \cdots, \lambda_{k-1})$ and the integer t satisfy the system (*), then $(\lambda_0, \cdots, \lambda_{k-1})$ "splits" into three parts: $h_1 = (\lambda_0, \cdots, \lambda_{p-1})$, (λ_p), $h_2 = (\lambda_{p+1}, \cdots, \lambda_{k-1})$ such that the pair $(h_1; h_2)$ is admissible for (*) and the pair (t, λ_p) satisfies the system

$$(11) \qquad [S^t P_i]_{i=1}^n = [S^{a\lambda} S^{a\delta_i} F_{p-1}]_{i=1}^m S^{a\lambda} \langle S^a, \quad F_{p-1}\rangle_{k-p}, \qquad t > N + 6M + 1,$$
$$a\lambda [S^\partial > (6M)^3, \qquad \lambda > 1,$$

Arguing in the reverse order, we see that if the procession $(\lambda_0, \cdots, \lambda_{p-1})$ satisfies (5), the procession $(\lambda_{p+1}, \cdots, \lambda_{k-1})$ satisfies (7), and the pair (t, λ_p)

satisfies (11), then the pair $(t, (\lambda_0, \cdots, \lambda_{k-1}))$ satisfies the equation of the system (*) and the condition $t > N+6M+1$. To each admissible pair $(h_i; g_i)$ corresponds its own system $(11)_i$, where $\delta_1 \cdots \delta_{m+1}$ is $\delta(g_i)$, a is defined by the equality $F(A,B,h_i) = S^a$, and F_{p-1} is $F(A,B,h_i')$, where h_i' is obtained from h_i by dropping the last component. By 2.3, the system $(11)_i$ is equivalent to some LDR $\mathscr{D}_i(t,\lambda)$. It follows that any solution of (*) satisfies the PDR (1) and that any solution (t,h) of the PDR (1) satisfies the equation of the system (*) and the condition $t > N+6M+1$. It remains to prove that if (t,h) is a solution of the PDR (1), then the procession h satisfies the condition $\neg\mathscr{B}([A,B,h],M)$. But this follows in an obvious way from condition 3) of admissibility and from the estimates $\lambda > 1$ and $a\lambda[S^\partial > (6M)^3$. We have proved that the system (*) is equivalent to the PDR (1). The proposition is proved.

The following assertion plays a fundamental role in solving equations of types ΘG and VG.

16.5. *Consider the system in the three unknowns σ, ν and h*

$$(1)\qquad \begin{gathered} [(S^\sigma C)^\nu S^\sigma A_i]_{i=1}^{n-1}(S^\sigma C)^{\nu+\epsilon} S^\sigma A_n = F(A, B, h), \\ \nu \geqslant N, \quad \sigma[S^\partial + [C^\partial > (6M)^3, \quad \neg\mathscr{B}([A, B, h], M), \end{gathered}$$

where σ and ν are positive unknowns, h is a processional unknown, $\epsilon = 0, \pm 1$; $n \geqq 2$, $N \geqq \max\{2n, 16\}$, $\mathscr{E}(A,B)$, $\mathscr{E}(S,C)$, S is a simple word, and

$$(2)\qquad [A^\partial,\ [B^\partial \leqslant M,\quad [S^\partial \leqslant 30M,\quad [C^\partial \leqslant 90M,\quad [A_1^\partial, \dots, [A_n^\partial \leqslant 91M.$$

The system (1) *is equivalent to some PDR. If $\epsilon = 0$, then* (1) *has no solutions.*

PROOF. The scheme of the argument is the same as in 16.4, but must be augmented because of the two-stage arrangement of the unknowns in the left-hand side of the equation of the system (1) and the analysis of the case $\epsilon = 0$.

We will say, with reference to the system (1), that we are in Case α (Case β) if there exist a procession $h = (\lambda_0, \cdots, \lambda_{p-1})$ and integers $a > 0$ and d such that $F_p = S^a$ and $S^d C \perp\!\!\!\perp F_{p-1}$ (such that $F_p = S^a$ and $S^d C \perp\!\!\!\perp F_{p-1} S^a$), the symbol $\perp\!\!\!\perp$ being the same as that defined after 1.20. It is easy to see that these cases are inconsistent and that the integers a and d and the procession h are uniquely determined by the given A, B, S and C. Indeed, if, for example, $F(A,B,h) = S^a$ and $F(A,B,g) = S^b$, then $F(A,B,h)F(A,B,g) = F(A,B,g)F(A,B,h)$. Using the condition $\mathscr{E}(A,B)$, we obtain $h = g$. Then $a = b$. If now $S^d C \perp\!\!\!\perp F_{p-1}$ and $S^e C \perp\!\!\!\perp F_{p-1} S^a$ for some e and d, then $S^e C \perp\!\!\!\perp S^d C S^a$ and $S^{e-d} C \perp\!\!\!\perp C S^a$. Since $a > 0$, we have $e - d > 0$ and $S^{e-d} C = C S^a$, which by 1.1 contradicts $\mathscr{E}(S,C)$.

A pair of processions $(h_1 = (\lambda_0, \cdots, \lambda_{p-1}), h_2)$ is called an α-pair (β-pair) of the system (1) if the following conditions are satisfied: 1) $F_p = S^a$ and $S^d C \perp\!\!\!\perp F_{p-1}$ ($S^d C \perp\!\!\!\perp F_{p-1} S^a$) for some a and d; 2) $h_2 \neq \Lambda$, $h_2 \neq (1)$ and the length of the δ-procession $\delta(h_2)$ is less than $n+1$. It is obvious that the number of α-pairs and β-pairs of a given system (1) is finite and that the system has only α-pairs or only β-pairs. It is also obvious how to find all α-pairs or β-pairs of the system (1).

Suppose there is given a system (1), and let $(h_i; g_i)$ $(i = 1, \cdots, l;\ l \geqq 0)$

be a list of all of its α-pairs, and $(\tilde{h}_i;\tilde{g}_i)$ $(i=1,\cdots,s;\ s\geqq 0)$ a list of all of its β-pairs. We will prove that (1) is equivalent to a PDR

$$\bigvee_{i=1}^{l} \exists\lambda\mu\,[h=(h_i,\lambda,\mu+1,g_i)\,\&\,\mathscr{D}_i(\sigma,\nu,\lambda,\mu)] \tag{3}$$
$$\bigvee_{i=1}^{s} \exists\lambda\mu\,[h=(\tilde{h}_i,\lambda,1,\mu,\tilde{g}_i)]\,\&\,\tilde{\mathscr{D}}_i(\sigma,\nu,\lambda,\mu)],$$

where each of $\mathscr{D}_i$ and $\tilde{\mathscr{D}}_i$ is a LDR.

Consider the solutions of (1). We will prove that if the system has a solution, then we are in either Case α or Case β. Suppose the triple $(\sigma,\nu,h=(\lambda_0,\cdots,\lambda_{k-1}))$ is a solution of (1). Then

$$[(S^\sigma C)^\nu S^\sigma A_i]_{i=1}^{n-1}(S^\sigma C)^{\nu+\varepsilon}S^\sigma A_n=F(A,B,\lambda_0,\ldots,\lambda_{k-1}), \tag{4}$$

$$\neg\mathscr{B}([A,B,\lambda_0,\ldots,\lambda_{k-1}],M), \tag{5}$$

$$\nu\geqslant N\geqslant 16,\qquad [S^\sigma C^\partial>(6M)^3. \tag{6}$$

Since $\nu\geqq 16$, the left-hand side of (4) begins with the word $(S^\sigma C)^{15}$, so that (4) is an equality of the form 16.3 (*). Applying 16.3 to (4) (the applicability of 16.3 to (4) obviously follows from our assumptions), we conclude that there exist integers a, p and q, $0<a<\sigma$, $0\leqq p<q\leqq k-1$, such that

$$F_p=S^a,\qquad [F_p^\partial\leqq 30M, \tag{7}$$

$$F_q=S^\sigma C, \tag{8}$$

and either $q=p+1$ or $q=p+2$, $\lambda_{p+1}=1$. In accordance with this we consider two cases.

Case 1. $q=p+1$. We will prove that for the system (1) we are in Case α, and the pair $((\lambda_0,\cdots,\lambda_{p-1}),(\lambda_{p+2},\cdots,\lambda_{k-1}))$ is an α-pair of (1). Since $q=p+1$, from (7) and (8) we have $S^\sigma C=F_q=F_{p+1}=F_p^{\lambda_p}F_{p-1}=S^{a\lambda_p}F_{p-1}$, which implies that there exists d such that $\sigma-a\lambda_p=d$ and $S^dC\amalg F_{p-1}$. From the latter relation and (7) we see that for the system (1) we are in Case α. We will now prove that the pair $((\lambda_0,\cdots,\lambda_{p-1}),(\lambda_{p+2},\cdots,\lambda_{k-1}))$ is an α-pair of (1). We know $0<p+1\leqq k-1$. Hence by 5.1

$$\begin{aligned}F(A,B,\lambda_0,\cdots,\lambda_{k-1})&=F(F_{p+1},F_p,\lambda_{p+1},\cdots,\lambda_{k-1})\\&=F(S^\sigma C,S^a,\lambda_{p+1},\cdots,\lambda_{k-1}).\end{aligned} \tag{9}$$

Since $[S^\partial\leqq 30M$, it follows from (7) and (5) that the procession $(\lambda_{p+2},\cdots,\lambda_{k-1})$ is nonempty and different from the procession "(1)". Then by 9.1.2 there exists a δ-procession $\delta_1\cdots\delta_{m+1}$ such that

$$\delta_1\cdots\delta_{m+1}=\delta(\lambda_{p+2},\cdots,\lambda_{k-1}),\qquad m>0,\quad \delta_1=0. \tag{10}$$

From (10) and 9.1.3 we have (for any λ_{p+1})

$$F(S^\sigma C,S^a,\lambda_{p+1},\ldots,\lambda_{k-1})=[(S^\sigma C)^{\lambda_{p+1}+\delta_i}S^a]_{i=1}^m(S^\sigma C)^{\lambda_{p+1}}\langle S^\sigma C,S^a\rangle_{k-p-1}. \tag{11}$$

From 5.1 we see that $\langle S^\sigma C,S^a\rangle_q=S^\sigma\langle C,S^a\rangle_q$ for any integer q. Taking this into account and putting $\mu=\lambda_{p+1}-1$, we have from (4), (9) and (11)

$$\begin{aligned}&[(S^\sigma C)^\nu S^\sigma A_i]_{i=1}^{n-1}(S^\sigma C)^{\nu+\varepsilon}S^\sigma A_n\\&\qquad=[(S^\sigma C)^{\mu+\delta_i}S^\sigma CS^a]_{i=1}^m(S^\sigma C)^{\mu+1}S^\sigma\langle C,S^a\rangle_{k-p-1}.\end{aligned} \tag{12}$$

We will now prove that $m < n$. We will first prove by contradiction that $\mu \geqq \nu - 1$. The left-hand side of (12) begins with the word $(S^\sigma C)^\nu S^\sigma$, the right-hand side with the word $(S^\sigma C)^{\mu+1} S^a S^\sigma C$ (since $\delta_1 = 0$, $m > 0$ and $\mu \geqq 0$). If $\nu > \mu + 1$, then, cancelling (12) on the left by $(S^\sigma C)^{\mu+1}$, we obtain $S^\sigma C S^\sigma V_1 = S^a S^\sigma C V_2$ for some V_1 and V_2. Cancelling on the left by S^σ and taking into account that $0 < a < \sigma$, we obtain $CS^a = S^a C$, $a > 0$, which by 1.2 implies that S and C are powers of some word, which contradicts $\mathscr{E}(S, C)$. From this contradiction we conclude that $\nu \leqq \mu + 1$, so that

$$\mu \geqq \nu - 1 \geqq N - 1 \geqq 2n - 1. \tag{13}$$

Comparing the lengths of the left- and right-hand sides of (12) and considering that $(\mu + 1)[S^\sigma C^\partial > n\sigma[S^\partial + [A_1 \cdots A_n^\partial$, we see that $m < n$. We have proved that the pair of processions $((\lambda_0, \cdots, \lambda_{p-1}), (\lambda_{p+2}, \cdots, \lambda_{k-1}))$ is an α-pair of (1). It follows that the quadruple $(\sigma, \nu, \lambda_p, \lambda_{p+1})$ satisfies the system

$$(14)\quad \begin{gathered} [(S^\sigma C)^\nu S^\sigma A_i]_{i=1}^{n-1} (S^\sigma C)^{\nu+\varepsilon} S^\sigma A_n = [(S^\sigma C)^{\mu+\delta_i} S^\sigma B_i]_{i=1}^{m+1}, \\ \sigma - a\lambda = d, \qquad \sigma[S^\partial + [C^\partial > (6M)^3, \qquad \mu \geqslant \nu - 1, \qquad \nu \geqslant N, \end{gathered}$$

where $\mu = \lambda_{p+1} - 1$, $\delta_{m+1} = 1$, $B_i = CS^a$ $(i = 1, \cdots, m)$ and $B_{m+1} = \langle C, S^a \rangle_{k-p-1}$.

Arguing in the reverse order, we see that if the pair of processions $(h_1; h_2)$ is an α-pair of (1), and the quadruple $(\sigma, \nu, \lambda_p, \lambda_{p+1})$ satisfies (14), then the triple $(\sigma, \nu, h = (h_1, \lambda_p, \lambda_{p+1}, h_2))$ satisfies the upper row of the system (1). To each α-pair $(h_i; g_i)$ corresponds its own system $(14)_i$ (the δ-procession $\delta_1 \cdots \delta_m$ and the integers a and d depend on i). Consider the system (14).

Estimating the lengths of the words A_i and B_j, we obtain (see (2) and (7))

$$\max_{i,j}\{[A_i^\partial, [B_j^\partial\} + 2[SC^\partial \leqq (6M)^3. \tag{15}$$

By 2.9, the system $(14)_i$ is equivalent to some LDR $\mathscr{D}_i(\sigma, \nu, \lambda, \mu)$. It follows that any solution of (1) satisfies the PDR (3) and any solution of (3) satisfies the upper row of the system (1) (in the case under consideration $s = 0$). It remains to prove that any solution (σ, ν, h) of (3) satisfies the condition $\neg\, \mathscr{B}([A, B, h], M)$. But this follows easily from the fact that $g_i \neq \Lambda$, $g_i \neq (1)$, and for the solution $(\sigma, \nu, \lambda, \mu)$ of (3) we have $\mu > 1$ and $a\lambda > 30M$ (the latter follows from the equality $\sigma - a\lambda = d$ after comparing the sizes of σ, a and d). We have proved that the system (1) is equivalent to the PDR (3).

Now consider the case $\epsilon = 0$. If in this case the system (1) has a solution, then, by what has been proved, the system (14) also has a solution. Then by 2.8 (the applicability of 2.8 is guaranteed by the estimates (6), (15) and the condition $\mathscr{E}(S, C)$) there exist sequences $(n_1, \cdots, n_k)$, $(m_1, \cdots, m_k)$ and $(\epsilon_1, \cdots, \epsilon_k)$ of integers such that $k > 0$, $1 \leqq n_1 < \cdots < n_k = n$, $1 \leqq m_1 < \cdots < m_k = m + 1$, $|\epsilon_1|, \cdots, |\epsilon_k| \leqq 1$ and $\epsilon_k = 0$; $B_i = C$ for any i different from $m_1, \cdots, m_k$; and we have the equalities

$$\lambda_1 + \cdots + \lambda_{n_i} - (\mu_1 + \cdots + \mu_{m_i}) = m_i - n_i + \epsilon_i \qquad (i = 1, \cdots, k), \tag{16}$$

where $\lambda_1 = \cdots = \lambda_n = \nu$ and $\mu_i = \mu + \delta_i$ $(i = 1, \cdots, m + 1)$. Since $a > 0$, we have $B_i \neq C$ $(i = 1, \cdots, m)$. Hence the sequence $(m_1, \cdots, m_k)$ agrees with the sequence $(1, \cdots, m + 1)$, so that $k = m + 1 > 1$ and $m_i = i$ $(i = 1, \cdots, k)$. Considering these relations and passing from $\lambda_1, \cdots, \lambda_n$ and $\mu_1, \cdots, \mu_{m+1}$

to ν and μ, we obtain from (16)

$$(17)\quad n_i\nu - (i\mu + \delta_1 + \ldots + \delta_i) = i - n_i + \varepsilon_i \quad (i = 1, \ldots, k-1),$$

$$(18)\quad n_k\nu - (k\mu + \delta_1 + \ldots + \delta_m + 1) = k - n_k + \varepsilon_k.$$

By simple transformations we can rewrite (17) and (18) as

$$(19)\quad n_i(\nu + 1) - i(\mu + 1) = \delta_1 + \ldots + \delta_i + \varepsilon_i \quad (i = 1, \ldots, k-1),$$

$$(20)\quad n_k(\nu + 1) - k(\mu + 1) = \delta_1 + \ldots + \delta_m + 1 + \varepsilon_k.$$

We have $n_k = n$, $\epsilon_k = 0$, $k = m+1$ and $\delta_1 = 0$. Now from (19) with $i = 1$ and from (20) we conclude that the system of linear equations

$$(21)\quad n_1x - y = \epsilon_1, \qquad nx = (m+1)y = \delta_2 + \cdots + \delta_m + 1$$

has a solution $(x = \nu + 1,\ y = \mu + 1)$. We will prove that this is impossible. If the determinant Δ of the system (21) is different from zero, then by Cramer's rule

$$|\nu + 1| = \frac{|\varepsilon_1(m+1) - (\delta_2 + \ldots + \delta_m + 1)|}{|\Delta|} \leqslant 2m + 1,$$

since $|\epsilon_1| \leqq 1$ and $\delta_2, \cdots, \delta_m$ are either 0 or 1. We know $m < n$; hence $\nu + 1 < 2n$, which contradicts the inequality $\nu \geqq N$ and the assumption $N \geqq 2n$. From this contradiction we deduce that $\Delta = 0$. Since the system (21) has a solution, we must have $\epsilon_1(m+1) = \delta_2 + \cdots + \delta_m + 1$, which is again impossible, since $|\epsilon_1| \leqq 1$, ϵ_1 is an integer, and $\delta_2, \cdots, \delta_m$ are either 0 or 1. Thus we have proved that if $\epsilon = 0$, $N \geqq 2n$, then the system (1) has no solutions. The consideration of Case 1 is completed.

Case 2. $q = p + 2$ and $\lambda_{p+1} = 1$. This case is handled by the same scheme as Case 1. We will prove that in Case 2 we are, with reference to the system (1), in Case β, and the pair of processions $((\lambda_0, \cdots, \lambda_{p-1}), (\lambda_{p+3}, \cdots, \lambda_{k-1}))$ is a β-pair of (1). We have (see (7) and (8)) $S^\sigma C = F_q = F_{p+2} = F_p^{\lambda_p} F_{p-1} F_p = S^{a\lambda_p} F_{p-1} S^a$. Therefore, $F_{p+1} = F_p^{\lambda_p} F_{p-1} = [S^\sigma C * S^a]$ and there exists d such that $\sigma - a\lambda_p = d$ and $S^d C \perp\!\!\!\perp F_{p-1} S^a$, so that (see (7)) with reference to (1) we are in Case β. We know $0 < p + 2 \leqq k - 1$. Hence by 5.1

$$(22)\quad \begin{aligned} F(A, B, \lambda_0, \ldots, \lambda_{k-1}) &= F(F_{p+2}, F_{p+1}, \lambda_{p+2}, \ldots, \lambda_{k-1}) \\ &= F(S^\sigma C, [S^\sigma C * S^a], \lambda_{p+2}, \ldots, \lambda_{k-1}). \end{aligned}$$

Since $[S^\partial \leqq 30M$ and $\lambda_{p+1} = 1$, from (7) and (5) it follows that the procession $(\lambda_{p+3}, \cdots, \lambda_{k-1})$ is nonempty and different from the procession "(1)". Then by 9.1.2 there exists a δ-procession $\delta_1 \cdots \delta_{m+1}$ such that

$$(23)\quad \delta_1 \cdots \delta_{m+1} = \delta(\lambda_{p+3}, \cdots, \lambda_{k-1}), \qquad m > 0, \quad \delta_1 = 0.$$

From (23) and 9.1.3 we have, putting $\mu = \lambda_{p+2}$ and $q = k - p - 2$,

$$(24)\quad \begin{aligned} &F(S^\sigma C, [S^\sigma C * S^a], \lambda_{p+2}, \ldots, \lambda_{k-1}) \\ &\quad = [(S^\sigma C)^{\mu+\delta_i} [S^\sigma C * S^a]]_{i=1}^m) S^\sigma C)^\mu \langle S^\sigma C, [S^\sigma C * S^a]\rangle_q. \end{aligned}$$

If q is even, then

$$\langle S^\sigma C, [S^\sigma C * S^a]\rangle_q = [S^\sigma C * S^a] S^\sigma C = S^\sigma C S^{\sigma - a} C = S^\sigma C S^{\sigma - a}[\langle S^a, C\rangle_q * S^a].$$

If q is odd, then

$$\langle S^\sigma C, [S^\sigma C * S]\rangle_q = S^\sigma C[S^\sigma C * S^\sigma] = S^\sigma C S^{\sigma - a}[\langle S^a, C\rangle_q * S^a].$$

Hence for any q we have

$$\langle S^\sigma C, [S^\sigma C * S^a]\rangle_q = S^\sigma C S^{\sigma-a}[\langle S^a, C\rangle_q * S^a].$$

Now from (4), (22) and (24) we have

$$(25)\quad \begin{aligned}&[(S^\sigma C)^\nu S^\sigma A_i]_{i=1}^{n-1}(S^\sigma C)^{\nu+\varepsilon}S^\sigma A_n\\ &\quad = [(S^\sigma C)^{\mu+\delta_i}[S^\sigma C * S^a]]_{i=1}^{m}(S^\sigma C)^{\mu+1}S^{\sigma-a}[\langle S^a, C\rangle_q * S^a].\end{aligned}$$

We will now prove that (13) holds. The left-hand side of (25) begins with the word $(S^\sigma C)^\nu$. Since $m > 0$, $\delta_1 = 0$, $\mu > 0$ and $S^{\sigma-a} \angle [S^\sigma C * S^a]$, the right-hand side of (25) begins with the word $(S^\sigma C)^\mu[S^\sigma C * S^a]S^\sigma CS^{\sigma-a}$, which is equal to the word $(S^\sigma C)^\mu S^\sigma CS^{\sigma-a}CS^{\sigma-a}$. If $\nu > \mu+1$, then, cancelling (25) on the left by $(S^\sigma C)^{\mu+1}S^{\sigma-a}$, we obtain $S^aCV_1 = CS^{\sigma-a}V_2$ for some V_1 and V_2, and therefore $S^aC = CS^a$ (obviously, $\sigma - a > a$). This, in view of 1.2, contradicts the condition $\mathscr{L}(S, C)$. From this contradiction we conclude that $\nu \leqq \mu+1$, which implies (13). From (25) and the inequality $(\mu+1)[S^\sigma C^\partial > n(\sigma[S^\partial + 151M)$ we see that $m < n$. We have proved that the pair

$$((\lambda_0, \cdots, \lambda_{p-1}), (\lambda_{p+3}, \cdots, \lambda_{k-1}))$$

is a β-pair of (1).

Let $D = S^aC$, $Q_i = S^aA_i$, $\tau = \sigma - a$, $B_i = [S^aC * S^a]$ $(i = 1, \cdots, m)$, $B_{m+1} = [\langle S^a, C\rangle_q * S^a]$ and $\delta_{m+1} = 1$. From the equalities $\sigma - a\lambda_p = d$ and (25) it follows that the quintuple $(\sigma, \tau, \nu, \lambda_p, \lambda_{p+2} = \mu)$ satisfies the system

$$(26)\quad \begin{aligned}&[(S^\tau D)^\nu S^\tau Q_i]_{i=1}^{n-1}(S^\tau D)^{\nu+\varepsilon}S^\tau Q_n = [(S^\tau D)^{\mu+\delta_i}S^\tau B_i]_{i=1}^{m+1},\\ &\tau + a = \sigma, \quad \sigma - a\lambda = d, \quad \tau[S^\partial + [D^\partial > (6M)^3, \quad \nu \geqslant N, \quad \mu \geqslant \nu - 1.\end{aligned}$$

Arguing in the reverse order, we see that if the pair of processions $(h_1; h_2)$ is a β-pair of (1), and the quintuple $(\sigma, \tau, \nu, \lambda, \mu)$ satisfies (26), then the triple $(\sigma, \nu, h = (h_1, \lambda, 1, \mu, h_2))$ satisfies the two upper rows of the system (1).

Estimating the lengths of the words Q_i and B_j, we obtain

$$\max_{i,j}\{[Q_i^\partial, [B_j^\partial\} + 2[SD^\partial \leqslant (6M)^3.$$

The rest of the argument on the system (26) is the same as in the first case on the system (14). Preparing for the application of 2.8, we note that $\mathscr{L}(S, D)$ and $B_i \neq C$ $(i = 1, \cdots, m)$. The first fact follows trivially from the relations $D = S^aC$ and $\mathscr{L}(S, C)$. The second is proved by contradiction: if $B_i = C$, then $[S^aC * S^a] = C$ and $S^aC = CS^a$, which contradicts the condition $\mathscr{L}(S, C)$. Applying 2.9 to (26) and repeating the corresponding argument from Case 1, we conclude that the system (1) is equivalent to the PDR (3) (now we have $l = 0$), and the LDR's $\widetilde{\mathscr{D}}_i$ are constructed starting from 2.9 and the relation $\tau = \sigma + a$. The condition $\urcorner\,\mathscr{B}([A, B, h], M)$ for solutions of the PDR (3) is verified as in Case 1. In the analysis of the case $\epsilon = 0$, the inequalities $B_i \neq C$ lead to the fact that the sequence $(m_1, \cdots, m_k)$ of 2.8 agrees with the sequence $(1, \cdots, m+1)$. This allows us, by repeating verbatim the corresponding argument from Case 1, to prove that if $\epsilon = 0$, then the system (26) has no solutions. This completes the consideration of Case 2. Thus, the proposition is proved.

We can now turn to equations containing a G-term. In §§13–15 we saw how the solution of equations was facilitated by the fact that the terms with lower index one had the form $\alpha^\lambda\beta$, $\lambda \geqq N$, where $N > 1$ and (α,β) is a normal pair of predicate terms. In solving equations the λ is eliminated by means of 13.1 and 14.2. In a certain sense the situation is the same for G-terms. Namely, in solving equations we may assume that a G-term has the form $\alpha^\lambda\beta$, $\lambda > 1$, where (α,β) is a normal pair of predicate terms. Indeed (using the usual abbreviations), we have $G_k = F_{k-1}^{\lambda_{k-1}} G_{k-2}$. By 5.12, the word F_{k-1} is simple in the region defined by the predicate $\urcorner \mathscr{B}$. Also, $G_{k-2} \angle F_{k-1}$ if $k > 2$, so certainly in the region $\urcorner \mathscr{B}$. It remains to ensure the inequality $\lambda_{k-1} > 1$. This is done with the aid of 5.11, and properties 3 and 4 of the predicate $\mathscr{B}$ also hold for predicate G-terms.

We need a relation between processions, which we will denote by an arrow. Suppose $h = (\lambda_0, \cdots, \lambda_{k-1})$ and $g = (\mu_0, \cdots, \mu_{s-1})$ are processions. By definition, we write $h \to g$ if $s > 0$, $k = s+1$, $\lambda_{k-1} = 1$, $\mu_{s-1} = \lambda_{k-2}+1$ and $(\mu_0, \cdots, \mu_{s-2}) = (\lambda_0, \cdots, \lambda_{k-3})$.

Denote by h^0 the last component of a nonempty procession h, and by $|h|$ the number of components of h. If $h = (\lambda_0, \cdots, \lambda_{k-1})$, then $|h| = k$; if $k > 0$, then $h^0 = \lambda_{k-1}$.

16.6. *Suppose x is a word unknown, h, h_1 and g are processional unknowns, $T(h)$ is a G^M-term, and P, Q and R are words in the alphabet* **И** *such that $[PQR^\partial \leqq M$, $\mathscr{L}(QP, QR)$ and $Q \neq \Lambda$. The following are true*:

(1) $$(h \to g) \supset (g^0 > 1),$$

(2) $$\exists g\,[(h_1 \to g) \,\&\, (h \to g)] \Leftrightarrow (h = h_1 \,\&\, h^0 = 1),$$

(3) $$\urcorner \mathscr{B}([QP, QR, h], M) \supset (|h| > 1),$$

(4) $$(x = T(h)) \supset (|h| > 1)$$

(5) $$(|h| > 1) \supset \exists g\,((h = g \vee h \to g) \,\&\, g^0 > 1),$$

(6) $$(h \to g) \,\&\, (x = G(QP, QR, h)) \Leftrightarrow (h \to g) \,\&\, (x = G(QP, QR, g)),$$

(7) $$(h \to g) \,\&\, \urcorner \mathscr{B}([QP, QR, h], M) \Leftrightarrow (h \to g) \,\&\, \urcorner \mathscr{B}([QP, QR, g], M),$$

(8) $$(x = T(h)) \,\&\, (h = g \vee h \to g) \Leftrightarrow (x = T(g)) \,\&\, (h = g \vee h \to g),$$

(9) $$x = T(h) \Leftrightarrow \exists g\,[x = T(g) \,\&\, g^0 > 1 \,\&\, (h = g \vee h \to g)].$$

PROOF. Statements (1), (2) and (5) follow immediately from the definition of the relation $\to$. Statement (3) follows from property 1, and (7) from property 3, of the predicate $\mathscr{B}$. Statement (6) follows from 5.11. The equations $x = T(h)$ and $x = T(g)$ expanded have the form

$$x = G(QP, QR, h)\,Q \,\&\, \urcorner \mathscr{B}([QP, QR, h], M),$$
$$x = G(QP, QR, g)\,Q \,\&\, \urcorner \mathscr{B}([QP, QR, g], M),$$

respectively. Now (8) follows in an obvious way from (6) and (7), and (4) from (3). Statement (9) follows from (4), (5) and (8). The proposition is proved.

16.7. *Suppose $T(h)$ is a G^M-term, (α,β) is a normal pair of predicate terms, and $N > 1$. Then the system*

(1) $$T(h) = \alpha^\lambda\beta, \quad h^0 > 1, \quad \lambda \geqq N$$

is equivalent to the predicate

(2) $$\exists fg[T(g) = \alpha^2\beta,\ g = (f, 2), h = (f, \lambda)] \,\&\, \lambda \geqq N.$$

PROOF. Write the systems (1) and (2) in expanded form. We obtain

$$(1') \quad A^{\lambda}B = G(QP,\ QR,\ h)\,Q, \quad h^0 > 1, \quad \lambda \geqslant N, \quad \mathcal{P}\ \&\ \neg\,\mathcal{B}([QP,\ QR,\ h],\ M),$$

$$(2') \quad \exists fg\,[A^2B = G(QP,\ QR,\ g)\,Q, \quad g = (f,\ 2), \quad h = (f,\ \lambda), \quad \lambda \geqslant N, \quad \mathcal{P}\ \&\ \neg\,\mathcal{B}([QP,\ QR,\ g],\ M),$$

where A and B are the left-hand sides, and $\mathcal{P}$ is the right-hand side, of the predicate terms α and β; P, Q and R are words in the alphabet И, $[PQR^{\partial} \leqq M$, $\mathcal{S}(QP, QR)$ and $Q \neq \Lambda$.

Suppose the unknowns occurring in the term $A^{\lambda}B$ comprise the list $\xi_1, \cdots, \xi_n$, and suppose $(\overline{\xi}_1, \cdots, \overline{\xi}_n, h = (\lambda_0, \cdots, \lambda_{k-1}))$ is a solution of (1′). Then in the alphabet И we have the equality

$$(3) \quad \overline{A}^{\overline{\lambda}}\overline{B} = G(QP, QR, \lambda_0, \cdots, \lambda_{k-1})\,Q,$$

where $\lambda, \lambda_{k-1} > 1$ and the condition $\neg\,\mathcal{B}([QP, QR, \lambda_0, \cdots, \lambda_{k-1}], M)$ is satisfied. This condition implies that $k > 2$; hence $G_k = F_{k-1}^{\lambda_{k-1}}G_{k-2}$. We have

$$(4) \quad \overline{A}^{\overline{\lambda}}\overline{B} = F_{k-1}^{\lambda_{k-1}}G_{k-2}Q, \quad \overline{\lambda} > 1, \quad \lambda_{k-1} > 1.$$

Since $k > 2$, $G_{k-2}Q \angle F_{k-2} \angle F_{k-1}$. Thus, $\Lambda \neq G_{k-2}Q \angle F_{k-1}$. Also, the condition $\neg\,\mathcal{B}([\cdots], M)$ implies that $[F_{k-1}^{\partial} > 30M$, whence by 5.12 the word F_{k-1} is simple. Because of the normality of the pair (α, β) we have $\Lambda \neq \overline{B} \angle \overline{A}$ and $\overline{A}$ is a simple word. Now from (4) and 1.7 we obtain $A = F_{k-1}$, $\overline{\lambda} = \lambda_{k-1}$ and $\overline{B} = G_{k-2}Q$, whence

$$(5) \quad \overline{A}^2\overline{B} = F_{k-1}^2G_{k-2}Q = G(QP,\ QR,\ \lambda_0,\ \ldots,\ \lambda_{k-2},\ 2)\,Q.$$

By property 4 of the predicate $\mathcal{B}$ we have $\neg\,\mathcal{B}([(QP, QR, \lambda_0, \cdots, \lambda_{k-2}, 2)], M)$. It is now obvious that the set $(\overline{\xi}_1, \cdots, \overline{\xi}_n, \overline{h})$ satisfies the predicate (2). Here $f = (\lambda_0, \cdots, \lambda_{k-2})$. We have proved that any solution of the predicate (1′) satisfies the predicate (2′).

We will now prove that any solution of (2′) is a solution of (1′). Suppose $(\overline{\xi}_1, \cdots, \overline{\xi}_n, \overline{h} = (\lambda_0, \cdots, \lambda_{k-1}))$ is a solution of (2′). This means that there exist processions f and g such that $f = (\lambda_0, \cdots, \lambda_{k-2})$ and $g = (\lambda_0, \cdots, \lambda_{k-2}, 2)$, where we have (5) and the condition

$$\neg\,\mathcal{B}([QP, QR, \lambda_0, \cdots, \lambda_{k-2}, 2], M).$$

Moreover, $\lambda_{k-1} = \overline{\lambda} \geqq N > 1$. From (5) and 1.7 we obtain $\overline{A} = F_{k-1}$ and $\overline{B} = G_{k-2}Q$. Since $\overline{\lambda} = \lambda_{k-1}$, we obtain (3) and (4). Also, by property 4 of the predicate $\mathcal{B}$, we have $\neg\,\mathcal{B}([QP, QR, \lambda_0, \cdots, \lambda_{k-1}], M)$. It is now obvious that the set $(\overline{\xi}_1, \cdots, \overline{\xi}_n, \overline{h})$ satisfies (1′). The proposition is proved.

16.7′. *Suppose $T(h)$ is a G^M-term and (α, β) is a normal pair of predicate terms. Then the system*

$$(*) \quad T(h) = \alpha^{\lambda}\beta, \quad \lambda \geqq N,$$

where $N > 1$, is equivalent to the predicate (f and g are processional unknowns)

$$\exists\, fg[\alpha^2\beta = T(g),\ g = (f, 2),\ h = (f, \lambda) \vee h = (f, \lambda - 1, 1)], \quad \lambda \geqq N.$$

PROOF. By 16.6 (9), the system (*) is equivalent to the predicate

$$\exists\, h_1[\alpha^{\lambda}\beta = T(h_1),\ h_1^0 > 1,\ h = h_1 \vee h \rightarrow h_1], \quad \lambda \geqq N.$$

It remains to apply 16.7 to the system $\alpha^\lambda\beta = T(h_1)$, $h_1^0 \geqq 1$, $\lambda \geqq N$. The proposition is proved.

We can now prove the following assertion.

16.8. *The equation $T_1 = T_2$, where T_1 is a Σ^M-term and T_2 a G^M-term, is equivalent to some PDR.*

PROOF. Let h denote the processional unknown occurring in T_2. We consider four cases.

Case 1. T_1 is a Θ_2^M-term. Let the triple $[P, Q, R]$ be the base of T_2. Then the expanded version of the equation $T_1 = T_2$ has the form

$$\text{(1)}\quad \begin{gathered} [(S^\sigma C)^\nu S^\sigma A_i]_{i=1}^{n-1}(S^\sigma C)^{\nu+\varepsilon} S^\sigma A = G(QP, QR, h)\,Q, \\ \sigma[S^\partial + [C^\partial > (6M)^3, \quad \nu > 4M^2, \quad \neg\mathscr{B}([QP, QR, h], M), \end{gathered}$$

where σ and ν are positive unknowns, $\epsilon = \pm 1$, $1 < n \leqq M^2$, S is a simple word, $\mathscr{L}(S, C)$, $\mathscr{L}(QP, QR)$, $[PQR^\partial \leqq M$, $Q \neq \Lambda$, $[S^\partial \leqq 30M$ and $[A^\partial, [C^\partial, [A_i^\partial \leqq 90M$.

It follows from the definitions of the functions F and G that if h is an even unknown, then $\forall h[G(QP, QR, h)QRQP = F(QP, QR, h)]$, and if h is an odd unknown, then $\forall h[G(QP, QR, h)QPQR = F(QP, QR, h)]$. Putting $A_n = ARQP$ if h is even, $A_n = APQR$ if h is odd, we see that the upper equation of the system (1) is equivalent to the equation $[(S^\sigma C)^\nu S^\sigma A_i]_{i=1}^{n-1}(S^\sigma C)^{\nu+\epsilon} S^\sigma A_n = F(QP, QR, h)$. The system consisting of this last equation and the second row of (1) is the system described in 16.5 with $N = 4M^2$, and our assertion follows from 16.5.

Case 2. T_1 is a Θ_1^M-term. Then by 13.2 the equation $T_1 = T_2$ is equivalent to the system $\alpha^{\lambda+a}\beta = T_2$, $\lambda \geqq M$, where λ is a positive unknown, $a \geqq 0$, and (α, β) is a normal pair of predicate terms; here the term $\alpha^2\beta$ is equivalent to some Θ_2^M-term T. By 16.7′ this last system is equivalent to the predicate

$$\text{(2)}\quad \exists fg[\alpha^2\beta = T_2(g), g = (f, 2), (h = (f, \lambda + a) \vee h = (f, \lambda + a - 1, 1))] \;\&\; \lambda \geqq M,$$

where f and g are processional unknowns. Since the term $\alpha^2\beta$ is equivalent to a Θ_2^M-term T, the equation $\alpha^2\beta = T_2(g)$ is equivalent to the equation $T = T_2$. Since T is a Θ_2^M-term, by Case 1 the equation $T = T_2$ is equivalent to some PDR $\mathscr{K}(g, \sigma)$, where σ is a list of the unknowns occurring in the term T. Then the predicate (2) is equivalent to the predicate

$$\text{(3)}\quad \exists fg[\mathscr{K}(g, \sigma), g = (f, 2), (h = (f, \lambda + a) \vee h = (f, \lambda + a - 1, 1))] \;\&\; \lambda \geqq M.$$

Remembering that this is a PDR, it is not difficult to see that the predicate (3) is equivalent to some PDR $\mathscr{K}'(h; \sigma, \lambda)$, which is the desired one.

Case 3. T_1 is a Ψ_2^M-term. We consider a more general case. Suppose T_1 is a $\overline{\Psi}_2^M$-term. Then the expanded version of the equation $T_1 = T_2$ has the form

$$\text{(4)}\quad [S^\sigma A_i]_{i=1}^{n-1} S^\sigma A = G(QP, QR, h)\,Q, \quad \sigma[S^\partial > N, \quad \neg\mathscr{B}([QP, QR, h], M),$$

where σ is a positive unknown, $S, A_1, \cdots, A_{n-1}$ and A are words in the alphabet $\mathbf{И}$, S is simple, N is a natural number, and $[S^\partial \leq (6M)^3$. Arguing as in Case 1, we see that the equation of the system (4) is equivalent to the equation $[S^\sigma A_i]_{i=1}^n = F(QP, QR, h)$, where $[A_n^\partial = [APQR^\partial$. We divide the domain of the variable σ into two parts: 1) $\sigma[S^\partial \leqq N_1$; 2) $\sigma[S^\partial > N_1$, where

N_1 is sufficiently large. In the region 1) the assertion is trivial, and in 2) it follows from 16.4.

Case 4. T_1 is a Ψ_1^M-term. This case is handled exactly like Case 2, only instead of 13.2 and Case 1 we refer to 13.3 and Case 3. The proposition is proved.

16.9. *Consider the system* $x=\alpha=\beta$, *where* α *is a* G^M-*term*, β *is a* Σ^M-*term and* x *is a word unknown. The system* $x=\alpha=\beta$ *is equivalent to the system* $(x=\alpha)\,\&\,\mathscr{K}$, *where* $\mathscr{K}$ *is a PDR. In this connection we have the implication* $[(x=\alpha^*)\,\&\,\mathscr{K}]\supset(x=\alpha^*=\beta^*)$, *where* α^* *and* β^* *are the left-hand sides of the predicate terms* α *and* β.

PROOF. Let $[P,Q,R]$ be the base of α, and h the processional unknown occurring in α. The system $x=\alpha=\beta$ expanded has the form

$$x=\alpha^*,\quad \alpha^*=\beta^*,\quad \neg\mathscr{B}([QP,QR,h],M),\mathscr{D}, \tag{1}$$

where $\mathscr{D}$ is the right-hand side of β. By 16.8, the subsystem $\alpha^*=\beta^*$, $\neg\mathscr{B}$, $\mathscr{D}$ of the system (1) is equivalent to some PDR $\mathscr{K}$. Then (1) is equivalent to the system $(x=\alpha^*)\,\&\,\mathscr{K}$, where we have the implication

$$\mathscr{K}\supset[\mathscr{D}\,\&\neg\mathscr{B}([QP,QR,h],M)].$$

The proposition is proved.

We now turn to equations of type VG. We will prove that such equations have no solutions. We require two lemmas.

LEMMA 1. *Suppose the integers* r *and* k *are such that* $0<r<k$, *and suppose in the alphabet* $\mathbf{H}\cup\{y\}$ *we have the equality*

$$[yQ_i]_{i=1}^{k-1}=[yQ_i]_{i=r+1}^{k}[yQ_i]_{i=1}^{r-1}, \tag{1}$$

where $Q_1,\cdots,Q_k$ *are words in* $\mathbf{H}$. *Then* $Q_k=Q_r$.

PROOF. We define words $Q_{k+1},\cdots,Q_{k^2}$ in the following manner. If $k+1\leqq j\leqq k^2$ and ν is the remainder of the division of j by k, then

$$Q_j=\begin{cases}Q_\nu, & \text{if } 0<\nu<k,\\ Q_k, & \text{if } \nu=0.\end{cases} \tag{2}$$

Then $Q_i=Q_{k+1}$ $(i=1,\cdots,r-1)$, $[yQ_i]_{i=1}^{r-1}=[yQ_i]_{i=k+1}^{k+r-1}$, and from (1) we have

$$Q_i=Q_{r+i}\quad(i=1,\cdots,k-1). \tag{3}$$

In the sequence $1,\cdots,k-1$ choose t such that $r,2r,\cdots,tr$ are not divisible by k, but $(t+1)r$ is divisible by k. Such a t obviously exists. Each of the integers $r,2r,\cdots,tr$ on division by k leaves as remainder one of the integers $1,\cdots,k-1$. Hence by (2) the words $Q_r,Q_{2r},\cdots,Q_{tr}$ are words in the list $Q_1,\cdots,Q_{k-1}$. Hence by (3) we have $Q_r=Q_{2r}=\cdots=Q_{tr}=Q_{(t+1)r}$. Since $(t+1)r$ is divisible by k, by definition $Q_{(t+1)r}=Q_k$. But then $Q_r=Q_k$, as required.

LEMMA 2. *Suppose* $m>1$, $\mathscr{L}(QP,QR)$, $[PQR^\partial\leqq M$, *and*

$$U=G(QP,QR,\lambda_0,\cdots,\lambda_{m-1})Q.$$

Then there exist words X, A *and* B *such that* $A\neq B$, $[A^\partial,[B^\partial\leqq M$, $[X^\partial>[U^\partial$, $XAU=UBX$ *and* $UB=F(QP,QR,\lambda_0,\cdots,\lambda_{m-1})$.

PROOF. Put $X = G(QP, QR, \lambda_0, \cdots, \lambda_{m-1}, 1)Q$. We define the words A and B according to the parity of m. Put $A = PQR$ and $B = RQP$ if m is even, $A = RQP$ and $B = PQR$ if m is odd. Obviously, $QA = \langle QP, QR\rangle_{m+1}$ and $QB = \langle QP, QR\rangle_m$. We will prove that the words X, A and B are the desired ones. We have $[A^\partial = [B^\partial = [PQR^\partial \leqq M$. Also, from $\mathscr{L}(QP, QR)$ we have $QPQR \neq QRQP$, whence $A \neq B$. The inequality $[U^\partial < [X^\partial$ follows from the definition of the function G. Since $QB = \langle QP, QR\rangle_m$, we have

$$UB = G(QP, QR, \lambda_0, \cdots, \lambda_{m-1}) \langle QP, QR\rangle_m = F(QP, QR, \lambda_0, \cdots, \lambda_{m-1}).$$

It remains to prove that $XAU = UBX$. But this follows from 5.9 b). The lemma is proved.

In order to prove that equations of type $V^M G^M$ have no solutions, we express y from the equation $[yQ_i]_{i=1}^{k-1} y = \alpha$, where α is a G^M-term.

16.10. *Suppose the integers $k > 1$, $M > 0$ and the words $P, Q, R, Y, Q_1, \cdots, Q_{k-1}$ satisfy the conditions $\mathscr{L}(QP, QR)$, $[PQR^\partial, [Q_1^\partial, \cdots, [Q_{k-1}^\partial \leqq M$, and $\omega(Y, M)$, and suppose $m > 1$ and*

$$[YQ_i]_{i=1}^{k-1} Y = G(QP, QR, \lambda_0, \ldots, \lambda_{m-1})\, Q. \tag{1}$$

Then there exist σ, ν, C, D and a simple word S such that $Y = (S^\sigma C)^\nu S^\sigma D$, $\mathscr{L}(S, C)$, $[S^\partial \leqq M$, $[C^\partial, [D^\partial \leqq 7M$, $[S^\sigma C^\partial > (6M)^3$ and $\nu \geqq 4M^2$.

PROOF. By Lemma 2, there exist words X, P_1 and Q_k (it is convenient with regard to subsequent notation to denote the words A and B of Lemma 2 by P_1 and Q_k) such that $P_1 \neq Q_k$, $[P_1^\partial, [Q_k^\partial \leqq M$, $[X^\partial > [[YQ_i]_{i=1}^{k-1} Y^\partial$ and

$$[YQ_i]_{i=1}^{k} = F(QP, QR, \lambda_0, \ldots, \lambda_{m-1}), \tag{2}$$

$$XP_1 [YQ_i]_{i=1}^{k-1} Y = [YQ_i]_{i=1}^{k} X. \tag{3}$$

Since by hypothesis $\omega(Y, M)$, we have

$$\neg\tau(Y, (6M)^3) \,\&\, \neg\rho(Y, M, 4M^2). \tag{4}$$

As is evident from (3), the pair (X, Y) is a solution of the equation

$$xP_1 [yQ_i]_{i=1}^{k-1} yxx = [yQ_i]_{i=1}^{k} xxx. \tag{5}$$

Equation (5) is a normal equation of the form 11.1 (*). By hypothesis, $[Q_1^\partial, \cdots, [Q_{k-1}^\partial \leqq M$; we know that $[P_1^\partial, [Q_k^\partial \leqq M$, so that the lengths of the coefficients of (5) do not exceed M. It follows from (4) that $[Y^\partial > 0$, and since $[X^\partial > [[YQ_i]^{k-1} Y^\partial$, we have $[X^\partial > [Y^\partial > 0$. Hence by 11.1, for any integers $N \geqq 15M$ and $L > 1$ we have one of 11.1 a) – 11.1 e). Choose $N = (6M)^3$, $L = 4M^2$ and consider each of these five cases.

Case 11.1 a). Impossible (see (4)).

Case 11.1 b). In this case $X = ([YQ_i]_{i=1}^{k})^\lambda [YQ_i]_{i=1}^{r-1} Y$, where $\lambda \geqq 0$, $0 < r \leqq k$. Since (X, Y) is a solution of (5),

$$([YQ_i]_{i=1}^{k})^\lambda [YQ_i]_{i=1}^{r-1} YP_1 [YQ_i]_{i=1}^{k-1} YXX = ([YQ_i]_{i=1}^{k})^{\lambda+1} [YQ_i]_{i=1}^{r-1} YXX.$$

Cancelling this equality on the left by $([YQ_i]_{i=1}^{k})^\lambda [YQ_i]_{i=1}^{r-1} Y$ and on the right by YXX, we obtain $P_1 [YQ_i]_{i=1}^{k-1} = Q_r [YQ_i]_{i=r+1}^{k} [YQ_i]_{i=1}^{r-1}$. Then from (4) and Lemma 3 of §11 we have in the alphabet $\mathbf{N} \cup \{y\}$ the equality

$$P_1 [yQ_i]_{i=1}^{k-1} = Q_r [yQ_i]_{i=r+1}^{k} [yQ_i]_{i=1}^{r-1}.$$

Hence $P_1 = Q_r$ and $[yQ_i]_{i=1}^{k-1} = [yQ_i]_{i=r+1}^{k}[yQ_i]_{i=1}^{r-1}$ in the alphabet $\mathbf{И} \cup \{y\}$. Since $P_1 = Q_r$ and $P_1 \neq Q_k$, we have $r \neq k$. Hence $0 < r < k$ and, by Lemma 1, $Q_r = Q_k$. The relations $P_1 = Q_r$, $Q_r = Q_k$ and $P_1 \neq Q_k$ constitute a contradiction. Thus Case 11.1 b) is impossible.

Case 11.1 c). The σ, ν, S, C and R appearing in this case are exactly those whose existence we claim (here $D = R$).

Case 11.1 e). In this case $Q_1 = \cdots = Q_k$. Since by hypothesis $k > 1$, (2) implies that the word $F(QP, QR, \lambda_0, \cdots, \lambda_{m-1})$ is not simple. On the other hand, since $[Y^{\partial} > (6M)^3$ (see (4)), we have (see (2)) $[F(QP, QR, \lambda_0, \cdots, \lambda_{m-1})^{\partial} \geqq (6M)^3 > 6M$. Since by hypothesis $[PQR^{\partial} \leqq M$ and $\mathscr{L}(QP, QR)$, by 5.12 the word $F(QP, QR, \lambda_0, \cdots, \lambda_{m-1})$ must be simple. From this contradiction we conclude that Case 11.1 e) is impossible.

Case 11.1 d). In this case $X = (ZP_1)^m Z$ and $Y = (ZP_1)^n Z$, where $m > n$. Since (X, Y) is a solution of (5), Z satisfies the equation

$$(6) \qquad (zP_1)^m zP_1\, [(zP_1)^n zQ_i]_{i=1}^{k-1} (zP_1)^n z = [(zP_1)^n zQ_i]_{i=1}^{k} (zP_1)^m z.$$

Assume that equation (6) is trivial, i.e. the equality (6) is graphical. Since $m > n$, the graphical equality (6) implies that $P_1 = Q_1$. Also, we obtain from (6) successively $Q_2 = P_1, \cdots, Q_k = P_1$, which contradicts the relation $P_1 \neq Q_k$. Hence (6) is not trivial. Since the lengths of the coefficients of (6) do not exceed M and Z is a solution of this equation, by Lemma 3 of §11 we have $Z = S^{\sigma} D$, where $[S^{\partial} \leqq M$, $D \angle S$ and S is a simple word. Then $Y = (S^{\sigma} DP)^n S^{\sigma} D$. It follows from (4) that DP_1 is not a power of S (so that $\mathscr{L}(S, DP_1)$ in view of the simplicity of S), $[S^{\sigma} DP_1^{\partial} > (6M)^3$ and $n \geqq 4M^2$. Putting $C = DP_1$ and $\nu = n$, we see that σ, ν, S, C and D are as desired. The proposition is proved.

16.11. *Suppose T_1 is a V^M-term, h is a processional unknown, and $T(h)$ is a G^M-term. Then the system*

$$(*) \qquad T_1 = T(h), \qquad h^0 > 1,$$

has no solutions.

PROOF. There are two possible cases.

Case 1. T_1 is a V_1^M-term. Then by 14.1 the system (*) has the form

$$(1) \qquad \alpha^{\lambda}\beta = T(h), \quad \lambda \geqq (4M)^6, \quad h^0 > 1,$$

where λ is a positive unknown, (α, β) is a normal pair of predicate terms, and $\alpha^2\beta$ is a V_2^M-term. By 16.7, if (1) has a solution, then the equation $\alpha^2\beta = T(h)$ also has a solution. The latter expanded has the form

$$(2) \qquad [yQ_i]_{i=1}^{k-1} y = G(QP, QR, h)\, Q, \quad \omega(y, M), \quad \neg\, \mathscr{B}([QP, QR, h], M),$$

where $1 < k \leqq 3M$; Q_i, P, Q and R are words in the alphabet $\mathbf{И}$; $[Q_i^{\partial}, [PQR^{\partial} \leqq M$, and $\mathscr{L}(QP, QR)$. We will prove, however, that (2) has no solutions. We argue by contradiction. We obtain in $\mathbf{И}$ the equality

$$[YQ_i]_{i=1}^{k-1} Y = G(QP, QR, \lambda_0, \cdots, \lambda_{m-1})\, Q,$$

where $\neg\, \mathscr{B}([QP, QR, \lambda_0, \cdots, \lambda_{m-1}], M)$ and $\omega(Y, M)$. From $\neg\, \mathscr{B}(\cdots)$ we obtain $m > 1$. Then by 16.10 there exist σ, ν, S, C and D such that $Y = (S^{\sigma} C)^{\nu} S^{\sigma} D$, $\mathscr{L}(S, C)$, $[S^{\partial} \leqq 30M$, $[C^{\partial}, [D^{\partial} \leqq 90M$, $[S^{\sigma} C^{\partial} > (6M)^3$, $\nu \geqq 4M^2$, and S is a simple word.

We define a word A. Put $A = RQP$ if m is even, $A = PQR$ if m is odd. Then $[A^\partial \leq M$, $QZ = \langle QP, QR\rangle_m$, and

$$F(QP, QR, \lambda_0, \cdots, \lambda_{m-1}) = G(QP, QR, \lambda_0, \cdots, \lambda_{m-1}) \langle QP, QR\rangle_m = [YQ_i]_{i=1}^{k-1} YA.$$

But $Y = (S^\sigma C)^\nu S^\sigma D$. Hence the triple $(\sigma, \nu, h' = (\lambda_0, \cdots, \lambda_{m-1}))$ satisfies the system

$$\text{(3)}\quad \begin{gathered}[(S^\sigma C)^\nu S^\sigma DQ_i]_{i=1}^{k-1}(S^\sigma C)^\nu S^\sigma DA = F(QP, QR, h),\\ \sigma[S^\partial + [C^\partial > (6M)^3, \quad \nu \geqslant 4M^2, \quad \urcorner \mathscr{B}([QP, QR, h], M).\end{gathered}$$

By 16.5 the system (3) has no solutions, since it is the system 16.5 (1) with $\epsilon = 0$, $N = 4M^2$ and $n = k$, and we have $k \leqq 3M \leqq 2M^2$. We have obtained the required contradiction.

Case 2. T_1 is a V_1^M-term. Then the system (*) in expanded form can be written as

$$\text{(4)}\quad \begin{gathered}(xA)^\lambda x = G(QP, QR, h)Q, \quad \mathscr{L}(x, A), \quad \urcorner \tau(x, M),\\ \urcorner \mathscr{B}([QP, QR, h], M), \quad h^0 > 1,\end{gathered}$$

where x is a word unknown, λ is a positive unknown, A is a word in the alphabet $\mathbf{И}$, $[A^\partial \leqq M$, and P, Q and R are the same as in Case 1. We will prove by contradiction that (4) has no solutions. Assume that the triple $(X, \lambda, h = (\lambda_0, \cdots, \lambda_{k-1}))$ is a solution of (4). Then

$$\text{(5)}\quad (XA)^\lambda X = G(QP, QR, \lambda_0, \cdots, \lambda_{k-1})\, Q,$$

$$\text{(6)}\quad \urcorner \mathscr{B}([QP, QR, \lambda_0, \cdots, \lambda_{k-1}], M),$$

$$\text{(7)}\quad \urcorner \tau(X, M),$$

where XA is a simple word. For brevity, put $F_i = F(QP, QR, \lambda_0, \cdots, \lambda_{i-1})$, $i = 0, 1, \cdots, k$; $G_i = G(QP, QR, \lambda_0, \cdots, \lambda_{i-1})$, $i = 1, \cdots, k$. It follows from (6) that $k > 2$. Then $G_k Q = F_{k-1}^{\lambda_{k-1}} G_{k-2} Q$. From (5) we have

$$\text{(8)}\quad (XA)^\lambda X = F_{k-1}^{\lambda_{k-1}} G_{k-2} Q.$$

Exactly as in the proof of 16.7, we establish that F_{k-1} is a simple word and $\Lambda \neq G_{k-2}Q \angle F_{k-1}$. It follows from (7) that $[X^\partial > M \geqq [A^\partial$. Also, from (6) we have $[F_{k-2}^\partial > 30M$; hence $[G_{k-2}Q^\partial > 29M$, since $[F_{k-2}^\partial = [G_{k-2}^\partial + [QPQR^\partial$ and $[PQR^\partial \leqq M$. Thus, $[G_{k-2}Q^\partial > 29M > [A^\partial$. Now from (8) and 1.26 we obtain $XA = F_{k-1}$ and $X = G_{k-2}Q$. But $[F_{k-2}^\partial = [G_{k-2}QPQR^\partial$; $[A^\partial, [PQR^\partial \leqq M$. Hence $|[F_{k-2}^\partial - [F_{k-1}^\partial| \leqq M$. Since $k > 2$, we have $F_{k-1} = F_{k-2}^{\lambda_{k-2}} F_{k-3}$. Hence if $\lambda_{k-2} = 1$, then $[F_{k-3}^\partial \leqq M$, and if $\lambda_{k-2} > 1$, then $[F_{k-2}^\partial \leqq M$. These deductions contradict (6). This contradiction means that (4) has no solutions. The proposition is proved.

16.12. *Equations of type $V^M G^M$ have no solutions.*

Proof. This follows from 16.11 and 16.6 (9).

§17. Equations of type *GG*

Any equation of type $G^M G^M$ when expanded has the form

$$\text{(1)}\quad \begin{gathered}G(QP, QR, h)\, Q = G(Q_1 P_1, Q_1 R_1, g)\, Q_1,\\ \urcorner \mathscr{B}([QP, QR, h], M) \,\&\, \urcorner \mathscr{B}([Q_1 P_1, Q_1 R_1, g], M),\end{gathered}$$

where h and g are processional unknowns, and P, Q, R, P_1, Q_1 and R_1 are words in the alphabet $\mathbf{H}$ such that $[PQR^\partial, [P_1Q_1R_1^\partial \leq M$, $\mathscr{S}(QP, QR)$, $\mathscr{S}(Q_1P_1, Q_1R_1)$, $Q, Q_1 \neq \Lambda$. If the pair of processions $h = (\lambda_0, \cdots, \lambda_{k-1})$, $g = (\mu_0, \cdots, \mu_{s-1})$ satisfies the system (1), then

$$(2) \qquad G(QP, QR, \lambda_0, \ldots, \lambda_{k-1})\, Q = G(Q_1P_1, Q_1R_1, \mu_0, \ldots, \mu_{s-1})\, Q_1,$$

$$(3) \qquad \neg \mathscr{B}([QP, QR, \lambda_0, \ldots, \lambda_{k-1}], M) \,\&\, \neg \mathscr{B}([Q_1P_1, Q_1R_1, \mu_0, \ldots, \mu_{s-1}], M).$$

The analysis of the equality (2) under the condition (3) is, in essence, the content of §17. Technically, it is more convenient to append the condition $\lambda_{k-1}, \mu_{s-1} > 1$, which we may easily get rid of by means of 16.6 (9). We will prove (17.5) that if (2) and (3) hold and $\lambda_{k-1}, \mu_{s-1} > 1$, then $P = P_1$, $Q = Q_1$, $R = R_1$, and $(\lambda_0, \cdots, \lambda_{k-1})$ is equal to $(\mu_0, \cdots, \mu_{s-1})$. Condition (3) plays a decisive role in this connection.

Throughout this section we use the following abbreviations. Suppose there are given the processions

$$(*) \qquad [QP, QR, \lambda_0, \cdots, \lambda_{k-1}], \quad [Q_1P_1, Q_1R_1, \mu_0, \cdots, \mu_{s-1}].$$

For brevity, put

$$F_{-1} = QR, \quad F_i = F(QP, QR, \lambda_0, \cdots, \lambda_{i-1}), \qquad i = 0, \cdots, k,$$
$$\widetilde{F}_{-1} = Q_1R_1, \quad \widetilde{F}_i = F(Q_1P_1, Q_1R_1, \mu_0, \cdots, \mu_{i-1}), \qquad i = 0, \cdots, s.$$

17.1. *Suppose there are given the processions* (*) *and suppose* $F_n = \widetilde{F}_m$ *and* $F_{n-1} = \widetilde{F}_{m-1}$, *where* $1 < n \leq k$ *and* $1 < m \leq s$. *Then* $\lambda_{n-1} = \mu_{m-1}$ *and* $F_{n-2} = \widetilde{F}_{m-2}$.

PROOF. Since $n, m > 1$, we have $F_n = F_{n-1}^{\lambda_{n-1}} F_{n-2}$, $\widetilde{F}_m = \widetilde{F}_{m-1}^{\mu_{m-1}} \widetilde{F}_{m-2}$. By hypothesis, $F_n = \widetilde{F}_m$. Therefore,

$$(1) \qquad F_{n-1}^{\lambda_{n-1}} F_{n-2} = \widetilde{F}_{m-1}^{\mu_{m-1}} \widetilde{F}_{m-2}.$$

Since $n, m > 1$, we have $[F_{n-2}^\partial < [F_{n-1}^\partial$ and $[\widetilde{F}_{m-2}^\partial < [\widetilde{F}_{m-1}^\partial$. By hypothesis, $F_{n-1} = \widetilde{F}_{m-1}$. Comparing the lengths of the left- and right-hand sides of (1), we conclude that $\lambda_{n-1} = \mu_{m-1}$ and $F_{n-2} = \widetilde{F}_{m-2}$, as required.

17.2. *Suppose there are given the processions* (*), *and suppose* $[Q_1P_1^\partial \geq [QP^\partial$. *Suppose also that the integers* n *and* m *are such that* $1 < n \leq k$, $1 < m \leq s$, $F_n = \widetilde{F}_m$ *and* $F_{n-1} = \widetilde{F}_{m-1}$. *Then there exists an integer* q, $0 \leq q < n-1$, *such that* $Q_1P_1 = F_q$, $(Q_1P_1)^{\mu_0} Q_1R_1 = F_{q+1}$ *and* $(\mu_1, \cdots, \mu_{m-1}) = (\lambda_{q+1}, \cdots, \lambda_{n-1})$.

PROOF. In view of 17.1, it follows from our assumptions that

$$(1) \qquad \lambda_{n-1} = \mu_{m-1},$$

$$(2) \qquad F_{n-2} = \widetilde{F}_{m-2}.$$

We now proceed by induction on $\min\{n, m\}$. By hypothesis, $\min\{n, m\} \geq 2$.

Basis of the induction. Suppose $\min\{n, m\} = 2$. We will prove by contradiction that $m \leq n$. Assume that $m > n$. Then $n = 2$ and from (2) we have $F_0 = \widetilde{F}_{m-2}$. Since $m > 2$, we have $[\widetilde{F}_{m-2}^\partial > [\widetilde{F}_0^\partial = [Q_1P_1^\partial$. Thus, $[Q_1P_1^\partial < [F_0^\partial = [QP^\partial$, which contradicts the assumption $[Q_1P_1^\partial \geq [QP^\partial$. It follows from this contradiction that $m \leq n$, so that $m = \min\{n, m\} = 2$. Then from (2) we have $F_{n-2} = \widetilde{F}_0 = Q_1P_1$; the condition $F_{n-1} = \widetilde{F}_{m-1}$ gives

$$F_{n-1} = \widetilde{F}_1 = (Q_1P_1)^{\mu_0} Q_1R_1;$$

(1) implies that $\mu_1 = \lambda_{n-1}$, and our assertion holds with $q = n-2$.

Induction step. Suppose $l > 2$ and the assertion being proved is true if $\min\{n, m\} < l$. Now assume that $\min\{n, m\} = l$. Then $n, m \geqq 3$. We have $1 < n-1 < k$, $1 < m-1 < s$ and $\min\{n-1, m-1\} < l$. Hence from $F_{n-1} = \widetilde{F}_{m-1}$, $F_{n-2} = \widetilde{F}_{m-2}$ and the induction assumption it follows that there exists an integer q, $0 \leqq q < n-2$, such that $Q_1P_1 = F_q$, $(Q_1P_1)^{\mu_0}Q_1R_1 = F_{q+1}$ and $(\mu_1, \cdots, \mu_{m-2}) = (\lambda_{q+1}, \cdots, \lambda_{n-2})$. Since $\lambda_{n-1} = \mu_{m-1}$, we have $(\mu_1, \cdots, \mu_{m-1}) = (\lambda_{q+1}, \cdots, \lambda_{n-1})$. The proposition is proved.

In Proposition 17.3 all of the conditions except $\neg \mathscr{B}$ are "natural" and arise in the analysis of the equality (2). The condition $\neg \mathscr{B}$ is supplementary and necessary in order to obtain the conclusions a) and b). Propositions 17.3 and 17.4 are the main reason for the introduction of the predicate $\mathscr{B}$.

17.3. *Suppose M is a positive integer and the processions* (*) *are such that* $k \geqq 4$, $s \geqq 3$, $[F^{\partial}_{k-4} \leqq 2M$, $[F^{\partial}_{k-3} > 30M$, $\lambda_{k-2} = 1$, $\mathscr{S}(QP, QR)$, $\mathscr{S}(Q_1P_1, Q_1R_1)$; $[QP^{\partial}, [QR^{\partial}, [Q_1P_1^{\partial}, [Q_1R_1^{\partial} \leqq M$, $\neg \mathscr{B}([Q_1P_1, Q_1R_1, \mu_0, \cdots, \mu_{s-1}], M)$, *and* $F_{k-1} = \widetilde{F}_{s-1}$. *Then*: a) *if* $\mu_{s-2} = 1$, *then* $F_{k-3} = \widetilde{F}_{s-3}$; b) *if* $\mu_{s-2} > 1$, *then* $F_{k-3} = \widetilde{F}_{s-2}$.

PROOF. Write the word F_{k-4} as a power of a simple word: $F_{k-4} = S^a$, where S is simple and $a > 0$. For brevity, put $d = \lambda_{k-3}$, $\lambda = \lambda_{k-4}$ and $X = F_{k-5}$ (we know $k \geqq 4$). Then $F_{k-3} = F^{\lambda_{k-4}}_{k-4}F_{k-5} = S^{a\lambda}X$, $F_{k-1} = F(F_{k-3}, F_{k-4}, \lambda_{k-3}, \lambda_{k-2}) = F(S^{a\lambda}X, S^a, d, 1) = (S^{a\lambda}X)^d S^a S^{a\lambda}X$. The equality $F_{k-1} = \widetilde{F}_{s-1}$ assumes the form

$$(S^{a\lambda}C)^d S^a S^{a\lambda} X = \widetilde{F}_{s-1}. \tag{1}$$

We apply 16.1 to (1). For this purpose we estimate $a\lambda[S^{\partial}$. By hypothesis, $[F^{\partial}_{k-3} > 30M$, so that $[S^{a\lambda}X^a > 30M$. Also, it is obvious that

$$[F^{\partial}_{k-5} \leqq \max\{[QR^{\partial}, [F^{\partial}_{k-4}\};$$

hence by our assumptions $[F^{\partial}_{k-5} \leqq 2M$, i.e. $[X^{\partial} \leqq 2M$. We obtain $a\lambda[S^{\partial} > 28M$. By hypothesis, $[F^{\partial}_{k-4} \leqq 2M$, but $F_{k-4} = S^a$, $a > 0$. Therefore, $[S^{\partial} \leqq 2M$. Finally, $a\lambda[S^{\partial} > 8[S^{\partial} + \max\{6M, 6[S^{\partial}\}$. Applying 16.1 to (1), we obtain $\widetilde{F}_q = S^b$ for some q and b, $0 \leqq q \leqq s-2$, $b > 0$. Now our aim is to prove that $F_{k-3} = \widetilde{F}_{q+1}$, from which it is not difficult to deduce a) and b). By 5.1 we have

$$\widetilde{F}_{s-1} = F(\widetilde{F}_q, \widetilde{F}_{q-1}, \mu_q, \cdots, \mu_{s-2}) = F(S^b, Y, \mu_q, \cdots, \mu_{s-2}), \tag{2}$$

where $Y = \widetilde{F}_{q-1}$. Since $[S^{\partial} \leqq 2M$ and $\widetilde{F}_q = S^b$, the condition

$$\neg \mathscr{B}([Q_1P_1, Q_1R_1, \mu_0, \cdots, \mu_{s-1}], M)$$

implies that the procession $(\mu_{q+1}, \cdots, \mu_{s-2})$ is nonempty and different from the procession "(1)". Then by 9.1.2 there exists a δ-procession $\epsilon_1 \cdots \epsilon_{m+1}$ such that

$$\epsilon_1 \cdots \epsilon_{m+1} = \delta(\mu_{q+1}, \cdots, \mu_{s-2}), \qquad \epsilon_1 = 0, \quad m > 0. \tag{3}$$

From (3) and 9.1.3 it follows that

$$F(S^b, Y, \mu_q, \ldots, \mu_{s-2}) = S^{b\mu_q} Y [S^{b(\mu_q + \varepsilon_i)} Y]_{i=2}^m S^{b\mu_q} \langle S^b, Y\rangle_{s-q-1}. \tag{4}$$

From (1), (2) and (4), putting $\mu = \mu_q$ and $t = s-q-1$, we have

$$(S^{a\lambda}X)^d S^a S^{a\lambda} X = S^{b\mu} Y [S^{b\mu + b\varepsilon_i} Y]_{i=2}^m S^{b\mu} \langle S^b, Y\rangle_t. \tag{5}$$

We know $F_{k-3}=S^{a\lambda}X$ and $\widetilde{F}_{q+1}=\widetilde{F}_q^{\mu_q}\widetilde{F}_{q-1}=S^{b\mu}Y$. The equality $F_{k-3}=\widetilde{F}_{q+1}$ assumes the form $S^{a\lambda}X=S^{b\mu}Y$. Let $\perp\!\!\!\perp$ denote equality in the free group whose positive alphabet coincides with **И**. Let $a\lambda-b\mu=p$. Then from (5) we have (we know $d>0$)

$$S^pX\,(S^{a\lambda}X)^{d-1}\,S^aS^{a\lambda}X\perp\!\!\!\perp Y\,[S^{b\mu+b\varepsilon_i}Y]_{i=2}^m\,S^{b\mu}\langle S^b,\,Y\rangle_t. \tag{6}$$

The left hand side of (6) begins with the word $S^pXS^{a\lambda}$, the right-hand side with the word $YS^{b\mu}$. We will apply Proposition 1.22 to (6). Let us estimate $b\mu[S^\partial$. From $\widetilde{F}_q=S^b$, $\widetilde{F}_{q-1}=Y$, $\mathscr{G}(Q_1P_1,Q_1R_1)$ and 5.3 we obtain $\mathscr{G}(Y,S)$. Analogously, we have $\mathscr{G}(X,S)$. If $a\lambda[S^\partial\geqq[S^{b\mu}YS^\partial$, then from (5) we obtain $S^{b\mu}YS\angle S^{a\lambda}$ and $YS\angle S^{a\lambda-b\mu}$, which by 1.4 is incompatible with $\mathscr{G}(Y,S)$. Hence $a\lambda[S^\partial<[S^{b\mu}YS^\partial$. Analogously, $b\mu[S^\partial<[S^{a\lambda}XS^\partial$. But we know $[S^\partial\leqq 2M$ and $\widetilde{F}_q=S^b$. Hence by 5.12′ we have $[\widetilde{F}_q^\partial\leqq 6M$, and then $[Y_q^\partial=[\widetilde{F}_{q-1}^\partial\leqq\max\{[Q_1R_1^\partial,[\widetilde{F}_q^\partial\}\leqq 6M$. Now the inequalities $[S^{b\mu}YS^\partial>a\lambda[S^\partial>28M$ imply that $b\mu[S^\partial>20M$. We have proved that $\mathscr{G}(X,S)$, $\mathscr{G}(Y,S)$, $a\lambda[S^\partial>28M>[Y^\partial+2[S^\partial$ and $b\mu[S^\partial>20M>[X^\partial+2[S^\partial$. Applying Proposition 1.22 to (6), we obtain $S^pX\perp\!\!\!\perp YS^c$ for some integer c. Since $a\lambda=b\mu+p$, we get $S^{a\lambda}X\perp\!\!\!\perp S^{b\mu}YS^c$. Now from (6) we have

$$S^c(S^{b\mu}YS^c)^{d-1}\,S^aS^{b\mu}YS^c\perp\!\!\!\perp[S^{b\mu+b\varepsilon_i}Y]_{i=2}^m\,S^{b\mu}\langle S^b,\,Y\rangle_t. \tag{7}$$

Multiplying (7) on the right by $S^{b\mu}$ and transforming the left-hand side of (7) in an obvious way, we obtain

$$(S^{b\mu+c}Y)^{d-1}\,S^{a+c+b\mu}\,YS^{c+b\mu}\perp\!\!\!\perp[S^{b\mu+b\varepsilon_i}Y]_{i=2}^m\,S^{b\mu}YS^{b+b\mu},\qquad\text{if } t \text{ is even,} \tag{8}$$

$$(S^{b\mu+c}Y)^{d-1}\,S^{a+c+b\mu}\,YS^{b\mu+c}\perp\!\!\!\perp[S^{b\mu+b\varepsilon_i}Y]_{i=2}^m\,S^{b\mu+b}\,YS^{b\mu},\qquad\text{if } t \text{ is odd.} \tag{9}$$

From the above inequalities $a\lambda[S^\partial<[S^{b\mu}YS^\partial$, $b\mu[S^\partial<[S^{a\lambda}XS^\partial$ it follows that $|[S^{a\lambda}X^\partial-[S^{b\mu}Y^\partial|\leqq[XYS^\partial\leqq 10M$. Also, we have proved that $b\mu[S^\partial>20M$ and $S^{a\lambda}X\perp\!\!\!\perp S^{b\mu}YS^c$. From this last equality it follows that $|c|[S^\partial=|[S^{a\lambda}X^\partial-[S^{b\mu}Y^\partial|$, so that $|c|[S^\partial\leqq 10M<b\mu[S^\partial$. Therefore, $b\mu+c>0$. Then the left- and right-hand sides of each of the equalities (8) and (9) are words in the alphabet **И**. Hence the pair $(X=S,Y)$ satisfies one of the word equations in two unknowns:

$$(x^{b\mu+c}y)^{d-1}\,x^{a+c+b\mu}yx^{b\mu+c}=[x^{b\mu+b\varepsilon_i}y]_{i=2}^m\,x^{b\mu}yx^{b+b\mu}, \tag{8′}$$

$$(x^{b\mu+c}y)^{d-1}\,x^{a+c+b\mu}yx^{b\mu+c}=[x^{b\mu+b\varepsilon_i}y]_{i=2}^m\,x^{b\mu+b}yx^{b\mu}. \tag{9′}$$

From $\mathscr{G}(Y,S)$ and 1.1 it follows that in the alphabet $\{x,y\}$ we have the (graphical) equalities (8′) and (9′). Let us consider these.

From (8′) it follows that $m=d$, $b\mu+c=b\mu+b\epsilon_i$ $(i=2,\cdots,m)$, $a+c+b\mu=b\mu$ and $b\mu+c=b\mu+b$. From the last two equalities we obtain $a+c=0$ and $c=b$, which is impossible, since $a,b>0$. Thus (8′) is impossible.

Consider (9′). Here we obtain $m=d$, $b\mu+c=b\mu+b\epsilon_i$ $(i=2,\cdots,m)$, $a+c+b\mu=b\mu+b$, $b\mu+c=b\mu$ and $c=0$. Since $c=0$, from $S^{a\lambda}X\perp\!\!\!\perp S^{b\mu}YS^c$ it follows that $S^{a\lambda}X=S^{b\mu}Y$. The equality $S^{a\lambda}X=S^{b\mu}Y$, and hence also the equality $F_{k-3}=\widetilde{F}_{q+1}$, is proved. Also, since $c=0$, from $b\mu+c=b\mu+b\epsilon_i$ $(i=2,\cdots,m)$ we have $b\epsilon_2=\cdots=b\epsilon_m=0$. Since $b>0$, $\epsilon_2=\cdots=\epsilon_m=0$. But we know $\epsilon_1=0$. Therefore, $\epsilon_1\cdots\epsilon_m=0^m$. Since $\epsilon_1\cdots\epsilon_{m+1}=\delta(\mu_{q+1},\cdots,\mu_{s-2})$, by 9.1.4 we have: α) if $\mu_{s-2}=1$, then $q+1=s-3$; β) if $\mu_{s-2}>1$, then $q+1=s-2$.

Now a) and b) follow from α) and β) and the equality $F_{k-3} = \widetilde{F}_{q+1}$. The proposition is proved.

17.4. *Suppose the integer $M > 0$ and the processions (*) are such that $s > 2$, $k > 4$, $[\widetilde{F}^{\partial}_{s-3} \leq 2M$, $[\widetilde{F}^{\partial}_{s-2} > 30M$, $\mu_{s-2} > 1$, $\mathscr{S}(QP, QR)$, $\mathscr{S}(Q_1P_1, Q_1R_1)$; $\lceil QP^{\partial}$, $[QR^{\partial}$, $\lfloor Q_1P_1^{\partial}$, $[Q_1R_1^{\partial} \leq M$, $\neg\, \mathscr{B}([QP, QR, \lambda_0, \cdots, \lambda_{k-1}], M)$, $F_{k-1} = \widetilde{F}_{s-1}$, $\lambda_{k-2} = 1$. Then $\widetilde{F}_{s-2} = F_{k-3}$.*

The proof is analogous to that of 17.3.

17.5. *Suppose the integer $M > 0$ and the words P, Q, R, P_1, Q_1 and R_1 satisfy the conditions $\mathscr{S}(QP, QR)$, $\mathscr{S}(Q_1P_1, Q_1R_1)$, $[PQR^{\partial} \leq M$, $[P_1Q_1R_1^{\partial} \leq M$, $Q \neq \Lambda$, $Q_1 \neq \Lambda$, and suppose $\lambda_{k-1} > 1$, $\mu_{s-1} > 1$ and*

(1) $$G(QP, QR, \lambda_0, \ldots, \lambda_{k-1})\, Q = G(Q_1P_1, Q_1R_1, \mu_0, \ldots, \mu_{s-1})\, Q_1,$$

where $\neg\, \mathscr{B}([QP, QR, \lambda_0, \cdots, \lambda_{k-1}], M)$ and $\neg\, \mathscr{B}([Q_1P_1, Q_1R_1, \mu_0, \cdots, \mu_{s-1}], M)$. Then $QP = Q_1P_1$, $QR = Q_1R_1$, $(\lambda_0, \cdots, \lambda_{k-1}) = (\mu_0, \cdots, \mu_{s-1})$ and $Q = Q_1$.

PROOF. We will use the abbreviations which precede 17.1 and some new ones as well. Put

$$G_i = G(QP, QR, \lambda_0, \ldots, \lambda_{i-1}), \qquad i = 1, \ldots, k,$$
$$\widetilde{G}_i = G(Q_1P_1, Q_1R_1, \mu_0, \ldots, \mu_{i-1}), \quad i = 1, \ldots, s.$$

Then (1) can be written in the form

(1′) $$G_kQ = \widetilde{G}_sQ_1.$$

In view of property 1 of the predicate $\mathscr{B}$ we have $k, s > 2$ and $[F^{\partial}_{k-2}$, $[F^{\partial}_{s-2} > 30M$. Since $k, s > 2$, we get $[F^{\partial}_{k-2} < [F^{\partial}_{k-1}$ and $[\widetilde{F}^{\partial}_{s-2} < [\widetilde{F}^{\partial}_{s-1}$. Thus,

$$[F^{\partial}_{k-1} > [F^{\partial}_{k-2} > 30M > 6M,$$
$$[\widetilde{F}^{\partial}_{s-1} > [\widetilde{F}^{\partial}_{s-2} > 30M > 6M.$$

By 5.12, the words F_{k-1}, F_{k-2}, $\widetilde{F}_{s-1}$ and $\widetilde{F}_{s-2}$ are simple. Since $k, s > 2$, we have $G_k = F^{\lambda_{k-1}}_{k-1}G_{k-2}$ and $\widetilde{G}_s = \widetilde{F}^{\mu_{s-1}}_{s-1}\widetilde{G}_{s-2}$. The equality (1′) assumes the form

(2) $$F^{\lambda_{k-1}}_{k-1}G_{k-2}Q = \widetilde{F}^{\mu_{s-1}}_{s-1}\widetilde{G}_{s-2}Q_1.$$

Since $k, s > 2$, we have $G_{k-2}Q \angle F_{k-2} \angle F_{k-1}$ and $\widetilde{G}_{s-2}Q_1 \angle \widetilde{F}_{s-2} \angle \widetilde{F}_{s-1}$. We have proved that F_{k-1} and $\widetilde{F}_{s-1}$ are simple, and by hypothesis λ_{k-1}, $\mu_{s-1} > 1$. Therefore, applying Proposition 1.7′ to (2), we obtain

(3) $$\lambda_{k-1} = \mu_{s-1},$$

(4) $$F_{k-1} = \widetilde{F}_{s-1}.$$

Since $k, s > 2$, we have $F_{k-1} = F^{\lambda_{k-2}}_{k-2}F_{k-3}$ and $\widetilde{F}_{s-1} = \widetilde{F}^{\mu_{s-2}}_{s-2}\widetilde{F}_{s-3}$. From (4) we have

(5) $$F^{\lambda_{k-2}}_{k-2}F_{k-3} = \widetilde{F}^{\mu_{s-2}}_{s-2}\widetilde{F}_{s-3}.$$

The following three cases are possible.

Case 1. $\lambda_{k-2}, \mu_{s-2} > 1$. Then 1.7′ is applicable to (5), since the words F_{k-2} and $\widetilde{F}_{s-2}$ are simple and, in addition, $F_{k-3} \angle F_{k-2}$ and $\widetilde{F}_{s-3} \angle \widetilde{F}_{s-2}$. From (5) and 1.7′ we obtain

(6) $$F_{k-2} = \widetilde{F}_{s-2}.$$

Two cases are possible: 1) $[PQ^\partial \geqq [P_1Q_1^\partial$; 2) $[P_1Q_1^\partial \geqq [PQ^\partial$. Without loss of generality, we consider the case $[P_1Q_1^\partial \geqq [PQ^\partial$. Then from (4), (6) and 17.2 (the roles of n and m in 17.2 are played here by $k-1$ and $s-1$) it follows that there exists q, $0 \leqq q < k-2$, such that

$$(7) \qquad Q_1P_1 = F_q, \qquad (Q_1P_1)^{\mu_0}Q_1R_1 = F_{q+1},$$

$$(8) \qquad (\mu_1, \cdots, \mu_{s-2}) = (\lambda_{q+1}, \cdots, \lambda_{k-2}).$$

Since $F_k = F_{k-1}^{\lambda_{k-1}}F_{k-2}$ and $\widetilde{F}_s = \widetilde{F}_{s-1}^{\mu_{s-1}}\widetilde{F}_{s-2}$, from (3), (4) and (6) it follows that $F_k = \widetilde{F}_s$. From this and the equalities $[F_k^\partial = [G_kQPQR^\partial$, $[\widetilde{F}_s^\partial = [\widetilde{G}_sQ_1P_1Q_1R_1^\partial$ and (1′) it follows that $[PQR^\partial = [P_1Q_1R_1^\partial$. This relation allows us to complete the proof as follows. By hypothesis, $Q \neq \Lambda$. If we assume that $q \geqq 1$, then we have $[F_q^\partial \geqq [F_1^\partial = [(QP)^{\lambda_0}QR^\partial \geqq [QPQR^\partial > [PQR^\partial = [P_1Q_1R_1^\partial \geqq [Q_1P_1^\partial$, i.e. $[F_q^\partial \geqq [Q_1P_1^\partial$, which contradicts (7). Consequently, $q = 0$. Then from (7) we obtain $Q_1P_1 = F_0 = QP$, $(Q_1P_1)^{\mu_0}Q_1R_1 = F_1 = (QP)^{\lambda_0}QR$. From these equalities, using the relation $[PQR^\partial = [P_1Q_1R_1^\partial$ and the fact that Q and Q_1 are nonempty, we conclude (by contradiction) that $\lambda_0 = \mu_0$ and $Q_1R_1 = QR$. Finally, from (3) and (8), using $q = 0$ and $\lambda_0 = \mu_0$, we obtain $(\mu_0, \cdots, \mu_{s-1}) = (\lambda_0, \cdots, \lambda_{k-1})$. We have proved that $QP = Q_1P_1$, $QR = Q_1R_1$ and $(\lambda_0, \cdots, \lambda_{k-1}) = (\mu_0, \cdots, \mu_{s-1})$. Then $G_k = \widetilde{G}_s$ and it follows from (1) that $Q = Q_1$. Our assertion is proved in Case 1.

In considering the remaining cases we put, for brevity, $A = \langle QP, QR\rangle_{k-3}$, $B = \langle Q_1P_1, Q_1R_1\rangle_{s-3}$. Then from the assumptions $[PQR^\partial \leqq M$ and $[P_1Q_1R_1^\partial \leqq M$ it follows that $[A^\partial, [B^\partial \leqq 2M$.

Case 2. Assume that

$$(9) \qquad \lambda_{k-2} = \mu_{s-2} = 1.$$

By property 1 of the predicate $\mathscr{B}$ we have $k, s > 3$ and $[F_{k-3}^\partial, [\widetilde{F}_{s-3}^\partial > 30M$. By 5.12, the words F_{k-3} and $\widetilde{F}_{s-3}$ are simple. Since $k, s > 3$, we have $F_{k-2} = F_{k-3}^{\lambda_{k-3}}F_{k-4}$, $\widetilde{F}_{s-2} = \widetilde{F}_{s-3}^{\mu_{s-3}}\widetilde{F}_{s-4}$. From this, (4) and (9) we have

$$(10) \qquad F_{k-3}^{\lambda_{k-3}}F_{k-4}F_{k-3} = \widetilde{F}_{s-3}^{\mu_{s-3}}\widetilde{F}_{s-4}\widetilde{F}_{s-3}.$$

We will prove that $F_{k-3} = \widetilde{F}_{s-3}$. Since $k, s > 3$, there are two possible subcases.

Subcase 2.1. $k, s > 4$. Then by 5.8 we have $F_{k-4}F_{k-3} = F_{k-3}G_{k-4}A$, $\widetilde{F}_{s-4}\widetilde{F}_{s-3} = \widetilde{F}_{s-3}\widetilde{G}_{s-4}B$, and (10) assumes the form

$$(11) \qquad F_{k-3}^{\lambda_{k-3}+1}G_{k-4}A = \widetilde{F}_{s-3}^{\mu_{s-3}+1}\widetilde{G}_{s-4}B.$$

Without loss of generality, we consider the case $[\widetilde{G}_{s-4}B^\partial \geqq [G_{k-4}A^\partial$. Then, cancelling (11) on the right by $G_{k-4}A$, we obtain

$$(12) \qquad F_{k-3}^{\lambda_{k-3}+1} = \widetilde{F}_{s-3}^{\mu_{s-3}+1}V,$$

where V satisfies the relation $VG_{k-4}A = \widetilde{G}_{s-4}B$.

2.1.1) Assume that $[V^\partial \leqq [\widetilde{G}_{s-4}^\partial$. Then $V \angle \widetilde{G}_{s-4}$. Since $s > 4$, we have $\widetilde{G}_{s-4} \angle \widetilde{F}_{s-4} \mathrel{\dot{\angle}} \widetilde{F}_{s-3}$, so that $V \mathrel{\dot{\angle}} \widetilde{F}_{s-3}$. Since $\lambda_{k-3}+1 > 1$, $\mu_{s-3}+1 > 1$ and the words F_{k-3} and $\widetilde{F}_{s-3}$ are simple, from (12) and 1.7′ it follows that $F_{k-3} = \widetilde{F}_{s-3}$.

2.1.2) Now suppose $[V^\partial > [\widetilde{G}_{s-4}^\partial$. Then we have $[G_{k-4}A^\partial < [B^\partial \leqq 2M$. On the other hand, $[F_{k-4}^\partial = [F_{k-4}A^\partial$. Hence $[F_{k-4}^\partial \leqq 2M$. It was noted above that $[F_{k-3}^\partial > 30M$. It follows from our assumptions and 17.3 a) that $F_{k-3} = \widetilde{F}_{s-3}$.

Subcase 2.2. One of the numbers k or s is equal to 4. Without loss of generality, we assume that $k=4$. Then $[F_{k-4}^{\partial}=[F_0^{\partial}=[QP^{\partial}\leqq M$. Also, $s>3$ and $[F_{k-3}^{\partial}>30M$. Now from our assumptions and 17.3 a) we conclude that $F_{k-3}=\widetilde{F}_{s-3}$.

We have proved that $F_{k-3}=\widetilde{F}_{s-3}$. Since $k,s>2$, we have $F_{k-1}=F_{k-2}^{\lambda_{k-2}}F_{k-3}$, $\widetilde{F}_{s-1}=\widetilde{F}_{s-2}^{\mu_{s-2}}\widetilde{F}_{s-3}$. From the last three equalities, (9) and (4) we conclude that (6) holds. Arguing now exactly as in Case 1 following (6), we see that our assertion is true.

Case 3. Now suppose that one of the numbers λ_{k-2} or μ_{s-2} is equal to 1 and the other is greater than 1. We will prove that this case is impossible. Without loss of generality, we assume that

$$\lambda_{k-2}=1, \qquad \mu_{s-2}>1. \tag{13}$$

Since $\lambda_{k-2}=1$, by property 1 of the predicate $\mathscr{B}$ we have $k>3$ and $[F_{k-3}^{\partial}>30M$. By 5.12, F_{k-3} is a simple word. Since $k>3$, we have $F_{k-2}=F_{k-3}^{\lambda_{k-3}}F_{k-4}$. From (4) and (13) we have

$$F_{k-3}^{\lambda_{k-3}}F_{k-4}F_{k-3}=\widetilde{F}_{s-2}^{\mu_{s-2}}\widetilde{F}_{s-3}. \tag{14}$$

We will prove that $F_{k-3}=\widetilde{F}_{s-2}$. Since $k>3$, there are two possible subcases.

Subcase 3.1. $k=4$. Then $[F_{k-4}^{\partial}=[F_0^{\partial}=[QP^{\partial}\leqq M$. Also, $s>2$ and $[F_{k-3}^{\partial}>30M$. An application of 17.3 b) gives $F_{k-3}=\widetilde{F}_{s-2}$.

Subcase 3.2. $k>4$. By 5.8, we have $F_{k-4}F_{k-3}=F_{k-3}G_{k-4}A$. From (14) we obtain

$$F_{k-3}^{\lambda_{k-3}+1}G_{k-4}A=\widetilde{F}_{s-2}^{\mu_{s-2}}\widetilde{F}_{s-3}. \tag{15}$$

3.2.1) Suppose $[G_{k-4}A^{\partial}<[\widetilde{F}_{s-3}^{\partial}$. Then, cancelling (15) on the right by $G_{k-4}A$, we obtain

$$\mathbf{F}_{k-3}^{\lambda_{k-3}+1}=\widetilde{F}_{s-2}^{\mu_{s-2}}V, \tag{16}$$

where $V\angle\widetilde{F}_{s-3}$. Since $s>2$, we have $F_{s-3}\underset{\cdot}{\angle}F_{s-2}$, whence $V\underset{\cdot}{\angle}F_{s-2}$. We have proved that the words F_{k-3} and $\widetilde{F}_{s-2}$ are simple. Hence from (16), (13) and 1.7′ it follows that $F_{k-3}=\widetilde{F}_{s-2}$.

3.2.2) Suppose $[G_{k-4}A^{\partial}>[\widetilde{F}_{s-3}^{\partial}$. Then, cancelling (15) on the right by $\widetilde{F}_{s-3}$, we obtain

$$F_{k-3}^{\lambda_{k-3}+1}V=\widetilde{F}_{s-2}^{\mu_{s-2}}, \tag{17}$$

where V satisfies the relation $G_{k-4}A=V\widetilde{F}_{s-3}$. If $V\angle G_{k-4}$, then, using $G_{k-4}\angle F_{k-4}\underset{\cdot}{\angle}F_{k-3}$, we obtain $V\underset{\cdot}{\angle}F_{k-3}$. Then from (17), (13) and 1.7′ it follows that $F_{k-3}=\widetilde{F}_{s-2}$. It remains to consider the case where V is not a beginning of G_{k-4}. In this case we must have $[V^{\partial}>[G_{k-4}^{\partial}$. Then $[\widetilde{F}_{s-3}^{\partial}<[A^{\partial}\leqq 2M$. An application of 17.4 gives $F_{k-3}=\widetilde{F}_{s-2}$.

Thus we have proved that $F_{k-3}=\widetilde{F}_{s-2}$. A contradiction is now obtained as follows. From (14) we have $V_1\widetilde{F}_{s-2}=V_2\widetilde{F}_{s-2}\widetilde{F}_{s-3}$ for some V_1, V_2. Since $s>2$, $\widetilde{F}_{s-2}$ ends with $\langle Q_1P_1,Q_1R_1\rangle_{s-2}$. The word $\widetilde{F}_{s-2}\widetilde{F}_{s-3}$ ends with $\langle Q_1P_1,Q_1R_1\rangle_{s-2}$. We obtain $\langle Q_1P_1,Q_1R_1\rangle_{s-2}=\langle Q_1R_1,Q_1P_1\rangle_{s-2}$, i.e. Q_1P_1 and Q_1R_1 commute, which contradicts the condition $\mathscr{L}(Q_1P_1,Q_1R_1)$. From this contradiction it follows that Case 3 is impossible. The proposition is proved.

17.6. *Consider the system*

$$x=\alpha(h)=\beta(g), \qquad h^0>1, \quad g^0>1, \tag{*}$$

where x is a word unknown, h and g are processional unknowns, and α and β are G^M-terms. One of the following two assertions holds: 1) *the system* (*) *has no solutions;* 2) *the system* (*) *is equivalent to the system $x=\alpha$, $h=g$, $h^0>1$, and we have the implication*

$$[(x=\alpha^*(h)) \,\&\, ((h=g) \vee (h\rightarrow g) \vee (g\rightarrow h))] \supset [x=\alpha^*(h)=\beta^*(g)],$$

where α^* and β^* are the left-hand sides of the predicate terms α and β.

PROOF. The system (*) expanded has the form

$$x = G(QP, QR, h)\,Q, \quad G(QP, QR, h)\,Q = G(Q_1P_1, Q_1R_1, g)\,Q_1,$$

$$\neg\mathcal{B}([QP, QR, h]\,M), \ \neg\mathcal{B}([Q_1P_1, Q_1R_1, g], M), \ h^0>1, \ g^0>1,$$

where P, Q, R, P_1, Q_1 and R_1 are words in the alphabet $\mathbf{И}$ such that

$$[PQR^\partial, [P_1Q_1R_1^\partial \leq M, \quad \mathscr{L}(QP, QR), \quad \mathscr{L}(Q_1P_1, Q_1R_1), \quad Q, Q_1 \neq \Lambda.$$

By 17.5, if even one of the equalities $QP=Q_1P_1$, $QR=Q_1R_1$ or $Q=Q_1$ does not hold, then we have 1). If these equalities do hold, then we have 2); we use the fact that from $F(A,B,h)=F(A,B,g)$ and $\mathscr{L}(A,B)$ follows $h=g$ (see the beginning of the proof of 16.5) and also 16.6 (8). The proposition is proved.

We will now get rid of the conditions $h^0, g^0>1$.

17.7. *Consider the system*

$$(*) \qquad x=\alpha(h)=\beta(g),$$

where x is a word unknown, h and g are processional unknowns, and α and β are G^M-terms. One of the following two assertions holds: 1) *the system* (*) *has no solutions;* 2) *the system* (*) *is equivalent to the system*

$$(x=\alpha) \,\&\, (h=g \vee h\rightarrow g \vee g\rightarrow h);$$

here we have the implication

$$[(x=\alpha^*(h)) \,\&\, (h=g \vee h\rightarrow g \vee g\rightarrow h] \supset [x=\alpha^*(h)=\beta^*(g)],$$

where α^* and β^* are the left-hand sides of the predicate terms α and β.

PROOF. By 16.6 (9) we have the equivalences

$$(1) \qquad x=\alpha(h) \Leftrightarrow \exists f\,[x=\alpha(f), \quad (h=f \vee h\rightarrow f), \quad f^0>1],$$

$$(2) \qquad x=\beta(g) \Leftrightarrow \exists \varphi\,[x=\beta(\varphi), \quad (g=\varphi \vee g\rightarrow \varphi), \quad \varphi^0>1],$$

where f and φ are processional unknowns. From (1) and (2) we conclude that (*) is equivalent to the system

$$(3) \qquad \exists f\varphi[x=\alpha(f)=\beta(\varphi), f^0>1, \varphi^0>1, (h=f \vee h\rightarrow f), (g=\varphi \vee g\rightarrow \varphi)].$$

By 17.6 one of the following two assertions holds: a) the predicate

$$(4) \qquad x=\alpha(f)=\beta(\varphi),\ f^0>1,\ \varphi^0>1$$

has no solutions; b) the predicate (4) is equivalent to the predicate $x=\alpha(f)$, $f=\varphi$, $f^0>1$, and we have the implication

$$[(x=\alpha^*(f)) \,\&\, (f=\varphi \vee f\rightarrow\varphi \vee \varphi\rightarrow f)] \supset [x=\alpha^*(f)=\beta^*(\varphi)].$$

In case a) we have 1). Consider case b). In this case the predicate (3) is equivalent to the predicate

(5) $\exists f\varphi[x=\alpha(f), f=\varphi, f^0>1, (h=f \vee h\to f), (g=\varphi \vee g\to\varphi)]$.

Dropping the conjunctive factor $f=\varphi$ and applying 16.6 (8), we obtain that (5) is equivalent to the predicate

$$\exists f[x=\alpha(h), f^0>1, h=f \vee h\to f, g=f \vee g\to f],$$

which, in turn, is equivalent to the predicate

(5′) $x=\alpha(h) \mathbin{\&} \exists f[f^0>1, h=f \vee h\to f, g=f \vee g\to f]$.

We will prove that we have the equivalence

(6) $\exists f[f^0>1, h=f \vee h\to f, g=f \vee g\to f] \Leftrightarrow [h=g \vee h\to g \vee g\to h]$.

By analyzing cases we establish that the left-hand side implies the right; here we use the fact that $\exists f[h\to f, g\to f]$ implies $h=g$. To prove the implication from right to left it suffices to use 16.6 (5) and 16.6 (1). Thus (6) is proved. It is now obvious that in case b) we have 2). The proposition is proved.

In solving systems there arise PDR's of a more general form than those defined in §16. The inductive definition of the "new" PDR's is as follows:

0) Any LDR is a PDR.

1) Suppose h, g and f are distinct processional unknowns, $\lambda_1, \cdots, \lambda_k$ and $\mu_1, \cdots, \mu_s$ are positive unknowns, and $\mathscr{D}(\lambda_1, \cdots, \lambda_k, \mu_1, \cdots, \mu_s)$ is a LDR. Then each of the predicates

$$\exists \lambda_1 \cdots \lambda_k[h=(\lambda_1, \cdots, \lambda_k) \mathbin{\&} \mathscr{D}(\lambda_1, \cdots, \lambda_k, \mu_1, \cdots, \mu_s)],$$
$$\exists f\lambda_1 \cdots \lambda_k\mu_1 \cdots \mu_s[h=(f, \lambda_1, \cdots, \lambda_k), g=(f, \mu_1, \cdots, \mu_s), \mathscr{D}(\lambda_1, \cdots, \lambda_k, \mu_1, \cdots, \mu_s)]$$

is a PDR.

2) If $\mathscr{K}_1$ and $\mathscr{K}_2$ are PDR's, then each of the predicates $\mathscr{K}_1 \vee \mathscr{K}_2$ and $\mathscr{K}_1 \mathbin{\&} \mathscr{K}_2$ is also a PDR.

Any PDR in the sense of §16 is equivalent to some "new" PDR. This is obvious. For example, the PDR $\exists \lambda\mu[h=(2,\lambda,\mu)]$ is equivalent to the PDR $\exists \lambda\mu\nu[h=(\nu,\lambda,\mu),\ \nu=2]$. In what is to follow, a PDR will be understood as a PDR defined in accordance with 0), 1) and 2).

We state in terms of PDR's the main result on equations which do not contain V-terms.

17.8. *Consider the system*

(*) $$x=\alpha=\beta,$$

where α and β are special predicate terms of characteristic M which are not V^M-terms. Then there exists a PDR $\mathscr{K}$ such that (*) is equivalent to the system $(x=\alpha) \mathbin{\&} \mathscr{K}$ and we have the implication

$$[(x=\alpha^*) \mathbin{\&} \mathscr{K}] \supset [(x=\alpha^*=\beta^*) \mathbin{\&} \mathscr{K}],$$

where α^* and β^* are the left-hand sides of the predicate terms α and β.

PROOF. There are three possible cases.

Case 1. α and β are G^M-terms. By 17.7, either (*) has no solutions or it is equivalent to the predicate

(1) $$x = \alpha(h), \quad [h = g \vee h \to g \vee g \to h],$$

where h and g are the unknowns occurring in α and β; here we have the implication $[(x = \alpha^*) \& (h = g \vee h \to g \vee g \to h)] \supset [x = \alpha^* = \beta^*]$. In the first alternative the assertion is trivial: as $\mathscr{K}$ we may take $0 = 1$. Consider the second alternative. From the definition of the relation $\to$ (preceding 16.6) it follows that the predicate contained in the square brackets in (1) is equivalent to the PDR

$$\exists f(h = f, g = f) \vee \exists f\lambda\mu\nu\,[h = (f, \lambda, \nu), g = (f, \mu), \mu = \lambda + 1, \nu = 1]$$
$$\vee \exists f\lambda\mu\nu\,[h = (f, \mu), g = (f, \lambda, \nu), \mu = \lambda + 1, \nu = 1].$$

Denoting this PDR by $\mathscr{K}$, we see that our assertion is true.

Case 2. One of the terms α or β is a G^M-term, the other is a Σ^M-term. Obviously, it doesn't matter which of α or β we take to be the G^M-term and which the Σ^M-term. Our assertion follows from 16.9.

Case 3. α and β are Σ^M-terms. By 13.8, the equation $\alpha = \beta$ is equivalent to some LDR $\mathscr{D}$. It is not difficult to see that our assertion is true with $\mathscr{K} = \mathscr{D}$. The proposition is proved.

In conclusion, we construct an algorithm recognizing whether or not a given PDR has a solution.

Let us begin with the notion of an ω-object. By definition, an ω-object is a nonempty ordered set of unknowns whose first component is either a processional or a positive unknown and whose other components (if there are any) are positive unknowns. We identify a processional (positive) unknown ξ with the ω-object consisting of one component, the processional (positive) unknown ξ. We say that an unknown ξ occurs in the ω-object $(\xi_1, \cdots, \xi_m)$ if ξ is one of the letters $\xi_1, \cdots, \xi_m$.

Suppose h is a processional unknown, and $\xi = (\xi_1, \cdots, \xi_m)$ and $\eta = (\eta_1, \cdots, \eta_k)$ are ω-objects. We define "the result of substituting η for h in ξ" (denoted by $\mathrm{S}^h_\eta \xi|$). By definition, $\mathrm{S}^h_\eta \xi|$ is ξ if h does not occur in ξ, and is the ω-object $(\eta_1, \cdots, \eta_k, \xi_2, \cdots, \xi_m)$ if ξ has the form $(h, \xi_2, \cdots, \xi_m)$. (Obviously, either h does not occur in ξ or ξ has the form $(h, \xi_2, \cdots, \xi_m)$.) Note that if $(\xi_1, \cdots, \xi_m)$ is an ω-object and $\bar{\xi}_1, \cdots, \bar{\xi}_m$ are values of the unknowns $\xi_1, \cdots, \xi_m$, then $(\bar{\xi}_1, \cdots, \bar{\xi}_m)$ is a procession. Suppose now that $\xi = (\xi_1, \cdots, \xi_m)$ and $\eta = (\eta_1, \cdots, \eta_k)$ are ω-objects. We call the equality $\xi = \eta$ an equation. By a solution of this equation we mean any set $(\bar{\xi}_1, \cdots, \bar{\xi}_m, \bar{\eta}_1, \cdots, \bar{\eta}_k)$ of values of the unknowns $\xi_1, \cdots, \xi_m, \eta_1, \cdots, \eta_k$ such that the procession $(\bar{\xi}_1, \cdots, \bar{\xi}_m)$ is equal to the procession $(\bar{\eta}_1, \cdots, \bar{\eta}_1, \cdots, \bar{\eta}_k)$.

We define elementary and standard PDR's. By definition, any standard PDR has the form

(*) $$\mathop{\&}_{i=1}^{m} (h_i = \omega_i) \& \mathscr{D},$$

where $m > 0$; $h_1, \cdots, h_m$ are processional unknowns, $\omega_1, \cdots, \omega_m$ are ω-objects, $\mathscr{D}$ is a LDR, and none of the letters $h_1, \cdots, h_m$ occurs in any of $\omega_1, \cdots, \omega_m$. By the left-hand side of the standard PDR (*) we mean a list $g_1, \cdots, g_n$ $(n \leq m)$ of all the different unknowns contained among $h_1, \cdots, h_m$. If all of the letters $h_1, \cdots, h_m$ are distinct, the standard PDR (*) is called elementary. Formally, standard PDR's are not PDR's in the sense of the definition preceding 17.8. However, the following is true.

17.9. *Any PDR is equivalent to a predicate of the form* $\exists \xi \mathfrak{A}(\xi, \eta)$, *where* ξ *and* η *are lists of unknowns, and* $\mathfrak{A}$ *is a disjunction of standard PDR's.*

PROOF. By definition, any PDR is obtained by means of the operations & and $\vee$ from basic PDR's of the form $\exists \xi \mathfrak{B}(\xi, \eta)$, where ξ and η are lists of unknowns and $\mathfrak{B}$ is an elementary PDR. Using the permutability of the symbols & and $\vee$, we can write any PDR as a disjunction of conjunctions of basic PDR's. It remains to place & and then $\vee$ inside the existential quantifiers. The first goal can be accomplished using the fact that the choice of variables related by the quantifier is at our disposal and using the equivalence $\mathscr{A} \& \exists x \mathscr{P}(x) \Leftrightarrow \exists x \mathscr{A} \& \mathscr{P}(x))$, where the predicate $\mathscr{A}$ does not depend on x. For the second it suffices to use the permutability of the symbols $\exists$ and $\vee$.

Standard PDR's reduce to elementary PDR's.

17.10. *Any standard PDR with left-hand side* $h_1, \cdots, h_n$ *is equivalent with respect to* $h_1, \cdots, h_n$ *to any elementary PDR with left-hand side* $h_1, \cdots, h_n$.

PROOF. This is a consequence of the following lemma.

LEMMA. *Suppose* $h_1, \cdots, h_n, \cdots, h_m$ $(1 \leq n \leq m)$ *are processional unknowns, where* $h_1, \cdots, h_n$ *are distinct and* $h_{n+1}, \cdots, h_m$ *are letters from the list* $h_1, \cdots, h_n$. *Suppose also that* $\omega_1, \cdots, \omega_n$ *are* ω*-objects such that none of the letters* $h_1, \cdots, h_m$ *occurs in any of* $\omega_1, \cdots, \omega_m$ *and that* $\mathscr{D}$ *is a LDR. Then the system*

$$\underset{i=1}{\overset{m}{\&}} (h_i = \omega_i) \& \mathscr{D} \tag{1}$$

is equivalent with respect to $h_1, \cdots, h_n$ *to an elementary PDR with left-hand side* $h_1, \cdots, h_n$.

PROOF OF THE LEMMA. We proceed by induction on the difference $m - n$. By hypothesis, this difference is nonnegative. If $m - n = 0$, then $m = n$ and the system (1) is an elementary PDR with left-hand side $h_1, \cdots, h_n$. The lemma is trivially true.

Suppose $k > 0$ and our assertion is true whenever $m - n < k$. Suppose there is given a system (1) and $m - n = k$. Write (1) in the form

$$\underset{i=1}{\overset{m-1}{\&}} (h_i = \omega_i) \& \mathscr{D}, \qquad h_m = \omega_m. \tag{2}$$

By hypothesis, h_m is a letter in the list $h_1, \cdots, h_n$. Without loss of generality, we may assume that h_m is h_n. Then (2) is equivalent to the system

$$\underset{i=1}{\overset{m-1}{\&}} (h_i = \omega_i) \& \mathscr{D}, \qquad \omega_n = \omega_m. \tag{3}$$

The rest of the proof is divided into three cases.

Case 1. $\omega_n = (\lambda_1, \cdots, \lambda_k)$ and $\omega_m = (\mu_1, \cdots, \mu_s)$ where $\lambda_1, \cdots, \lambda_k$ and $\mu_1, \cdots, \mu_s$ are positive unknowns. If $k \neq s$, then the equation $\omega_n = \omega_m$, and hence also the system (3), has no solutions. Then (3) is equivalent with respect to $h_1, \cdots, h_n$ to the elementary PDR $(h_1 = \omega_1) \& (0 = 1)$, which also has no solutions. If $k = s$, then the equation $\omega_n = \omega_m$ is equivalent to the system $\&_{i=1}^{k} (\lambda_i = \mu_i)$. Then (3) is equivalent to the system

$$\underset{i=1}{\overset{m-1}{\&}} (h_i = \omega_i) \& \mathscr{D}_1, \tag{4}$$

where $\mathscr{D}_i$ is the LDR $\mathscr{D}$ & $\&_{i=1}^{k}(\lambda_i = \mu_i)$. Applying the induction assumption to (4), we see that our assertion is true.

Case 2. $\omega_n = (h, \lambda_1, \cdots, \lambda_k)$ and $\omega_m = (g, \mu_1, \cdots, \mu_s)$, where h and g are processional unknowns and $\lambda_1, \cdots, \lambda_k$ and $\mu_1, \cdots, \mu_s$ are positive unknowns. Without loss of generality, we may assume that $k \leqq s$. Then the equation $\omega_n = \omega_m$ is equivalent to the system $h = (g, \mu_1, \cdots, \mu_{s-k})$ & $\&_{i=1}^{k}(\lambda_i = \mu_{s-k+i})$. Since h and g do not occur in the list $h_1, \cdots, h_n$, (3) is equivalent with respect to $h_1, \cdots, h_n$ to the system

$$(5) \qquad \mathop{\&}_{i=1}^{m-1} (h_i = \tilde{\omega}_i) \& \mathscr{D}_1,$$

where $\mathscr{D}_1$ is the LDR $\mathscr{D}$ & $\&_{i=1}^{k}(\lambda_i = \mu_{s-k+i})$, and $\tilde{\omega}_i = \mathrm{S}^{h}_{(g,\mu_1,\cdots,\mu_{s-k})}\omega_i|$. Applying the induction assumption to (5), we see that our assertion is true.

Case 3. One of the ω-objects ω_n or ω_m has the form $(h, \lambda_1, \cdots, \eta_k)$, and the other the form $(\mu_1, \cdots, \mu_s)$, where h is a processional unknown and $\lambda_1, \cdots, \lambda_k$ and $\mu_1, \cdots, \mu_s$ are positive unknowns. It is clear that if $k \geqq s$, then the equation $\omega_n = \omega_m$, and hence also the system (3), has no solutions and the lemma is true. Suppose $k < s$. In this case the equation $\omega_n = \omega_m$ is equivalent to the system $h = (\mu_1, \cdots, \mu_{s-k})$ & $\&_{i=1}^{k}(\lambda_i = \mu_{s-k+i})$. Then (3) is equivalent with respect to $h_1, \cdots, h_n$ to (5), where $\mathscr{D}_1$ is the LDR $\mathscr{D}$ & $\&_{i=1}^{k}(\lambda_i = \mu_{s-k+i})$, and $\tilde{\omega}_i = \mathrm{S}^{h}_{(\mu_1,\cdots,\mu_{s-k})}\omega_i|$. Applying the induction assumption to (5), we see that our assertion is true. The lemma is proved. Thus 17.10 is also proved.

17.11. *There exists an algorithm which determines whether or not a given PDR has a solution.*

Proof. In view of 17.9 and 17.10, it suffices to prove our assertion for elementary PDR's. But for elementary PDR's it is almost obvious. To see this we need only introduce the notion of the "parity of an ω-object". Suppose there is given an ω-object $(\xi_1, \cdots, \xi_n)$. Its parity $\gamma(\omega)$ is defined as follows: 1) $\gamma(\omega) = 0$ if ξ_1 is a processional unknown and m is odd, or if ξ_1 is a positive unknown and m is even; 2) $\gamma(\omega) = 1$ if ξ_1 is a processional unknown and m is even, or if ξ_1 is a positive unknown and m is odd. It is clear that the elementary PDR (*) has a solution if and only if $\gamma(h_i) = \gamma(\omega_i)$ $(i = 1, \cdots, m)$ and the LDR $\mathscr{D}$ has a solution. It remains to apply 2.3′. The proposition is proved.

§18. *T*-systems

Consider the system

$$(*) \qquad \begin{gathered} x_1 = \alpha_{11} = \ldots = \alpha_{1q_1}, \\ \cdots\cdots\cdots\cdots \\ x_n = \alpha_{n1} = \ldots = \alpha_{nq_n}, \\ \mathscr{K} \& \Omega, \end{gathered}$$

where $n > 1$; $q_1, \cdots, q_n \geqq 1$; $x_1, \cdots, x_n$ are distinct word unknowns, the α_{ij} are predicate terms, $\mathscr{K}$ is a PDR, and Ω is a conjunction of factors of the form $\omega(x, M)$, where x is a letter in the list $x_1, \cdots, x_n$ and M is a positive integer. The list $x_1, \cdots, x_n$ is called the left-hand side of the system (*). By the height of the system (*) we mean the integer $h = q_1 + \cdots + q_n$. We call (*) canonical if $q_1 = \cdots = q_n = 1$.

The system (*) is called a *T*-system of characteristic M if the following conditions are satisfied:

1) If the ith row of (*) contains more than one term ($q_i > 1$), then each term in this row is a special predicate term of characteristic M.

2) The word unknowns occurring in the V_1^M-terms of (*) are letters in the list $x_1, \cdots, x_n$.

3) $\widetilde{V}_1^M$-terms of (*) containing the same word unknown have the same base.

4) If (*) contains a V^M-term $\alpha(x_k, \lambda)$, where $1 \leq k \leq n$ and λ is a positive unknown, then the PDR $\mathscr{K}$ contains the factor $\lambda \geq 9M^2$, and the conjunction Ω the factor $\omega(x_k, M)$. There are no other factors in Ω.

5) If the ith row of (*) contains a Ψ_1^M-term α and a V^M-term in the word unknown x_k ($1 \leq k \leq n$), then there exists $j \neq k$ such that the jth row of (*) contains a Ψ_2^M-term with the same base as α.

For any T-system $\mathfrak{A}$ we define a system $\mathfrak{A}^*$. If the T-system $\mathfrak{A}$ has the form (*), then $\mathfrak{A}^*$ is the system

$$(**)\qquad \begin{array}{c} x_1 = \overset{*}{\alpha}_{11} = \ldots = \overset{*}{\alpha}_{1q_1}, \\ \cdots\cdots\cdots\cdots \\ x_n = \overset{*}{\alpha}_{n1} = \ldots = \overset{*}{\alpha}_{nq_n}, \\ \mathscr{K}_3 \end{array}$$

where each α_{ij}^* is the left-hand side of the predicate term α_{ij}.

We say that an unknown ξ occurs in the system (*) (in the system (**)) if ξ is one of the letters $x_1, \cdots, x_n$ or occurs in some predicate term α_{ij} (in some term α_{ij}^*) or occurs freely in the PDR $\mathscr{K}$ (i.e. the occurrences of ξ in $\mathscr{K}$ are not related by an existential quantifier). It is clear that the unknowns occurring in $\mathfrak{A}$ are exactly those occurring in $\mathfrak{A}^*$.

We write $\mathfrak{A}(x_1, \cdots, x_n, \xi)$ if $\mathfrak{A}$ is a T-system with left-hand side of $x_1, \cdots, x_n$ and ξ is a list of unknowns occurring in $\mathfrak{A}$ and different from $x_1, \cdots, x_n$. The notation $\mathfrak{A}^*(x_1, \cdots, x_n, \xi)$ has a similar meaning.

A solution $(X_1, \cdots, X_n)$ of the system (*) is called an $\bar{\epsilon}_M$-solution if $\neg\epsilon(x_1, \cdots, x_n, M)$ (the definition of the predicate ϵ follows 15.2′).

The following theorem gives an explicit representation for the $\bar{\epsilon}_M$-solutions of T-systems of characteristic M.

18.1. Theorem. *Suppose $\mathfrak{A}(x_1, \cdots, x_n, \xi)$ is a T-system of characteristic M with left-hand side $x_1, \cdots, x_n$. Then it is possible to construct a canonical T-system $\mathfrak{B}(x_1, \cdots, x_n, \eta)$ of characteristic M such that*: 1) *the systems $\mathfrak{A}$ and $\mathfrak{B}$ are equivalent in the region $\neg\epsilon(x_1, \cdots, x_n, M)$*; 2) *we have the implication*

$$\mathfrak{B}^*(x_1, \cdots, x_n, \eta) \supset \exists \xi \mathfrak{A}^*(x_1, \cdots, x_n, \xi).$$

Proof. If the system $\mathfrak{A}$ has no $\bar{\epsilon}_M$-solutions, the theorem is trivially true: as $\mathfrak{B}$ we may take any canonical T-system for which the PDR $\mathscr{K}$ contains the factor $0 = 1$.

To prove the theorem we proceed by induction on the height of $\mathfrak{A}$. Obviously, the height of $\mathfrak{A}$ is at least n. If the height of $\mathfrak{A}$ is equal to n, then $\mathfrak{A}$ is canonical and the theorem is true: as $\mathfrak{B}$ we may take $\mathfrak{A}$.

Suppose $h > n$ and our assertion is true whenever the height of $\mathfrak{A}$ is less than h, and suppose there is given a T-system (*) of characteristic M and height h. We will prove our assertion true for (*). Since the height of (*) is greater than n, there exists j, $1 \leq j \leq n$, such that $q_j > 1$. Without loss of generality we may assume that $j = 1$, so that $q_1 > 1$. Then (*) can be written in the form

$$
\begin{aligned}
&x_1 = \alpha = \beta = \alpha_{13} = \cdots = \alpha_{1q_1}, \\
(1) \qquad &x_i = \alpha_{i1} = \cdots = \alpha_{iq_i} \quad (i = 2, \cdots, n), \\
&\mathscr{K} \,\&\, \Omega,
\end{aligned}
$$

where, in view of condition 1), α and β are special predicate terms of characteristic M. Let us consider the possibilities for α and β.

Case 1. One of the terms α or β is a $\tilde{\Psi}_1^M$-term, and the other a V^M-term whose word unknown is x_k, where $1 \leq k \leq n$. Then in accordance with conditions 4) and 5) the system (*) contains the subsystem (if α is a V^M-term)

$$
\begin{aligned}
(2) \qquad &(\alpha = \beta) \,\&\, \omega(x_h, M), \\
&x_j = \gamma,
\end{aligned}
$$

where $x_j \neq x_k$ and γ is a Ψ_2^M-term with the same base as β. Assume that $(X_1, \cdots, X_n, \cdots)$ is a solution of (*). Then $(X_k, X_j, \cdots)$ is a solution of (2). By 14.4 we have $\epsilon(X_k, X_j, M)$, whence $\epsilon(X_1, \cdots, X_n, M)$. Hence in Case 1 the system (*) has no $\bar{\epsilon}_M$-solutions and the theorem is true.

REMARK 1. In considering the remaining cases we may assume that (1) contains no rows in which there occur a $\tilde{\Psi}_1^M$-term and a V^M-term whose word unknown is a letter in the list $x_1, \cdots, x_n$. If we do this, then 5) is trivially satisfied for the system (1) and any T-subsystem with left-hand side $x_1, \cdots, x_n$.

Case 2. Neither of the terms α or β is a V^M-term. Then by 17.8 there exists a PDR $\mathscr{K}_1$ such that the system $x_1 = \alpha = \beta$ is equivalent to the system $(x_1 = \alpha) \,\&\, \mathscr{K}_1$, and we have the implication

$$((x_1 = \alpha^*) \,\&\, \mathscr{K}_1) \supset ((x_1 = \alpha^* = \beta^*) \,\&\, \mathscr{K}_1).$$

Then (1) is equivalent to the system

$$
\begin{aligned}
&x_1 = \alpha = \alpha_{13} = \cdots = \alpha_{1q_1}, \\
(3) \qquad &x_i = \alpha_{i1} = \cdots = \alpha_{iq_i} \quad (i = 2, \cdots, n), \\
&\mathscr{K} \,\&\, \mathscr{K}_1 \,\&\, \Omega,
\end{aligned}
$$

and we have the implication $(3)^* \supset (1)^*$.

The system (3) is a T-system of characteristic M with left-hand side $x_1, \cdots, x_n$ and height $h-1$. Applying the induction assumption to (3) and taking into account the equivalence of (1), (3) and the implication $(3)^* \supset (1)^*$, we see that the theorem is true.

It remains to consider the situation where at least one of the terms α or β is a V^M-term. Without loss of generality, we will assume that α is a V^M-term. For β we have the following possibilities.

Case 3. β is a G^M-term. Then by 16.12 the equation $\alpha = \beta$, and hence also the system (1), has no solutions. Then certainly (1) has no $\bar{\epsilon}_M$-solutions, and the theorem is true.

Case 4. β is a Θ^M-term. Then by 14.7 the equation $\alpha = \beta$ has no solutions. The rest of the argument is the same as in Case 3.

Case 5. β is a V^M-term. Suppose α is $\alpha(u, \mu)$, and β is $\beta(v, \mu_1)$, where u and v are word unknowns, and μ and μ_1 are positive unknowns.

Assume that $u = v$. Then by 15.5 there exists a LDR $\mathscr{K}_1$ such that the system $x_1 = \alpha = \beta$ is equivalent to the system $(x_1 = \alpha) \,\&\, \mathscr{K}_1$, and the system $x_1 = \alpha^* = \beta^*$ to the system $(x_1 = \alpha^*) \,\&\, \mathscr{K}_1$. The conjunction $\mathscr{K} \,\&\, \mathscr{K}_1$ is a

PDR. Arguing as in Case 2 following the formula $(x_1 = \alpha^* = \beta^*) \,\&\, \mathscr{K}_1$, we see that the theorem is true. From now on, without loss of generality, we will assume that no row of the system (1) contains two V^M-terms with the same word unknown. In particular, $u \neq v$. Since α and β are V^M-terms, there are three possible subcases.

Subcase 5.1. α and β are V_1^M-terms. Then, in accordance with condition 2) in the definition of a T-system, u and v are letters in the list $x_1, \cdots, x_n$. Since $u \neq v$, we have $u = x_i$ and $v = x_j$, where $i \neq j$. If $(X_1, \cdots, X_n, \mu, \mu_1, \cdots)$ is a solution of (1), then (X_i, X_j, μ, μ_1) is a solution of the equation $\alpha = \beta$. By 15.1 we have $\epsilon(X_i, X_j, M)$, and hence $\epsilon(X_1, \cdots, X_n, M)$. Thus in Subcase 5.1 the system (1) has no $\bar{\epsilon}_M$-solutions and the theorem is true.

Subcase 5.2. α and β are $\widetilde{V}_1^M$-terms. Here we consider two subcases.

5.2.1) The letters u and v are letters in the list $x_1, \cdots, x_n$. Arguing as in Subcase 5.1 (only instead of 15.1 we use 15.2′), we see that (1) has no $\bar{\epsilon}_M$-solutions and the theorem is true.

5.2.2) At least one of the letters u or v does not occur in the list $x_1, \cdots, x_n$. Suppose v does not occur in $x_1, \cdots, x_n$. Look at the rows of the system (1) in which v does not appear. Suppose these to be the first k rows $(1 \leqq k \leqq n)$. Since v does not occur in $x_1, \cdots, x_n$, in accordance with condition 2) the letter v can occur only in the $\widetilde{V}_1^M$-terms of (1). By condition 3), the $\widetilde{V}_1^M$-terms of (1) containing v have the same base. Such $\widetilde{V}_1^M$-terms can differ from one another only in their positive unknowns. Furthermore, as we have agreed, no row of (1) contains two V^M-terms with the unknown v. Hence, without loss of generality, we may assume that the system (1) has the form

$$
\begin{aligned}
&x_1 = \alpha(u, \mu) = \beta(v, \mu_1) = \alpha_{13} = \ldots \alpha_{1q_1}, \\
(4)\qquad &x_i = \beta(v, \mu_i) = \alpha_{i2} = \ldots = \alpha_{iq_i} \qquad (i = 2, \ldots, k), \\
&x_i = \alpha_{i1} = \ldots = \alpha_{iq_i} \qquad (i = k+1, \ldots, n), \\
&\mathscr{K} \,\&\, \Omega,
\end{aligned}
$$

where the terms with double indices do not contain v. Consider the subsystem of (4)

$$
(5)\qquad
\begin{aligned}
&x_1 = \alpha(u, \mu) = \beta(v, \mu_1), \\
&x_i = \beta(v, \mu_i) \qquad (i = 2, \cdots, k).
\end{aligned}
$$

By 15.4, the system (5) either has no solutions or is equivalent in the region $\neg\, \epsilon(x_1, \cdots, x_k, M)$ to the system

$$
(6)\qquad x_i = \alpha(u, \mu_i) \qquad (i = 1, \cdots, k), \quad \mu = \mu_1,
$$

and we have the implication $(6)^* \supset \exists\, v\, (5)^*$.

In the first alternative the system (4) has no solutions and the theorem is true. Consider the second alternative. In this case (4) is equivalent in the region $\neg\, \epsilon(x_1, \cdots, x_n, M)$ to the system

$$
\begin{aligned}
&x_1 = \alpha = \alpha_{13} = \ldots = \alpha_{1q_1}, \\
(7)\qquad &x_i = \alpha(u, \mu_i) = \alpha_{i2} = \ldots = \alpha_{iq_i} \qquad (i = 2, \ldots, k), \\
&x_i = \alpha_{i1} = \ldots = \alpha_{iq_i} \qquad (i = k+1, \ldots, n), \\
&\mathscr{K} \,\&\, \Omega \,\&\, \mu = \mu_1.
\end{aligned}
$$

It is also obvious that we have the implication

$$
(8)\qquad (7)^* \supset \exists\, v\, (4)^*.
$$

The system (7) is a T-system of characteristic M with left-hand side $x_1, \cdots, x_n$ and height $h-1$. Applying the induction assumption to (7) and taking into account that (7) is equivalent in the region $\urcorner \epsilon(x_1, \cdots, x_n, M)$ to (4), and also using (8), we see that the theorem is true.

Subcase 5.3. One of the terms α or β is a V_1^M-term, and the other is a $\widetilde{V}_1^M$-term. Suppose $\alpha(u, \mu)$ is the V_1^M-term, and $\beta(v, \mu_1)$ is the $\widetilde{V}_1^M$-term. By condition 2), the unknown u occurs in the list $x_1, \cdots, x_n$. There are two possibilities for v.

5.3.1) n is a letter in the list $x_1, \cdots, x_n$. Then, arguing as in Subcase 5.1 (only instead of 15.1 we now use 15.3′), we see that (1) has no $\bar{\epsilon}_M$-solutions.

5.3.2) v does not occur in the list $x_1, \cdots, x_n$. Arguing as in Subcase 5.2.2), we write (1) in the form (4), where the terms with double indices do not contain v. Since u occurs in the list $x_1, \cdots, x_n$, by 15.3 (the role of x in 15.3 is now played by u) one of the following two assertions holds: 1) the system (5) has no solutions; 2) the system (5) is equivalent in the region

$$\urcorner \epsilon(x_1, \cdots, x_n, M)$$

to the system (6), and we have the implication $(6)^* \supset \exists v (5)^*$. The rest of the argument is the same as in 5.2.2) following (6). This completes the consideration of Case 5.

Case 6. β is a Ψ^M-term. This case is split into two subcases.

Subcase 6.1. The word unknown of the term α occurs in the list $x_1, \cdots, x_n$. Then $\alpha = \alpha(x_k, \lambda)$, where $1 \leqq k \leqq n$ and λ is a positive unknown. In view of condition 4) in the definition of a T-system, (1) contains the subsystem

$$\alpha(x_h, \lambda) = \beta \,\&\, \omega(x_k, M) \,\&\, \lambda \geqq 9M^2. \tag{9}$$

Since β is a Ψ^M-term, it follows that β is a $\overline{\Psi}_1^M$-, $\tilde{\Psi}_1^M$- or Ψ_2^M-term. By Remark 1, β is a $\overline{\Psi}_1^M$-term or a Ψ_2^M-term. Hence, by 14.3, the system (9), and so also the system (1), has no solutions. Certainly (1) has no $\bar{\epsilon}_M$-solutions and the theorem is true.

Subcase 6.2. The word unknown of the term α does not occur in the list $x_1, \cdots, x_n$. Then, by condition 2), α is a $\widetilde{V}_1^M$-term. Thus, the still unresolved Subcase 6.2 is characterized by the fact that one of the terms α or β is a Ψ^M-term, the other a $\widetilde{V}_1^M$-term whose word unknown does not occur in the list $x_1, \cdots, x_n$.

We now summarize the discussion of the induction step. We started with the fact that in some row of the system (*) there occur at least two terms α and β (we assumed, without loss of generality, that the first row is such a row), and proved that our assertion is true in all cases with respect to α and β except, possibly, where one of the terms α or β is a $\widetilde{V}_1^M$-term, the other a Ψ^M-term. We are justified in making the following

REMARK 2. If some row of the system (*) contains two terms for which it is not true that one is a $\widetilde{V}_1^M$-term and the other a Ψ^M-term, then the assertion being proved is true. In particular, if some row of (*) contains three terms, the assertion is true, since among three terms, obviously, there are two for which it is not true that one is a $\widetilde{V}^M$-term and the other a Ψ^M-term.

We now return to Subcase 6.2. In this subcase, α is a $\widetilde{V}^M$-term whose word unknown does not occur in the list $x_1, \cdots, x_n$, and β is a Ψ^M-term. Suppose $\alpha = \alpha(v, \mu_1)$, where v is a word unknown and μ_1 a positive unknown. Arguing as in Subcase 5.2.2), we write (1) in the form

$$
\begin{aligned}
&x_1 = \alpha(v, \mu_1) = \alpha_{12} = \ldots = \alpha_{1q_1}, \\
(1') \qquad &x_i = \alpha(v, \mu_i) = \alpha_{i2} = \ldots = \alpha_{iq_i} \qquad (i = 2, \ldots, k), \\
&x_i = \alpha_{i1} = \ldots = \alpha_{iq_i} \qquad (i = k+1, \ldots, n) \\
&\mathscr{K} \,\&\, \Omega,
\end{aligned}
$$

where the terms with double indices do not contain v. In view of Remark 2, we may assume that $1 \leq q_i \leq 2$ $(i = 1, \cdots, n)$. Without loss of generality, we will assume that among the integers $q_1, \cdots, q_k$ the first t are equal to 2 $(1 \leq t \leq k)$, those remaining (if there are any) being equal to 1. Let A be the base of the term α. Then, expressing α in more detail, we can write the system (1′) in the form

$$
\begin{aligned}
&x_i = (vA)^{\mu_i} v = \alpha_{i2} \qquad (i = 1, \ldots, t), \\
(1'') \qquad &x_i = (vA)^{\mu_i} v \qquad (i = t+1, \ldots, k), \\
&x_i = \alpha_{i1} = \ldots = \alpha_{iq_i} \qquad (i = k+1, \ldots, n), \\
&\mathscr{E}(vA), \ \neg\tau(v, M), \\
&\mathscr{K} \,\&\, \Omega.
\end{aligned}
$$

If among the terms $\alpha_{12}, \cdots, \alpha_{t2}$ there is at least one which is not a Ψ^M-term, then, by Remark 2, the assertion being proved is true. We will assume that $\alpha_{12}, \cdots, \alpha_{t2}$ are Ψ^M-terms. Then by 14.5 there exist $\overline{\Psi}_2^M$-terms $\beta_1, \cdots, \beta_t$ such that the system $(vA)^{\mu_i} v = \alpha_{i2}$ $(i = 1, \cdots, k)$, $\mathscr{E}(vA)$, $\neg\tau(v, M)$ is equivalent to the system $(v = \beta_1 = \cdots = \beta_t) \,\&\, \mathscr{D}$, where $\mathscr{D}$ is a LDR. By 2.3, the system $v = \beta_1 = \cdots = \beta_t$ is equivalent to a system $(v = \beta_1) \,\&\, \mathscr{D}_1$. Then (1″) is equivalent to the system

$$
\begin{aligned}
&x_i = (vA)^{\mu_i} v \qquad (i = 1, \ldots, k), \\
(10) \qquad &v = \beta_1, \\
&x_i = \alpha_{i1} = \ldots = \alpha_{iq_i} \qquad (i = k+1, \ldots, n), \\
&\mathscr{K} \,\&\, \Omega \,\&\, \mathscr{D}_1 \,\&\, \mathscr{D}.
\end{aligned}
$$

The equation $v = \beta_1$ expanded has the form

$$v = [S^\sigma A_i]_{i=1}^m B, \qquad \sigma[S^\partial > N,$$

where m and N are natural numbers, and $S, A_1, \cdots, A_m$ and B are words in the alphabet И. We define predicate terms $\alpha_1, \cdots, \alpha_k$. Put

$$\alpha_i = [(S^\sigma A_i]_{i=1}^m BA)^{\mu_i} [S^\sigma A_i]_{i=1}^m B; \ \sigma[S^\partial > N] \qquad (i = 1, \ldots, k).$$

The unknown v does not occur among the terms of the system (10) having double indices; v does not occur in the list $x_1, \cdots, x_n$; hence, in view of condition 4), it does not occur in the predicate Ω. Finally, v does not occur in the PDR $\mathscr{K} \,\&\, \mathscr{D}_1 \,\&\, \mathscr{D}$. Hence (10) is equivalent with respect to the unknowns $x_1, \cdots, x_n$ to the system

$$
\begin{aligned}
&x_i = \alpha_i \qquad (i = 1, \ldots, k), \\
(11) \qquad &x_i = \alpha_{i1} = \ldots = \alpha_{iq_i} \qquad (i = k+1, \ldots, n), \\
&\mathscr{K} \,\&\, \Omega \,\&\, \mathscr{D}_1 \,\&\, \mathscr{D}.
\end{aligned}
$$

It follows that (1) is equivalent with respect to the unknowns $x_1, \cdots, x_n$, so certainly in the region $\neg\epsilon(x_1, \cdots, x_n, M)$, to the system (11). The system (11) is a T-system of characteristic M with left-hand side $x_1, \cdots, x_n$; the height of (11) is less than that of (1), i.e. less than h. Hence the induction assumption is applicable to (11). It also follows that we have the implication

(12) $$(11)^* \supset \exists v(1)^*.$$

Applying the induction assumption to (11) and taking into account that (11) is equivalent to (1) in the region $\urcorner \epsilon(x_1, \cdots, x_n, M)$, and also the implication (12), we see that our assertion is true. The theorem is proved.

§19. Normalization of W_2-systems

An equation which is normal in the sense of §11 differs from equations of general form in that the left- and right-hand sides of a normal equation begin with different unknowns. It is easy to reduce an equation of general form to a normal equation and to equations in a smaller number of unknowns (see the proof of 4.10). For example, in the equation $Px \cdots = y \cdots$, where P is a coefficient, it suffices to make the substitution $y = Py_1$. But in the case of systems, because of occurrences of y in other equations of the system, these other equations "get worse", since the coefficient P appears in them.

Let us consider an example. We try to normalize the system

(1) $$Px \cdots = y \cdots, \qquad y \cdots = z \cdots, \qquad z \cdots = x \cdots.$$

The change of variable $y = Py_1$ transforms (1) into the system

(2) $$x \cdots = y_1, \cdots, \qquad Py_1 \cdots = z \cdots, \qquad z \cdots = x \cdots.$$

The change of variable $z = Pz_1$ transforms (2) into the system

$$x \cdots = y_1 \cdots, \qquad y_1 \cdots = z_1 \cdots, \qquad Pz_1 \cdots = x \cdots.$$

Finally, the change of variable $x = Px_1$ returns us to the system (1) with x_1, y_1 and z_1 in place of x, y and z (under these changes we lose only solutions with restricted components, and this loss can be disregarded). Hence it is impossible to normalize the system (1) by the change of variables indicated. The nonnormalizability of this system, roughly speaking, is of the same nature as the nonnormalizability of the equation $Px \cdots = x \cdots$. Namely, for sufficiently long solutions (x, y, z) of (1) we have from the first equation $P \angle y$, from the second $P \angle z$, from the third $P \angle x$, and again from the first $P^2 \angle y$, i.e. for any solution of the system (1) one of the components (the shortest) is a periodic word with period P. This component (say, y) satisfies, for some A, the equation $Py = yA$ (A being a coefficient). This property allows us to indicate the change of variable by which the coefficient P can be removed from (1).

We will not apply the normalization procedure to every W_2-system. A word equation is called a left type I (right type I) equation if it has the form $P\xi\Phi = \eta\Psi$ (the form $\Phi\xi P = \Psi\eta$), where P is a coefficient, ξ and η are distinct unknowns, and Φ and Ψ are words in the alphabet $\text{И} \cup \text{X}$. An equation is called a type I equation if it is a left type I equation and a right type I equation. A system of word equations is called a type I system (left type I system, right type I system) if each equation of the system is a type I equation (left type I equation, right type I equation). It is not difficult to show (§20) that any W_2-system reduces to type I W_2-systems. We apply the normalization procedure to type I W_2-systems.

A word equation is called left normal (right normal) if it has the form

$\xi\Phi = \eta\Psi$ (the form $\Phi\xi = \Psi\eta$), where ξ and η are distinct unknowns, and Φ and Ψ are words in the alphabet $\text{И} \cup \mathrm{X}$. A system of word equations is called left normal (right normal) if each equation of the system is left normal (right normal).

With each W_2-system of word equations in the n unknowns $x_1, \cdots, x_n$ we associate a graph in the following way. The vertices of the graph are the unknowns $x_1, \cdots, x_n$. The pair (x_i, x_j) is an edge of the graph if and only if the system contains a (left or right) type I equation in the unknowns x_i and x_j. A system of equations is called connected if its associated graph is connected, i.e. any two vertices are joined by some path.

Suppose there is given a W_2-system $\mathfrak{A}$, and let $\mathfrak{A}^*$ be its associated graph. Let x be a vertex of $\mathfrak{A}^*$. By the closure of x in the system $\mathfrak{A}$ we mean the set of all vertices of $\mathfrak{A}^*$ which are joined to x by some path and the edges joining these vertices. If $\mathfrak{A}$ is connected, then the closure of any vertex of $\mathfrak{A}^*$ coincides with $\mathfrak{A}^*$. We will identify a system with its associated graph.

19.1. *Suppose $\mathfrak{A}$ is a connected left normal (right normal) W_2-system in the n unknowns $x_1, \cdots, x_n$, and suppose $(X_1, \cdots, X_n)$ is a solution of $\mathfrak{A}$. Then among the words $X_1, \cdots, X_n$ there is one which is a beginning (end) of the others.*

PROOF. We restrict ourselves to the case where the system $\mathfrak{A}$ is left normal. The proof is by induction on the number N of equations in $\mathfrak{A}$. If $N = 1$, then $\mathfrak{A}$ consists of one left normal equation in the two unknowns x_1 and x_2. Our assertion is obvious.

Suppose $N > 1$ and the assertion is true for systems consisting of $N - 1$ equations. Suppose $\mathfrak{A}$ is a connected left normal W_2-system in the n unknowns $x_1, \cdots, x_n$ consisting of N equations. We will prove that our assertion is true for $\mathfrak{A}$. Select some equation in the system $\mathfrak{A}$. We obtain a system $\mathfrak{A}' \,\&\, \xi\Phi = \eta\Psi$, where ξ and η are unknowns and $\mathfrak{A}'$ is a left normal W_2-system consisting of $N - 1$ equations. Without loss of generality, we will assume that $\xi = x_1$ and $\eta = x_n$. Suppose now that $(X_1, \cdots, X_n)$ is a solution of $\mathfrak{A}$. Then $X_1 \sim X_n$, i.e. one of the words X_1 or X_n is a beginning of the other. Moreover, the set $(X_1, \cdots, X_n)$ is a solution of the systems $\mathfrak{A}'$. We will prove that among the words $X_1, \cdots, X_n$ there is one which is a beginning of the others. There are two cases to consider.

FIGURE 6

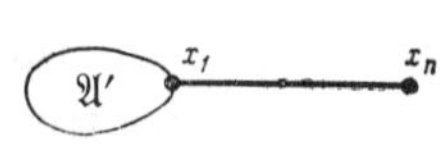

FIGURE 7

1) The unknowns x_1 and x_n occur in the system $\mathfrak{A}'$. Then $\mathfrak{A}'$ is a left normal W_2-system in n unknowns. If $\mathfrak{A}'$ is connected, the assertion follows from the induction assumption. If $\mathfrak{A}'$ is not connected, then consider the closures $\mathfrak{A}_1$ and $\mathfrak{A}_2$ in the system $\mathfrak{A}'$ of the unknowns x_1 and x_n (Figure 6). Each of $\mathfrak{A}_1$ and $\mathfrak{A}_2$ is a connected left normal W_2-system. Suppose $x_1, \cdots, x_k$ are the unknowns occurring in $\mathfrak{A}_1$, and $x_{k+1}, \cdots, x_n$ the unknowns occurring in $\mathfrak{A}_2$. By the induction assumption, there is among $X_1, \cdots, X_k$ a word X_ν $(1 \leq \nu \leq k)$ which is a beginning of the others, and there is among $X_{k+1}, \cdots, X_n$ a word X_μ $(k + 1 \leq \mu \leq n)$ which is a beginning of the others. We have $X_\nu \angle X_1$, $X_\mu \angle X_n$ and $X_1 \sim X_n$. The cases $X_1 \angle X_n$ and $X_n \angle X_1$ are handled in a similar way. We restrict ourselves to the case $X_1 \angle X_n$. Then $X_\nu \angle X_n$. We

had above $X_\mu \angle X_n$. Hence of the words X_ν and X_μ one is a beginning of the other. If $X_\nu \angle X_\mu$, then X_ν is a beginning of the words $X_1, \cdots, X_n$, and if $X_\mu \angle X_\nu$, then X_μ is a beginning of $X_1, \cdots, X_n$.

2) One of the unknowns (say, x_n) does not occur in the system $\mathfrak{A}'$. Then by the connectedness of $\mathfrak{A}$ the unknown x_1 occurs in $\mathfrak{A}'$ (Figure 7). The system $\mathfrak{A}'$ is a connected left normal W_2-system in the $n-1$ unknowns $x_1, \cdots, x_{n-1}$. By the induction assumption, there is among the words $X_1, \cdots, X_n$ a word X_ν $(1 \leqq \nu \leqq n-1)$ which is a beginning of the others. We know $X_1 \sim X_n$. If $X_1 \angle X_n$, then X_ν is a beginning of $X_1, \cdots, X_n$. If $X_n \angle X_1$, then, since $X_\nu \angle X_1$, we have $X_\nu \sim X_n$. If $X_\nu \angle X_n$, then X_ν is a beginning of $X_1, \cdots, X_n$. If $X_n \angle X_\nu$, then X_n is a beginning of $X_1, \cdots, X_n$.

The proposition is proved.

Let us turn to the normalization of W_2-systems.

Suppose there is given a W_2-system $\mathfrak{A}$ in the n unknowns $x_1, \cdots, x_n$ consisting of N equations. Let M denote the maximum length of the extreme left coefficients of the equations of the system. Put $l(\mathfrak{A}) = M(2^N - 1)$. A solution $(X_1, \cdots, X_n)$ of the system $\mathfrak{A}$ is called an l-solution if $[X_1^\partial, \cdots, [X_n^\partial \geqq l(\mathfrak{A})$. A set $\langle \mathfrak{A}_1, \cdots, \mathfrak{A}_k \rangle$ of systems of equations is called a left normalization of $\mathfrak{A}$ if the following conditions are satisfied:

a) If $k = 0$, then $\mathfrak{A}$ has no l-solutions.

b) Each of the systems $\mathfrak{A}_1, \cdots, \mathfrak{A}_k$ has the form

$$x_i = A_i y_i \quad (i = 1, \ldots, k),$$

$$\mathfrak{B}(y_1, \ldots, y_n), \ \mathfrak{C}(y_1, \ldots, y_n),$$

where $y_1, \cdots, y_k$ are unknowns different from $x_1, \cdots, x_n$; $A_1, \cdots, A_n$ are words in the alphabet И, and $[A_1^\partial, \cdots, [A_n^\partial \leqq l(\mathfrak{A})$; $\mathfrak{B}$ is a connected left normal W_2-system in the n unknowns $y_1, \cdots, y_n$ which is right normal if $\mathfrak{A}$ is right normal, and $\mathfrak{C}$ is a W_1-system.

c) For any l-solution $(X_1, \cdots, X_n)$ of $\mathfrak{A}$ there exist j, $1 \leqq j \leqq k$, and a set $(Y_1, \cdots, Y_n)$ such that $(X_1, \cdots, X_n, Y_1, \cdots, Y_n)$ is a solution of $\mathfrak{A}_j$.

d) For any j, $1 \leqq j \leqq k$, and any solution $(X_1, \cdots, X_n, Y_1, \cdots, Y_n)$ of $\mathfrak{A}_j$, the set $(X_1, \cdots, X_n)$ is a solution of $\mathfrak{A}$.

In a similar way we define the notion of a "right normalization of the system $\mathfrak{A}$". It is obtained from the definition of a left normalization by replacing each of the words "left" and "right" by the word "right" and "left", respectively, and replacing $x_i = A_i y_i$ by $x_i = y_i A_i$.

We will prove two assertions about normalization.

19.2. *Suppose $\mathfrak{A}$ is a connected left type* I *W_2-system, and suppose the left- and right-hand sides of each equation of $\mathfrak{A}$ contain more than one occurrence of the unknowns. Then $\mathfrak{A}$ has a left normalization.*

PROOF (BY INDUCTION ON THE NUMBER OF EQUATIONS IN $\mathfrak{A}$).

Basis of the induction. Suppose $\mathfrak{A}$ consists of one equation. It is a left type I equation in the two unknowns x_1 and x_2. The system $\mathfrak{A}$ has the form

$$A_2 x_1 \Phi(x_1, x_2) = A_1 x_2 \Psi(x_1, x_2), \tag{1}$$

where Φ and Ψ are words in the alphabet $\text{И} \cup \{x_1, x_2\}$; A_1 and A_2 are words in the alphabet И, one of A_1 or A_2 is empty and $[A_1^\partial, [A_2^\partial \leqq M$. Consider the system

$$
(2) \qquad \begin{aligned} x_i &= A_i y_i \quad (i = 1, 2), \\ y_1 \Phi (A_1 y_1, A_2 y_2) &= y_2 \Psi (A_1 y_1, A_2 y_2). \end{aligned}
$$

We will prove that (2) is a left normalization of (1). We will verify that conditions a) – d) are satisfied. We restrict ourselves to the case $A_2 = \Lambda$.

Condition a) holds: we have $k = 1$. Condition b) is obvious. Condition d) is almost obvious: if (X_1, X_2, Y_1, Y_2) is a solution of (2), then $Y_1 \Phi(X_1, X_2) = Y_2 \Psi(X_1, X_2)$. Since $A_2 = \Lambda$, after multiplying on the right by A_1 we obtain $X_1 \Phi(X_1, X_2) = A_1 X_2 \Psi(X_1, X_2)$, so that (X_1, X_2) is a solution of (1). We will prove that c) holds. Suppose (X_1, X_2) is an l-solution of (1). Then $X_1 \Phi(X_1, X_2) = A_1 X_2 \Psi(X_1, X_2)$, where $[X_1^\partial, [X_2^\partial \geqq M$. Hence $X_1 = A_1 Y_1$ for some Y_1. Putting $Y_2 = X_2$, we see that the set (X_1, X_2, Y_1, Y_2) satisfies (2). Thus c) holds also. The basis of the induction is proved.

Induction step. Suppose $\mathfrak{A}$ is a connected left type I W_2-system in the n unknowns $x_1, \cdots, x_n$ consisting of N equations, $N > 1$. We will prove that $\mathfrak{A}$ has a left normalization. Write $\mathfrak{A}$ in the form $\mathfrak{A}' \,\&\, \Phi' = \Psi'$, where $\mathfrak{A}'$ is a subsystem of $\mathfrak{A}$ consisting of $N - 1$ equations, and $\Phi' = \Psi'$ is an equation of $\mathfrak{A}$. Without loss of generality, we will assume that in the equation $\Phi' = \Psi'$ occur the unknowns x_1 and x_2. Then $\mathfrak{A}$ has the form

$$
(3) \qquad \begin{gathered} \mathfrak{A}', \\ R_1 x_1 \Phi(x_1, x_2) = R_2 x_2 \Psi(x_1, x_2), \end{gathered}
$$

where R_1 and R_2 are coefficients, $[R_1^\partial, [R_2^\partial \leqq M$ and one of R_1 or R_2 is empty. We consider three cases.

Case 1. One of the unknowns x_1 or x_2 (say, x_1) does not occur in $\mathfrak{A}'$. Then by the connectedness of $\mathfrak{A}$ the unknown x_2 occurs in $\mathfrak{A}'$ (Figure 8). The

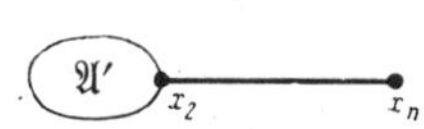

FIGURE 8

system $\mathfrak{A}'$ is a connected left type I W_2-system in the $n - 1$ unknowns $x_2, \cdots, x_n$ consisting of $N - 1$ equations. By the induction assumption, $\mathfrak{A}'$ has a left normalization $\langle \mathfrak{A}_1, \cdots, \mathfrak{A}_s \rangle$, $s \geqq 0$. Starting from this normalization we construct a set Γ of systems in $2n$ unknowns. If $s = 0$, Γ is the empty set. Suppose $s > 0$. Then, in view of b) in the definition of a normalization, each of the systems $\mathfrak{A}_1, \cdots, \mathfrak{A}_s$ has the form

$$
(4) \qquad \begin{gathered} x_i = A_i y_i \quad (i = 2, \ldots, n), \\ \mathfrak{D} (y_2, \ldots, y_n), \; \mathfrak{C} (y_2, \ldots, y_n), \end{gathered}
$$

where $A_2, \cdots, \mathfrak{A}_n$ are coefficients, $[A_2^\partial, \cdots, [A_n^\partial \leqq M(2^{N-1} - 1)$; $\mathfrak{C}$ is a W_1-system, and $\mathfrak{D}$ is a connected left normal W_2-system which is right normal if $\mathfrak{A}'$ is right normal. We select those systems from among $\mathfrak{A}_1, \cdots, \mathfrak{A}_s$ for which $R_1 \sim R_2 A_2$. Suppose the selected systems comprise the list $\mathfrak{A}_1, \cdots, \mathfrak{A}_k$, $k \geqq 0$. For each $\mathfrak{A}_i$ $(i = 1, \cdots, k)$ we construct a system of equations $\mathfrak{D}_i$ as follows. Suppose that $\mathfrak{A}_i$ is a selected system and has the form (4). Then $R_1 \sim R_2 A_2$, i.e. for some C_1 and C_2, one of which is empty, we have the equality

$$
(5) \qquad R_1 C_1 = R_2 A_2 C_2.
$$

By definition, $\mathfrak{D}_i$ is the system

$$
\begin{aligned}
&x_1 = C_1 v_1,\\
&x_i = A_i C_2 v_i \quad (i = 2, \ldots, n),\\
(6)\qquad &v_1 \Phi (C_1 v_1, A_2 C_2 v_2) = v_2 \Psi (C_1 v_1, A_2 C_2 v_2),\\
&[\mathfrak{D} (C_2 v_2, \ldots, C_2 v_n)], \quad \mathfrak{C} (C_2 v_2, \ldots, C_2 v_n),
\end{aligned}
$$

where the system $[\mathfrak{D}]$ is obtained from the system $\mathfrak{D}(C_2v_2, \cdots, C_2v_n)$ by carrying out all possible left cancellations. We will prove that the set $\Gamma = \langle \mathfrak{D}_1, \cdots, \mathfrak{D}_k \rangle$ is a left normalization of $\mathfrak{A}$. We will verify conditions a) – d).

Condition a). By construction, the list Γ is empty in only two cases: α_1) $s = 0$; α_2) $s > 0$ and for all j in the sequence $1, \cdots, s$ we have $\urcorner (R_1 \sim R_2 A_2)$. If $s = 0$, then, by a) in the definition of a normalization, the system $\mathfrak{A}'$ has no $l(\mathfrak{A}')$-solutions. But then $\mathfrak{A}$ has no $l(\mathfrak{A}')$-solutions. Then $\mathfrak{A}$ certainly has no $l(\mathfrak{A})$-solutions, since $l(\mathfrak{A}) > l(\mathfrak{A}')$. Thus, if $s = 0$, the system $\mathfrak{A}$ has no l-solutions. We will show below that in the case α_2) the system $\mathfrak{A}$ also has no l-solutions.

Condition b). Each of the systems $\mathfrak{D}_1, \cdots, \mathfrak{D}_k$ has the form (6). Each equation of the system $\mathfrak{D}(C_2v_2, \cdots, C_2v_n)$ can be cancelled on the left by C_2, after which there results a left normal system (because of the left normality of $\mathfrak{D}$). Hence the system $[\mathfrak{D}(C_2v_2, \cdots, C_2v_n)]$ is left normal. It is now obvious that b) holds.

Condition c). Suppose $(X_1, \cdots, X_n)$ is an $l(\mathfrak{A})$-solution of $\mathfrak{A}$. Then (see (3))

$$
(7)\qquad R_1 X_1 \Phi(X_1, X_2) = R_2 X_2 \Psi(X_1, X_2),
$$

where $[X_1^\partial, \cdots, [X_n^\partial \geq M(2^N - 1)$ and the set $(X_2, \cdots, X_n)$ is an $l(\mathfrak{A}')$-solution of $\mathfrak{A}'$. By condition c) for the normalization $\langle \mathfrak{A}_1, \cdots, \mathfrak{A}_s \rangle$ there exist an integer j, $1 \leq j \leq s$, and a set $(Y_2, \cdots, Y_n)$ such that $(X_2, \cdots, X_n, Y_2, \cdots, Y_n)$ is a solution of $\mathfrak{A}_j$. We will assume that $\mathfrak{A}_j$ has the form (4). Then we have

$$
(8)\qquad X_i = A_i Y_i \quad (i = 2, \cdots, n),
$$

$$
(9)\qquad \mathfrak{D}(Y_2, \cdots, Y_n), \quad \mathfrak{C}(Y_2, \cdots, Y_n).
$$

From (7) and (8) we have

$$
(10)\qquad R_1 X_1 \Phi(X_1, A_2 Y_2) = R_2 A_2 Y_2 \Psi(X_1, A_2 Y_2),
$$

from which we obtain $R_1 \sim R_2 A_2$. It is now clear that in case α_2) the system $\mathfrak{A}$ has no solutions, so certainly no l-solutions. Condition a), therefore, is satisfied.

Furthermore, since $R_1 \sim R_2 A_2$, we have (5), where either $C_1 = \Lambda$ or $C_2 = \Lambda$. From (5) and (10) we have (analyzing the cases $C_1 = \Lambda$ and $C_2 = \Lambda$)

$$
(11)\qquad C_2 X_1 \Phi(X_1, A_2 Y_2) = C_1 Y_2 \Psi(X_1, A_2 Y_2).
$$

From (5), (8) and the estimates for $[R_i^\partial$, $[A_2^\partial$, $[X_1^\partial$ and $[X_2^\partial$ we obtain

$$
[C_1^\partial, [C_2^\partial \leq \min\{ [X_1^\partial, [Y_2^\partial \}.
$$

Now from (11), inasmuch as $C_1 = \Lambda$ or $C_2 = \Lambda$, we obtain for some V_1 and V_2

$$
(12)\qquad X_1 = C_1 V_1, \qquad Y_2 = C_2 V_2.
$$

Since $C_1C_2 = C_2C_1$, from (11) and (12) we have

$$V_1\Phi(C_1V_1, A_2C_2V_2) = V_2\Psi(C_1V_1, A_2C_2V_2). \tag{13}$$

Since $(Y_2, \cdots, Y_n)$ is a solution of the connected left normal W_2-system $\mathfrak{D}$ (see (9)), by 19.1 one of the words $Y_2, \cdots, Y_n$ is a beginning of the others. Since $\min\{[Y_2^\partial, \cdots, [Y_n^\partial\} \geq M \geq [C_2^\partial$, it follows that C_2 is a beginning of $Y_2, \cdots, Y_n$. Therefore

$$Y_i = C_2V_i \qquad (i = 2, \cdots, n) \tag{14}$$

for some $V_2, \cdots, V_n$. Then we have the equalities

$$[\mathfrak{D}(C_2V_2, \cdots, C_2V_n)], \qquad \mathfrak{C}(C_2V_2, \cdots, C_2V_n). \tag{15}$$

From (12), (14), (8), (13) and (15) we conclude that $(X_1, \cdots, X_n, Y_1, \cdots, V_n)$ is a solution of (6). Thus condition c) holds.

Condition d). Here we argue in the reverse order. Suppose the set $(X_1, \cdots, X_n, Y_1, \cdots, Y_n)$ is a solution of (6). Then we have (13), (15), (14), (12), (8) and (9) (if we put $Y_i = C_2V_i$ $(i = 1, \cdots, n)$). Multiplying (13) on the left by $R_2A_2C_2$ and using (5), (12) and (18), we obtain (7). From (8) and (9) it follows that $(X_2, \cdots, X_n, Y_2, \cdots, Y_n)$ is a solution of (4). Then by condition d) for the normalization $\langle \mathfrak{A}_1, \cdots, \mathfrak{A}_s \rangle$ the set $(X_2, \cdots, X_n)$ satisfies $\mathfrak{A}'$. Now from (3) and (7) we conclude that $(X_1, \cdots, X_n)$ is a solution of $\mathfrak{A}$. Thus condition d) holds. The set $\langle \mathfrak{D}_1, \cdots, \mathfrak{D}_k \rangle$ is a left normalization of $\mathfrak{A}$.

Case 2. The unknowns x_1 and x_2 occur in $\mathfrak{A}'$, but $\mathfrak{A}'$ is not connected. Then we consider the closures $\mathfrak{A}_1$ and $\mathfrak{A}_2$ in the system $\mathfrak{A}'$ of the unknowns x_1 and x_2 (Figure 9): it is more convenient to consider x_n instead of x_2. The

FIGURE 9

systems $\mathfrak{A}_1$ and $\mathfrak{A}_2$ are connected left type I W_2-systems. Since $\mathfrak{A}'$ is not connected, the systems $\mathfrak{A}_1$ and $\mathfrak{A}_2$ have no unknowns in common. Suppose $\mathfrak{A}_1 = \mathfrak{A}_1(x_1, \cdots, x_k)$, $\mathfrak{A}_2 = \mathfrak{A}_2(x_{k+1}, \cdots, x_n)$. The induction assumption is applicable to $\mathfrak{A}_1$ and $\mathfrak{A}_2$; hence each of them has a left normalization. Let $\langle \alpha_1, \cdots, \alpha_s \rangle$ be a normalization of $\mathfrak{A}_1$, and $\langle \beta_1, \cdots, \beta_q \rangle$ a normalization of $\mathfrak{A}_2$. Each of the systems α_j has the form

$$\begin{gathered} x_i = A_iy_i \qquad (i = 1, \cdots, k), \\ \mathfrak{b}'(y_1, \cdots, y_k) \qquad \mathfrak{l}'(y_1, \cdots, y_k), \end{gathered} \tag{16}$$

and each of the systems β_j the form

$$\begin{gathered} x_i = A_iy_i \qquad (i = k+1, \cdots, n), \\ \mathfrak{b}''(y_{k+1}, \cdots, y_n), \qquad \mathfrak{l}''(y_{k+1}, \cdots, y_n), \end{gathered} \tag{17}$$

where $A_1, \cdots, A_k$ are coefficients, $[A_1^\partial, \cdots, [A_n^\partial \leq M(2^{N-1} - 1)$; the y_i are new unknowns, $\mathfrak{b}'$ and $\mathfrak{b}''$ are connected left normal W_2-systems which are right normal if $\mathfrak{A}_1$ and $\mathfrak{A}_2$ are right normal, and $\mathfrak{l}'$ and $\mathfrak{l}''$ are W_1-systems.

We construct a set of systems Γ in the following way. We select pairs $(\mathfrak{D}_i, \beta_j)$ of systems (16) and (17) for which $R_1A_1 \sim R_2A_n$ (see (3)). Suppose the selected pairs comprise the list $\pi_1, \cdots, \pi_m$, $m \geq 0$. For each pair π_i we construct a system π_i' as follows. Suppose $R_1A_1 = R_2A_nP$. Then π_i' is the system

$$
(18)\qquad
\begin{gathered}
x_i = A_iPv_i \quad (i = 1, \dots, k),\\
x_i = A_iv_i \quad (i = k+1, \dots, n),\\
[\mathfrak{b}'(Pv_1, \dots, Pv_k)], \quad \mathfrak{b}''(v_{k+1}, \dots, v_n)\\
\mathfrak{c}'(Pv_1, \dots, Pv_k), \quad \mathfrak{c}''(v_{k+1}, \dots, v_n),\\
v_1\Phi(Pv_1, v_n) = v_n\Psi(Pv_1, v_n),
\end{gathered}
$$

where the system $[\mathfrak{b}']$ is defined as in Case 1. The situation is analogous if $R_1A_1P = R_2A_n$, only now the groups of indices $i = 1, \cdots, k$ and $i = k+1, \cdots, m$ are interchanged. The set $\Gamma = \langle \pi_1', \cdots, \pi_m' \rangle$ is a left normalization of $\mathfrak{A}$.

The argument (the verification of conditions a) – d)) is completely analogous to that of Case 1.

Case 3. The unknowns x_1 and x_2 occur in $\mathfrak{A}'$, and $\mathfrak{A}'$ is connected. Then, by the induction assumption, $\mathfrak{A}'$ has a left normalization $\langle \mathfrak{A}_1, \cdots, \mathfrak{A}_s \rangle$. Starting from this normalization we construct a set of systems Γ in $2n$ unknowns as follows. If $s = 0$, then Γ is the empty set. Suppose $s > 0$. By b) in the definition of a normalization, each of the systems $\mathfrak{A}_1, \cdots, \mathfrak{A}_s$ has the form

$$
(19)\qquad
\begin{gathered}
x_i = A_iy_i \quad (i = 1, \cdots, n),\\
\mathfrak{B}(y_1, \cdots, y_n), \quad \mathfrak{C}(y_1, \cdots, y_n),
\end{gathered}
$$

where $A_1, \cdots, A_n$ are coefficients, $[A_1^\partial, \cdots, [A_n^\partial \leq M(2^{N-1} - 1)$; $\mathfrak{C}$ is a W_1-system, and $\mathfrak{B}$ is a connected left normal W_2-system in the n unknowns $x_1, \cdots, x_n$ which is right normal if $\mathfrak{A}'$ is right normal. We select those systems from among $\mathfrak{A}_1, \cdots, \mathfrak{A}_s$ for which $R_1A_1 \sim R_2A_2$. Suppose the selected systems comprise the list $\mathfrak{A}_1, \cdots, \mathfrak{A}_k$, $k \geqq 0$. For each of the selected systems $\mathfrak{A}_i$ we construct a list Γ_i of systems of equations as follows. Consider the $n!$ ordered sets formed from the list $y_1, \cdots, y_n$. Each such set κ defines an order κ among $y_1, \cdots, y_n$. Namely, y_i is "shorter" than y_j in the sense of the order κ if y_i stands to the left of y_j in the set κ. Fix a system $\mathfrak{A}_i$ and an order κ. We will construct a set of systems $\Gamma_{i,\kappa}$. Suppose $\mathfrak{A}_i$ has the form (19). We have $R_1A_1 \sim R_2A_2$. Consider the case $R_2A_2 \angle R_1A_1$. Then

$$
(20)\qquad R_1A_1 = R_2A_2P
$$

for some P. We divide the set $\{y_1, \cdots, y_n\}$ of vertices of the graph of the system $\mathfrak{B}$ into three sets $\mathfrak{M}_1$, $\mathfrak{M}_2$ and $\mathfrak{M}_3$ as follows (depending on the order κ).

Suppose v is a vertex of the graph $\mathfrak{B}$. By definition:

1) $v \in \mathfrak{M}_1$ if v is the shortest (in the sense of the fixed order κ) vertex of some path joining y_1 and y_2. The possibilities $v = y_1$ and $v - y_2$ are not cxcluded.

2) $v \in \mathfrak{M}_2$ if $v \neq y_2$ and there exists a path joining v and y_2 in which no vertex belongs to $\mathfrak{M}_1$. If $y_2 \notin \mathfrak{M}_1$, then by definition $y_2 \in \mathfrak{M}_2$.

3) $v \in \mathfrak{M}_3$ if $v \notin \mathfrak{M}_1 \cup \mathfrak{M}_2$.

The sets $\mathfrak{M}_1$, $\mathfrak{M}_2$ and $\mathfrak{M}_3$ are mutually disjoint, and their union is the set $\{y_1, \cdots, y_n\}$ of all vertices of $\mathfrak{B}$. It is easy to see that this partition possesses the following properties: α) $y_1 \notin \mathfrak{M}_2$; β) $y_2 \notin \mathfrak{M}_3$; γ) no edge can join a vertex of $\mathfrak{M}_2$ to a vertex of $\mathfrak{M}_3$.

The systems $\Gamma_{i,\kappa}$ are constructed as follows. If $y_1 \in \mathfrak{M}_1$, then the set $\Gamma_{i,\kappa}$ consists of the systems of the form

$$\begin{aligned} &x_i = A_i y_i \quad (i = 1, \ldots, n), \\ (21)\qquad &y_1 Q \Phi(A_1 y_1, A_2 y_2) = y_2 \Psi(A_1 y_1, A_2 y_2), \\ &\mathfrak{C}(y_1, \ldots, y_n), \quad \mathfrak{B}(y_1, \ldots, y_n), P y_1 = y_1 Q, \end{aligned}$$

where the Q are all possible words in the alphabet И of length $[P^\partial$. If $y_1 \notin \mathfrak{M}_1$, then $\Gamma_{i,\kappa}$ consists of the systems of the form

$$\begin{aligned} &x_i = A_i P v_i \quad (y_i \in \mathfrak{M}_1 \cup \mathfrak{M}_2), \\ &x_i = A_i v_i \quad (y_i \in \mathfrak{M}_3), \\ (22)\qquad &P v_i = v_i Q_i \quad (y_i \in \mathfrak{M}_1), \\ &v_1 \Phi(A_1 v_1, A_2 P v_2) = v_2 \Psi(A_1 v_1, A_2 P v_2), \\ &\mathfrak{B}^*(v_1, \ldots, v_n), \quad \mathfrak{C}^*(v_1, \ldots, v_n), \end{aligned}$$

where the Q_i are words in И of length $[P^\partial$, and the system $\mathfrak{C}^*$ is obtained from $\mathfrak{C}$ by substituting Pv_i for y_i if $y_i \in \mathfrak{M}_1 \cup \mathfrak{M}_2$, and by substituting v_i for y_i if $y_i \in \mathfrak{M}_3$. The system $\mathfrak{B}^*$ is constructed as follows. Choose an equation of the system $\mathfrak{B}$. It has the form

$$(23)\qquad y_i \Phi(y_i, y_j) = y_j \Psi(y_i, y_j),$$

since $\mathfrak{B}$ is left normal. The following cases are possible:

1) $y_i, y_j \in \mathfrak{M}_1 \cup \mathfrak{M}_2$. Then we replace (23) by the equation

$$(24)\qquad v_i \Phi(Pv_i, Pv_j) = v_j \Psi(Pv_i, Pv_j);$$

2) $y_i, y_j \in \mathfrak{M}_3$. Then we replace (23) by the equation

$$(25)\qquad v_i \Phi(v_i, v_j) = v_j \Psi(v_i, v_j);$$

3) $y_i \in \mathfrak{M}_1 \cup \mathfrak{M}_2$ and $y_j \in \mathfrak{M}_3$. Then we replace (23) by the equation

$$(26)\qquad v_i Q_i \Phi(Pv_i, v_j) = v_j \Psi(Pv_i, v_j),$$

where Q_i is the same as in the third row of the system (22).

Replacing each equation of the system $\mathfrak{B}$ as indicated, we obtain a system $\mathfrak{B}^*(v_1, \cdots, v_n)$. The set $\Gamma_{i,\kappa}$ is defined. Similarly, interchanging y_1 and y_2, we define the $\Gamma_{i,\kappa}$ in the case $R_1 A_1 \angle R_2 A_2$. Now put $\Gamma_i = \bigcup_\kappa \Gamma_{i,\kappa}$ and $\Gamma = \bigcup_{i=1}^k \Gamma_i$, where the symbol $\bigcup_\kappa$ denotes the (set-theoretic) union over all orders κ. The set Γ has been constructed. We will prove that Γ is a left normalization of $\mathfrak{A}$. We will verify the conditions a) – d).

Condition a). As in Case 1, we distinguish two cases: α_1) $s = 0$; α_2) $s > 0$ and for all of the systems $\mathfrak{A}_1, \cdots, \mathfrak{A}_s$ we have $\neg(R_1 A_1 \sim R_2 A_2)$. In case α_1) the system $\mathfrak{A}$ has no l-solutions (the argument is the same as in Case 1). We will show below that $\mathfrak{A}$ has no solutions in case α_2).

Condition b). This condition is satisfied by construction. Only the right normality in the system (21) and equation (26) can be spoiled, but by the hypothesis of the theorem the word Φ contains occurrences of the unknowns.

Condition c). We will prove that this condition is satisfied. Suppose $(X_1, \cdots, X_n)$ is an $l(\mathfrak{A})$-solution of $\mathfrak{A}$. Then

$$(27)\qquad R_1 X_1 \Phi(X_1, X_2) = R_2 X_2 \Psi(X_1, X_2).$$

Moreover, $[X_1^\partial, \cdots, [X_n^\partial \geq M(2^N - 1)$ and the set $(X_1, \cdots, X_n)$ is an $l(\mathfrak{A}')$-solution of $\mathfrak{A}'$. Hence by c) for the normalization $\langle \mathfrak{A}_1, \cdots, \mathfrak{A}_s \rangle$ there exist an integer j and set $(Y_1, \cdots, Y_n)$ such that $(X_1, \cdots, X_n, Y_1, \cdots, Y_n)$ is a solution of $\mathfrak{A}_j$. Suppose the system $\mathfrak{A}_j$ has the form (19). We have

$$X_i = A_i Y_i \qquad (i = 1, \cdots, n), \tag{28}$$

$$\mathfrak{B}(Y_1, \cdots, Y_n), \qquad \mathfrak{C}(Y_1, \cdots, Y_n). \tag{29}$$

From (28) and (27) we have

$$R_1 A_1 Y_1 \Phi(x_1, x_2) = R_2 A_2 Y_2 \Psi(x_1, x_2) \tag{30}$$

from which we obtain $R_1 A_1 \sim R_2 A_2$. It is now clear that in case α_2) the system $\mathfrak{A}$ has no solutions. Thus condition a) holds.

Furthermore, $R_1 A_1 \sim R_2 A_2$ implies that either $R_1 A_1 = R_2 A_2 P$ or $R_2 A_2 = R_1 A_1 P$ for some P. Both cases are handled in a similar way. We consider the case $R_1 A_1 = R_2 A_2 P$. From (30) we have

$$PY_1 \Phi(X_1, X_2) = Y_2 \Psi(X_1, X_2). \tag{31}$$

We arrange $Y_1, \cdots, Y_n$ in order of increasing length. If the lengths of Y_i and Y_j are equal, we place them side by side (in either order). This ordering defines in a natural way an order κ among the unknowns $y_1, \cdots, y_n$. In accordance with κ we construct the sets $\mathfrak{M}_1$, $\mathfrak{M}_2$ and $\mathfrak{M}_3$. Suppose $y_1 \in \mathfrak{M}_1$. Then $\Gamma_{i,\kappa}$ consists of systems of the form (21). Since $y_1 \in \mathfrak{M}_1$, in some path Δ joining y_1 and y_2 the vertex y_1 is shorter than the others. To the path Δ there corresponds a connected subsystem Δ of the system $\mathfrak{B}$. By 19.1, among the components $Y_1, Y_2, \cdots$ occurring in Δ there is one which is a beginning of the others. Since Y_1 is shorter than the others, $Y_1 \angle Y_2$. From (31) we have $PY_1 = Y_1 Q$ for some Q. Obviously, $(X_1, \cdots, X_n, Y_1, \cdots, Y_n)$ is a solution of (21). Thus c) holds if $y_1 \in \mathfrak{M}_1$.

Suppose $y_1 \notin \mathfrak{M}_1$. Then by α), $y_1 \in \mathfrak{M}_3$. By β), $y_2 \in \mathfrak{M}_1 \cup \mathfrak{M}_2$. Choose from among $y_1, \cdots, y_n$ that y_j which is shorter than the others. Applying 19.1 to the system $\mathfrak{B}$, we obtain that Y_j is a beginning of the words $Y_1, \cdots, Y_n$; in particular, $Y_j \angle Y_2$. Estimating $[P^\partial$ and $[Y_j^\partial$ by means of (20), (28) and the estimates for $[A_j^\partial$, $[X_j^\partial$ and $[R_i^\partial$, we obtain

$$[P^\partial \leqq [R_1 A_1^\partial \leqq M \cdot 2^{N-1},$$

$$[Y_j^\partial = [X_j^\partial - [A_j^\partial \geqq M(2^N - 1) - M(2^{N-1} - 1) \geqq M \cdot 2^{N-1},$$

so that $[P^\partial \leqq [Y_j^\partial$. From (31) and the relation $Y_j \angle Y_2$ we obtain $P \angle Y_j$. Then certainly $P \angle Y_i$ for $i = 1, \cdots, n$. We can now define the words $V_1, \cdots, V_n$ by $V_i = Y_i$ if $y_i \in \mathfrak{M}_3$; $Y_i = PV_i$ if $y_i \in \mathfrak{M}_1 \cup \mathfrak{M}_2$. From (28) we have

$$X_i = A_i P V_i \qquad (y_i \in \mathfrak{M}_1 \cup \mathfrak{M}_2), \tag{32}$$

$$X_i = A_i V_i \qquad (y_i \in \mathfrak{M}_3). \tag{33}$$

Since $y_1 \in \mathfrak{M}_3$ and $y_2 \in \mathfrak{M}_1 \cup \mathfrak{M}_2$, we have $Y_1 = V_1$ and $Y_2 = PV_2$. From (31) we have $V_1 \Phi(A_1 V_1, A_2 P V_2) = V_2 \Psi(A_1 V_1, A_2 P V_2)$. We will prove: δ) if $y_i \in \mathfrak{M}_1$, then $PV_i = V_i Q$ for some Q. Indeed, if $y_i \in \mathfrak{M}_1$, then the vertex y_i is the shortest in some path Δ joining y_1 and y_2. By 19.1, the component Y_i, being the shortest in the connected subsystem Δ of $\mathfrak{B}$, is a beginning of Y_1 and Y_2. Then from (31) we have $PY_i = Y_i Q$ for some Q. Since $y_i \in \mathfrak{M}_1$, we have $Y_i = PV_i$, $P^2 V_i = PV_i Q$ and $PV_i = V_i Q$. Thus δ) is proved. It is now obvious that the set $(X_1, \cdots, X_n, V_1, \cdots, V_n)$ satisfies the upper four rows of (22) and the system $\mathfrak{C}^*$. We will prove that $(V_1, \cdots, V_n)$ is a solution of $\mathfrak{B}^*$. It suffices to prove that if $(Y_1, \cdots, Y_n)$ is a solution of (23), then $(V_1, \cdots, V_n)$ is a solution of the replacing equation. If $y_i, y_j \in \mathfrak{M}_1 \cup \mathfrak{M}_2$, then $Y_i = PV_i$, $Y_j = PV_j$ and the pair (V_i, V_j) satisfies (24). If $y_i, y_j \in \mathfrak{M}_3$, then $Y_i = V_i$, $Y_j = V_j$ and the pair (V_i, V_j) satisfies (25). There remains the case $y_i \in \mathfrak{M}_1 \cup \mathfrak{M}_2$, $y_j \in \mathfrak{M}_3$. In view of γ) we have $y_i \in \mathfrak{M}_1$. By δ) we have $PV_i = V_iQ$ for some Q. Hence $Y_i = PV_i = V_iQ$ and the pair (V_i, V_j) satisfies (26). Thus c) holds.

Condition d). Arguing in the reverse order, we see that if $(X_1, \cdots, X_n, V_1, \cdots, V_n)$ is a solution of some system $\Gamma_{i,\varkappa}$, then the set $(X_1, \cdots, X_n)$ is a solution of $\mathfrak{A}$. The proposition is proved.

19.3. *Suppose $\mathfrak{A}$ is a connected right type* I *W_2-system, and suppose the left- and right-hand sides of each equation of $\mathfrak{A}$ contain more than one occurrence of the unknowns. Then $\mathfrak{A}$ has a right normalization.*

The proof is analogous to that of 19.2.

§20. The main theorem

We will prove a theorem describing the general solution of W_2-systems by means of terms.

Suppose $\mathfrak{A}$ is a system of word equations in the n unknowns $x_1, \cdots, x_n$, and $\mathscr{P}(x_1, \cdots, x_n)$ is a decidable predicate. A solution $(X_1, \cdots, X_n)$ of $\mathfrak{A}$ is called a $\mathscr{P}$-solution if $\mathscr{P}(X_1, \cdots, X_n)$ is true.

A system $\Gamma_1, \cdots, \Gamma_m$ of sets,

$$\Gamma_i = [T_{1i}(\xi_1, \cdots, \xi_k), \cdots, T_{ni}(\xi_1, \cdots, \xi_k), \mathscr{P}_i(\xi_1, \cdots, \xi_k)], \quad i = 1, \cdots, m,$$

where the T_{ij} are terms and $\mathscr{P}_1, \cdots, \mathscr{P}_m$ are nonempty decidable predicates, is called a representation ($\mathscr{P}$-representation) of $\mathfrak{A}$ if the following conditions are satisfied:

a) if $m = 0$, then $\mathfrak{A}$ has no solutions ($\mathscr{P}$-solutions);

b) for any solution ($\mathscr{P}$-solution) $(X_1, \cdots, X_n)$ of $\mathfrak{A}$ there exist an integer j, $1 \leq j \leq m$, and a set $(\bar{\xi}_1, \cdots, \bar{\xi}_k)$ satisfying $\mathscr{P}_j$ such that

$$(1) \qquad X_i = T_{ij}(\bar{\xi}_1, \cdots, \bar{\xi}_k) \qquad (i = 1, \cdots, n);$$

c) for any integer j, $1 \leq j \leq m$, and any set $(\bar{\xi}_1, \cdots, \bar{\xi}_k)$ satisfying $\mathscr{P}_j$ the system of words $(X_1, \cdots, X_n)$ in (1) is a solution of $\mathfrak{A}$.

We say that "the solutions ($\mathscr{P}$-solutions) of $\mathfrak{A}$ are representable by terms" if $\mathfrak{A}$ has a representation ($\mathscr{P}$-representation). Obviously, if $\mathfrak{A}$ has no $\mathscr{P}$-solutions, then the $\mathscr{P}$-solutions of $\mathfrak{A}$ are representable by terms (with $m = 0$).

We will also use the following obvious assertion.

20.1. *Suppose we have an equivalence $\mathscr{P} \Leftrightarrow \bigvee_1^n \mathscr{P}_i$, where $n \geq 0$ and $\mathscr{P}$, $\mathscr{P}_1, \cdots, \mathscr{P}_n$ are predicates, and suppose the $\mathscr{P}_i$-solutions of a system $\mathfrak{A}$ are representable by terms ($i = 1, \cdots, n$). Then the $\mathscr{P}$-solutions of $\mathfrak{A}$ are representable by terms. In particular, if a system $\mathfrak{A}$ is equivalent to a disjunction of systems $\mathfrak{A}_1, \cdots, \mathfrak{A}_n$ and the solutions of $\mathfrak{A}_1, \cdots, \mathfrak{A}_n$ are representable by terms, then the solutions of $\mathfrak{A}$ are also representable by terms (the solutions of the system $\mathfrak{A}_i$ can be viewed as $\mathfrak{A}_i$-solutions of $\mathfrak{A}$)*

Main Theorem. *The solutions of any W_2-system in n unknowns are representable by terms.*

PROOF (BY INDUCTION ON n).

Basis of the induction. Suppose $n=1$. By 1.28, we may assume that the system consists of a single equation. If the equation is trivial, then a representation consists of the one set $\Gamma_1=[x,0=0]$. If the equation is nontrivial, then it has the form (after left cancellation) $A[xP_i]_{i=1}^k=[xQ_i]_{i=1}^l$, where $k,l\geq 1$ and $A,P_1,\cdots,P_k,\ Q_1,\cdots,Q_l$ are coefficients, $A\neq\Lambda$. By 1.16, any solution X of the equation has the form $X=A^tA_1$, where $A_1\angle A$ and $t\geqq 0$. Arguing as in 1.19, we conclude that a representation consists of a finite number ($\geqq 0$) of sets of the form $[A^\lambda A_1,0=0]$, where $A_1\angle A$ and λ is a natural unknown, and a finite number ($\geqq 0$) of words of the form A^mA_1, where $A_1\angle A$ and $0\leqq m\leqq G$, G being the number in the proof of 1.19.

Induction step. Suppose $n>1$ and $\mathfrak{A}(x_1,\cdots,x_n)$ is a W_2-system. Without loss of generality, we will assume that each equation of $\mathfrak{A}$ is noncancellable from the left and right and that $\mathfrak{A}$ contains no trivial equations. A word equation is called degenerate if it has the form $P_0[\xi_iP_i]_{i=1}^n=Q$, where $n>0$; P_i and Q are coefficients, and the ξ_i are unknowns. The system $\mathfrak{A}$ may contain degenerate equations. If it does, then in any solution $(X_1,\cdots,X_n)$ of $\mathfrak{A}$ one of the components, say X_n, is bounded by the absolute constant $[Q^\partial$. Then the system $\mathfrak{A}(x_1,\cdots,x_n)$ is equivalent to the disjunction of the systems $\mathfrak{A}(x_1,\cdots,x_{n-1},A)$, $x_n=A$, taken over all words A in the alphabet И such that $[A^\partial\leqq[Q^\partial$. By the induction assumption, the solutions of each of the systems $\mathfrak{A}(x_1,\cdots,x_{n-1},A)$ are representable by terms. Hence the solutions of $\mathfrak{A}$ are also representable by terms.

We will assume that each equation of $\mathfrak{A}$ is nontrivial and nondegenerate. The system $\mathfrak{A}$ may contain equations in two unknowns of the form $Px\cdots=x\cdots$, where P is a nonempty coefficient and x is an unknown. We call these left type II equations. Suppose $\mathfrak{B}(x,y)$ is a left type II equation. From the argument in 4.10 concerning the equation 4.10 (1) it follows that there are three possible cases.

1) In any solution of the equation $\mathfrak{B}$ one of the components (say y) is bounded in length by an absolute constant.

2) The equation $\mathfrak{B}$ is equivalent to a disjunction of systems consisting of equations in one unknown.

3) The equation $\mathfrak{B}$ is equivalent to a disjunction of left type I equations.

In case 1) the theorem follows from the induction assumption. In case 2) the system $\mathfrak{A}$ is equivalent to a disjunction of systems, each of which is obtained from $\mathfrak{A}$ by replacing $\mathfrak{B}$ by a system consisting of equations in one unknown. Thus, including case 3), if $\mathfrak{A}$ contains left type II equations, then the theorem is true, or $\mathfrak{A}$ is equivalent to a disjunction of W_2-systems in which any equation is either a left type I equation or an equation in one unknown. Arguing analogously in the case where the system contains right type II equations (i.e. equations of the form $\cdots xP=\cdots x$, where P is a nonempty coefficient and x is an unknown), we obtain that $\mathfrak{A}$ is equivalent to a disjunction of W_2-systems consisting of type I equations and equations in one unknown. It remains to prove the theorem for W_2-systems consisting of type I equations and equations in one unknown.

Suppose $\mathfrak{A}$ is such a system. Consider the subsystem of $\mathfrak{A}$ consisting of the type I equations. We call this subsystem the kernel of $\mathfrak{A}$ and denote it by $\mathfrak{A}^*$. It may happen that $\mathfrak{A}^*$ is not connected. Then $\mathfrak{A}$ consists of two not empty subsystems $\mathfrak{A}_1$ and $\mathfrak{A}_2$ with no common unknowns, so that each of $\mathfrak{A}_1$ and $\mathfrak{A}_2$ is a W_2-system in a number of unknowns less than n. Applying the induction assumption to $\mathfrak{A}_1$ and $\mathfrak{A}_2$, we see that the theorem is true.

Suppose $\mathfrak{A}^*$ is connected. It may happen that $\mathfrak{A}^*$ contains an equation in which one of its sides contains only one occurrence of an unknown. We may assume that this equation has the form $P_0[xP_i]_{i=1}^n = AyB$, where $n \geqq 1$; $P_0, \cdots, P_n$, A and B are coefficients, and x and y are unknowns. In the remaining case one of the components x or y is bounded in length by an absolute constant, and for this case the theorem is true. This equation can be solved for x and y. For example, let us consider the case $P_0 = B = \Lambda$. We have $[xP_i]_{i=1}^n = Ay$. We divide the domain of $(x_1, \cdots, x_n)$ into two parts: 1) $\partial(x) \leqq [A^\partial$; 2) $\partial(x) > [A^\partial$. In 2) we have for x and y the expressions $x = Au$ and $y = [uP_iA]_{i=1}^{n-1}uP_n$. By the induction assumption, the solutions of $\mathfrak{A}$ are representable by terms in each of the regions 1) and 2). By 20.1, the solutions of $\mathfrak{A}$ are representable by terms.

Thus, we may assume that the left- and right-hand sides of each equation of $\mathfrak{A}^*$ contain more than one occurrence of the unknowns. Then, by 19.3, $\mathfrak{A}^*$ has a right normalization. From the definition of a right normalization we conclude that the representability by terms of the solutions of $\mathfrak{A}$ will be proved if we prove that the solutions of any W_2-system in n unknowns with a connected right normal and left type I kernel are representable by terms. Suppose $\mathfrak{A}(x_1, \cdots, x_n)$ is a W_2-system with a connected right normal and left type I kernel $\mathfrak{A}^*$. By 19.2, $\mathfrak{A}^*$ has a left normalization. From the definition of a left normalization we conclude that the representability by terms of the solutions of $\mathfrak{A}$ will be proved if we prove that the solutions of any W_2-system in n unknowns with a connected left and right normal kernel are representable by terms.

Suppose $\mathfrak{A}(x_1, \cdots, x_n)$ is a W_2-system with a connected left and right normal kernel. We will prove that the solutions of $\mathfrak{A}$ are representable by terms. If the kernel of $\mathfrak{A}$ contains two equations in the same pair of unknowns, we can fuse them into one equation by means of 1.28. The resulting equation will be left and right normal. By 1.28, each equation in one unknown can be attached on the right to some equation of the kernel. The resulting equation will be left normal and right type II. Thus, $\mathfrak{A}$ is equivalent to a system in which each equation is left normal, and right normal or right type II. But these are exactly the equations which are normal in the sense of §11. Thus, $\mathfrak{A}$ is equivalent to a system of the form

(1) $$\mathop{\&}_{\omega(i,j)} \mathfrak{A}_{ij}(x_i, x_j),$$

where the $\mathfrak{A}_{ij}$ are normal equations in the sense of §11, and the symbol $\omega(i,j)$ means that the conjunction extends over some set of pairs (i,j) such that (for definiteness) $i > j$. Here all of the pairs (i,j) are distinct. We will prove that the solutions of (1) are representable by terms.

Choose an integer $M > 2$ such that it is a characteristic of each of the equations $\mathfrak{A}_{ij}$. Following 15.2′ we defined the predicate $\epsilon(x_1, \cdots, x_n, M)$. The representability by terms of the solutions of $\mathfrak{A}$ in the region $\epsilon(x_1, \cdots, x_n, M)$ follows from the definition of ϵ and the induction assumption. By 20.1, it remains to prove that the solutions of (1) in the region $\neg\epsilon(x_1, \cdots, x_n, M)$ are representable by terms. By Theorem 12.2, each of the equations $\mathfrak{A}_{ij}$

either has no ϵ_M-solutions or is equivalent in the region $\neg\epsilon(x_i, x_j, M)$ to a disjunction of the form

$$\bigvee_{\nu=1}^{N_{ij}} (x_i = \alpha_{ij\nu}, x_j = \beta_{ij\nu}, \mathscr{P}_{ij\nu}), \tag{2}$$

where $N_{ij} \geqq 1$ and $[\alpha_{ij\nu}, \beta_{ij\nu}, \mathscr{P}_{ij\nu}]$ is a special triple of characteristic M. Here we have the implications

$$(x_i = \alpha^*_{ij\nu}, x_j = \beta^*_{ij\nu}, \mathscr{P}^*_{ij\nu}) \supset \mathfrak{A}_{ij}(x_i, x_j), \qquad \nu = 1, \cdots, N_{ij}. \tag{3}$$

We have arrived at the following alternatives: 1) at least one of the systems $\mathfrak{A}_{ij}$ has no ϵ_M-solutions; 2) each of the systems $\mathfrak{A}_{ij}$ is equivalent in the region $\neg\epsilon(x_i, x_j, M)$ to a disjunction (2), and we have the implications (3). In the first case, obviously, (1) has no $\neg\epsilon(x_1, \cdots, x_n, M)$-solutions and the theorem is trivially true. In the second case (1) is equivalent in the region $\neg\epsilon(x_1, \cdots, x_n, M)$ to the conjunction

$$\mathop{\&}_{\omega(i,j)} \bigvee_{\nu=1}^{N_{ij}} (x_i = \alpha_{ij\nu}, x_j = \beta_{ij\nu}, \mathscr{P}_{ij\nu}), \tag{4}$$

where $N_{ij} > 0$ for any i and j such that $\omega(i,j)$. Consider the second case.

Using the permutability of the symbols & and $\vee$, we conclude that the predicate (4) is equivalent to a disjunction of systems of the form

$$\mathop{\&}_{\omega(i,j)} (x_i = \alpha_{ij\nu}, x_j = \beta_{ij\nu}, \mathscr{P}_{ij\nu}), \tag{5}$$

where ν runs from 1 to N_{ij}. From (3) we have the implications

$$\mathop{\&}_{\omega(i,j)} (x_i = \alpha^*_{ij\nu}, x_j = \beta^*_{ij\nu}, \mathscr{P}^*_{ij\nu}) \supset \mathfrak{A}(x_1, \cdots, x_n) \tag{6}$$

for any set of values of ν. It is not difficult to see (using the definition of special triples in §12 and of T-systems in §18) that the conjunction (5) is a T-system of characteristic M with left-hand side $x_1, \cdots, x_n$. We write it as $\pi(x_1, \cdots, x_n, \xi)$. The implication (6) takes the form

$$\pi^*(x_1, \cdots, x_n, \xi) \supset \mathfrak{A}(x_1, \cdots, x_n). \tag{7}$$

By Theorem 18.1, the system π is equivalent in the region $\neg\epsilon(x_1, \cdots, x_n, M)$ to a canonical T-system $\mathfrak{B}(x_1, \cdots, x_n, \eta)$ of characteristic M with left-hand side $x_1, \cdots, x_n$; we have the implication

$$\mathfrak{B}^*(x_1, \cdots, x_n, \eta) \supset \exists \xi \pi^*(x_1, \cdots, x_n, \xi). \tag{8}$$

The system $\mathfrak{B}$ has the form

$$\mathop{\&}_{i=1}^{n} (x_i = \gamma_i) \mathbin{\&} \mathscr{K} \mathbin{\&} \Omega,$$

where the γ_i are predicate terms, $\mathscr{K}$ is a PDR, and Ω is a conjunction of ω-predicates. The system $\mathfrak{B}^*$ has the form

$$\mathop{\&}_{i=1}^{n} (x_i = \beta_i^*) \mathbin{\&} \mathscr{K}, \tag{9}$$

where γ_i^* is the left-hand side of the predicate term γ_i. From the implications (7) and (8) and from (9) we have

$$\left[\mathop{\&}_{i=1}^{n}(x_i=\gamma_i^*)\,\&\,\mathscr{K}\right]\supset\mathfrak{A}(x_1,\cdots,x_n). \tag{10}$$

We select from the systems $\mathfrak{B}$ those for which the PDR $\mathscr{K}$ has solutions. This can be done (see 17.11). Suppose the selected systems comprise the list $\mathfrak{B}_1,\cdots,\mathfrak{B}_m$, $m\geqq 0$. Each $\mathfrak{B}_i$ has the form $\&_{j=1}^{n}(x_j=\gamma_{ij})\,\&\,\mathscr{K}_i\,\&\,\Omega_i$, where the γ_{ij} are predicate terms of characteristic M, $\mathscr{K}_j$ is a PDR, and Ω_j is a conjunction of ω-predicates. We construct the system of sets $\Gamma_1,\cdots,\Gamma_m$,

$$\Gamma_i=[\gamma_{i1}^*,\cdots,\gamma_{in}^*,\mathscr{K}_i],$$

where γ_{ij}^* is the left-hand side of the predicate term γ_{ij}. It follows (see the implication (10)) that the system of sets $\Gamma_1,\cdots,\Gamma_m$ is a $\urcorner\epsilon(x_1,\cdots,x_n,M)$-representation of $\mathfrak{A}$. Thus, the $\urcorner\epsilon(x_1,\cdots,x_n,M)$-solutions of $\mathfrak{A}$ are representable by terms, as required. The theorem is proved.

From the main theorem we obtain the following

THEOREM. *There exists an algorithm which determines whether or not a given W_2-system of word equations has solutions.*

Indeed, we construct a representation $\Gamma_1,\cdots,\Gamma_m$ of $\mathfrak{A}$ in accordance with the main theorem. If $m=0$, then $\mathfrak{A}$ has no solutions, and if $m>0$, then $\mathfrak{A}$ has solutions. The theorem is proved.

CHAPTER IV

COEFFICIENT-FREE EQUATIONS

The subject of the first and second parts of this monograph has been the decision problem for equations. This problem was solved for equations in three unknowns and for W_2-systems. Another very interesting problem related to equations is that of describing the structure of the solutions of an equation. In §12 we obtained a description of the general solution of normal equations in two unknowns. The main theorem of §20 contains a description of the general solution of W_2-systems.

In the investigation of word equations coefficient-free equations play an important role. One use of coefficient-free equations can be found in §6, where the reduction of the decision problem for equations in three unknowns is carried out in terms of coefficient-free equations (cf. 6.23). The latter form a "skeleton" of arbitrary word equations. Another use is related to obtaining formulas describing the structure of the solutions of coefficient-free equations. For example, in §1 the premises of almost all the assertions involved solutions of coefficient-free equations, and the conclusions asserted something about the structure of the solutions (cf. 1.1 – 1.7, 1.10 – 1.12, 1.16, etc.).

In this chapter we study the structure of the solutions of coefficient-free word equations in three unknowns. The simplest such equations were considered in 1.23 – 1.25 and 1.16. For example, by 1.16″ the general solution of the equation $xy = yz$ can be written in the form $x = uv$, $z = vu$, $y = (uv)^{\lambda}u$, where u and v are word unknowns, and λ is a natural unknown. By 1.23, the general solution of the equation $xyz = zyx$ has the form $x = (uv)^{\lambda}u$, $y = v(uv)^{\mu}$, $z = (uv)^{\nu}u$, where u and v are word unknowns and λ, μ and ν are natural unknowns. These and other examples suggest that the situation is analogous for any coefficient-free equation in three unknowns. Namely, the general solution of any coefficient-free equation in three unknowns can be expressed by a finite number of formulas whose right-hand sides involve only multiplication and raising to a natural power of word unknowns. This conjecture turns out to be true, and its proof is given in §§21 and 22. In order to formulate here, in the introduction, the main result of these sections, we will make some necessary definitions.

Parametric words will be understood in the sense of §3, i.e. words obtained by the inductive rules 1), 2) and 3) of §2 and the rule 1′) of §3. Parametric words not containing letters of the alphabet $\mathbf{И}$ will be called parametric words without constants.

A set $(X_1, \cdots, X_n)$ of words in $\mathbf{И}$ is called homogeneous if there exist a word S and natural numbers $a_1, \cdots, a_n$ such that $X_i = S^{a_i}$ $(i = 1, \cdots, n)$. The determination of the homogeneous solutions of coefficient-free equations obviously reduces to the determination of the natural solutions of linear Diophantine equations. Our main result concerns the determination of the nonhomogeneous solutions and can be stated as follows.

MAIN THEOREM. *Suppose $\mathfrak{A}$ is a coefficient-free word equation in the three unknowns x_1, x_2, x_3. Then it is possible to construct a finite list of L quadruples* ($L \geqq 1$)

1) $T_{11}(u, v, \lambda_1, \ldots, \lambda_m)$, $T_{12}(u, v, \lambda_1, \ldots, \lambda_m)$, $T_{13}(u, v, \lambda_1, \ldots, \lambda_m)$, $D_1(\lambda_1, \ldots, \lambda_m)$,

. . . .

L) $T_{L1}(u, v, \lambda_1, \ldots, \lambda_m)$, $T_{L2}(u, v, \lambda_1, \ldots, \lambda_m)$, $T_{L3}(u, v, \lambda_1, \ldots, \lambda_m)$, $D_L(\lambda_1, \ldots, \lambda_m)$,

where the T_{ij} are parametric words without constants, u and v are word variables, $\lambda_1, \cdots, \lambda_m$ are natural variables, and the $\mathscr{D}_i$ are LDR's, in such a way that the following conditions hold:

a) *For any nonhomogeneous solution (X_1, X_2, X_3) of $\mathfrak{A}$ there exist an integer k, $1 \leqq k \leqq L$, and a set $(U, V, \overline{\lambda}_1, \cdots, \overline{\lambda}_m)$ of values of the variables $u, v, \lambda_1, \cdots, \lambda_m$ such that $UV \neq VU$, $\mathscr{D}_k(\overline{\lambda}_1, \cdots, \overline{\lambda}_m)$, and*

$$X_i = T_{ki}(U, V, \overline{\lambda}_1, \cdots, \overline{\lambda}_m) \qquad (i = 1, 2, 3); \tag{1}$$

b) *For any integer k, $1 \leqq k \leqq L$, and any set $(U, V, \overline{\lambda}_1, \cdots, \overline{\lambda}_m)$ of values of the variables $u, v, \lambda_1, \cdots, \lambda_m$ such that $UV \neq VU$ and $\mathscr{D}_k(\overline{\lambda}_1, \cdots, \overline{\lambda}_m)$, the triple (X_1, X_2, X_3) of (1) is a nonhomogeneous solution of $\mathfrak{A}$.*

In other words (in the language of the predicate calculus), if $\mathfrak{A}(x_1, x_2, x_3)$ denotes the predicate "(x_1, x_2, x_3) is a nonhomogeneous solution of $\mathfrak{A}$", then we have the equivalence

$$\mathfrak{A}(x_1, x_2, x_3) \Leftrightarrow \bigvee_{k=1}^{L} \exists u v \lambda_1 \ldots \lambda_m \left(\mathop{\&}_{i=1}^{3} (x_i = T_{ki}) \,\&\, uv \neq vu \,\&\, \mathscr{D}_k \right). \tag{2}$$

If for an equation $\mathfrak{A}$ we have the equivalence (2), we say that the nonhomogeneous solutions of $\mathfrak{A}$ are representable by parametric words.

Our investigation of coefficient-free equations in three unknowns is carried out according to the plan for equations with coefficients. The essential difference is the problem of eliminating the parameter of transformations. It is now necessary to know for which values of the parameter of transformations there will exist nonhomogeneous solutions of the images. We will prove the eliminability of the parameter in an appropriate form (§21). This, however, forces us to change somewhat the notion of a complete system of images and to reconsider certain proofs of §6. As in Chapter II, we introduce basic equations and prove that the nonhomogeneous solutions of any coefficient-free equation in three unknowns can be obtained by means of a finite number of formulas from the nonhomogeneous solutions of basic equations. The nonhomogeneous solutions of these basic equations are explicitly written out in §22. There the main theorem will be proved.

In connection with the main theorem, it is interesting that the set of nonhomogeneous solutions (and also the set of all solutions) of a coefficient-free equation in four unknowns is, in general, not representable (in the corresponding precise sense) by means of parametric words. We will show this in §23 by the example of the equation $xyz = zvx$.

In conclusion, we mention that along with equations we can consider systems of coefficient-free word equations. There is a theorem which reduces the consideration of a system of coefficient-free equations in n unknowns to the consideration of a single coefficient-free equation in n unknowns. The proof of this theorem is given in §23.

Throughout this entire chapter, unless it is explicitly stated otherwise, word equations are understood to be coefficient-free.

§21. Elimination of parameters in transformations of coefficient-free equations in three unknowns

In this chapter we change somewhat the notion of a solution, and also the notions of a transformation and an image, of a one-sided equation. In Chapter II, when looking for a decision algorithm, we were interested in infinite images of equations. The finite images were equations in two unknowns, and for these a decision procedure was known. Now the finite images can be (and, in fact, are) a source of the nonhomogeneous solutions of equations and must be taken into consideration. It is technically more convenient to deal only with infinite transformations and images. This can be done by changing the notion of a solution of a one-sided equation. By a solution of the one-sided equation $x_1\Phi \to x_2\Psi$ we will mean any solution $(X_1, X_2, \cdots)$ of the equation $x_1\Phi = x_2\Psi$ such that $[X_1^\partial \geqq [X_2^\partial$ (including the possibilities $[X_1^\partial = [X_2^\partial$ and $X_i = \Lambda$, which earlier were "pushed" into the finite images). We have the equivalence

$$(\Phi = \Psi) \Leftrightarrow (\Phi \to \Psi) \vee (\Psi \to \Phi).$$

We now define transformations and images of coefficient-free equations. Suppose there is given a one-sided coefficient-free equation $x_1 \cdots \to \cdots$ in n unknowns. Write it in the form

$$x_1\Phi(x_1, \cdots, x_n) \to D(x_2, \cdots, x_n)x_1\Psi(x_1, \cdots, x_n), \tag{1}$$

where $D \neq \Lambda$ and does not contain x_1. A transformation of (1) is any equality of the form

$$x_1 = D^tAy, \tag{2}$$

where $t \geqq 0$, A is a proper beginning of D, $D^tA \neq \Lambda$, y is a new unknown, and t is the parameter of the transformation. The image of (1) under the transformation (2) is the equation

$$y\Phi(D^tAy, x_2, \cdots, x_n) \leftarrow BAy\Psi(D^tAy, x_2, \cdots, x_n), \tag{3}$$

where B is defined by the (graphical) equality $AB = D$.

The following proposition is analogous to Lemma 3.1.

21.1. a) *If $(X_1, \cdots, X_n)$ is a nonhomogeneous solution of the equation* (1) *and $X_2, \cdots, X_n \neq \Lambda$, then there exist an integer $t \geqq 0$, a proper beginning A of the word D, and a word Y such that*

$$X_1 = \mathrm{S}_{Y, X_2, \cdots, X_n}^{y, x_2, \cdots, x_n} D^tAy|, \tag{1'}$$

where $(Y, X_2, \cdots, X_n)$ is a nonhomogeneous solution of the image of (1) *under the transformation $x_1 = D^tAy$.*

b) *If $(Y, X_2, \cdots, X_n)$ is a nonhomogeneous solution of the image of* (1) *under the transformation $x_1 = D^tAy$, then the set $(X_1, \cdots, X_n)$, where X_1 is defined by* (1′), *is a nonhomogeneous solution of* (1).

PROOF. It is obvious that if X_1 is defined by (1′), then the sets $(X_1, \cdots, X_n)$ and $(Y, X_2, \cdots, X_n)$ are either both homogeneous or both nonhomogeneous.

PROOF OF a). Suppose $(X_1, \cdots, X_n)$ is a solution of (1). Then we have in the alphabet И the equality

$$X_1\Phi(X_1,\cdots,X_n)=D(X_2,\cdots,X_n)X_1\Psi(X_1,\cdots,X_n).$$

Since $D\neq\Lambda$ and $X_2,\cdots,X_n\neq\Lambda$, we have $D(X_2,\cdots,X_n)\neq\Lambda$.

Then by 1.16 there exist an integer $t\geqq 0$ and a word P such that $X_1=(D(X_2,\cdots,X_n))^tP$, $P\mathrel{\dot{\angle}} D(X_2,\cdots,X_n)$.

The word $D(X_2,\cdots,X_n)$ has the form $X_{i_1}\cdots X_{i_k}$, where $k\geqq 1$ and $2\leqq i_j\leqq n$ $(j=1,\cdots,k)$. Since $P\mathrel{\dot{\angle}} D(X_2,\cdots,X_n)$, there exist an integer s, $0\leqq s<k$, and a word Y such that $P=X_{i_1}\cdots X_{i_s}Y$ and $Y\mathrel{\dot{\angle}} X_{i_{s+1}}$. If $s=t=0$, then $X_1=P=Y\mathrel{\dot{\angle}} X_{i_1}$, which is impossible, since $[X_1^\partial\geqq[X_{i_1}^\partial$ because of the one-sidedness of (1). Hence if $t=0$, then $s>0$. Put $A=x_{i_1}\cdots x_{i_s}$ and $B=x_{i_{s+1}}\cdots x_{i_k}$. Then $D^tA\neq\Lambda$, $A\angle D$, $AB=D$, and the word X_1 is expressed by (1′). Substituting the expression for X_1 given by (1′) into equation (1) and cancelling the resulting equality on the left by

$$(D(X_2,\cdots,X_n))^tA(X_2,\cdots,X_n),$$

we obtain

$$\text{(2′)}\qquad S^{y,\,x_2,\,\ldots,\,x_n}_{Y,\,X_2,\,\ldots,\,X_n}\,y\Phi\,(D^tAy,x_2,\ldots,x_n)\,|$$

$$=S^{y,\,x_2,\,\ldots,\,x_n}_{Y,\,X_2,\,\ldots,\,X_n}\,BAy\Psi\,(D^tAy,x_2,\ldots,x_n)|.$$

Since $Y\mathrel{\dot{\angle}} X_{i_{s+1}}$, we have $[Y^\partial\leqq[X_{i_{s+1}}^\partial$. Since $x_{i_{s+1}}\angle B$, it follows from (2′) that the set $(Y,X_2,\cdots,X_n)$ satisfies (3). Thus a) is proved.

The assertion b) becomes obvious if we multiply (2′) on the left by $(D(X_2,\cdots,X_n))^tA(X_2,\cdots,X_n)$ and consider that $[X_1^\partial\geqq[X_{i_1}^\partial$ if X_1 is obtained from (1′) with nonempty D^tA, and that $x_{i_1}\angle D^tA$. The proposition is proved.

Lemma 2.1 outlines the following path for determining the nonhomogeneous solutions of coefficient-free equations in n unknowns. The determination of the nonhomogeneous solutions with an empty component reduces to solving equations in $n-1$ unknowns. The nonhomogeneous solutions with nonempty components can be obtained by means of (2) from the nonhomogeneous solutions of the images of an equation. It remains to solve the images of an equation. But since $0\leqq t<\infty$, it will be necessary to look at infinitely many images, and there will be infinitely many formulas (2). There arises the question of whether it is possible to restrict ourselves to a finite number of formulas and a finite number of images. We can make this more precise in connection with the problem of describing the set of nonhomogeneous solutions. Consider the equation

$$\text{(4)}\qquad x_1\Phi(x_1,\cdots,x_n,t)\to x_2\Psi(x_1,\cdots,x_n,t),$$

where Φ and Ψ are parametric words (without constants), $x_1,\cdots,x_n$ are word variables, and t is a natural variable. Denote the predicate defined by (4) by $\mathfrak{A}(x_1,\cdots,x_n,t)$. A number N is called an elimination bound of the parameter t in equation (4) if for any nonhomogeneous set $(x_1,\cdots,x_n)$ we have the implication

$$(\exists t\geqslant N)\,\mathfrak{A}(x_1,\ldots,x_n,t)\supset(\forall t\geqslant N)\,\mathfrak{A}(x_1,\ldots,x_n,t).$$

In other words, for any $x_1,\cdots,x_n$, if $(x_1,\cdots,x_n)$ is a nonhomogeneous solution of (4) for some $t \geqq N$, then $(x_1, \cdots, x_n)$ is a solution of (4) for any $t \geqq N$. Thus, the eliminability considered in this chapter is the strong eliminability in the sense of §3 in the region of nonhomogeneous solutions.

A number N is called an elimination bound of the parameter in transformations of (1) if for any A such that $A \measuredangle D$ the number N is an elimination bound of the parameter t in (3). If N is an elimination bound of the parameter in transformations of (1), then the set of images (3) for which $t \leqq N$ is called a complete system of images of equation (1). We say that the parameter is eliminable in transformations of equation (1) (in equation (4)) if there exists an elimination bound of the parameter in transformations of (1) (in (4)).

Now consider the equation in three unknowns

$$x\Phi(x,y,z) \to D(y,z)x\Psi(x,y,z), \tag{5}$$

where $D \neq \Lambda$ and does not contain x. Denote this equation by $\mathfrak{A}$. Let D_i denote the beginning of length i of the word D, and let $\mathfrak{A}_{it}$ denote the image of $\mathfrak{A}$ under the transformation $x = D^tD_iv$ (v being a new unknown). Let us also agree that if π is an equation in the three unknowns x, y and z, then $\pi(x,y,z)$ is the predicate "(x,y,z) is a nonhomogeneous solution of π". Put $d = [D^\partial$. From 21.1 and the definitions of elimination we obtain the following assertion.

21.2. *If N is an elimination bound of the parameter in transformations of an equation $\mathfrak{A}$, then we have the equivalence*

$$\mathfrak{A}(x,y,z) \Leftrightarrow [y = \Lambda, \mathfrak{A}(x,\Lambda,z)] \vee [z = \Lambda, \mathfrak{A}(x,y,\Lambda)] \vee \bigvee_{k=0}^{N-1}\bigvee_{i=0}^{d-1}$$

$$\times\ \exists v\,[x = D^kD_iv,\quad \mathfrak{A}_{ik}(v,y,z)] \vee \bigvee_{i=0}^{d-1} \exists t \exists v\,[x = D^tD_iv, t \geqslant N, \mathfrak{A}_{iN}(v,y,z)].$$

This brings us to the notion of parametric expressibility.

We will use the same symbol to denote a predicate and its truth set. Suppose $\mathscr{P}, \mathscr{P}_1, \cdots, \mathscr{P}_n$ are ternary relations among words (predicates). We say that the set $\mathscr{P}$ is parametrically expressed in terms of the sets $\mathscr{P}_1, \cdots, \mathscr{P}_n$ if it is possible to indicate lists $\{\alpha_1, \cdots, \alpha_N\}$ and $\{\beta_1, \cdots, \beta_M\}$, $N+M > 0$, of triples of parametric words

$$(\alpha_i)\quad [\alpha_{i1}(x,y,z,\lambda_1,\ldots,\lambda_n), \alpha_{i2}(x,y,z,\lambda_1,\ldots,\lambda_n), \alpha_{i3}(x,y,z,\lambda_1,\ldots,\lambda_n)],$$

$$(\beta_i)\quad [\beta_{i1}(u,v,\mu_1,\ldots,\mu_m), \beta_{i2}(u,v,\mu_1,\ldots,\mu_m), \beta_{i3}(u,v,\mu_1,\ldots,\mu_m)].$$

and lists $\{\mathscr{D}_1, \cdots, \mathscr{D}_N\}$ and $\{\mathscr{B}_1, \cdots, \mathscr{B}_M\}$ of LDR's such that we have the equivalence

$$\mathscr{P}(x_1,x_2,x_3) \Leftrightarrow \bigvee_{i=1}^{N} \exists xyz\lambda_1\ldots\lambda_n\,[\mathop{\&}_{i=1}^{3}(x_i = \alpha_{ij})\,\&\,\mathscr{P}_i(x,y,z)\,\&\,(\&\,\mathscr{D}_i(\lambda_1,\ldots,\lambda_n)]$$

$$\vee \bigvee_{i=1}^{M} \exists uv\mu_1\ldots\mu_m\,[\mathop{\&}_{i=1}^{3}(x_i = \beta_{ij})\,\&\,uv \neq vu\,\&\,\mathscr{B}_i(\mu_1,\ldots,\mu_m)]$$

(x,y,z,u and v are word unknowns, and λ_i and μ_j are natural unknowns).

Thus, besides the expression of the elements of $\mathscr{P}$ in terms of the elements of the sets $\mathscr{P}_i$ (the summand $\bigvee_{i=1}^{N}$) there is assumed a parametric representation of the nonhomogeneous elements of the set $\mathscr{P}$ (the summand $\bigvee_{i=1}^{M}$). It is possible to get along without the latter summand, but this would lengthen the argument.

It is obvious that if a set $\mathscr{P}$ is parametrically expressed in terms of the sets $\mathscr{P}_1, \cdots, \mathscr{P}_n$, and each of the sets $\mathscr{P}_i$ is parametrically expressed in terms of the sets $\mathscr{P}_{ij}$, then $\mathscr{P}$ is parametrically expressed in terms of the $\mathscr{P}_{ij}$. It is also obvious that if the set of nonhomogeneous solutions of an equation $\mathfrak{A}$ is parametrically expressed in terms of the sets of nonhomogeneous solutions of the equations $\mathfrak{A}_1, \cdots, \mathfrak{A}_n$, and the nonhomogeneous solutions of each of the equations $\mathfrak{A}_1, \cdots, \mathfrak{A}_n$ are representable by parametric words, then the nonhomogeneous solutions of $\mathfrak{A}$ are representable by parametric words.

By 1.1, an equation in two unknowns has nonhomogeneous solutions if and only if its left-hand side is graphically equal to its right-hand side. It follows that the nonhomogeneous solutions of an équation in three unknowns having exactly one empty component are representable by parametric words in accordance with the formulas (for the case $y = \Lambda$) $x = u$, $y = \Lambda$, $z = v$ and $uv \neq vu$.

Proposition 21.2 can now be stated in the following way.

21.2′. *If the parameter is eliminable in transformations of an equation $\mathfrak{A}$, then the set of nonhomogeneous solutions of $\mathfrak{A}$ is parametrically expressed in terms of the sets of nonhomogeneous solutions of the equations comprising a complete system of images of $\mathfrak{A}$.*

Equation (5) is called a type I equation if both of the unknowns y and z occur in the word D. By 4.1, the parameter is eliminable (in the present sense) in transformations of type I equations, since the present elimination is strong elimination in the sense of §3. Thus we obtain

21.2″. *For any type* I *equation $\mathfrak{A}$ it is possible to construct a finite number of images $\mathfrak{A}_1, \cdots, \mathfrak{A}_n$ (a complete system of images) in such a way that the set of nonhomogeneous solutions of $\mathfrak{A}$ is parametrically expressed in terms of the sets of nonhomogeneous solutions of the equations $\mathfrak{A}_1, \cdots, \mathfrak{A}_n$.*

The main difficulty regarding elimination involves the case where one of the unknowns y or z does not occur in D. In this case the parameter is generally not eliminable in the strong sense in transformations of the equation (5). This is well illustrated in the following example. Consider the equation $xy \rightarrow zx^2$. Any image of this equation has the form $vy \leftarrow zvz^tv$, where v is a new unknown and t is the parameter of the transformation. Eliminability of the parameter in transformations of the equation $xy \rightarrow zx^2$ means eliminability of the parameter t in the equation $zvz^tv \rightarrow vy$. It is not difficult to verify that the triple $v = a$, $z = ab$, $y = ba^2(ba)^t$ (a and b are letters of the alphabet **И**) is a nonhomogeneous solution of the equation $zvz^tv \rightarrow vy$. If we now assume that N is an elimination bound of the parameter t in the equation $zvz^tv \rightarrow vy$, then the triple $v = a$, $z = ab$, $y = ba^2(ba)^N$, being a solution of the equation $zvz^Nv \rightarrow vy$, must satisfy the equation $zvz^tv \rightarrow vy$ for any $t \geqq N$, which is obviously not so. The parameter is not eliminable in transformations of the equation $xy \rightarrow zx^2$. A simple argument shows that the nonhomogeneous solutions of the equation $xy \rightarrow zx^2$ are obtained by means of the formulas $x = (vu)^tv$ and $z = vu$ from the nonhomogeneous solutions of the equation $y \rightarrow uv(vu)^tv$, which is an image of the second order of the equation $xy \rightarrow zx^2$. This property turns out to be typical. We will prove that if one of the unknowns y or z does not occur in D, then the nonhomogeneous solutions of (5) (under certain easily removed restrictions) are parametrically expressed in terms of the nonhomogeneous solutions of a finite number of images of second and third order of (5).

For definiteness, we will assume that y does not occur in D. Then equation (5) has the form $x\Phi \to z^c x\Psi$, where $c \geqq 1$. We may assume that $\Phi \neq \Lambda$, multiplying, if necessary, the equation on the right by an unknown. If $x \angle \Phi$ or $z \angle \Phi$, then the solutions of the equation can be found without resorting to images. These cases are considered in §22. There remains the case $y \angle \Phi$. Multiplying, if necessary, the equation $x\Phi \to z^c x\Psi$ on the right by suitable unknowns, we can write it in the form

$$(*) \qquad xy^b A(x,y,z) \to z^c x B(x,z) y C(x,y,z) \qquad (b, c \geqq 1),$$

where $A \neq \Lambda$, the first letter of the word A is different from y, and B does not contain y. Any transformation of (*) has the form $x = z^t u$, where u is a new unknown and t is a positive integer. Correspondingly, any image of (*) has the form

$$(**) \qquad uy^b A(z^t u, y, z) \leftarrow z^c u B(z^t u, z) y C(z^t u, y, z).$$

Eliminability of the parameter in transformations of (*) means eliminability of the parameter t in (**). As the above example shows, the parameter t in transformations of (*) is not always eliminable.

We will begin, nevertheless, with a case where the parameter is eliminable in transformations of (*).

21.3. *If $B \neq \Lambda$ and does not contain x, then the parameter is eliminable in transformations of* (*).

PROOF. We must prove that the parameter t is eliminable in the equation (**). Consider a nonhomogeneous solution (U, Y, Z) of (**). Since $[Z^\partial \geqq [U^\partial$ (because of the arrow in (**)), there exists V such that $Z = UV$. Since $B \neq \Lambda$ and does not contain x, B has the form z^d, where $d \geqq 1$. Substituting Y, U and Z into (**) and cancelling the resulting equality on the left by U, we obtain $YT_1 = (VU)^c(UV)^d YT_2$ for some T_1 and T_2. Since (U, Y, Z) is a nonhomogeneous solution and $Z = UV$, we have $UV \neq \Lambda$. Then from the next to last equality and 1.16 we have $Y = ((VU)^c(UV)^d)^\lambda P$, where $\lambda \geqq 0$ and $P \angle (VU)^c(UV)^d$. For brevity, put $E = (VU)^c(UV)^d$. Then $Y = E^\lambda P$ and $P \angle E$. Substituting the expressions for Y and Z into (**), we obtain in the alphabet $\mathbf{И}$ the equality

$$(1) \qquad U(E^\lambda P)^b A((UV)^t U, E^\lambda P, UV) = UEE^\lambda PC((UV)^t U, E^\lambda P, UV).$$

It is obvious that if (1) holds, then the triple $(U,\ Y = E^\lambda P,\ Z = UV)$ satisfies (**). Consequently, the eliminability of the parameter t in (**) will be proved if we prove the existence of a number N which does not depend on λ, U, V or P and satisfies property α: if (1) holds for some $t \geqq N$, then it holds for any $t \geqq N$ (λ, P, U and V are fixed). Put

$$l = \max\{b + c + d + [A^\partial, c + d + [C^\partial, 8\}.$$

We will prove that the number $N = 4l^5$ possesses property α. The idea of the proof is very simple. For small λ we view (1) as an exponential equation in the one unknown t and apply 2.7. For large λ we view (1) as an exponential equation in the two unknowns t and λ, and apply 2.6.

Assume that $\lambda < 2l^2$. Then, viewing (1) as an exponential equation in the one unknown t, we obtain for the height h and characteristic H of this equation the following estimates: $h \leqq l$, $H \leqq l(c+d)(\lambda+1) \leqq 2l^4$, $2hH \leqq 4l^5$, and, by 2.7, the number N possesses property α.

Suppose $\lambda \geqq 2l^2$. Having in mind an application of 2.6, we represent the words E and UV as powers of simple words. We know $UV \neq \Lambda$. We have

$$UV = S^a \quad (a > 0), \tag{2}$$

$$E = (VU)^c(UV)^d = T^k \quad (c, d, k > 0), \tag{3}$$

where S and T are simple words. We will prove that $[S^\partial \neq [T^\partial$. We argue by contradiction. Assume that $[S^\partial = [T^\partial$. Then it follows from (2) and (3) that $S = T$ and also that the words UV and VU are powers of S. The latter conclusion implies that $UV = VU$. By 1.2, the words U and V are powers of a simple word W. But then Z is a power of W. It is now evident from (**) that the pair (Y, W) satisfies a nontrivial coefficient-free equation in two unknowns, and hence Y is a power of W. The words U, Z and Y are powers of W, which contradicts the nonhomogeneity of the solution (U, Y, Z). Thus we have proved that $[S^\partial \neq [T^\partial$.

We now regard (1) as an exponential equation in the two unknowns t and λ. Since $[S^\partial \neq [T^\partial$, this is the equation described in 2.6. For the height h and characteristic H of this equation we obtain the following estimates: $h \leqq \max\{b + [A^\partial, 1 + [C^\partial, 8\} \leqq l$,

$$H \leqslant \frac{(c+d)[UV^\partial + \max\{[A^\partial, [C^\partial\}[UV^\partial}{\min\{[UV^\partial, [E^\partial\}} \leqslant \frac{l[UV^\partial}{[UV^\partial} = l, \qquad 2Hh \leqslant 2l^2,$$

and the assertion concerning N follows from 2.6. Thus we have shown that $N = 4l^5$ is an elimination bound of the parameter t in (**). The proposition is proved.

In what is to follow we will need the notion of conditional elimination. Suppose $\mathscr{P}(x_1, \cdots, x_n, t)$ is a decidable predicate. Let $\mathfrak{A}(x_1, \cdots, x_n, t)$ denote the predicate defined by (4). We say that the parameter t is eliminable in (4) under the condition $\mathscr{P}$ if there exists a number N such that for any nonhomogeneous set $(x_1, \cdots, x_n)$ we have the implication

$$(\exists t \geqslant N)(\mathfrak{A}(x_1, \ldots, x_n, t) \& \mathscr{P}) \supset (\forall t \geqslant N)\mathfrak{A}(x_1, \ldots, x_n, t). \tag{6}$$

If (6) holds, then N is called an elimination bound of the parameter t in the equation (4) under the condition $\mathscr{P}$.

A number N is called an elimination bound of the parameter in transformations of (1) under the condition $\mathscr{P}(x_1, \cdots, x_n)$ if N is an elimination bound of the parameter t in (3) under the condition $\mathscr{P}(D^tAy, x_2, \cdots, x_n)$ for all A such that $A \angle D$. We say that the parameter is eliminable in transformations of (1) under the condition $\mathscr{P}(x_1, \cdots, x_n)$ if there exists an elimination bound of the parameter in transformations of (1) under the condition $\mathscr{P}(x_1, \cdots, x_n)$.

We introduce a binary predicate $\partial(x) < \partial(y)$, where x and y are word unknowns. By definition, a pair of words (X, Y) satisfies this predicate if and only if $[X^\partial < [Y^\partial$. We define the predicate $\partial(x) \leqq \partial(y)$ analogously.

21.4. *The parameter t is eliminable in (**) under the condition $\partial(y) \leqq \partial(z)$.*

PROOF. For nonhomogeneous solutions (u,y,z) of (**) satisfying the condition $\partial(y) \leqq \partial(z)$ we have

$$\partial(z) \geqq \max\{\partial(u), \partial(y)\} > 0 \tag{1}$$

(because of the arrow in (**)). If we fix u, y and z and view (**) as an exponential equation in the one unknown t, then (1) implies that the characteristic of the equation is bounded by a constant which does not depend on (u,y,z). This is trivially true of the height. Now our assertion follows from 2.7.

A sequence $\mathfrak{A}_0, \cdots, \mathfrak{A}_k$ of word equations is called a chain of images of $\mathfrak{A}_0$ if for any i, $1 \leqq i \leqq k$, the equation $\mathfrak{A}_i$ is an image of the equation $\mathfrak{A}_{i-1}$. Degenerate and nondegenerate transformations, images and chains of images are defined exactly as in §5, omitting the word "infinite".

Suppose $\mathfrak{A}_0, \cdots, \mathfrak{A}_k$ is a chain of images. Then we say that $\mathfrak{A}_k$ is an image of order k of $\mathfrak{A}_0$. If the chain $\mathfrak{A}_0, \cdots, \mathfrak{A}_k$ is degenerate, then $\mathfrak{A}_k$ is a degenerate image of order k of $\mathfrak{A}_0$, and if the chain $\mathfrak{A}_0, \cdots, \mathfrak{A}_k$ is nondegenerate, then $\mathfrak{A}_k$ is a nondegenerate image of order k of $\mathfrak{A}_0$.

21.5. *Among the degenerate images of second order and the nondegenerate images of third order of the equation (*) it is possible to find finitely many $\mathfrak{A}_1, \cdots, \mathfrak{A}_k$ such that the set of nonhomogeneous solutions of (*) satisfying the condition $\partial(y) \leqq \partial(z)$ is parametrically expressed in terms of the sets of nonhomogeneous solutions of the equations $\mathfrak{A}_1, \cdots, \mathfrak{A}_k$.*

PROOF. Nonhomogeneous solutions of the equations (*) and (**) satisfying the condition $\partial(y) \leqq \partial(z)$ will be called τ-solutions. By 1.1, τ-solutions of (*) and (**) have nonempty components. Denote the equation (**) with fixed t by $\mathfrak{A}_t$. By 21.4, the parameter t is eliminable in (**) under the condition $\partial(y) \leqq \partial(z)$. Let N be a bound of this elimination. Then the τ-solutions of (*) are parametrically expressed in terms of the τ-solutions of the equations $\mathfrak{A}_0, \cdots, \mathfrak{A}_N$. We divide the τ-solutions of $\mathfrak{A}_t$ into two sets: $\mathfrak{M}_t$ and $\mathfrak{N}_t$. To $\mathfrak{M}_t$ we assign the τ-solutions of $\mathfrak{A}_t$ such that $\partial(z) \geqq \partial(uy)$. To $\mathfrak{N}_t$ we assign the τ-solutions for which $\partial(y) \leqq \partial(z) \leqq \partial(uy)$. The elements of $\mathfrak{N}_t$ are obtained by the formula $z = uv$ from the solutions (u,y,v) of the equation

$$y^b A((uv)^t u, y, uv) \to (vu)^c B((uv)^t u, uv) y C((uv)^t u, y, uv) \tag{1}$$

such that $\partial(y) \leqq \partial(vu)$. But in the region $\partial(y) \leqq \partial(vu)$ the solutions of (1) are obtained by the formula $y = v\eta$ from the solutions of the image of (1) under a transformation $y = v\eta$. Denote this image by $\mathfrak{b}_t$. The equation $\mathfrak{b}_t$ is a nondegenerate image of the third order of (*). We have proved that the set $\mathfrak{N}_t$ is parametrically expressed in terms of the set of nonhomogeneous solutionsf $\mathfrak{b}_t$.

Consider the set $\mathfrak{M}_t$. By 4.1, the parameter is eliminable in transformations of (**). Hence the set of nonhomogeneous solutions of (**) is parametrically expressed in terms of the sets of nonhomogeneous solutions of equations comprising a complete system Σ_t of images of this equation. Drop from Σ_t the image of $\mathfrak{A}_t$ under the transformation $z = uv$ and denote the resulting set by Σ_t^*. If we recall the condition $\partial(z) \geqq \partial(uy)$ defining $\mathfrak{M}_t$, we obtain that $\mathfrak{M}_t$ is parametrically expressed in terms of the sets of nonhomogeneous solutions of the equations comprising Σ_t^*. The elements of Σ_t^* are images of the second order of (*). Let Σ_t' and Σ_t'' denote the subsets of Σ_t^* consisting of the degenerate and nondegenerate images, respectively, of the second order. The elements of Σ_t'', as is easily computed, have the form $y \cdots \to vu \cdots$, where v is a new unknown, i.e. they are type I equations. For each equation $\pi \in \Sigma_t''$ we denote by Ω_π a complete system of images of π. The elements of Ω_π are nondegenerate images of the third order of (*). It follows that the set of nonhomogeneous solutions of (*) satisfying the condition $\partial(y) \leqq \partial(z)$ are parametrically expressed in terms of the sets of nonhomogeneous solutions of the equations of the set $\bigcup_{t=0}^N [\mathfrak{b}_t \cup \Sigma_t' \cup \bigcup_\pi \Omega_\pi]$, where $\pi \in \Sigma_t''$. The proposition is proved.

In view of 21.5, there remains the problem of determining the nonhomogeneous solutions of (*) satisfying the condition $\partial(y) \geqq \partial(z)$. The case where $B = z^d$, $d > 0$, was considered in 21.3. We will consider the case $B = \Lambda$ in §22. There remains the case where x occurs in B. We now take up this case.

21.6. *Suppose the unknown x occurs in B. Then among the nondegenerate images of the third order of (*) it is possible to find finitely many $\mathfrak{A}_1, \cdots, \mathfrak{A}_k$ such that the set of nonhomogeneous solutions of (*) satisfying the condition $\partial(y) \geqq \partial(z) > 0$ is parametrically expressed in terms of the sets of nonhomogeneous solutions of the equations $\mathfrak{A}_1, \cdots, \mathfrak{A}_k$.*

The proof is preceded by a lemma.

LEMMA. *Suppose $A(\alpha,\beta,\gamma)$ is a word in the alphabet $\{\alpha,\beta,\gamma\}$ and that A contains each of the letters α, β, γ and begins with α or γ. Suppose also that B and C are words in the alphabet $\{x,y,z\}$, and that λ, t, a, c, p, q and $a_1, \cdots, a_n$ are integers such that $a, p \geqq 0$, $c > 0$, $0 < q \leqq n$, $a_1, \cdots, a_n \geqq 0$ and $a_q + c + 2 \leqq t \leqq \lambda - c - [A^\partial$. Put*

$$D_1(x,z) = (zx)^c\,[(xz)^{\lambda + a_i}x]_{i=1}^{q-1}\,xz,$$

$$D_2(x,z) = (xz)^{\lambda - t + a_q}x\,[(xz)^{\lambda + a_i}x]_{i=q+1}^{n}\,(xz)^p.$$

Then the equations in three unknowns

$$y\,(D_1B)^a\,A\,((xz)^\lambda x,\, D_1B,\, xz) \leftarrow zD_2D_1C\,(x, y, z), \tag{1}$$

$$y\,(D_1B)^a\,A\,((xz)^\lambda x,\, D_1B,\, xz) \leftarrow D_2D_1C\,(x, y, z) \tag{2}$$

have no nonhomogeneous solutions.

PROOF OF THE LEMMA. Consider equation (1). Suppose (X,Y,Z) is a solution of (1). Then $Z = YV$ for some V. Substituting into (1) the words X, Y and YV for the unknowns x, y and z and cancelling the resulting equality on the left by Y, we have for some P and Q

$$(D_1(X, YV)P)^a\,A((XYV)^\lambda X,\, D_1(X, YV)P,\, XYV) = VD_2(X, YV)D_1(X, YV)Q. \tag{3}$$

Let us consider (3). The cases $a = 0$ and $a > 0$ will be treated separately.

Suppose $a > 0$. Then the left-hand side of (3) begins with the word $(YVX)^cXYV$, the right-hand side with $(VXY)^{\lambda - t}$. Since by hypothesis $c > 0$ and $\lambda - t \geqq c + 1$, (3) implies that $YVX = VXY$ and $XYV = VXY$. By 1.25, the triple (X, Y, V) is homogeneous. Hence the solution (X,Y,Z) is homogeneous.

Suppose $a = 0$. Here there are two possible subcases.

1) $\gamma^k\beta \angle A$, where $k > 0$. Then the left-hand side of (3) begins with $(XYV)^kYVX$, the right-hand side with $(VXY)^{\lambda-t}$. By hypothesis, $\lambda - t \geqq k+1$. From (3) we obtain $XYV = VXY = YVX$, so that (see the first case) the solution (X, Y, Z) is homogeneous.

2) $\gamma^k\alpha \angle A$, where $k \geqq 0$. Then the left-hand side of (3) begins with $(XYV)^{k+\lambda}$. The right-hand side of (3), as is easily seen from the expressions for D_1 and D_2, begins with the word $(VXY)^{\lambda-t+a_q}VXXYV$ if $n > q$ or $p > 0$, and with the word $(VXY)^{\lambda-t+a_q}VX(YVX)^cXYV$ if $n = q$ and $p = 0$. By hypothesis, $\lambda - t + a_q > 0$ and $\lambda > \lambda - t + a_q + c$. Hence $XYV = VXY$, and (3) can be cancelled on the left by $(VXY)^{\lambda-t+a_q}$. After cancellation, we obtain for some R_1, R_2, R_3 and R_4

$$(XYV)^{t-a_q}R_1 = VXXYVR_2, \tag{4}$$

if $n > q$ or $p > 0$, and

$$(XYV)^{t-a_2-c}R_3 = VXXYVR_4, \tag{5}$$

if $n = q$ and $p = 0$. By hypothesis, $t \geqq a_q + c + 2$. We have proved that $XYV = VXY$. Hence each of the equalities (4), (5) implies that $VXYVXY = VXXYV\cdots$. From this last equality we have $VXYVX = VXXYV$; hence $YVX = XYV$. We have proved that $XYV = VXY$. Hence, $XYV = VXY = YVX$ and, as in the first case, (X, Y, Z) is a homogeneous solution. The lemma is proved for equation (1).

The argument is analogous for equation (2). If (X, Y, Z) is a solution of (2), then $X = YV$, and after left cancellation by Y we have

$$\begin{aligned}(D_1(YV,\ Z)P)^aA\,((YVZ)^\lambda\, YV,\ D_1(YV,\ Z)\,P,\ YVZ)\\ = V\,(ZYV)^{\lambda-t+a_q}[(YVZ)^{\lambda+a_i}YV]_{i=q+1}^{n}(YVZ)^p\ D_1(YV, Z)Q.\end{aligned} \tag{6}$$

It suffices to prove that the triple (V, Y, Z) is homogeneous.

We now list the possible cases.

1) $a > 0$. Then the left-hand side of (6) begins with the word $(ZYV)^cYVZ$, the right-hand side with $(VZY)^{\lambda-t}$. By hypothesis, $\lambda - t \geqq c+1$. Hence $VZY = ZYV = YVZ$ and the triple (V, Y, Z) is homogeneous.

2) $a = 0$, $\gamma^k\beta \angle A$ and $k > 0$. Then the left-hand side of (6) begins with $(YVZ)^kZYV$, the right-hand side with $(VZY)^{\lambda-t}$. By hypothesis, $\lambda - t \geqq k+1$. Hence $VZY = YVZ = ZYV$ and the triple (V, Y, Z) is homogeneous.

3) $a = 0$, $\gamma^k\alpha \angle A$ and $k \geqq 0$. Then the left-hand side of (6) begins with $(YVZ)^{k+\lambda}$, the right-hand side with $(VZY)^{\lambda-t+a_q}V(YVZ)^p(ZYV)^cYVZ$ if $n = q$ and with $(VZY)^{\lambda-t+a_q}V(YVZ)^\lambda$ if $n < q$. By hypothesis, $\lambda - t + a_q > 0$. Hence $YVZ = VZY$, and (6) can be cancelled on the left by $(VZY)^{\lambda-t+a_q}$. After cancellation we obtain for some R_1, R_2, R_3 and R_4

$$(YVZ)^{t-a_q}R_1 = V(YVZ)^p(ZYV)^cYVZR_2, \quad \text{if } n = q, \tag{7}$$

$$(YVZ)^{t-a_q}R_3 = V(YVZ)^\lambda R_4, \quad \text{if } n < q. \tag{8}$$

Let us consider (7). Since $t - a_q \geqq 2$ and $YVZ = VZY$, (7) implies that $VZYV = VYVZ$ if $p > 0$. Also, $ZYV = YVZ$, and the solution (V, Y, Z) is homogeneous. If $p = 0$, we have $(VZY)^{t-a_q}R_1 = (VZY)^cVYVZR_2$. Since $t - a_q \geqq c+2$, after cancelling on the left by $(VZY)^c$ we obtain $VZYV = VYVZ$ and $ZYV = YVZ$, and the triple (V, Y, Z) is homogeneous. The treatment of (8) is even simpler. Since $t-a_q \geqq 2$, $\lambda > 0$ and $YVZ = VZY$, from (8) we have $VZYV = VYVZ$ and $ZYV = YVZ$, and the triple (V, Y, Z) is homogeneous. The lemma is proved.

PROOF OF 21.6. The nonhomogeneous solutions of (*) satisfying the condition $\partial(y) \geqq \partial(z) > 0$ are obtained by the formula $x = z^t u$, where $t > 0$, from the nonhomogeneous solutions (u, y, z) of (**) such that $\partial(y) \geqq \partial(z) > 0$. The latter are obtained by the formula $z = uv$ from the nonhomogeneous solutions (u, y, v) of the equation

$$y^b A((uv)^t u, y, uv) \to (vu)^c B((uv)^t u, uv) y C((uv)^t u, y, uv) \tag{1}$$

such that $\partial(y) \geqq \partial(uv)$ (the solutions of (1) such that $\partial(uv) = 0$ are homogeneous). Equation (1) is the image of (**) under the transformation $z = uv$ and, consequently, is a nondegenerate image of the second order of (*).

Consider the images of (1). These are nondegenerate images of the third order of (*). Put $D(t) = (vu)^c B((uv)^t u, uv)$. Then any transformation of (1) has the form $y = D^\lambda E w$, where $\lambda \geqq 0$, $E \mathrel{\dot{\angle}} D(t)$, $D^\lambda E \neq \Lambda$, and w is a new unknown. The images of (1) have the form

$$w(D^\lambda E w)^{b-1} A((uv)^t u, D^\lambda E w, uv) \leftarrow E_1 E w C((uv)^t u, D^\lambda E w, uv), \tag{2}$$

where E_1 is defined by the graphical equality $EE_1 = D$.

The solutions (u, y, v) of (1) having at least one empty component are homogeneous (this follows easily from 1.1). Hence by 21.2 the nonhomogeneous solutions of (1) are obtained by the formula $y = D^\lambda E w$ from the nonhomogeneous solutions (u, v, w) of (2). We obtained above expressions for x and z: $x = z^t u$ and $z = uv$. Hence the nonhomogeneous solutions of (*) satisfying the condition $\partial(y) \geqq \partial(z) > 0$ are obtained by the formulas

$$x = (uv)^t u, \quad y = (D(t))^\lambda E w, \quad z = uv, \quad t > 0, \quad \lambda \geqq 0 \tag{3}$$

from the nonhomogeneous solutions (u, v, w) of (2) such that $\partial(y) \geqq \partial(uv)$. The latter condition imposes on $D^\lambda E$ the restriction $vu \angle D^\lambda E$, which is satisfied for all transformations of (1) except the transformation $y = uv$. This transformation is excluded from the following discussion.

Let us see how the word E cuts the word $D(t)$. The word $D(t)$ has the form $D(t) = (vu)^c [(uv)^{t+a_i} u]_{i=1}^{n} (uv)^p$, where $n \geqq 1$, $p \geqq 0$ and $a_1, \cdots, a_n \geqq 0$ (n is the number of occurrences of x in B, and $a_1, \cdots, a_n$ and p correspond to occurrences of y in B). We are interested in the cuts by E of the occurrences of $(uv)^t$ in $D(t)$. The lemma implies that the cuts generated by nonhomogeneous solutions (x, y, z) of (*) cannot be arbitrarily deep. Indeed, put $M = \max_{1 \leqq i \leqq n}\{a_i + c + 1 + [A^\partial\}$. If E cuts occurrences of $(uv)^t$ in D, then for E and E_1 we have one of the following two representations:

$$E = (vu)^c\, [(uv)^{t+a_i} u]_{i=1}^{q-1}\, (uv)^k, \qquad E_1 = (uv)^{t+a_q-k}\, [(uv)^{t+a_i} u]_{i=q+1}^{n}\, (uv)^p,$$
$$E = (vu)^c\, [(uv)^{t+a_i} u]_{i=1}^{q-1}\, (uv)^{k-1} u, \quad E_1 = v\, (uv)^{t+a_q-k}\, [(uv)^{t+a_i} u]_{i=q+1}\, (uv)^p,$$

where $0 < k \leqq t$ and $0 < q \leqq n$. If $M \leqq k \leqq t - M$, then by the lemma (the roles of D_2, y, z, x, λ and t in the lemma are played by E_1, w, v, u, t and k, and the role of D_1 by the beginning $(vu)^c[\cdots]_{i=1}^{q-1} uv$ of E) equation (2) has only homogeneous solutions and the corresponding solutions of (*) are also homogeneous. Hence for the nonhomogeneous solutions of (2) either $k < M$ or $k > t - M$.

Let $\mathfrak{A}$ denote the set of nonhomogeneous solutions of (*) satisfying the condition $\partial(y) \geqq \partial(z) > 0$. As we noted above, the elements of $\mathfrak{A}$ are obtained by the formulas (3) from the nonhomogeneous solutions (u,v,w) of (2). Let $\mathfrak{M}$ and $\mathfrak{N}$ denote the subsets of $\mathfrak{A}$ consisting of the triples (x,y,z) for which in (3) we have $t \leqq 2M$ and $t \geqq 2M$, respectively. Then $\mathfrak{A} = \mathfrak{M} \cup \mathfrak{N}$. We introduce the predicate $\mathfrak{A}_E(u,v,w,t,\lambda)$, which denotes "the quintuple (u,v,w,t,λ) is a solution of (2) and (u,v,w) is a nonhomogeneous triple". Since $(uv)^t$ occurs in D, $E \angle D$ and $\partial(uv) \geqq \partial(w)$ for any solution (u,v,w) of (2) (because of the arrow in (2)), by using 2.7 it is easy to prove the existence of an L, which does not depend on u, v, w, t and E, such that for any u, v, w, t and E we have the implication

(4) $$(\exists\lambda \geqslant L)\, \mathfrak{A}_E\,(u,\ v,\ w,\ t,\ \lambda) \supset (\forall\lambda \geqslant L)\, \mathfrak{A}_E\,(u,\ v,\ w,\ t,\ \lambda).$$

For this purpose it suffices to view (2) with fixed u, v, w, t and E as an exponential equation in the one unknown λ and to verify, using $E \angle D$, $\partial(uv) \geqq \partial(w)$ and the fact that $(uv)^t$ occurs in D, that the characteristic G of this equation is bounded by a constant which does not depend on u, v, w, t or E. Then the existence of L follows from 2.7.

It follows from (3) and (4) that the set $\mathfrak{M}$ is parametrically expressed in terms of the sets of nonhomogeneous solutions of the equations obtained from (2) with $t \leqq 2M$ and $\lambda \leqq L$.

We now turn to the set $\mathfrak{N}$. Put $c_i = a_i + 2M$ and

$$F(\sigma) = (vu)^c[(uv)^{\sigma+c_i}u]_{i=1}^n (uv)^p.$$

Obviously, the words $D(t)$ and $F(\sigma)$ are equivalent in the region $t = \sigma + 2M$. The elements of $\mathfrak{N}$ are obtained by the formulas

(5) $$x = (uv)^{c+2M}u, \qquad y = (F(\sigma))^\lambda Hw, \qquad z = uv, \qquad \sigma \geqq 0, \qquad \lambda \geqq 0,$$

where $H \underset{\cdot}{\angle} F(\sigma)$ and $F^\lambda H \neq \Lambda$, from the solutions of the equation $\mathfrak{B}_{H,\sigma,\lambda}$, which is obtained from (2) by substituting $(uv)^{M+\sigma+M}$, $F(\sigma)$, H and H_1 in place of $(uv)^t$, $D(t)$, E and E_1, respectively. Here $HH_1 = F(\sigma)$. In view of what has been said about the cuts by E of the occurrences of $(uv)^t$ in $D(t)$, we may assume that H does not cut occurrences of $(uv)^\sigma$ in $F(\sigma)$ (we have in mind the occurrences $(uv)^M * (uv)^\sigma * (uv)^{M+a_i}$; see [**9**], Chapter I, §4.2). Otherwise, the equation $\mathfrak{B}_{H,\sigma,\lambda}$ has only homogeneous solutions (u,v,w) and the corresponding solutions of (*) are homogeneous. We define the predicate $\mathfrak{B}_H(u,v,w,\sigma,\lambda)$ as follows: $\mathfrak{B}_H(u,v,w,\sigma,\lambda)$ is true if and only if (u,v,w,σ,λ) is a solution of $\mathfrak{B}_{H,\sigma,\lambda}$ such that (u,v,w) is a nonhomogeneous triple. Viewing $\mathfrak{B}_{H,\sigma,\lambda}$ with fixed u, v, w, σ and H as an exponential equation in the one unknown λ and arguing as above on $\mathfrak{A}_E$ with an application of 2.7, we see that there exists a constant l, not depending on u, v, w, σ or H, such that for any u, v, w, σ and H we have the implication

(6) $$(\exists\lambda \geqq l)\,\mathfrak{B}_H(u,v,w,\sigma,\lambda) \supset (\forall\,\lambda \geqq l)\,\mathfrak{B}_H(u,v,w,\sigma,\lambda).$$

Viewing $\mathfrak{B}_{H,\sigma,\lambda}$ with fixed u, v, w, λ and H as an exponential equation in the one unknown σ and considering that H does not cut occurrences of $(uv)^\sigma$ in F, we see with the aid of 2.7 that there exists N_λ, not depending on u, v, w or H, such that for any u, v, w, H and λ we have the implication

(7) $$(\exists\,\sigma \geqq N_\lambda)\,\mathfrak{B}_H(u,v,w,\sigma,\lambda) \supset (\forall\,\sigma \geqq N_\lambda)\,\mathfrak{B}_H(u,v,w,\sigma,\lambda).$$

Put $N = N_l$. We will prove that for any u, v and w we have

$$\mathfrak{B}_H(u,v,w,N,l) \supset (\forall \sigma \geqq N)(\forall \lambda \geqq l)\mathfrak{B}_H(u,v,w,\sigma,\lambda). \tag{8}$$

Indeed, suppose $\mathfrak{B}_H(u,v,w,N,l)$ holds. Then from (7) with $\sigma = N$ and $\lambda = l$ we obtain $(\forall \sigma \geqq N)\mathfrak{B}_H(u,v,w,\sigma,l)$. Hence by (6) with $\lambda = l$ we have $(\forall \sigma \geqq N)(\forall \lambda \geqq l)\mathfrak{B}_H(u,v,w,\sigma,\lambda)$. Thus (8) is proved.

Using (6), (7) and (8), it is not difficult to prove that the set $\mathfrak{N}$ is parametrically expressed in terms of the sets of nonhomogeneous solutions of the equations $\mathfrak{B}_{H,\sigma,\lambda}$ for which $\sigma \leqq N$ and $\lambda \leqq l$. Indeed, the elements of $\mathfrak{N}$ are obtained by the formulas (5) from the solutions of the predicate $\mathfrak{B}_H(u,v,w,\sigma,l)$ in the sense that we have the equivalence

$$\mathfrak{N}(x,y,z) \Leftrightarrow \exists uvw\sigma\lambda H[x = (uv)^{\sigma+2M}u, y = F^{\lambda}Hw, z = uv, \mathfrak{B}_H(u,v,w,\sigma,\lambda)]. \tag{9}$$

Furthermore, since H does not cut occurrences of $(uv)^{\sigma}$ in the word F, there are for H only finitely many distinct representations and the existential quantifier $\exists H$ in (9) can be replaced by a disjunction. The desired parametric representation for $\mathfrak{N}$ follows from (9) and (6) – (8) and has the form

$$\begin{aligned} ((x,y,z) \in \mathfrak{N}) \Leftrightarrow \bigvee_H \{ &\exists uvw\sigma\lambda\, [x = (uv)^{\sigma+2M}u, \quad y = (F(\sigma))^{\lambda} Hw, \quad z = uv, \\ &\sigma \geqslant N, \lambda \geqslant l, \quad \mathfrak{B}_H(u,v,w,N,l)] \vee \bigvee_{m=0}^{l-1} \exists uvw\sigma\, [x = (uv)^{\sigma+2M}u, \\ &y = (F(\sigma))^m Hw, \quad z = uv, \quad \sigma \geqslant N_m, \quad \mathfrak{B}_H(u,v,w,N_m,m)] \vee \\ &\vee \bigvee_{k=0}^{N-1} \exists uvw\lambda\, [x = (uv)^{k+2M}u, \quad y = (F(k))^{\lambda} Hw, \quad z = uv, \quad \lambda \geqslant l, \\ &\mathfrak{B}_H(u,v,w,k,l)] \vee \bigvee_{k=0}^{N-1} \bigvee_{m=0}^{l-1} \exists uvw\, [x = (uv)^{k+2M}u, \; y = (F(k))^m Hw, \quad z = vu, \\ &\mathfrak{B}_H(u,v,w,k,m)] \}, \end{aligned} \tag{10}$$

where the summation $\bigvee_H$ extends over all possible beginnings H of the word F which do not cut occurrences of $(uv)^{\sigma}$ in F.

Considering the connection between the equations $\mathfrak{B}_{H,\sigma,\lambda}$ and (2) and the fact that $\mathfrak{A} = \mathfrak{M} \cup \mathfrak{N}$, we obtain that the set $\mathfrak{A}$ is parametrically expressed in terms of the sets of nonhomogeneous solutions of (2) for which $t \leqq N + 2M$ and $\lambda \leqq \max\{L, l\}$. We need only recall that (2) with any fixed $t > 0$ and $\lambda \geqq 0$ is a nondegenerate image of the third order of (*). The proposition is proved.

From 21.5 and 21.6 obviously follows

21.7. *If x occurs in B, then among the degenerate images of the second order and the nondegenerate images of the third order of (*) it is possible to find finitely many $\mathfrak{A}_1, \cdots, \mathfrak{A}_k$ such that the set of nonhomogeneous solutions of (*) is parametrically expressed in terms of the sets of nonhomogeneous solutions of the equations $\mathfrak{A}_1, \cdots, \mathfrak{A}_k$.*

An analogous result holds in the case where $B \neq \Lambda$ and does not contain x.

21.8. *If $B \neq \Lambda$ and does not contain x, then among the degenerate images of the second order and the nondegenerate images of the third order of (*) it is possible to find finitely many $\mathfrak{A}_1, \cdots, \mathfrak{A}_k$ such that the set of nonhomogeneous solutions of (*) is parametrically expressed in terms of the sets of nonhomogeneous solutions of the equations $\mathfrak{A}_1, \cdots, \mathfrak{A}_k$.*

PROOF. The images of the first order and the nondegenerate images of the second order of (*) are type I equations. By 21.3, 21.2′ and 21.2″ the nonhomogeneous solutions of (*) are parametrically expressed in terms of the nonhomogeneous solutions of finitely many images of the second order, and the nonhomogeneous solutions of nondegenerate images of the second order are parametrically expressed in terms of the nonhomogeneous solutions of finitely many of their images. It remains to observe that the latter are nondegenerate images of the third order of (*). The proposition is proved.

Equations of the form (*), excluding the case $B = \Lambda$, $b = 1$, will be called type II equations. We define θ-images for type I and type II equations. If $\mathfrak{A}$ is a type II equation, the θ-images of $\mathfrak{A}$ are the degenerate images of the second order and the nondegenerate images of the third order of $\mathfrak{A}$. The degenerate images of the second order of $\mathfrak{A}$ are called degenerate θ-images, and the nondegenerate images of the third order are called nondegenerate θ-images of $\mathfrak{A}$. A sequence $\mathfrak{A}_0, \cdots, \mathfrak{A}_k$ of word equations is called a chain of θ-images (of the equation $\mathfrak{A}_0$) if for any i, $1 \leqq i \leqq k$, $\mathfrak{A}_i$ is a θ-image of $\mathfrak{A}_{i-1}$. A chain $\mathfrak{A}_0, \cdots, \mathfrak{A}_k$ of θ-images is called degenerate if for any i, $1 \leqq i \leqq k$, $\mathfrak{A}_i$ is a degenerate θ-image of $\mathfrak{A}_{i-1}$. A chain of θ-images which is not degenerate is called nondegenerate. A complete system of θ-images of an equation $\mathfrak{A}$ is the set of equations $\mathfrak{A}_1, \cdots, \mathfrak{A}_k$ of which we spoke in 21.5 if $B = \Lambda$ and $b > 1$, in 21.7 if x occurs in B, and in 21.8 if $B \neq \Lambda$ and does not contain x. It follows from these definitions that any type II equation has a complete system of θ-images, and a complete system of θ-images of $\mathfrak{A}$ consists of its θ-images.

If $\mathfrak{A}$ is a type I equation, then by definition the θ-images of $\mathfrak{A}$ are its images. All of the notions pertaining to images of type I equations carry over verbatim to θ-images (complete system, chains, degeneracy, nondegeneracy).

The following proposition is the goal of §21.

21.9. *If $\mathfrak{A}$ is a type* I *or type* II *equation with $B \neq \Lambda$, then the set of nonhomogeneous solutions of $\mathfrak{A}$ is parametrically expressed in terms of the sets of nonhomogeneous solutions of the equations comprising a complete system of θ-images of $\mathfrak{A}$. If $\mathfrak{A}$ is a type* II *equation with $B = \Lambda$, then the set of nonhomogeneous solutions of the equation such that $\partial(y) \leqq \partial(z)$ is parametrically expressed in terms of the sets of nonhomogeneous solutions of the equations comprising a complete system of θ-images of $\mathfrak{A}$.*

PROOF. For type I equations the assertion follows from 21.2″, and for type II equations from 21.7, 21.8 and 21.5. The proposition is proved.

§22. Formular representation of the general solution of coefficient-free equations in three unknowns

Proposition 21.9 suggests that the θ-images of equations play the same role as the images in §6. Instead of trees we will consider θ-trees. We define θ-trees exactly as we did the trees of coefficient-free equations in §6 (including the "replacement rule"), only the word "image" is replaced everywhere by the word "θ-image". Moreover, in constructing θ-trees it is permissible to discard equations which do not have nonhomogeneous solutions, since such vertices of a tree cannot generate nonhomogeneous solutions of the original vertex.

Our aim is to prove the analogue of Theorem 6.17 for θ-trees (Theorem 22.0). We must reconsider the proof in §6, since not every tree is a θ-tree. For example, trees in whose construction the rules of admissible permutation, inversion or substitution are applied to images of type II equations are not θ-trees (in constructing θ-trees these rules may be applied to θ-images of equations, but not to images). In particular, we will change somewhat the notion of a basic equation. By basic equations we will mean the equations (6.1) – (6.7) of §6 and equations of the form

$$xy\alpha \cdots \to z^c xy \cdots \quad (\alpha \neq y,\ c \geqq 1), \tag{22.1}$$

$$xyxz \cdots \to zx^2y \cdots. \tag{22.2}$$

Note that (22.2) is the equation considered in 6.4, but the tree constructed there is not a θ-tree.

It is not difficult to see that any coefficient-free equation is a type I, type II or basic equation. Thus, any equation either has a complete system of θ-images or is basic.

A θ-tree is called basic if its final vertices are basic equations.

22.0. Theorem. *Any coefficient-free equation in three unknowns possesses a basic θ-tree.*

The proof of this theorem is obtained from the proof of Theorem 6.17 as follows. In the proof of 6.17, which consists of Propositions 6.4 – 6.17, we make the following alterations. The word "image" is replaced everywhere by the word "θ-image", "tree" by "θ-tree", and the words "α-tree" and "β-tree" by "basic θ-tree". Then the analogues of 6.4 – 6.7 and 6.9 – 6.17 are true, and 6.8 is replaced by 22.0.1, below. The analogue of 6.17 is 22.0. The proofs of 6.5 – 6.7, 6.10, 6.12 and 6.14 – 6.17 are the same as in §6 with the alterations indicated above. The equation of 6.4 is basic of the form (22.2), and the equation of 6.13 with $B = \Lambda$ and $b = 1$ (which does not have a complete system of θ-images) is basic of the form (22.1). The proofs of 6.9 and 6.11 require the changes indicated in 22.0.2 and 22.0.3.

22.0.1. *Any nondegenerate θ-image of the equation*

$$xyz\eta A(x,y,z) \to zx^2yB(x,y,z), \qquad \eta \neq z,$$

possesses a basic θ-tree.

Proof. The given equation is of type II. Any image of the equation $xyz \cdots \to zx \cdots$ is degenerate and has the form $xyz \cdots \leftarrow zx \cdots$. The degenerate images of the latter equation again have the form $xyz \cdots \to zx \cdots$. It follows that the nondegenerate θ-images of the equation $xyz\eta \cdots \to zx^2y \cdots$ are images of nondegenerate images of the second order of this equation. Nondegenerate images of the second order have the form 6.8 (2). The proof for the case $t = 0$ is exactly the same as in 6.8. The argument of 6.8 for $t > 0$ is not suitable, since substitutions were applied to images of the second order there. For $t > 0$ the proof is as follows. Nondegenerate images of the second order of the equation $xyz\eta A \to zx^2yB$ have the form (see 6.8 (2))

(1) $$yvuS^{z}_{(vy)^t vu}C(z^\lambda v, y, z)\,| \to uv((vy)^t vu)^\lambda vyS^{z}_{(vy)^t vu}B(z^\lambda v, y, z)\,|.$$

We will prove that any image of (1) is a basic equation or has no nonhomogeneous solutions. There are two cases to consider.

1) $x \angle C$. Then the left-hand side of (1) begins with the word $yvu(vy)^t vuv$. Since $t, \lambda > 0$, the right-hand side of (1) begins with $uv(vy)^t vuvy$. Equation (1) has the form

(2) $$yvuv(yv)^{t-1}yvvu \cdots \to uvvy(vy)^{t-1}vuvy \cdots.$$

Any solution of (2) satisfies the equations $yvuv = uvvy$ and $yvuv = vuvy$. Substituting in the left-hand side of the second equation $uvvy$ in place of $yvuv$, we obtain $uvvy = vuvy$; hence $uv = vu$. Thus any solution of (2) satisfies the system $yvuv = uvvy$, $uv = vu$. By 1.2, this system has only homogeneous solutions. We have proved that (2) has only homogeneous solutions. By 21.1, if an equation has only homogeneous solutions, then any of its images has only homogeneous solutions. Thus, our assertion concerning the images of (1) is proved for the case $x \angle C$.

2) $y \angle C$. Then (1) has the form $yvuy \cdots \to uv^2y(vy)^{t-1}vu \cdots$. Any image of this equation has one of the following three forms:

$$\xi v \cdots \leftarrow v^2 \cdots, \qquad \xi vu \cdots \leftarrow vuv \cdots, \qquad \xi vuuv^2 \cdots \leftarrow uv^2\xi vu \cdots.$$

The first two equations are basic of the forms (6.2), (6.3). The last equation has no nonhomogeneous solutions. Indeed, any solution of this equation satisfies the equation $\xi vu^2v^2 = uv^2\xi vu$. On the ends u and v "commute". Hence by the analogue of 1.17 for ends, any solution of the last equation satisfies the system $\xi vu^2v^2 = uv^2\xi vu$, $uv = vu$. By 1.2, this system has only homogeneous solutions. Thus, our assertion concerning the images of (1) is proved for the case $y \angle C$. The proposition is proved.

22.0.2. *Proposition* 6.9 *with the word "tree" replaced by "θ-tree" is true.*

The proof is preceded by a lemma.

LEMMA. *Any nondegenerate θ-image of the equation $xy^2\xi \cdots \to zxy\eta \cdots$, where $\xi, \eta \neq y$, possesses a basic θ-tree. Any nondegenerate θ-image of the equation $xy^2z \cdots \to zx^2y \cdots$ possesses an h-tree.*

PROOF OF THE LEMMA. For the equation $xy^2z \cdots \to zx^2y \cdots$ the assertion follows from the definitions of nondegenerate θ-images, h-trees and from 6.3.3), b). As for the equation $xy^2\xi \cdots \to zxy\eta \cdots$, by 6.3.2), b) its nondegenerate θ-images either are basic equations or have the form $yxzy \cdots \to zxzy \cdots$. The latter equation is reduced by the substitution $y = zv$ to the equation $vxz^2 \cdots = xz^2v \cdots$, which by 6.5 possesses a basic θ-tree. The lemma is proved.

PROOF OF 22.0.2. The scheme of the argument is the same as in 6.9, only now we construct a tree of θ-images. A tree of θ-images of height l is defined in the same way as a tree of images of height l in the proof of 6.9, except that the word "image" is replaced by "θ-image". Equation (5) (numerals in parentheses denote equations in 6.9) is considered as in 6.9. If $\mathfrak{A}_0$ is an equation in the list (1) – (4), we write $\mathfrak{A}_0$ in the form $xy^azy^p\xi \cdots \to zy^bxy^q\eta \cdots$, where p, q, ξ, η, a and b are the same as in 6.9, and we construct the tree Д of θ-images of $\mathfrak{A}_0$ of height l, where l is even, $l \geqq 8 + |p - q|$. It follows from the definitions that a degenerate chain of θ-images of an equation is a degenerate chain of images. Hence the analysis of degenerate chains is exactly the same as in 6.9, except that "β-tree" is replaced by "θ-tree". Consider a nondegenerate chain $\mathfrak{A}_0, \cdots, \mathfrak{A}_l$. Then for some j, $1 \leq j \leq l$, the chain $\mathfrak{A}_0, \cdots, \mathfrak{A}_{j-1}$ is degenerate and $\mathfrak{A}_j$ is a nondegenerate θ-image of $\mathfrak{A}_{j-1}$. If $\mathfrak{A}_0$ is equation (4), then all of the equations in the chain $\mathfrak{A}_0, \cdots, \mathfrak{A}_{j-1}$ are of type I and the argument of 6.9 carries over without change. If $\mathfrak{A}_0$ has the form (1), (2) or (3), then, as is easily calculated, $\mathfrak{A}_{j-1}$ has the form (1), (2) or (3), respectively. Hence if $\mathfrak{A}_0$ has the form (1) or (3), then $\mathfrak{A}_j$ possesses the desired tree by the lemma, and if $\mathfrak{A}_0$ has the form (2), then $\mathfrak{A}_j$ possesses it by 22.0.1. The proposition is proved.

22.0.3. *Proposition* 6.11 *with the word "tree" replaced by "θ-tree" is true.*

PROOF. The proof of the fact that the equation $zv^2xz \cdots \to xz^2v^2 \cdots$ possesses a basic θ-tree requires alteration. We will prove that the equation

$$zv^2xz \cdots \to xz^2v^2 \cdots \tag{1}$$

possesses a basic θ-tree.

Consider the θ-images of (1). Equation (1) is of type II. Hence its θ-images are degenerate images of the second order or nondegenerate images of the third order of (1). Any image of (1) has the form

$$uv^2xx^tu \cdots \leftarrow zux^tuv^2 \cdots, \tag{2}$$

where $t > 0$ and u is a new unknown. The images of the second order of (1) are the images of (2). Under the transformations $x = uz$, $x = uvz$ of (2) we obtain images of the form

$$v^2(uz)^{t+1} \cdots \to zu(uz)^tuv^2 \cdots, \tag{3}$$

$$vuv \cdots \to zu^2v \cdots. \tag{4}$$

We will prove that the images of (2) obtained under the other transformations have only homogeneous solutions. The "other" transformations have the form $x = uv^2A(u,v)z$, where A is a word in the alphabet $\{u,v\}$ and z is a new unknown. Put $y = A(u,v)z$. Then $x = uv^2y$. The triple (u,v,y) satisfies the equation

$$uv^2y(uv^2y)^tu \cdots = yu(uv^2y)^tuv^2 \cdots. \tag{5}$$

Any solution of (5) satisfies the system $uv^2yu = yu^2v^2$, $v^2yu = yuv^2$. Substituting in the first equation yuv^2 in place of v^2yu, we obtain $uyuv^2 = yu^2v^2$, whence $uy = yu$. Now by 1.1 we see that the above system, hence also equation (5), has only homogeneous solutions. But if u, v and $A(u,v)z$ are powers of some word S, then z is a power of S and the assertion about the "other" images is proved.

Hence among the θ-images of (1) nonhomogeneous solutions can exist only for those which are images of (3) or (4), since the transformations $x = uz$ and $x = uvz$ are nondegenerate. We will prove that the images of (3) and (4) have basic θ-trees. By 6.2, the images of (4) are σ-supporting and the assertion follows from 6.10. Consider (3). By 6.2, any image of (3) is either σ-supporting or is obtained under the transformation $v = zx$ and has the form $xzx(uz)^{t+1}u \cdots \to u^2(zu)^t(zx)^2 \cdots$. The latter equation is of type I. By

6.2, any of its images (and, therefore, any of its θ-images) either is σ-supporting or is obtained under the transformation $u = xv$ and has the form

$$zx(xvz)^{t+1}xv \cdots \to vxv(zxv)^t(zx)^2 \cdots.$$

The latter equation is reduced by the substitution $z = vy$ to the equation $yx(xv^2y)^{t+1}xvP = xv(vyxv)^t(vyx)^2Q$; an admissible permutation of this last equation is the basic equation $yxvPyx^2v^2(yxv^2)^t = vyxQxv^2yx(v^2yx)^t$. We have shown that (3) possesses a basic θ-tree. The proposition is proved.

Thus, Theorem 22.0 is proved.

Now our aim is to prove that the nonhomogeneous solutions of basic equations are representable by parametric words. Below, p and q are word unknowns, and λ, μ and ν are natural unknowns.

22.1. *The nonhomogeneous solutions of equation* (6.6) *are obtained by the formulas* $x = (pq)^\lambda p$, $y = q(pq)^\mu$, $z = (pq)^\nu p$ *and* $\mathscr{E}(p,q)$.

PROOF. One aspect of the assertion follows from 1.23. Let us prove the other. Obviously, the triple $x = (pq)^\lambda p$, $y = q(pq)^\mu$, $z = (pq)^\nu p$ is a solution of the equation $xyz = zyx$ for any values of the unknowns p, q, λ, μ and ν. It is nonhomogeneous if $\mathscr{E}(p,q)$. Indeed, if for some p, q, λ, μ and ν the triple (x,y,z) is homogeneous, then $xy = yx$. Then, obviously, $(pq)^{\lambda+\mu+1} = (qp)^{\lambda+\mu+1}$, whence $qp = pq$, which contradicts the condition $\mathscr{E}(p,q)$. The proposition is proved.

22.2. *The nonhomogeneous solutions of the equation* $xyz = zxy$ *are obtained by the formulas* $x = (pq)^\lambda p$, $y = q(pq)^\mu$, $z = (pq)^\tau$ *and* $\mathscr{E}(p,q)$.

The proof is analogous to that of 22.1 (see 1.24).

22.3. *Suppose* $a \geqq 2$. *Then the nonhomogeneous solutions of the equation* $xy^az = zy^ax$ *are obtained by the formulas*

$$\begin{aligned} &x = (pq^a)^\lambda p, \quad y = q, \quad z = (pq^a)^\mu p, \quad \mathscr{E}(p,q); \\ &x = qp\,((pq)^{\nu+1}p)^{a-2}pq\,[((pq)^{\nu+1}p)^{a-1}pq]^\lambda, \end{aligned} \tag{1}$$

$$\begin{aligned} &y = (pq)^{\nu+1}p, \\ &z = qp\,((pq)^{\nu+1}p)^{a-2}pq\,[((pq)^{\nu+1}p)^{a-1}pq]^\mu, \quad \mathscr{E}(p,q). \end{aligned} \tag{2}$$

PROOF. 1) We can verify directly that the triples (x,y,z) in (1) and (2) satisfy $xy^az = zy^ax$ for any p, q, λ, μ and ν. Arguing by contradiction, we see that under the additional condition $\mathscr{E}(p,q)$ a solution obtained by (1) or (2) is nonhomogeneous.

2) Now suppose that (X,Y,Z) is a nonhomogeneous solution of $xy^az = zy^ax$. Then $XY^aZ = ZY^aX$ and $Y^aXY^aZ = Y^aZY^aX$, where $XY \neq \Lambda$ and $XZ \neq \Lambda$. Thus Y^aX and Y^aZ commute. By 1.2, we have

$$Y^aX = S^n, \qquad Y^aZ = S^m \qquad (n,m > 0), \tag{3}$$

where S is a simple word. It follows from (3) that there exist $k,c,d \geqq 0$ and U and V such that

$$Y^a = (UV)^cU, \tag{4}$$

$$X = V(UV)^d, \qquad Z = V(UV)^k, \tag{5}$$

where $UV = S$. Consider (4). The following cases are possible.

Case 1. $c > 1$. From (4) and 1.7 it follows, in view of the simplicity of UV, that Y is a power of UV, i.e. a power of S. From (3) it is evident that X and Z are also powers of S, which contradicts the nonhomogeneity of the set (X, Y, Z). Hence the case $c > 1$ is impossible.

Case 2. $c = 0$. Then $U = Y^a$ and (see (5)) the solution (X, Y, Z) is obtained by means of (1).

Case 3. $c = 1$. Assume that $[U^\partial \geqq [Y^\partial$. From (4) we have $U = YQ$ and $Y^{a-1} = QVYQ$ for some Q. By 1.4′, the words QV and Y are powers of a simple word T. Then Q, V and U are powers of T and (see (5)) the solution (X, Y, Z) is homogeneous. Consequently, $[Y^\partial > [U^\partial$. It follows from (4) that there exists $R \neq \Lambda$ such that $Y = UR$ and $R(UR)^{a-1} = VU$. From the latter equality we deduce that there exists $B \neq \Lambda$ such that $UR = BU$ and $R(UR)^{a-2}B = V$. From $UR = BU$ and 1.16 there exist $t \geqq 0$ and P and Q such that $B = PQ$ and $U = (PQ)^tP$. Then $R = QP$. Substituting the expressions for B, U and R into the formulas for V, Y, X and Z, we see that the solution (X, Y, Z) is obtained by means of (2). The proposition is proved.

22.4. *The nonhomogeneous solutions of the equation $xyxz \to zx^2y$ are obtained by the formulas*

(1) $x = (pq)^\nu p,\quad z = pq,\quad y = qp^2q((pq)^\nu p^2q)^\lambda,\quad \mathscr{E}(p; q),\qquad \nu > 0.$

PROOF. 1) We see that the triple (x, y, z) in (1) is a nonhomogeneous solution of the equation $xyxz \to zx^2y$ for any p, q, ν and λ such that $\mathscr{E}(p, q)$ holds. 2) Suppose (X, Y, Z) is a nonhomogeneous solution of $xyxz \to zx^2y$. We have $[X^\partial \geqq [Z^\partial$ and $XYXZ = ZXXY$. If $[X^\partial = [Z^\partial$ or $Z = \Lambda$, by 1.2 we obtain that (X, Y, Z) is a homogeneous solution. Therefore, $[X^\partial > [Z^\partial > 0$. By 1.23, $XY = Q(PQ)^a$, $X = (PQ)^bP$ and $Z = Q(PQ)^c$ for some $a, b, c \geqq 0$ and P and Q such that $\mathscr{E}(P, Q)$. We have $XY = (PQ)^bPY = Q(PQ)^a$. If $a, b > 0$, then $PQ = QP$, which contradicts the condition $\mathscr{E}(P, Q)$. If $a = 0$, then $b = 0$, since $P \neq \Lambda$ in view of $\mathscr{E}(P, Q)$. We have $Q = PY$, $X = P$, $Z = XY(X^2Y)^c$, which is impossible, since $[X^\partial > [Z^\partial$. There remains the case $a > 0$, $b = 0$. Then $PY = (QP)^aQ$, whence $PA = QP$ and $Y = A(QP)^{a-1}Q$ for some A. Solving the equation $QP = PA$ with the aid of 1.16, we obtain $A = UV$, $Q = VU$ and $P = (VU)^tV$ for some U and V and some $t \geqq 0$. Substituting these expressions in place of P, Q and A in the formulas for X, Y and Z, we see that the solution (X, Y, Z) is obtained by means of (1). We need only observe that $b = 0$ and $[X^\partial > [Z^\partial$ imply $c = 0$ and $t > 0$, and that $\mathscr{E}(V, U)$ holds because of the nonhomogeneity of (X, Y, Z). The proposition is proved.

22.5. *Suppose* $A(x, y, z)$ and $B(x, y, z)$ *are words in the alphabet* $\{x, y, z\}$. *Let* $\bar{x}$, $\bar{y}$ *and* $\bar{z}$ *denote the parametric words appearing to the right of* x, y *and* z *in* 22.3 (2). *View the equation* $A(\bar{x}, \bar{y}, \bar{z}) = B(\bar{x}, \bar{y}, \bar{z})$ *as an exponential equation in the three unknowns* λ, μ *and* ν *and with the alphabet* $\{p, q\}$ *in the role of* И. *This equation is equivalent to some LDR.*

PROOF. Suppose $Q(x, y, z)$ is a word in the alphabet $\{x, y, z\}$. Let us agree to write $\bar{Q}$ instead of $Q(\bar{x}, \bar{y}, \bar{z})$. We will regard $\bar{Q}$ as a function of the three variables λ, μ and ν. The value of this function on the set $(\tilde{\lambda}, \tilde{\mu}, \tilde{\nu})$ will be denoted by $\tilde{Q}$. Note that $\tilde{Q}$ is a word in the alphabet $\{p, q\}$.

Words of the form $[\xi_i y^a]_{i=1}^k \xi$, where $k \geqq 0$ and ξ_i and ξ are letters in the list $\{x, z\}$, will be called elementary. We will assume that the word equation $A = B$ is left and right noncancellable. If the exponential equation $\overline{A} = \overline{B}$ has a solution, then, obviously, one of A or B begins with x, the other with z, i.e. with elementary words. We now represent A and B in the form $A = UA_1$ and $B = VB_1$, where U and V are elementary words which are not beginnings of other elementary beginnings of A and B. It is obvious that such a representation is possible. We will prove that the equation $\overline{U}\overline{A}_1 = \overline{V}\overline{B}_1$ is equivalent to $(\overline{U} = \overline{V}) \& (\overline{A}_1 = \overline{B}_1)$. Put $T = ((pq)^{\nu+1}p)^{a-1}pq$. Obviously, $T = \bar{y}^{a-1}pq$. It is easily verified that $\bar{y}^a\bar{x} = T^2T^\lambda$ and $\bar{y}^a\bar{z} = T^2T^\mu$. It follows that if C is an elementary word, then $\bar{y}^a\overline{C}$ has the form $T^{\varphi(\lambda,\mu)}$, where φ is an α-polynomial with positive free term. Along with $\overline{U}\overline{A}_1 = \overline{V}\overline{B}_1$ we consider the equivalent equation $\bar{y}^a\overline{U}\overline{A}_1 = \bar{y}^a\overline{V}\overline{B}_1$. Since U and V are elementary, the latter equation has the form $T^{\alpha(\lambda,\mu)}\overline{A}_1 = T^{\beta(\lambda,\mu)}\overline{B}_1$, where α and β are α-polynomials. If $(\tilde\lambda, \tilde\mu, \tilde\nu)$ is a solution of $\overline{U}\overline{A}_1 = \overline{V}\overline{B}_1$, then $(\tilde\lambda, \tilde\mu, \tilde\nu)$ is a solution of $T^{\alpha(\lambda,\mu)}\overline{A}_1 = T^{\beta(\lambda,\mu)}\overline{B}_1$, and we have in the alphabet $\{p, q\}$ the equality

$$\widetilde{T}^{\alpha(\tilde\lambda,\tilde\mu)}\widetilde{A}_1 = \widetilde{T}^{\alpha(\tilde\lambda,\tilde\mu)}\widetilde{B}_1.$$

We will prove that $\alpha(\tilde\lambda, \tilde\mu) = \beta(\tilde\lambda, \tilde\mu)$. Suppose, for example, that $\alpha(\tilde\lambda, \tilde\mu) > \beta(\tilde\lambda, \tilde\mu)$. Then from (1) we have $\widetilde{T}^c\widetilde{A}_1 = \widetilde{B}_1$, where $c > 0$. We have $y^{a-1}pq \angle T \angle T^c$. Then $y^{a-1}pq \angle \widetilde{B}_1$. This last relation implies that $\bar{y}^{a-1}pq \angle \overline{B}_1$. But the words $\bar{x}$ and $\bar{z}$ begin with qp. Therefore $\bar{y}^a \angle \overline{B}_1$, whence $y^a \angle B_1$, so that $B_1 = y^a B_2$ for some B_2. Cancelling the equality $\widetilde{T}^c\widetilde{A}_1 = \widetilde{B}_1$ on the left by $y^{a-1}pq$, we have $\widetilde{T}^{c-1}\widetilde{A}_1 = (pq)^\nu p\widetilde{B}_2$. Hence $p \angle \widetilde{T}^{c-1}\widetilde{A}_1$, and then either $c > 1$ or $y \angle A_1$. In both cases $(pq)^{\nu+1}p \angle \widetilde{T}^{c-1}\widetilde{A}_1$. Now from $\widetilde{T}^{c-1}\widetilde{A}_1 = (pq)^\nu p\widetilde{B}_2$ it is clear that $pq \angle \widetilde{B}_2$, which is possible only if B_2 begins with x or z. Then $y^a\xi \angle B_1$, where ξ is x or z, so that B begins with the word $Vy^a\xi$, which is elementary. This is impossible by definition of V. From this contradiction we conclude that $\alpha(\tilde\lambda, \tilde\mu) = \beta(\tilde\lambda, \tilde\mu)$. Then (1) implies that $\widetilde{A}_1 = \widetilde{B}_1$. Thus, any solution of $\overline{U}\overline{A}_1 = \overline{V}\overline{B}_1$ satisfies the system $(\overline{U} = \overline{V}) \& (\overline{A}_1 = \overline{B}_1)$. The converse is obvious. Consequently, the equation $\overline{U}\overline{A}_1 = \overline{V}\overline{B}_1$ is equivalent to the system $(\overline{U} = \overline{V}) \& (\overline{A}_1 = \overline{B}_1)$. We have shown that if the equation $\overline{A} = \overline{B}$ has a solution, then the words A and B are representable in the forms $A = UA_1$ and $B = VB_1$, where U and V are elementary words; in this regard, the equation $\overline{A} = \overline{B}$ is equivalent to the system $(\overline{U} = \overline{V}) \& (\overline{A}_1 = \overline{B}_1)$. By an obvious induction we conclude from this that if the equation $\overline{A} = \overline{B}$ has a solution, then A and B are representable in the forms $A = [U_iC_i]_{i=1}^n$ and $B = [V_iC_i]_{i=1}^n$, where $n \geqq 1$, U_i and V_i are elementary words, and the equation $\widetilde{A} = \widetilde{B}$ is equivalent to the system $\&_1^n(\overline{U}_i = \overline{V}_i)$. We have seen above that if C is an elementary word, then $\bar{y}^a\overline{C}$ has the form $T^{a\lambda + b\mu + c}$. Multiplying each of the equations $\overline{U}_i = \overline{V}_i$ on the left by $\bar{y}^a$, we see that the equation $\overline{A} = B$ is equivalent to some LDR. The proposition is proved.

22.6. *The nonhomogeneous solutions of basic equations are representable by parametric words.*

Proof. Equation (6.2) is reduced by the substitution $x = yv$ to the equation $vy \cdots = y^cv \cdots$, which is basic of the form (6.1). Equations (6.3) and (6.4) are reduced by the substitution $x = zv$ to the respective equations $vyz \cdots = zvy \cdots$ and $vyz \cdots = yzv \cdots$, which are basic of the form (6.5). Equation

(22.1) is reduced by the substitution $x=zv$ to the equation $vyzA=z^c vyB$. By 1.17, any solution of this equation satisfies $vyz=zvy$. Then the equation $vyzA=z^c vyB$ is equivalent to the equation $(vyz)^2A=zvyz^c vyB$, which is basic of the form (6.5). It remains to prove that the nonhomogeneous solutions of the basic equations (6.1), (6.5) – (6.7) and (22.2) are representable by parametric words. By 1.17 and 1.2, all nonhomogeneous solutions of (6.1) are contained among the triples $x=p^\lambda$, $y=p^\mu$, $z=q$, $\mathscr{B}(p,q)$ where λ and μ are natural unknowns, and p and q are word unknowns. Obviously, equation (6.5) is equivalent to the system $xyz=zxy$, $A=B$, equations (6.6), (6.7) are equivalent to the system $xy^a z=zy^a x$, $A=B$, and equation (22.2) is equivalent to the system $xyxz\rightarrow zx^2y$, $A=B$, where A and B are the ends of the left- and right-hand sides of the equation. Thus (see 22.1 – 22.4), the nonhomogeneous solutions of the basic equations (6.1), (6.5) – (6.7) and (22.2) are contained among the triples of the form

$$(1)\qquad x=T_1(p,q,\lambda,\mu,\nu),\quad y=T_2(p,q,\lambda,\mu,\nu),\quad z=T_3(p,q,\lambda,\mu,\nu),\quad \mathscr{B}(p,q),$$

where T_1, T_2 and T_3 are appropriate parametric words, p and q are word unknowns, and λ, μ and ν are natural unknowns. If $\Phi=\Psi$ ($\Phi\rightarrow\Psi$) is a basic equation of one of the forms (6.1), (6.5), (6.6), (6.7) or (22.2), then the determination of the nonhomogeneous solutions of this equation reduces to the determination of the solutions of the equation in five unknowns

$$(2)\qquad \Phi(T_1,T_2,T_3)=\Psi(T_1,T_2,T_3)$$

such that $\mathscr{B}(p,q)$. If we fix λ, μ and ν, then (2) becomes a coefficient-free word equation in the two unknowns p and q. By 1.1, it will have solutions such that $\mathscr{B}(p,q)$ only if its left-hand side is graphically equal to its right-hand side. Consequently, the nonhomogeneous solutions of the basic equation $\Phi=\Psi$ ($\Phi\rightarrow\Psi$) will be obtained by the formulas (1) only with those values of λ, μ and ν for which in the alphabet $\{p,q\}$ we have the graphical equality (2), i.e. with those λ, μ and ν which satisfy the exponential equation (2), if as the alphabet И we take $\{p,q\}$. We will prove that this exponential equation is equivalent to some LDR. If the formulas (1) are those of 22.1, 22.2 or 22.3 (1), then we have a one-stage exponential equation, and the assertion follows from 2.3. If the formulas (1) are those of 22.3 (2), then the assertion follows from 22.5. There remains the case where the formulas (1) are those of 22.4 (1). Here we denote equation (2) by $\mathfrak{A}(\nu,\lambda)$. Fix ν and view $\mathfrak{A}(\nu,\lambda)$ as an exponential equation in the one unknown λ. Considering the occurrence of the word $(pq)^\nu$ in the base of the power with exponent λ, it is not difficult to calculate that the characteristic of $\mathfrak{A}(\nu,\lambda)$ is bounded by a constant which does not depend on ν. The height does not exceed $[\Phi\Psi^\partial$. In view of 2.7, this implies the existence of L, not depending on ν, such that we have the implication

$$(\exists\lambda\geqslant L)\,\mathfrak{A}(\nu,\lambda)\supset(\forall\lambda\geqslant L)\,\mathfrak{B}(\nu,\lambda).$$

This implication implies the equivalence

$$\mathfrak{A}(\nu,\lambda)\Leftrightarrow\bigvee_{i=0}^{L-1}[\mathfrak{A}(\nu,i),\lambda=i]\vee[\mathfrak{A}(\nu,L),\lambda\geqslant L].$$

It remains to observe that $\mathfrak{A}(\nu,i)$ is a one-stage exponential equation in the one unknown ν and to apply 2.7. The proposition is proved.

22.7. *Suppose* $\mathfrak{A}$ *is the equation* (*) *preceding* 21.3. *If* $B = \Lambda$ *and* $b > 1$, *then the nonhomogeneous solutions of* $\mathfrak{A}$ *satisfying the condition* $\partial(y) \geqq \partial(z)$ *are representable by parametric words.*

PROOF. It is easy to see that the solutions of $\mathfrak{A}$ are obtained by the formulas $x = (uv)^t u$ and $z = uv$ from the solutions (y, u, v, t) of the equation

$$y^h A((uv)^t u, y, uv) = (vu)^c y C((uv)^t u, y, uv). \tag{1}$$

The condition $\partial(y) \geqq \partial(z)$ assumes the form $\partial(y) \geqq \partial(uv)$. Such solutions of (1) are obtained by the formula $y = vus$ from the solutions (v, u, s) of the equation $svu \cdots = (vu)^c s \cdots$ (since $b > 1$). By 1.17, any solution of this last equation satisfies the equation $svu = vus$. By 22.2, the nonhomogeneous solutions of the latter are obtained by the formulas $s = (pq)^a$, $v = (pq)^b p$, $u = q(pq)^c$ and $\mathscr{B}(p, q)$. Substituting the expressions for s, u and v into the formulas for x, y and z, we obtain $x = (qp)^\lambda q$, $z = (qp)^\mu$ and $y = (pq)^\nu$ for some λ, μ, ν, p and q such that $\mathscr{B}(p, q)$. We have proved that any nonhomogeneous solution of $\mathfrak{A}$ satisfying the condition $\partial(y) \geqq \partial(z)$ is representable in the form

$$x = (qp)^\lambda q, \quad z = (qp)^\mu, \quad y = (pq)^\nu, \quad \mathscr{B}(p, q) \tag{2}$$

for some λ, μ, ν, p and q. Denoting the right-hand sides of the equalities in (2) by T_1, T_2 and T_3, we see that 22.6 (2) is a one-stage exponential equation. The assertion being proved obviously follows from 2.3.

22.8. MAIN THEOREM. *The nonhomogeneous solutions of any coefficient-free equation in three unknowns are representable by parametric words.*

PROOF. We will first prove that if the equations $\mathfrak{A}_1, \cdots, \mathfrak{A}_k$ form a neighborhood of the equation $\mathfrak{A}$ and the nonhomogeneous solutions of each of $\mathfrak{A}_1, \cdots, \mathfrak{A}_k$ are representable by parametric words, then the nonhomogeneous solutions of $\mathfrak{A}$ are also representable by parametric words. In accordance with items 1)–5) in the definition of a neighborhood we consider five cases. For item 1) (with θ-images instead of images) the assertion follows from 21.9 and 22.7. For item 3) the assertion follows from the equivalence $(\Phi = \Psi) \Leftrightarrow (\Phi \rightarrow \Psi) \vee (\Psi \rightarrow \Phi)$. For item 5) the assertion follows from the fact that the equation is equivalent to any of its admissible permutations. For item 2) the assertion follows from the equivalence (see the discussion following the proof of 6.3)

$$\mathfrak{A}(x, y, z) \Leftrightarrow \bigvee_{i=1}^{n} \exists v\, [(x = \xi_1 \ldots \xi_i v) \,\&\, \mathfrak{A}_i(v, y, z)].$$

With regard to item 4), we define inductively the inversion $\overline{T}$ of a parametric (as described at the beginning of this chapter) word T. 1) If T is a word in the alphabet X and $T = \xi_1 \xi_2 \cdots \xi_k$, then $\overline{T} = \xi_k \cdots \xi_2 \xi_1$. 2) If λ is a natural unknown and T is a parametric word, then $\overline{T^\lambda} = (\overline{T})^\lambda$. 3) If T_1 and T_2 are parametric words, then $\overline{T_1 T_2} = \overline{T_2}\, \overline{T_1}$. Obviously, the inversion of a parametric word is a parametric word. It is not difficult to check that formula (2) following the statement of the main theorem at the beginning of this chapter remains true if we replace the words T_{ij} and the equation $\mathfrak{A}$ by their inversions. This implies the assertion for item 4). Thus the statement concerning a neighborhood is proved. In constructing a θ-tree, besides passing from equations to their neighborhoods, we also had the replacement rule. But by 5.6 the replacing and replaced equations have the same nonhomogeneous solutions. Hence if we have a θ-tree of $\mathfrak{A}$ with final vertices $\mathfrak{A}_1, \cdots, \mathfrak{A}_k$ and the nonhomogeneous solutions of each of $\mathfrak{A}_1, \cdots, \mathfrak{A}_k$ are representable by parametric words, then the nonhomogeneous solutions of $\mathfrak{A}$ are also representable by parametric words. The assertion being proved now follows from 22.0 and 22.6. The theorem is proved.

§23. On equations in more than three unknowns

In conclusion, we consider coefficient-free equations in more than three unknowns. In this area very little is known. It is known, for example, that even for equations in four unknowns parametric words are insufficient for a description of the general solution. We will show this by the example of the equation $xyz = zvx$ and at the same time we will give formulas allowing us to find all solutions of this equation (cf. 5.9).

It can also be proved that any system of coefficient-free equations in n unknowns is equivalent to a single coefficient-free equation in n unknowns. This will be proved at the end of this section.

It is not difficult to construct an algorithm recognizing the existence of solutions of coefficient-free equations in n unknowns. Namely, if an equation has a solution $(X_1, \cdots, X_n)$, then it has a homogeneous solution $(a^{[X_1^\partial}, \cdots, a^{[X_n^\partial})$, where a is a letter of the alphabet $И$. However, an algorithm recognizing the existence of nonhomogeneous solutions is unknown even for $n = 4$. For $n = 3$ such an algorithm follows from the main theorem of §22.

Still more difficult is the rank problem, which consists of determining an algorithm which calculates for a given coefficient-free equation in n unknowns the largest integer r, $1 \leqq r \leqq n-1$, such that the equation has a solution of rank r. The rank of a solution is defined as follows. It can be shown (see [**12**]) that for any solution $(X_1, \cdots, X_n)$ of a nontrivial coefficient-free equation there exist words $\varphi_1, \cdots, \varphi_n$ in an alphabet $\{v_1, \cdots, v_r\}$, $1 \leqq r \leqq n-1$, and nonempty words $V_1, \cdots, V_r$ in the alphabet $И$ such that $X_i = \mathrm{S}^{v_1, \cdots, v_r}_{V_1, \cdots, V_r} \varphi_i|$, where $(V_1, \cdots, V_r)$ does not satisfy any nontrivial coefficient-free equation in r unknowns. The integer r is uniquely determined by the solution $(X_1, \cdots, X_n)$ and is called the rank of this solution. Solutions of rank $r = 1$ are homogeneous; those of rank $r > 1$ are nonhomogeneous. For $n = 3$ the rank problem is equivalent to the recognition of the existence of nonhomogeneous solutions. The rank problem was investigated for equations in $n+1$ unknowns of the form $\mathfrak{A}$: $\varphi(x_1, \cdots, x_n) = y^m$, $n, m > 1$, in [**10**], and for equations of the form $\mathfrak{B}$: $x_1^m x_2^m \cdots x_n^m = y^m$ in [**11**]. In particular, it was proved in [**11**] that when $m \geqq n$ the equation $\mathfrak{B}$ has no solutions of rank > 1 (i.e. no nonhomogeneous solutions). The main result of [**10**] implies that the rank of solutions of $\mathfrak{A}$ is "almost always" less than n. The rank problem is not completely solved even for $\mathfrak{B}$. It is also not known for which φ the equation $\mathfrak{A}$ has nonhomogeneous solutions and for which it does not.

We will prove that the solutions of the equation $xyz = zvx$ cannot be described by means of finitely many quadruples of parametric words. Let us make this more precise.

Parametric words are to be understood in the sense of §3. A set of quadruples of words in the alphabet $И$ is called parametrizable if it coincides with the set of values of some quadruple of parametric words. Finite (set-theoretic) sums of parametrizable sets are also regarded as parametrizable sets. Our aim is to prove that the set of all solutions (all nonhomogeneous solutions) of the equation $xyz = zvx$ is nonparametrizable.

We first give formulas allowing us to find all solutions of $xyz = zvx$. The notation $h \sim g$ means that one of the processions h or g is obtained from the other by dropping the last component.

23.1. *The solutions of the equation $xyz = zvx$ are obtained by the formulas*

$$(0) \qquad x = p^{\lambda}, \quad y = v = p^{\mu}, \quad z = p^{\nu},$$

$$(1) \qquad x = (pq)^{\lambda}p, \quad y = v = q(pq)^{\nu}, \quad z = (pq)^{\mu}p, \qquad \mathscr{E}(p, q),$$

$$(2) \quad x = G(qp, qr, \xi)q, \ z = G(qp, qr, \zeta)q, \ y = pqr, \ v = rqp, \ y \neq v, \ \xi \sim \zeta,$$

$$(2') \quad x = G(qr, qp, \eta)q, \ z = G(qr, qp, \zeta)\, q, \ y = pqr, \ v = rqp, \ y \neq v, \ \eta \sim \zeta,$$

where p, q and r are word unknowns, λ, μ and ν are natural unknowns, and ξ, η and ζ are processional unknowns with ξ odd and η even.

PROOF. The homogeneous solutions are obtained by means of the formulas (0). We divide the nonhomogeneous solutions into two regions: $u = v$ and $u \neq v$. In the region $u = v$ the nonhomogeneous solutions are obtained by means of (1) (see 22.1). In the region $u \neq v$, as a consequence of 5.9, the nonhomogeneous solutions are obtained by means of (2) and (2′). The proposition is proved.

The proof of the nonparametrizability of the equation $xyz = zvx$ is preceded by two lemmas.

23.2. *Suppose a and b are different letters and that*

$$(1) \qquad AS^tV = G(a, b, \lambda_0, \cdots, \lambda_{k-1}),$$

where $S \neq \Lambda$, $k > 1$ and $\lambda_0 = \cdots = \lambda_{k-1} = 1$. Then $t \leqq 6 + 6[A^{\partial}$.

PROOF. It suffices to prove the assertion for S simple, since if $S = T^m$, where $m \geqq 1$ and T is simple, then $mt \leqq 6 + 6[A^{\partial}$ implies $t \leqq 6 + 6[A^{\partial}$. Hence we assume that S is simple. For brevity, put $F_{-1} = b$, $F_0 = a$ and $F_i = F(a, b, \lambda_0, \cdots, \lambda_{i-1})$, $i = 1, \cdots, k$. We prove the assertion by contradiction. Assume that $t > 6 + 6[A^{\partial}$. Multiplying (1) on the right by $\langle a, b\rangle_k$ and recalling the definition of the function G, we obtain

$$(2) \qquad AS^tV_1 = F_k.$$

We have $1 = [F_0^{\partial} \leqq [AS^{\partial} < [F_k^{\partial}$. It is obvious that $[F_{-1}^{\partial} = [F_0^{\partial} < [F_1^{\partial} < \cdots < [F_k^{\partial}$. Hence there must exist q, $1 \leqq q \leqq k$, such that $[F_{q-1}^{\partial} \leqq [AS^{\partial} < [F_q^{\partial}$. Obviously, $F_q \angle F_k$. From (2) we have $F_q = ASB$, $B \neq \Lambda$. Since $\lambda_{q-1} = 1$, we have $F_q = F_{q-1}F_{q-2}$. From the inequalities $[F_{q-2}^{\partial} \leqq [F_{q-1}^{\partial} \leqq [AS^{\partial}$ we obtain $[F_q^{\partial} \leqq 2[AS^{\partial}$. By 5.1, $F_k = F(F_q, F_{q-1}, \lambda_q, \cdots, \lambda_{k-1})$. From (2) we have

$$(3) \qquad AS^tV_1 = F(F_q, F_{q-1}, \lambda_q, \cdots, \lambda_{k-1}).$$

We know $\lambda_q = \cdots = \lambda_{k-1} = 1$. Simple computations show that if $k - q \leqq 2$, then the length of the right-hand side of (3) does not exceed $[F_qF_{q-1}F_q^{\partial}$, i.e. does not exceed $5[AS^{\partial}$, which contradicts the assumption $t > 6'6[A^{\partial}$. Consequently, $k - q > 2$. Then the right-hand side of (3) begins with the word $F(F_q, F_{q-1}, 1, 1, 1)$, which is equal to $F_qF_{q-1}F_qF_qF_{q-1}$. We know that $t[S^{\partial} > 6[AS^{\partial}$. Moreover, $F_q = ASB$, $[F_q^{\partial} \leqq 2[AS^{\partial}$ and $[F_{q-1}^{\partial} \leqq [AS^{\partial}$. We have from (3) that $AS^t = ASBF_{q-1}ASBASZ$ for some Z; hence

$$S^{t-1} = BF_{q-1}ASBASZ.$$

In view of 1.4′, this last equality implies that the words $BF_{q-1}A$ and BA are powers of S, so that for some $n, m > 0$ we have

$$BF_{q-1}A = S^n, \tag{4}$$

$$BA = S^m. \tag{5}$$

It follows from (5) that there exist $c, d \geqq 0$ and P and Q such that $S = PQ$, $B = (PQ)^c P$ and $A = Q(PQ)^d$. Then (4) implies that F_{q-1} is a power of QP. Since $F_q = ASB$, F_q is also a power of QP. Thus, F_q and F_{q-1} are powers of QP, from which, using 5.3, we obtain the required contradiction. The proposition is proved.

Suppose K is a set $(T_1, \cdots, T_n)$ of parametric words. Let $[K]$ denote the set of all possible values of $(T_1, \cdots, T_n)$.

23.3. *Suppose $T_1, \cdots, T_N$ are parametric words which do not contain word unknowns. Let N be the set of words of the form $G(a, b, 1, \cdots, 1)$, where a and b are letters of the alphabet $\mathrm{И}$. Then $\mathfrak{N}$ contains an element not belonging to*

$$\bigcup_{i=1}^{N} [T_i].$$

PROOF. Suppose T is a parametric word which does not contain word unknowns. Using 23.2 and applying induction on the number of occurrences in T of natural unknowns, it is not difficult to prove that if an element X of the set $[T]$ belongs to $\mathfrak{N}$, then $[X^\partial$ does not exceed some constant which is independent of the values of the unknowns occurring in T. Now our assertion is obvious, since $\mathfrak{N}$ contains arbitrarily long words.

23.4. THEOREM. *The set $\mathfrak{M}$ of solutions of the equation $xyz = zvx$ is nonparametrizable.*

PROOF. We define three kinds of quadruples of parametric words: 1) quadruples (T_1, T_2, T_3, T_4) such that any word unknown contained in the word T_1T_3 occurs in the word T_2T_4; 2) quadruples (T_1, T_2, T_3, T_4) such that $T_2 = T_4$ for any values of the unknowns occurring in T_2T_4; 3) all remaining quadruples of parametric words. We will assume that natural unknowns assume only positive values. This can always be done at the expense of increasing the number of quadruples of parametric words.

Now suppose K is the set of quadruples $K_1, \cdots, K_s$ $(s > 0)$ of parametric words. The following cases are possible.

Case 1. K consists only of quadruples of the first and second kinds. We will prove that in this case $\mathfrak{M}$ contains an element which is not contained in $\bigcup_1^s [K_i]$.

Let a and b be distinct letters of the alphabet $\mathrm{И}$. By 5.9 a), for any odd $k > 2$ the quadruple

$$(G(a, b, \underbrace{1, \ldots, 1}), ab, G(a, b, \underbrace{1, \ldots, 1}_{k-1}), ba)$$

belongs to $\mathfrak{M}$. Consider those quadruples in $\bigcup_1^s [K_i]$ which have the form (X, ab, Z, ba). It is clear that they are values of those quadruples in the list

$K_1, \cdots, K_s$ which are of the first kind. In view of the assumptions about the word unknowns occurring in quadruples of the first kind and the positivity of the natural unknowns, the values of the word unknowns occurring in T_1 have length $\leqq 2$. Then the values assumed by the component X are contained in the set $\bigcup_1^N [T_i]$, where $N > 0$ and $T_1, \cdots, T_N$ are parametric words which do not contain word unknowns. It remains to apply 23.3.

Case 2. K contains at least one quadruple of the third kind. Suppose this quadruple to be K_1. We will prove that $[K_1]$ contains an element which is not contained in $\mathfrak{M}$. Suppose K_1 has the form (T_1, T_2, T_3, T_4). Since K_1 is a quadruple of the third kind, there exists a set $(\bar{\xi}_1, \cdots, \bar{\xi}_m)$ of values of the unknowns occurring in $T_2 T_4$ such that $T_2(\bar{\xi}_1, \cdots, \bar{\xi}_m) \neq T_4(\bar{\xi}_1, \cdots, \bar{\xi}_m)$. Put $A = T_2(\bar{\xi}_1, \cdots, \bar{\xi}_m)$ and $B = T_4(\bar{\xi}_1, \cdots, \bar{\xi}_m)$. We have $A \neq B$, and A and B are words in the alphabet $И$. Put $\bar{T}_i = \mathrm{S}_{\xi_1, \cdots, \xi_m}^{\bar{\xi}_1, \cdots, \bar{\xi}_m} T_i |$ $(i = 1, 3)$. Since K_1 is a quadruple of the third kind, it follows, in view of the assumptions on the word unknowns and the natural unknowns, that a word unknown occurs in at least one of the words $\bar{T}_1$ or $\bar{T}_3$. Suppose the word is $\bar{T}_1$ and that the word unknown v occurs in $\bar{T}_1$. Fix values of all the unknowns except v which occur in $\bar{T}_1$. We obtain parametric words $\tilde{T}_1$ and $\tilde{T}_3$. Here $\tilde{T}_1$ has the form $A_0[vA_i]_{i=1}^n$, where the A_i are words in $И$, and $n \geqq 1$ because of the positivity of the natural unknowns. Let L denote the quadruple $(\tilde{T}_1, A, \tilde{T}_3, B)$. Obviously, $[L] \subset [K_i]$. We will construct an element of $[L]$ which does not belong to $\mathfrak{M}$. Let $\mathfrak{N}$ denote the set of elements of $\mathfrak{M}$ having the form (X, A, Z, B). Since $A \neq B$, by 5.9 the set $\{X\}$ of values assumed in $\mathfrak{N}$ by the component X is contained in the set of words of the form $G(QP, QR, \lambda_0, \cdots, \lambda_{k-1})Q$, where $\lambda_0, \cdots, \lambda_{k-1}$ are positive integers, $0 < k < \infty$, and $[P^\partial, [Q^\partial, [R^\partial \leqq [A^\partial$. To construct an element of $[L]$ not belonging to $\mathfrak{M}$ it suffices to choose V such that the word $A_0[VA_i]_{i=1}^n$ is not representable in the form

$$G(QP, QR, \lambda_0, \cdots, \lambda_{k-1})Q$$

with the indicated P, Q, R, k and $\lambda_0, \cdots, \lambda_{k-1}$. The construction of such an element, starting from the definition of the function G, presents no difficulties. Thus, there exists an element of $[L]$, hence also an element of $[K_1]$, which does not belong to $\mathfrak{M}$. The theorem is proved.

REMARK. Theorem 23.4, as is easily seen, remains true if the natural unknowns range not over all possible values, but over the values of a specific set; in particular, if they satisfy some LDR.

We will prove that for any system of coefficient-free word equations in n unknowns it is possible to construct a coefficient-free equation in n unknowns which is equivalent to the system. The proof depends on several lemmas.

23.5. *Suppose $A \neq B$, $[A^\partial = [B^\partial$ and $[X^\partial \geqq [AB^\partial$. Then $XAXB$ is a simple word.*

PROOF (BY CONTRADICTION). Assume that the word $XAXB$ is not simple. Then

$$XAXB = S^a, \tag{1}$$

where $a > 1$ and S is simple. Since $[A^\partial = [B^\partial$, $A \neq B$ and $a > 1$, the situations $[S^\partial > [XA^\partial$ and $[S^\partial = [XA^\partial$ are impossible. Therefore, $[S^\partial < [XA^\partial$, and then $a \geqq 3$. By hypothesis $[X^\partial \geqq [AB^\partial$, and we have proved that $a \geqq 3$. Then (1) implies that $[S^\partial \leqq [X^\partial$, so that $X = SV$ for some V. Now from (1) we have

(2) $$VASVB = S^{a-1}.$$

From (2) and 1.4 it follows that VA is a power of S. Then, obviously, VB is also a power of S. Since $[A^\partial = [B^\partial$, this is possible only if $A = B$. We have obtained the required contradiction. The proposition is proved.

Suppose $n > 1$. We write $\mathscr{E}(X_1, \cdots, X_n)$ if the words $X_1, \cdots, X_n$ are not powers of some word. It is obvious that if $\mathscr{E}(X_1, \cdots, X_n)$, then at least one of $X_1, \cdots, X_n$ is nonempty.

23.6. *Suppose* $\mathscr{E}(X_1, \cdots, X_n)$ *and* $M \geqq n > 1$, *and let* $X = (X_1 \cdots X_n)^{4M}$ *and*

$$A = [X_1X_i]_{i=1}^n [X_2X_i]_{i=1}^n \ldots [X_nX_i]_{i=1}^n,$$

$$B = [X_iX_1]_{i=1}^n [X_iX_2]_{i=1}^n \ldots [X_iX_n]_{i=1}^n.$$

Then the word $XAXB$ *is simple.*

PROOF. In accordance with 23.5, it suffices to prove the following three assertions: 1) $[A^\partial = [B^\partial$; 2) $[X^\partial \geqq [AB^\partial$; 3) $A \neq B$. Assertion 1) is obvious. Assertion 2) follows from the fact that $[AB^\partial = 4n[X_1 \cdots X_n^\partial \leqq 4M[X_1 \cdots X_n^\partial$. We will prove 3) by contradiction. Assume that $A = B$. Then for any i and j we have $X_iX_j = X_jX_i$. Since $\mathscr{E}(X_1, \cdots, X_n)$, at least one of $X_1, \cdots, X_n$ is nonempty. Without loss of generality, we will assume that $X_1 \neq \Lambda$. We know that $X_1X_2 = X_2X_1, \cdots, X_1X_n = X_nX_1$. Applying Proposition 1.2 to each of these equalities, we conclude that there exist natural numbers $a_2, \cdots, a_n$, $b_2, \cdots, b_n$ and simple words $S_2, \cdots, S_n$ such that $X_i = S_i^{a_i}$ and $X_1 = S_i^{b_i}$ $(i = 2, \cdots, n)$. If $n = 2$, then X_1 and X_2 are powers of S_2 despite the assumption $\mathscr{E}(X_1, X_2)$. If $n > 2$, then we have $X_1 = S_2^{b_2} = \cdots = S_n^{b_n}$. Since $X_1 \neq \Lambda$, we have $b_2, \cdots, b_n > 0$. Hence, by 1.3, $S_2 = \cdots = S_n$, and $X_1, \cdots, X_n$ are powers of S_2 despite the assumption $\mathscr{E}(X_1, \cdots, X_n)$. The contradiction shows that 3) is true. The proposition is proved.

Suppose $\mathfrak{A}$ and $\mathfrak{B}$ are systems of coefficient-free equations in n unknowns. We say that $\mathfrak{A}$ and $\mathfrak{B}$ are equivalent with respect to nonhomogeneous solutions if any nonhomogeneous solution of $\mathfrak{A}$ is a solution of $\mathfrak{B}$ and any nonhomogeneous solution of $\mathfrak{B}$ is a solution of $\mathfrak{A}$.

23.7. THEOREM. *Suppose* $N \geqq 2$ *and that*

(*) $$\begin{gathered} \Phi_1 = \Psi_1, \\ \cdots\cdots \\ \Phi_N = \Psi_N \end{gathered}$$

is a system of coefficient-free word equations in the n *unknowns* $x_1, \cdots, x_n$. *Put*

$$M = \max_{i,j}\{[\Phi_i^\partial, [\Psi_j^\partial, n\}, \qquad X = (x_1 \cdots x_n)^{4M},$$

$$A = [x_1x_i]_{i=1}^n[x_2x_i]_{i=1}^n \cdots [x_nx_i]_{i=1}^n,$$

$$B = [x_ix_1]_{i=1}^n[x_ix_2]_{i=1}^n \cdots [x_ix_n]_{i=1}^n,$$

$$\Omega = XAXB.$$

Then the system (*) *is equivalent with respect to nonhomogeneous solutions to the equation*

(1) $$\Phi_1[\Omega^2\Phi_i]_{i=2}^N = \Psi_1[\Omega^2\Psi_i]_{i=2}^N.$$

Proof. Obviously, any solution (in particular, any nonhomogeneous solution) of (*) is a solution of (1). We will prove that any nonhomogeneous solution of (1) is a solution of (*).

Suppose $(X_1, \cdots, X_n)$ is a nonhomogeneous solution of (1). We have $\mathscr{B}(X_1, \cdots, X_n)$ and

$$(2) \qquad \mathrm{S}_{X_1,\ldots,X_n}^{x_1,\ldots,x_n}\Phi_1\,[\Omega^2\Phi_i]_{i=2}^N\,| = \mathrm{S}_{X_1,\ldots,X_n}^{x_1,\ldots,x_n}\Psi_1\,[\Omega^2\Psi_i]_{i=2}^N\,|.$$

For brevity, put

$$(3) \qquad S = \mathrm{S}_{X_1,\ldots,X_n}^{x_1,\ldots,x_n}\Omega\,|, \qquad P_i = \mathrm{S}_{X_1,\ldots,X_n}^{x_1,\ldots,x_n}\Phi_i\,|, \qquad Q_i = \mathrm{S}_{X_1,\ldots,X_n}^{x_1,\ldots,x_n}\Psi_i\,|.$$

From (2) and (3) we have

$$(4) \qquad P_1[S^2P_i]_{i=2}^N = Q_1[S^2Q_i]_{i=2}^N.$$

By 23.6, the word S is simple. Also, it is obvious that

$$[P_1^\partial, \cdots, [P_N^\partial, [Q_1^\partial, \cdots, [Q_N^\partial < [S^\partial.$$

We will prove that $P_1 = Q_1$. Without loss of generality, we consider the case $[P_1^\partial \geqq [Q_1^\partial$. Then $P_1 = Q_1R$ for some R. We have $[R^\partial \leqq [P_1^\partial < [S^\partial$. From (4) we obtain $RS \angle S^2$. By 1.4, the word R is a power of S. Since $[R^\partial < [S^\partial$, we have $R = \Lambda$, and then $P_1 = Q_1$. We have proved that $P_1 = Q_1$. By an obvious induction we obtain $P_i = Q_i$ $(i = 1, \cdots, N)$, and this means that $(X_1, \cdots, X_n)$ is a solution of (*). We have proved that any nonhomogeneous solution of (1) satisfies (*). This establishes the equivalence with respect to nonhomogeneous solutions of the systems (*) and the equation (1). The proposition is proved.

BIBLIOGRAPHY

[1] V. K. Detlovs, *The equivalence of normal algorithms and recursive functions,* Trudy Mat. Inst. Steklov. **52** (1958), 75-139, especially §7; English transl., Amer. Math. Soc. Transl. (2) **23** (1963), 43-47. MR **21** # 5.

[2] Ju. I. Hmelevskiĭ, *Solution of word equations in three unknowns,* Dokl. Akad. SSSR **177** (1967), 1023-1025 = Soviet Math. Dokl. **8** (1967), 1554-1556. MR **36** # 3899.

[3] ———, *The solution of certain systems of word equations,* Dokl. Akad. Nauk SSSR **156** (1964), 749-751 = Soviet Math. Dokl. **5** (1964), 724-727. MR **29** # 1145.

[4] ———, *Word equations without coefficients,* Dokl. Akad. Nauk SSSR **171** (1966), 1047-1049 = Soviet Math. Dokl. **7** (1966), 1611-1613. MR **35** # 1685.

[5] S. C. Kleene, *Introduction to metamathematics,* Van Nostrand, Princeton, N. J., 1952. MR **14**, 525.

[6] R. C. Lyndon, *Groups with parametric exponents,* Trans. Amer. Math. Soc. **96** (1960), 518-533. MR **27** # 1487.

[7] Ju. L. Eršov, et al., *Elementary theories,* Uspehi Mat. Nauk **20** (1965), no. 4 (124), 37-108, especially §5 = Russian Math. Surveys **20** (1965), no. 4, 52-56. MR **32** # 4012.

[8] A. A. Buhštab, *Number theory,* Gos. Uč.-Ped.-Izdat., Moscow, 1960. (Russian) MR **22** # 10938.

[9] A. A. Markov, *The theory of algorithms,* Trudy Mat. Inst. Steklov. **42** (1954); English transl., Israel Program for Scientific Translations, Jerusalem, 1961. MR **17**, 1038; **24** # A2527.

[10] G. Baumslag, *Residual nilpotence and relations in free groups,* J. Algebra **2** (1965), 271-282. MR **31** # 3487.

[11] K. I. Appel and F. M. Djorup, *On the equation* $z_1^n z_2^n \cdots z_k^n$ *in a free semigroup,* Trans. Amer. Math. Soc. **134** (1968), 461-470. MR **37** # 6385.

[12] D. Skordev and Bl. Sendov, *On equations in words,* Z. Math. Logik Grundlagen Math. **7** (1961), 289-297. (Russian) MR **31** # 57.

[13]* J. Nielsen, *Die Gruppe der dreidimensionalen Gittertransformationen,* Danske Vid. Selsk., Math.-Fys. Medd. **5** (1924), no. 12.

*Editor's note. This reference does not appear in the original Russian text. It is the basis of §7 of Detlovs' paper cited above.